6th International Conference on Ship and Offshore Technology 2019

Kharagpur, India
7-8 November 2019

ISBN: 978-1-7138-0867-1

Printed from e-media with permission by:

Curran Associates, Inc.
57 Morehouse Lane
Red Hook, NY 12571

Some format issues inherent in the e-media version may also appear in this print version.

The Royal Institution of Naval Architects
8-9 Northumberland Street
London, WC2N 5DA
United Kingdom

Phone: 020 7235 4622
Fax: 020 7259 5912

publications@rina.org.uk

Additional copies of this publication are available from:

Curran Associates, Inc.
57 Morehouse Lane
Red Hook, NY 12571 USA
Phone: 845-758-0400
Fax: 845-758-2633
Email: curran@proceedings.com
Web: www.proceedings.com

6th International Conference on Ship & Offshore Technology

7th-8th November 2019, IIT Kharagpur, India

PAPERS

In association with the

Department of Ocean Engineering and Naval Architecture, IIT Kharagpur

Sponsored by:

THE ROYAL INSTITUTION OF NAVAL ARCHITECTS

8-9 Northumberland Street, London, WC2N 5DA, Telephone: +44 (0)20 7235 4622

The Royal Institution of
Naval Architects

INTERNATIONAL CONFERENCE

6th International Conference on Ship & Offshore Technology

7th-8th November 2019, IIT Kharagpur, India

Papers published in these conference Proceedings are peer reviewed.

THE ROYAL INSTITUTION OF NAVAL ARCHITECTS
8-9 Northumberland Street
London WC2N 5DA

Telephone: 020 7235 4622
Fax: 020 7259 5912

Preface

The sixth **International Conference on Ship & Offshore Technology**, ICSOT-INDIA 2019 was organized by the Department of Ocean Engineering & Naval Architecture at IIT Kharagpur in collaboration with the Royal Institution of Naval Architects on **7th-8th November 2019** at **IIT Kharagpur**.

This conference seeks to explore technological advancements related to marine vessels and structures for sustainable marine operations and generation of renewable energy. The objective is to disseminate outcomes from research and development activities concerning marine structures focused on innovative and practical solutions for energy efficiency and economy.

The **theme** of **ICSOT-INDIA 2019** is *"Dynamics of Ocean Structures"*. The aim is to bring together members of the international maritime industry to present and discuss the latest developments in design, analysis and construction of ships and offshore structures which will contribute towards sustainable growth and development of the global marine industry.

The technical presentations for ICSOT 2019 included **5** Keynote/ Special Lectures and **32** Technical Papers. The technical papers covered a wide range of topics related to naval architecture, offshore and structural engineering, marine hydrodynamics, coastal protection, autonomous operations, applications of machine learning, deep learning and big data to marine problems etc. All papers published in the conference proceedings have been peer reviewed. The organizing committee would like to express their sincere thanks and gratitude to the sponsors, reviewers, keynote speakers, members of the International Technical Committee, authors, delegates and volunteers for their active participation in ICSOT-India 2019.

Secretary, Organizing Committee

Dr. Anirban Bhattacharyya, IIT Kharagpur

Chairmen, Organizing Committee

Prof. Prasad K Bhaskaran / Prof. H.V. Warrior, IIT Kharagpur
Mr. Trevor Blakeley, The Royal Institution of Naval Architects

Organizing Committee

Patron

Director, IIT Kharagpur

Chairmen

Prof. Prasad K Bhaskaran / Prof. H.V. Warrior, IIT Kharagpur
Mr. Trevor Blakeley, The Royal Institution of Naval Architects

Secretary

Dr. Anirban Bhattacharyya, IIT Kharagpur

Members

Prof. Debabrata Sen, IIT Kharagpur
Prof. Om Prakash Sha, IIT Kharagpur
Dr. Ashoke Bhar, IIT Kharagpur
Prof. Trilochan Sahoo, IIT Kharagpur
Dr. Vishwanath Nagarajan, IIT Kharagpur
Dr. Nabanita Datta, IIT Kharagpur
Dr. Ranadev Datta, IIT Kharagpur
Dr. Mohammed Rabius Sunny, IIT Kharagpur
Dr. Arunjyoti Sarkar, IIT Kharagpur
Dr. Kiran Vijayan, IIT Kharagpur
Dr. Ritwik Ghoshal, IIT Kharagpur
Dr. Dasharatha Achani, IIT Kharagpur

International Technical Committee

Prof. P. K. Das, Director, ASRANet Ltd., United Kingdom
Prof. Sverre Steen, Department of Marine Technology, NTNU, Norway
Prof. C. Guedes Soares, CENTEC, Technical University of Lisbon, Portugal
Prof. T. Ray, Australian Defence Force Academy, The University of New South Wales, Australia
Prof. G. Bulian, Department of Engineering and Architecture, University of Trieste, Italy
Prof. R. P. Gokarn, (Retd), Department of Ocean Engg & Naval Architecture, IIT Kharagpur, India
Prof. S. C. Misra, (Retd), Department of OE & NA, IIT Kharagpur, & Chairman, NRB, India
Prof. P. Krishnankutty, Department of Ocean Engineering, IIT Madras, India
Mr. A. R. Kar, Indian Register of Shipping, Mumbai, India
Prof. B.N. Singh, Department of Aerospace Engineering, IIT Kharagpur, India
Prof. D. Sen, Department of Ocean Engg & Naval Architecture, IIT Kharagpur, India
Prof. O. P.Sha, Department of Ocean Engg & Naval Architecture, IIT Kharagpur, India
Prof. T. Sahoo, Department of Ocean Engg & Naval Architecture, IIT Kharagpur, India
Prof. P. K. Bhaskaran, Department of Ocean Engg & Naval Architecture, IIT Kharagpur, India

SPONSORS

LARSEN & TOUBRO

IRCLASS
Indian Register of Shipping

Shoft Shipyard
We Promise. We Deliver.

Datasol
Innovative Labs

GRSE

ClassNK

Keynotes/ Special Lectures

'Hydroelasticity and Resonant Behaviour of Very Large Container Ships'

by Prof. Mike Meylan, *The University of Newcastle, Australia*

'Present and Future perspectives of Naval Shipbuilding'

by Rear Admiral G. K. Harish, DGND(SSG), India

'Ship Design: Where we are and where we need to go'

by Capt. KJH Christie (IN Retd.) *Vice President and Head Design, L&T Shipbuilding, India*

'Biomimetic Underwater Vehicles'

by Prof. RN Govardhan, *Department of Mechanical Engineering, IISC Bangalore, India*

'Design and Life Cycle Management of Marine Hoses as Engineered Consumables'

by Dr. Satish Balasubramanian, *ExxonMobil Services & Technology Pvt. Ltd., India*

CONTENTS

Proceedings of the 6th International Conference on Ship and Offshore Technology
ICSOT 2019
November 7-8, 2019, IIT Kharagpur, India

ICSOT INDIA 2019-01

SPH-DEM MODELLING OF TSUNAMI DRIVEN FLOATING DEBRIS AND ITS TRAPPING BY COASTAL FOREST

Aditya Gupta,
Research Scholar
IIT Bombay-Monash
Research Academy,
Mumbai, Maharashtra,
400076 India
adityagupta1906@gmail.com

Manasa R. Behera,
Associate Professor
Department of Civil Engineering,
Indian Institute of Technology
Bombay, Mumbai-400076, India
manasa.rb@iitb.ac.in

Amin Heidaropur
Professor
Department of Civil Engineering,
Monash University, Melbourne,
Victoria-3800, Australia
amin.heidarpour@monash.edu

ABSTRACT

This paper applied a three-dimensional approach for Smoothed Particle Hydrodynamics (SPH) based on particle approximation for modelling an inland surge due to aftermath of tsunami and Discrete element method (DEM) representing solids like trees and floating debris. The aim is to predict the behaviour of particle-fluid interaction and understand the trapping action of debris by finite-length forest considering the effects of forest density and aspect ratio of the forest by SPH-DEM coupling. In order to investigate the behaviour of floating debris and their interaction with forest, the model was validated and the flow pattern of the debris simulated by the model is in good agreement with the experimental study. Further, the debris flow behaviour pattern was explored for different type of debris such as shipping container. The simulation results based on the open source code DualSPHysics of SPH-DEM model were capable of capturing the debris flow pattern and identified the most critical cases which exerted the overturning pressure to trees. Practical application of this study will further help to design the inland forests that can trap debris or slow their propagation in front of a refuge.

Keywords: *SPH-DEM, Tsunami inland surge, Debris flow, Coastal Forest*

1 INTRODUCTION

Humans are always drawn towards the coastal zones as it has abundance of natural or subsistence resources and offer a great access for marine trade and transport. The population density is high in coastal areas compared to the hinterland which is expected to grow even further in the near future [1, 2]. The increase in the population growth puts pressure on the coastal ecosystem [3, 4] which significantly increases the level of risk along coastlines. One of the major threats that humans face from the wrath of ocean is Tsunami. In the last 20 years alone, tsunami events such as in Indian Ocean (2004). South of Java Island (2006), Solomon Island (2007), Chile (2007), Samoa (2009), Pacific coast of Japan (2011) have caused heavy destruction of life and property. There has been extensive studies done on the countermeasures to protect coastal zone from tsunami effects [5-7]. Inoue et al. [8, 9] investigation showed the poor hydrodynamic performance of the existing coastal protection structures and concluded that there is a need to improve the efficiency of these structures. One such countermeasure that is provided by nature itself is 'Coastal forests' which has played a very crucial role as a countermeasure against catastrophic tsunami events. The investigations done by Danielsen et al.; Kathiresan and Rajendran [10-11] have shown the importance of coastal forest in reducing the damage done by tsunami inundation. The water level and the velocity behind the vegetation can be controlled by changing the density of coastal forest [12]. When tsunami inundates inland it also carries with it debris in the form of broken trees, ship containers, automobiles, ships or a broken concrete structure from a damaged building. These debris along with tsunami surge collide with the building on its way and cause a secondary damage which may result in the fire outbreak or complete destruction of the building. Although there have been several experimental studies done which investigates the debris flow path and its flow pattern but there is very little or no work on developing a numerical model which can exactly predict the behaviour of floating debris in a water surge. An accurate numerical model will help us in predicting the debris behaviour in any condition which is sometimes difficult to carry in an experiment due to high randomness. In this paper, different types of forest arrangement which varies from sparse to dense adopted from Pasha and Tanaka (2016) [13] are modelled. The effect of forest density and aspect ratio on accumulation height has been studied by applying a 3 Dimensional approach of SPH-DEM coupling to represent trees and debris. This numerical study also extends the behaviour of flow pattern during the introduction of debris other than wooden logs such as shipping container so as to differentiate the flow pattern between circular cross section and rectangular cross section respectively. The numerical results are analysed and validated with experimental data.

2 NUMERICAL MODEL SETUP

Smoothed Particle Hydrodynamics (SPH) is a lagrangian meshless method. SPH uses a technique in which a continuum, which in this case is Navier Stokes equation for fluid simulation gets discretized using set of particles and these set of particles are determined by distance based function with n associated smoothening length 'h'. The new physical quantities are calculated for each neighboring particles at every time step which updates and move accordingly. The discrete element method (DEM) is used to model trees and debris in this case which computes rigid particle dynamics and account for interaction forces by considering contact laws. The couples SPH-DEM are lagrangian-lagrangian scheme which are capable of simulating the free surface flow and dynamic boundary [14]. To simulate the numerical model an open-source DualSPHysics code originated from SPHysics [15] is used which is implemented in C++ and CUDA language to carry out simulations with millions of particles on CPU.

2.1 SPH model of Fluid

Smoothed Particle Hydrodynamics (SPH) initially was used in astrophysical simulation as a lagrangian method [16] and later its application were used in fluid dynamics [17].In SPH the partial differential form of continuum fluid dynamics are transferred to particle based simulation using interpolation function referred to as Kernel function (W) which can be either in the form of cubic or quantic. It represents a function F(r) in region r' by integral approximation

$$F(r) = \int F(r')\, W(r - r', h)dr' \quad (1)$$

The above equation (1) can be discretized to a set of particles interpolated at particle 'a' by defining smoothening length 'h' that falls in a region of compact support of particle 'b' of neighboring volume ∇v_b as shown in equation (2).

$$(r_a) = \sum_b F(r_b) W(r_a - r_{b,h}) \nabla v_b \quad (2)$$

2.2 The Smoothing Kernel

Kernels are expressed as a function of non-dimensional distance between particles 'x', which is given by equation (3).

$$x = \frac{r}{h} \quad (3)$$

Where r is the distance between any two given particles 'a' and 'b' and 'h' is the smoothening length. Several smoothening kernel have been developed but the most used ones Cubic Spline [18] and Wendland [19]. The simulation using Wendland kernel show more orderly distribution of particle than cubic spline [20]. Hence, Wendland kernel shown in equation (4) is chosen for the following study.

1) Quintic/Wendland [19]

$$W(r,h) = \emptyset_d \left(1 - \frac{x}{2}\right)^4 (2x + 1) \qquad 0 \le x \le 2 \quad (4)$$

Where $\emptyset_d = \frac{21}{16\pi h^3}$ *in 3D*

2.3 Mass and Momentum Conservation equation

The mass and momentum conservation equations are written in Lagrangian form in equation (5):

$$\frac{D\rho}{Dt} = -\rho\nabla \cdot v; \qquad \frac{Dv}{Dt} = -\frac{1}{\rho}\nabla p + g + \theta \quad (5)$$

Where v is the velocity vector, p is the pressure, and θ refers to the viscous diffusion term.

Fluid in DualSPHysics is treated as weakly compressible and hence an equation of state is used which connects pressure and density is shown in equation (6):

$$p = B\left[\left(\frac{\rho}{\rho^0}\right)^\gamma - 1\right] \quad (6)$$

Where ρ0 is the reference density, γ is a dimensionless parameter typically assumed equal to 7, and $B = {c_0^2 \rho_0}/{\gamma}$ in which c_0 represents the speed of sound at the reference density.

The equation (7) is obtained by discretising equation (6) using SPH.

$$\frac{d\rho_a}{dt} = \sum_b m_b (v_a - v_b) \cdot \nabla_a W_{ab}\,; \quad \frac{dv_a}{dt} = -\sum_b m_b \left(\frac{p_b}{\rho_b^2} + \frac{p_a}{\sigma_a^2} + \Pi_{ab}\right) \cdot \nabla_a W_{ab} + g \tag{7}$$

Where Π_{ab} is the viscosity term that in this work is expressed using the classic artificial viscosity formulation as shown in equation (8).

$$\Pi_{ab} = \begin{cases} \dfrac{-\alpha c_0 \mu_{ab}}{\rho_{ab}} & \text{if } v_{ab} \cdot r_{ab} < 0 \\ 0 & \text{if } v_{ab} \cdot r_{ab} \geq 0 \end{cases} \tag{8}$$

Where $\mathbf{r}_{ab} = \mathbf{r}_a - \mathbf{r}_b$ and $\mathbf{v}_{ab} = \mathbf{v}_a - \mathbf{v}_b$, being $\mathbf{r}_k$ the position vector of the generic particle k and α a free parameter which is set to 0.01 in the present study as it gives the best results in wave loading onto coastal structure [21].

2.4 Time Stepping

There are several numerical time stepping integration schemes, one of which is Verlet Method [22] where the algorithm is split into two parts so as to reduce the computational effort. The variable is calculated in predictor part and the divergence of integrated values are controlled in second part, which results in in the stability of the scheme.

The other scheme which is also used in this study is the Symplectic Method [23] which is an explicit – second order with an accuracy in time of $O(\Delta t^2)$. Let momentum, density and position equations written as:

$$\frac{dr_a}{dt} = V_a \tag{9}$$

$$\frac{d\rho_a}{dt} = D_a \tag{10}$$

$$\frac{dr_a}{dt} = V_a \tag{11}$$

The predictor stage in Symplectic method estimates the value of acceleration and density at middle of time step:

$$r_a^{n+\frac{1}{2}} = r_a^n + \frac{\Delta t}{2} v_a^n \tag{12}$$

$$\rho_a^{n+\frac{1}{2}} = \rho_a^n + \frac{\Delta t}{2} D_a^n \tag{13}$$

The corrector stage uses $\frac{dv_a^{n+\frac{1}{2}}}{dt}$ to calculate corrected velocity and position of particle at the end of time step.

$$v_a^{n+1} = v_a^{n+\frac{1}{2}} + \frac{\Delta t}{2} F_a^{n+\frac{1}{2}} \tag{14}$$

$$r_a^{n+1} = r_a^{n+\frac{1}{2}} + \frac{\Delta t}{2} v_a^{n+\frac{1}{2}} \tag{15}$$

The corrected value of density $\frac{d\rho_a^{n+1}}{dt} = D_a^{n+1}$ is calculated using he updated vales of v_a^{n+1} and r_a^{n+1} [24]

3. COUPLING WITH DEM

When a solid object interacts with another solid or fluid particle, force arises. These interaction force are accounted by contact laws for computation of rigid particles. The contact force (F_c) is resolved into normal (F_n) and tangential components (F_t) given by equation (16):

$$F_C = F_n + F_t \tag{16}$$

Normal and tangential force components includes viscous dissipation effects, as the deformation due to collision between particles will be somewhere in between perfectly elastic and perfectly inelastic. The quantification of the deformation is done by the normal restitution coefficient as shown in equation (17) which is the ratio of final to initial relative velocity between two bodies after they collide

$$e_n = \frac{v_n|_{t=t^n}}{v_n|_{t=0}}, e \in [0,1] \tag{17}$$

The normal force (F_n) and Tangential force (F_t) as shown in equation (18) and (19) are decomposed into repulsion force (F_r) and damping force (F_d) respectively arising due to elastic deformation of material and viscous dissipation of energy. Detailed information on DEM implementation can be referred in Canelas et al. [25, 26]

$$F_{n,ij} = F_n^r + F_n^d = k_{n,ij}\delta_{ij}^{3/2} e_{ij}^n - \gamma_{n,ij}\delta_{ij}^{1/2}\dot{\delta}_{ij} e_{ij}^n \quad \textbf{(18)}$$

$$F_{t,ij} = F_t^r + F_t^d = k_{t,ij}\delta_{ij}^{3/2} e_{ij}^t - \gamma_{t,ij}\delta_{ij}^{1/2}\dot{\delta}_{ij} e_{ij}^t \quad \textbf{(19)}$$

Where, $e_{ij}^{n,t}$ is the unit vector between centers of particle i and j and stiffness for normal and tangential is given by equation (20) and (21) respectively:

$$k_{n.ij} = \frac{4}{3} E^* \sqrt{R^*} \quad \textbf{(20)}$$

$$k_{t.ij} = \frac{2}{7} k_{n.ij} \quad \textbf{(21)}$$

Where, R^* is the reduced radius and E^* is the reduced Young modulus

$$R^* = \left(\frac{1}{R_1} + \frac{1}{R_2}\right)^{-1}; \; E^* = \left(\frac{1-\nu_{p1}^2}{E_1} + \frac{1-\nu_{p2}^2}{E_2}\right)^{-1}$$

Where, ν_p is the Poisson coefficient of material. The values of Poisson and Young's Modulus in this study are discussed in the later section of this paper

And the damping coefficient is given by equation (22)

$$\gamma_{n,ij} = -\frac{\log e_{ij}}{\sqrt{\pi^2 + \log^2 e_{ij}}} \quad \textbf{(22)}$$

3.2 Boundary Condition

In SPH the particles are divided into two sets: Fluid particle and a Boundary particle. The boundary particles are the virtual particles that characterize the system limits. Dynamic boundary condition due to dynamic particle was first presented by Dalrymple and Knio, (2000) [27] and further used by Gómez-Gesteira and Dalrymple (2004) for waves and coastal structures. In this boundary condition mass conservation, momentum balance and energy conservations are obeyed by the particles located at the boundary. The boundary particles can be fixed or moving and when the distance between the fluid particle and boundary particle is less than '2h' ('h' being smoothening length), the density and pressure of the affected boundary particle increases. This increase in P/ρ^2 term in momentum equation results in the generation of repulsion force on the fluid particle. The expression given by Monaghan (1994) [28] as shown in equation (23) is used to analyzed the energy conservation.

$$\frac{du_a}{dt} = \frac{1}{2}\sum_b m_b \left(\frac{P_a}{\rho_a} + \frac{P_b}{\rho_b} + \Pi_{ab}\right) \vec{v}_{ab} \vec{\nabla}_a W_{ab} \quad \textbf{(23)}$$

Where, 'a' is the moving particle, 'b' is the boundary particle, P is pressure, ρ is the density, Π is the viscosity, W is the kernel function. In this study to generate the water surge Dynamic boundary condition is applied which has also been tested by Crespo et al., 2007 [29] and concluded that the particle does not leave the container and guarantees a proper movement of particles around the walls.

4 DESCRIPTION OF PHYSICAL EXPERIMENT

The experiments were carried out by Pasha et. al. (2016) [14] in laboratory at Saitama University in Japan. The dimensions of water flume was $5 \times 0.7 \times 0.5\, m$ with constant bed slope of $1/2400$. The scaling factor (L_s) was considered as 100. The forests were modelled as cylinder mounted on the flume bed in a rectangular patter at about 2.8 m from the upstream of channel start. In the study forest is set to be located inland to trap the debris. Froude number of 0.7 was chosen according to the study by Spiske et al. (2010) [30] and hence water depth of 4.5 cm was selected at start of experiment. The velocity was measured at 80 % of flow depth by Particle Image Velocimetry (PIV). Video analysis was used for the motion of the wooden debris.

Forests were modelled on $1/100$ scale of Japanese Pin tree having diameter of 0.004 m. The density of forest were calculated against G/D ratio where G is spacing between the cylinders in cross stream direction & D is the diameter of forest. Five values of G/D 2.13, 1.5, 1.09, 0.58, 0.25 were selected where 2.13 and 0.25 represents the sparse and dense arrangement respectively. Three aspect ratio W/L, (W is the forest width and L is the length) of 1, 1.7, 2.4 were selected to represent the inland forest. Wooden sticks of diameter 0.5 cm and lengths 12.5, 10, 7.5 & 5 cm was considered for the debris with specific gravity ranging from 0.90 to 1.05 and density (ρ_d) smaller than water density (ρ_w) so that debris can float. Their study identified four different patterns of debris interaction with forest: a) Trapped b) Held parallel to trees c) Pivoted with outer tree d) Passed through after hitting outer trees

4.1 Description of Computational Domain

The configuration of experimental flume is shown in the fig.1. The flume of dimension $5 \times 0.7 \times 0.5\ m$ is created as a particle representing an object in a Face mode, where particles are generated only as an encasing leaving the inside hollow. The fluid particles are generated by using an inlet/outlet condition where the fluid particles are pumped continuously from an opening or an inlet to maintain the continuous supply of water in a flume. The properties of the inlet condition is determined such that the Froude number of 0.7 as proposed in the experimental study could be maintained.

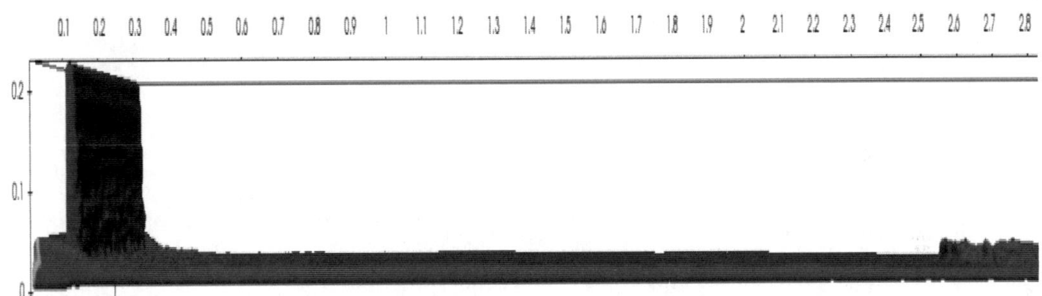

Fig. 1. CROSS SECTIONAL VIEW OF THE COMPUTATIONAL DOMAIN WITH VIRTUAL PUMP AT THE CHANNEL START. GRID SIZE IN METER.

4.2 Forest Modelling

The forests are modelled using DEM. In the current study the trees are considered to be rigid and hence no deflection occurs when water and debris hit them. Table 1. shows the property of the trees used for the simulation. The trees are modelled as cylinders with a height of 150 mm and diameter of 4 mm. Table 2 shows two cases that are considered in this study adopted from the experimental study of Pasha et. al [14]. The first case represents a sparse forest and the second case represents a dense forest condition arranged in staggered manner.

Table 1. MATERIAL PROPERTIES OF TREES

Young's Modulus ($^{kN}/_{m^2}$)	200000
Poissons Ratio	0.2
Restitution Coefficient	0.6
Kinetics Friction coefficient	0.7

Table 2. ARRANGEMENT OF FOREST IN COMPUTATIONAL DOMAIN

	Case1: SPARSE FOREST		**Case2: DENSE FOREST**	
Aspect Ratio ($^{W_y}/_{W_x}$)	1.70	2.4	0.98	2.4
Forest Density ($^{cyl}/_{cm^2}$)	0.24	0.24	1.22	1.22
G/D	2.10	2.13	0.25	0.25
L/D	5.25	5.25	1.5	1.5

Figure 3 shows the arrangement pattern for the modelling of forest that are considered in this numerical study transitioning from dense to sparse condition with respect to their aspect ratio.

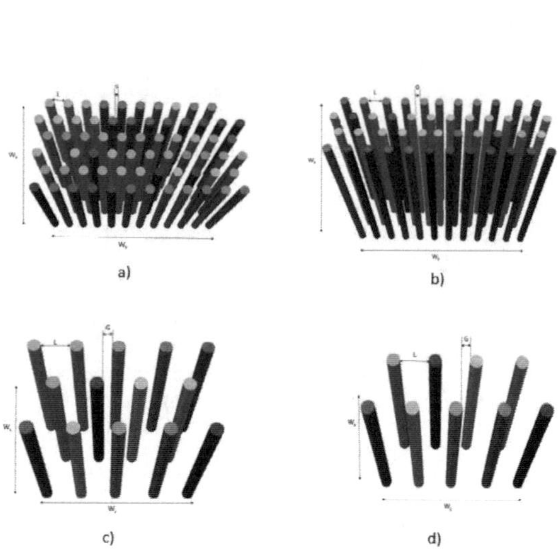

Fig. 3. ARRANGEMENT OF FOREST IN COMPUTATIONAL DOMAIN a) DENSE WITH A.R 2.4 b) DENSE WITH A.R 0.98 c) SPARSE WITH A.R 2.4 d) SPARSE WITH A.R 1.7

4.3 Debris Modelling

It was observed in the aftermath of tsunami 2011 that one of the main cause of severe destruction was the impact of debris such as cars, ships, uprooted tree, building material, containers etc. with the structures, which are often not designed for such extreme loads. Therefore, it is important to study the behavior of debris interaction with inland forest which could help the engineers to understand the most effective ways for an inland forest to trap the debris and save the humankind from catastrophe. In this study the debris were modelled using Discrete Element Method. In this numerical study only the behavior of wooden debris of a circular c/s and a shipping container of a rectangular c/s is considered. To replicate the behavior of full length uprooted trees a cylinder of constant diameter 0.5 cm and length of 12.5 cm was selected as shown in figure 4. As per ISO the standard size of a 20 ft. steel shipping container is $6.06 \times 2.43 \times 2.6\ m$. The container was modelled as a rectangular block of dimension $6 \times 2.4 \times 2.6\ cm$ as shown in figure 4 b).

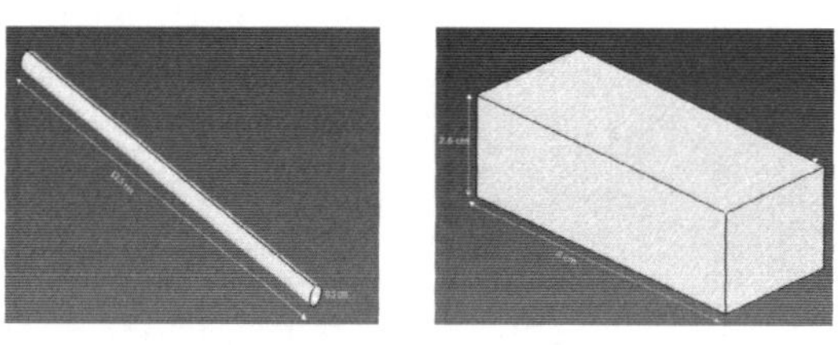

Fig. 4. TYPE OF DEBRIS. a) WOODEN LOG WITH CYLINDRICAL C/S b) STEEL SHIPPING CONTAINER WITH RECTANGULAR C/S

Table 3 shows the following properties that were assigned to the debris to study the floating properties in this numerical study.

Table 3. PROPERTIES OF DEBRIS CONSIDERED FOR THE NUMERICAL MODEL

PROPERTIES	WOODEN LOG	SHIPPING CONTAINER (EMPTY)
TYPE	Hard wood	Steel
Density (kg/m^3)	750	920
Young's Modulus (GN/m^2)	0.2	210.0
Poisson Ratio	0.20	0.30
Restitution Coefficient	0.60	0.80
Kinetic Friction coefficient	0.70	0.45

5 SENSITIVITY ANALYSIS

The model was calibrated against the water level measured along the center line of the flume. In the experiment the channel had a continuous and steady supply of water with a depth of 4.5 cm at the start. To generate a continuous supply of water in numerical model a virtual rectangular shape pump was designed which can maintain the flow depth of 0.45 m and Froude number of 0.70. Finally a pump with a c/s of 0.0975 m^2 which generates fluid particle at a uniform velocity of 0.11 m/s was considered in this study. Figure 5 a) represents the calibration study of the channel for pump uniform imposed velocity to achieve the required Froude number. Table 4 shows the following input parameters considered to generate fluid particles in the SPHysics.

Table 4. FLUID PARAMETERS

Parameter	Value
1) Density of fluid (ρ)	1000 kg/m^3
2) Politropic constant (γ)	7
3) Coefficient to calculate smoothing length $h = coef\ (h)\sqrt{3dp^2}$	1
4) Viscosity Value (ν)	0.01

The water height in the numerical model was calibrated against the experimental results by changing the initial inter particle distance ('dp') in meters. Table 5 shows that when the model was run against different values of 'dp' increasing from 0.005 to 0.007 m the number of total particle generated decreases and simulation time reduces which is shown in Figure 5 b) resulting in an inaccurate simulation as shown in Figure 6.

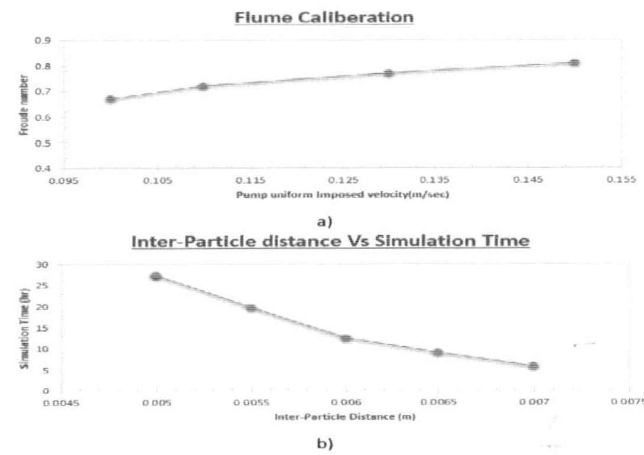

Fig. 5. .a) FLUME CALIBRATION FOR FROUDE NUMBER b) COMPUTATIONAL COST FOR NUMERICAL MODEL

TABLE 5. TOTAL PARTICLES GENERATED FOR DIFFERENT INTER-PARTICLE DISTANCE

Inter-Particle Distance (m)	Number of particles generated
0.005	876569
0.0055	723905
0.006	508504
0.0065	459676
0.007	443894

Figure 6 shows the graph for water level along the center of the flume. It can be observed that that when the 'dp' is set to 0.007 the water height is very unstable as the greater inter-particle distance tends to vanish the weighting function. The fluctuations tends to subside as the inter particle distance is reduced to 0.006. The results with particle distance 0.0055 and 0.005 are of greater accuracy with an average relative error of 2.3 and 0.88 % respectively and observe the same trend as experimental results. In this study all the simulation has been performed by considering 'dp' as 0.005 so as to balance the simulation time and desired fidelity.

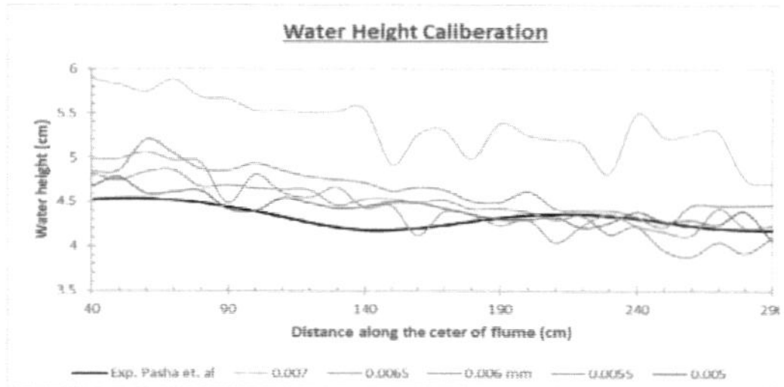

Fig. 6. SENSITIVITY ANALYSIS FOR WATER LEVEL BY CHANGING THE SIZE OF INTER-PARTICLE DISTANCE

6 NUMERICAL RESULTS AND DISCUSSION

6.1 Effect on water surge due to change in forests parameter

The inland forest was placed at 2.8 m upstream of channel start. Figure 7 shows the water height measured along the center line of the flume in front of forest for different cases. It was observed that the water level in front of forest increases when the density and aspect ratio increases. This is due to increase in resistance by increase in density of trees which makes it difficult for water surge to pass through it [12]. The numerical model was able to capture the similar trend with respect to experimental study. It was observed that the maximum peak of water level in front of forest was overestimated by numerical model, it may be due to the upward splashing of fluid particles when colliding with trees contributing in the addition of water height. The model was also able to capture the sudden increase in water surface slope inside the forest as the density increases from 0.24 to 1.22 cylinders per cm^2 as investigated by Iimura and Tanaka [12].

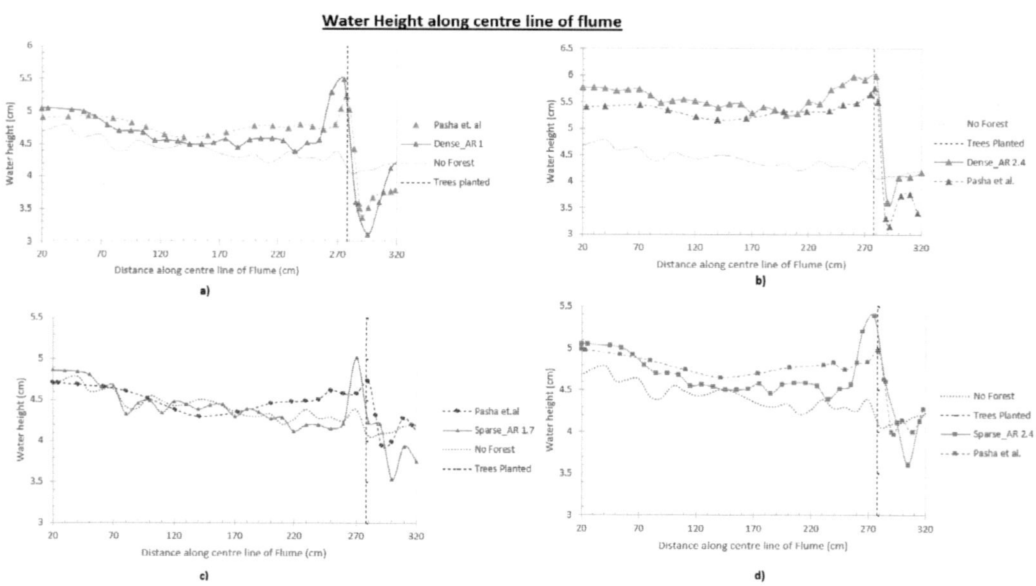

Fig. 7. VARIATION IN WATER LEVEL ALONG THE CENTER LINE IN FRONT OF THE FOREST WHEN a) DENSE FOREST WITH ASPECT RATIO 1 b) DENSE FOREST WITH ASPECT RATIO 2.4 c) SPARSE FOREST WITH ASPECT RATIO 1.7 d) SPARSE FOREST WITH ASPECT RATIO 2.4

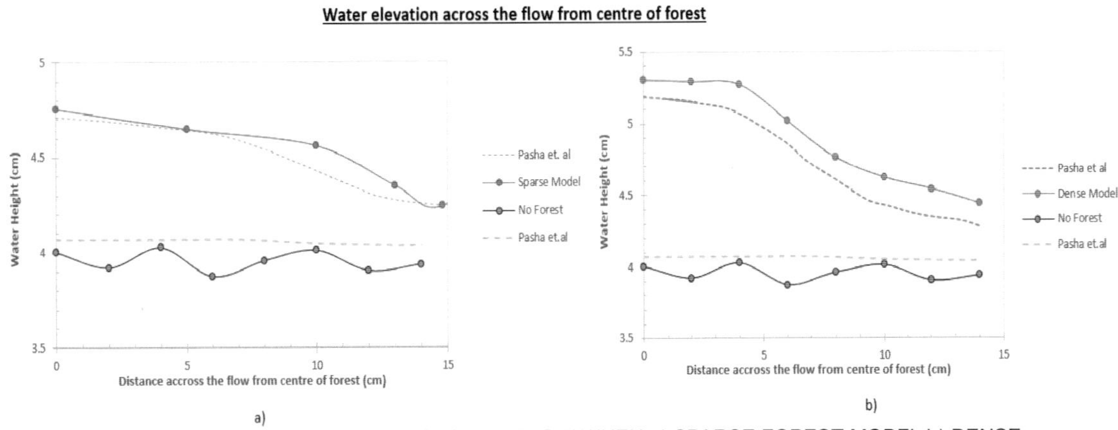

Fig. 8. VARIATION IN WATER LEVEL ACROSS THE FLOW WHEN a) SPARSE FOREST MODEL b) DENSE FOREST MODEL

Figure 8 shows the level of water across the flow measured from the center of forest. As expected the water level decreases as we move away from the forest area since there is no restriction for water to pass through.

6.2 Effect on flow velocity due to change in forests parameter

The velocities in the model were measured at two strips, Strip 1 that is 1 cm across the flow in front of forest while Strip 2 is 7 cm across the flow direction in front of forest. X-axis of Figure 9 shows the distance from center of forest towards the forest edge and Y-axis denotes the ratio of Velocity of water in presence of forest V_{fo} to the Velocity of water in absence of forest V_i. The numerical model was able to capture the trend as observed in experimental study. It was observed that as the Aspect ratio increases the fluctuations in front of forest (Strip 1) increases as more volume of water is resisted by trees which also results in the reduction of velocity. The velocity effect at Strip 2 has a smooth transition from center to edge as the effect of turbulence diminishes away from the forest area. In both the cases the velocity ratio increases as we move away from the center of forest as the blockage due to forest decreases.

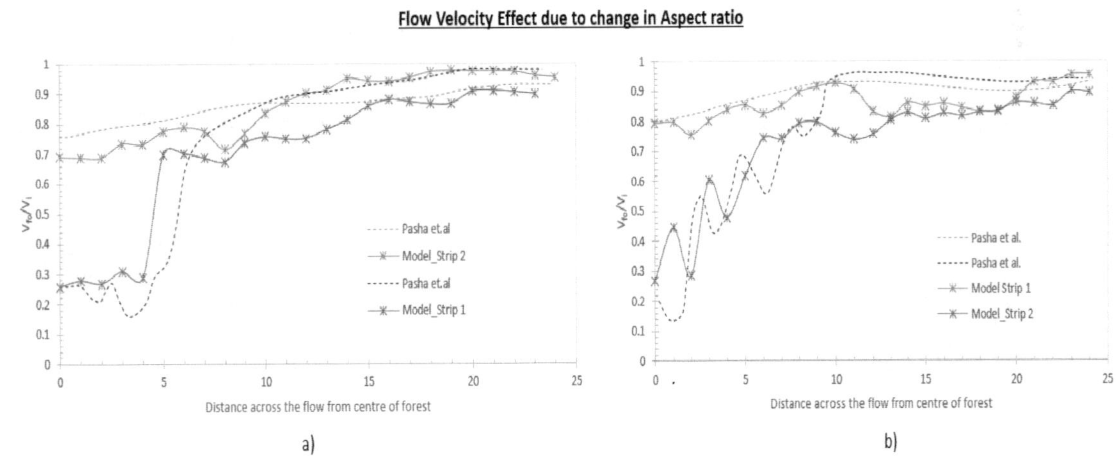

Fig. 9. VARIATION IN FLOW VELOCITY ACROSS THE FOREST WHEN a) FOREST WITH ASPECT RATIO 1.0 b) FOREST WITH ASPECT RATIO 1.7

6.2 Debris Interaction with Inland Forest

The debris were kept at a distance of 1 m upstream of inland forest as shown in Figure 10. The details of the two debris can be found in Table 3. In order to investigate the flow pattern of these debris, two conditions were considered a) Wooden log interaction with inland Forest b) Steel shipping container interaction with inland Forest.

Figure 11 and 12 shows the interaction of wooden log and shipping container with sparse forest. Flow behavior of both debris will be discussed for sparse forest as there was no visible change in flow pattern for dense forest compared to sparse forest.

It can be observed that the numerical study was accurately able to capture the flow pattern of a floating debris. For Figure 11 at 4.1 sec of simulation, the log collides with the sparse forest and gets trapped there till 4.5 sec. The continuous water surge from behind tilts the log at around 4.6 seconds which makes the wooden log's long face to orient itself perpendicular to the flow at 4.8 sec. After that the wooden log slides along the forest and pivot from the outer tree at around 5.1 seconds and then pass by the forest. The same flow pattern was also observed in the experimental studies of Tanaka and Ogino (2017) [31].

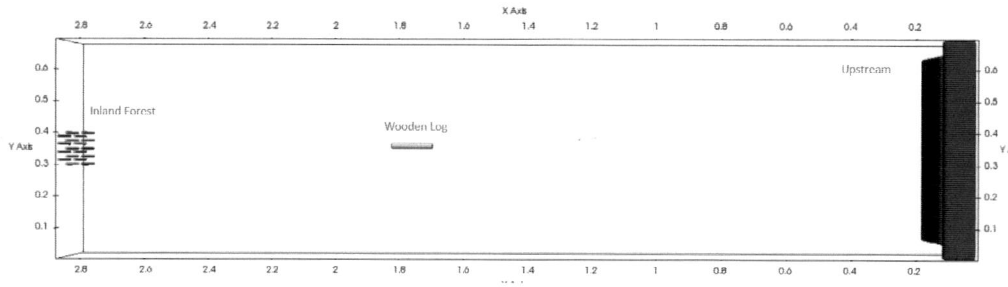

Fig. 10. TOP VIEW OF COMPUTATIONAL DOMAIN SHOWING THE POSITION OF DEBRIS WITH RESPECT TO FOREST

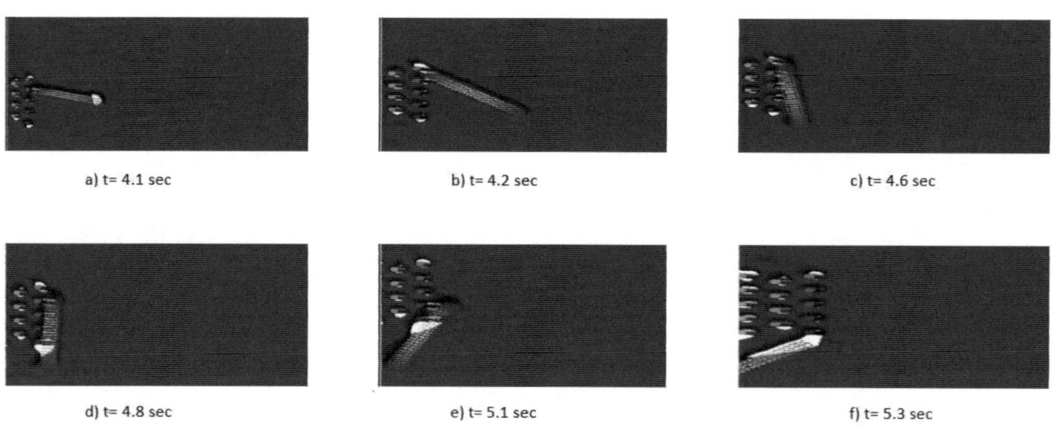

Fig. 11. FLOW PATTERN AND INTERACTION OF A WOODEN LOG WITH INLAND FOREST

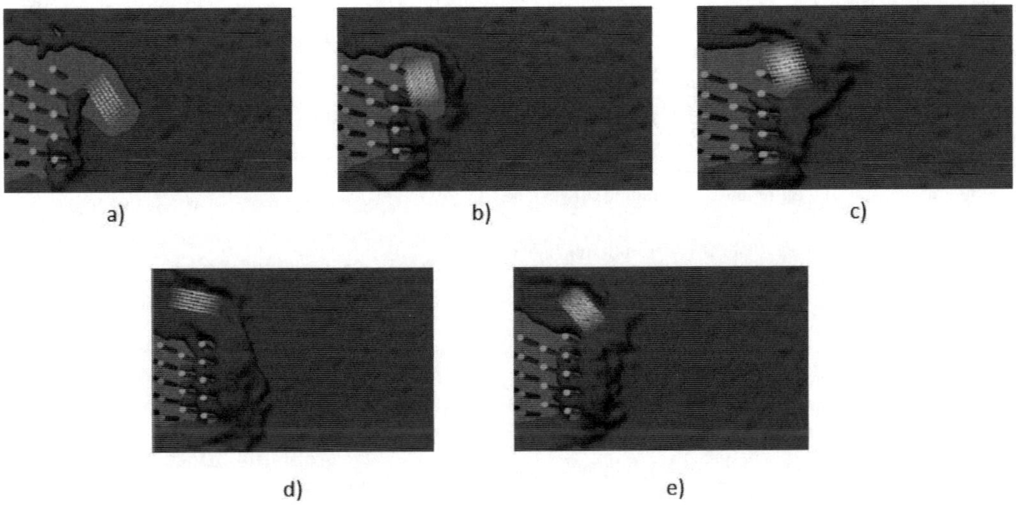

Fig. 12. FLOW PATTERN AND INTERACTION OF A SHIPPING CONTAINER WITH INLAND FOREST

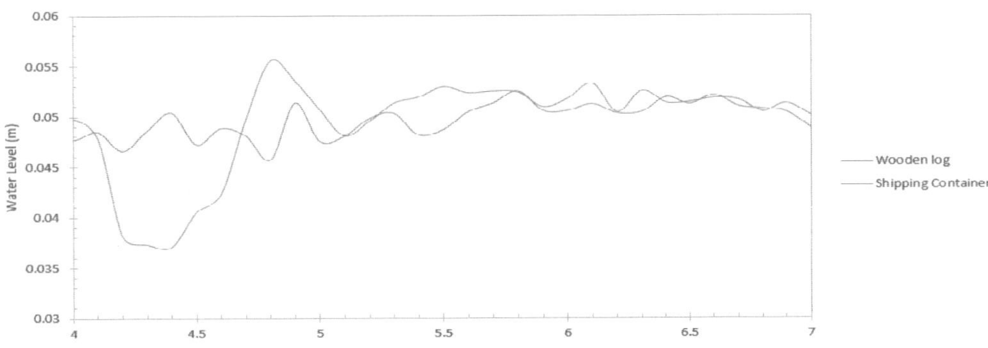

Fig. 13. WATER LEVEL IN FRONT OF FOREST DURING DEBRIS INTERACTION

Figure 12 shows the shipping container flow and its interaction with sparse forest. As the shipping container has a bigger cross section than a tree the water dragged the container and collided it with inland forest. The container did not get trapped in the trees due to its broader dimension but followed the same flow pattern as of the wooden log. In Figure 13, the difference between the water level variations for the two debris in front of forest can be observed. As wooden log has a smaller cross section and small density compared to container it was able to float on water thus not affecting the water height that much. The fluctuations are due to the turbulence created by debris in front of the forest. The increase in water level from 4 to 5.8 seconds is due to the debris changing its orientation from longitudinal axis parallel to perpendicular to the flow and thus in process blocking the water behind inland forest and increasing its level. This orientation is of grave concern as the increase in water level in front of forest also increases the overturning moment on the roots of trees thus resulting in its breakage [32]. The shipping container due to its bigger cross section and high density does not float easily and is rather dragged towards forest which is also the reason for major water level drop at the initial stage after impacting with forest. As soon as the container stops in front of forest container begins to float and the water level rises much greater than the wooden log case as expected. Once the container is pushed away by the water surge towards the edge of forest the water level comes to level and then flow uniformly through the forest.

5 CONCLUSION

In this study, a 3 Dimensional SPH-DEM modelling based on particle approximation is applied to study the i) Effect of variation of water level and flow velocity around the inland forest with respect to parameters like, Aspect Ratio and trees density and ii) Capture and understand the flow pattern of debris floating in tsunami surge. It was observed that when the aspect ratio and tree densities were increased the water level and fluctuations in flow velocity in front of inland forest increases as more volume of water is resisted by forest. The trend was captured by model with an error 0f $\pm$ 1% in water level and $\pm$ 2% in velocity variation.

To understand the interaction of debris floating and interacting with inland forest, two types of debris were considered for this study: i) Wooden log replicating as a full length uprooted tree ii) ISO standard empty shipping container. The flow pattern of the wooden log debris was captured by SPH-DEM model and was validated. Exploring further, shipping container was modelled for the floating case and its flow pattern was observed. Different patterns of debris interaction with sparse forests were identified. First, the debris interacts with forest and there is a chance of wooden log getting trapped in between trees but this was not in case of container due to its bigger dimensions. Second, after the initial impact and sudden rest of debris in front of forest the water from behind further pushes the debris, thus changing its orientation to the longitudinal axis perpendicular to the flow. Third, due to build-up of water behind them results in sliding of debris until its pivot to the outer tree and then pass along the forest. Overall, the model was also able to capture the critical point when the longitudinal axis of the debris is perpendicular to flow that tends to build-up water level behind forest in turn resulting in greater overturning moment and uprooting the trees.

References:

[1] Balk D, Montgomery MR, McGranahan G, Kim D, Mara V, Todd M, et al. Mapping urban settlements and the risks of climate change in Africa, Asia and South America. In: Guzmán JM, Martine G, McGranahan G, Schensul D and Tacoli C, editors. Population Dynamics and Climate Change. New York, London: United Nations Population Fund (UNFPA), International Institute for Environment and Development (IIED); 2009. pp. 80–103.

[2] Small C, Nicholls RJ. A global analysis of human settlement in coastal zones. Journal of Coastal Research 2003; 19: 584–599.

[3] Patterson M, Hardy D. Economic Drivers of Change and their Oceanic-Coastal Ecological Impacts. In: Patterson M, Glavovic BC, editors. Ecological economics of the oceans and coasts. Edward Elgar Publishing; 2008. pp. 187–209.

[4] Crossland C, Baird D, Ducrotoy J-P, Lindeboom H, Buddemeier R, Dennison W et al. The Coastal Zone—a Domain of Global Interactions. In: Crossland C, Kremer H, Lindeboom H, Marshall Crossland J, Tissier MA, editors. Coastal Fluxes in the Anthropocene. Berlin, Heidelberg: Springer; 2005. pp. 1–37

[5] Daigle, R. (2006). Impacts of sea level rise and climate change on the coastal zone of southeastern New Brunswick. Environment Canada.

[6] Matsui, T., Tateishi, H., Isobe, M., Watanabe, A., Mimura, N., & Shibazaki, A. (1992). The vulnerability assessment of sea coastal zone in Japan under the sea-level rise. In Proceedings of Coastal Engineering, JSCE (Vol. 39, pp. 1031-1035).

[7] Hosomi, Y, Kadoyu, K, Uchida, S, Gomi, H, Itahashi, H, and Mimura, N, (2005). Public Awareness toward Sea Level Rise Caused by Global Warming and Future Directions of Coastal Management. J Coastal Eng, JSCE, Vol 52, pp 1291-1295.

[8] Inoue, M, Shimada, H, and Sakamoto, Y, (1993). Effect of Sea Level Rise on Wave Overtopping over a Seawall. Proc Coastal Eng, JSCE, Vol. 40, pp 696-700.

[9] Inoue, M, Shimada, H, and Sakamoto, Y, (1994). Malfunction of Wave overtopping over Seawall due to Sea Level Rise. Proc Coastal Eng, JSCE, Vol. 41, pp 676-680.

[10] Danielsen, F., Sorensen, M. K., Olwig, M. F., Selvam, V., Parish, F., Burgess, N. D., Hiraishi, T., Karunagaran, V. M., Rasmussen, M. S., Hansen, L. B., Quarto, A. and Suryadiputra, N. [2005] "The Asian tsunami: A protective role for coastal vegetation," Science 310, 643.

[11] Kathiresan, K. and Rajendran, N. [2005] "Coastal mangrove forests mitigated tsunami," Estuar. Coast. Shelf Sci. 65(3), 601–606

[12] 12. Iimura, K. and Tanaka, N. [2012] "Numerical simulation estimating the effects of tree density distribution in coastal forest on tsunami mitigation," Ocean Eng. 54, 223– 232

[13] Pasha, G. A., & Tanaka, N. (2016). Effectiveness of finite length inland forest in trapping tsunami-borne wood debris. Journal of Earthquake and Tsunami, 10(04), 1650008

[14] Crespo AJC, Domínguez JM, Rogers BD, Gómez-Gesteira M, Longshaw S, Canelas R, Vacondio R, Barreiro A, García-Feal O. 2015. DualSPHysics: open-source parallel CFD solver on Smoothed Particle Hydrodynamics (SPH). Computer Physics Communications, 187: 204-216. doi: 10.1016/j.cpc.2014.10.004.

[15] Gingold RA, Monaghan JJ. Smoothed particle hydrodynamics: theory and application to non-spherical stars. Mon Not R Astron Soc 1977;181:375–89

[16] Gingold RA, Monaghan JJ. Smoothed particle hydrodynamics: theory and application to non-spherical stars. Mon Not R Astron Soc 1977;181:375–89.

[17] Monaghan JJ, Kocharyan A. SPH simulation of multi-phase flow. Comput Phys Commun 1995;87:225–35

[18] Monaghan JJ and Lattanzio JC. 1985. A refined method for astrophysical problems. Astron. Astrophys, 149, 135–143.

[19] Wendland H. 1995. Piecewiese polynomial, positive definite and compactly supported radial functions of minimal degree. Advances in Computational Mathematics 4, 389-396.

[20] Wu, K., Yang, D., & Wright, N. (2016). A coupled SPH-DEM model for fluid-structure interaction problems with free-surface flow and structural failure. Computers & Structures, 177, 141-161.

[21] Altomare C, Crespo AJC, Domínguez JM, Gómez-Gesteira M, Suzuki T, Verwaest T. 2015a. Applicability of Smoothed Particle Hydrodynamics for estimation of sea wave impact on coastal structures. Coastal Engineering 96, 1-12.

[22] Verlet L. 1967. Computer experiments on classical fluids. I. Thermodynamical properties of Lennard-Jones molecules. Physical Review, 159, 98-103.
[23] Leimkuhler BJ, Reich S, Skeel RD. 1996. Integration Methods for Molecular
[25] Canelas RB, Domínguez JM, Crespo AJC, Gómez-Gesteira M, Ferreira RML. 2015. A Smooth Particle Hydrodynamics discretization for the modelling of free surface flows and rigid body dynamics. International Journal for Numerical Methods in Fluids, 78, 581-593.
[26] Canelas RB, Crespo AJC, Domínguez JM, Ferreira RML and Gómez-Gesteira. 2016. SPH-DCDEM model for arbitrary geometries in free surface solid-fluid flows. Computer Physics Communications, 202, 131-140. doi:10.1016/j.cpc.2016.01.006.
[27] Dalrymple, R.A. and Knio, O., (2000): SPH Modelling of Water Waves. Proc. Coastal Dynamics, Lund, 779-787
[28] Monaghan, J. J. (1994): Simulating free surface flows with SPH. Journal Computational Physics, 110: 399- 406
[29] Gómez-Gesteira and Dalrymple, R. (2004): Using a 3D SPH Method for Wave Impact on a Tall Structure, J. Wtrwy. Port, Coastal and Ocean Engrg., 130(2), 63-69.
[30] Spiske, M., Weiss, R., Bahlburg. H., Roskosch, J. and Amijaya, H. [2010] "The TsuSedMod inversion model applied to the deposits of the 2004 Sumatra and 2006 Java tsunami and implications for estimating flow parameters of palaeo–tsunami," Sediment. Geol. 224, 29–37.
[31] Tanaka, N., & Ogino, K. (2017). Comparison of reduction of tsunami fluid force and additional force due to impact and accumulation after collision of tsunami-produced driftwood from a coastal forest with houses during the Great East Japan tsunami. Landscape and Ecological Engineering, 13(2), 287-304.
dynamic IMA Volume in Mathematics and its application. Springer
[24] Monaghan JJ. 2005. Smoothed Particle Hydrodynamics. Rep. Prog. Phys, 68, 1703-1759

Proceedings of the 6th International Conference on Ship and Offshore Technology
ICSOT 2019
November 7-8, 2019, IIT Kharagpur, India

ICSOT INDIA 2019-02

CHALLENGES IN MAINTAINING STRUCTURAL INTEGRITY OF AGED FIXED OFFSHORE PLATFORMS IN WESTERN OFFSHORE OF INDIA

Amit Jain
Institute of Engineering and Ocean technology
(IEOT)-ONGC, Panvel,
Navi Mumbai, Maharashtra-410206, India
E-Mail: jain_amit @ongc.co.in

S. Karunanithi
Institute of Engineering and Ocean technology
(IEOT)-ONGC, Panvel,
Navi Mumbai, Maharashtra-410206, India
E-Mail: karunanithi_s@ongc.co.in

ABSTRACT

ONGC is a national E&P company in India. Presently, ONGC is operating more than 300 fixed offshore platforms in western offshore of India. These platforms are supported by a lattice type of steel tubular structures. In general, these platforms are secured to seabed by steel tubular piles which are driven below seabed up to depth of 80m to 120m depending on the soil condition prevailing at the location. These platforms are designed with production facilities for a service life of 25 years. ONGC takes up number of re-development projects for enhanced production which requires post installation modifications such as addition of clamp-on conductor wells, additional risers, new facilities and deck extensions resulting in increase of hydrodynamic as well as gravity loads on these structures. Structural integrity of these platforms need to be documented with respect to prevailing provisions of the design code. It is imperative to carry out the re-assessment of these structures for extending their use.

This case study discusses the reassessment of a well platform for additional facilities and the challenges involved for documenting the structural integrity. Higher level non-linear ultimate strength analysis was also carried out for understanding the failure mechanism of the jacket. Based on the higher level analysis, mitigation such as load reduction and underwater strengthening of X-brace member have been recommended.

NOMENCLATURE

GoM-Gulf of Mexico
RSR-Reserve Strength Ratio
UC-Unity Check

1. INTRODUCTION

Many existing platforms in Indian offshore region have undergone various modifications during their service life for additional facilities such as clamp-on conductors & risers for accommodating the changing production requirements, deck extension for addition of associated equipments, revamp or modification of jacket platforms for field redevelopment projects, new discoveries in the region resulting in increase of loading on the structure. Reassessment of the jacket structure has been carried out in view of modifications and for extended use beyond the designed life to fulfill the operational and regulatory requirements (Ref. iii). Significant changes have been incorporated in the design code provisions of API RP 2A (WSD) i.e. increased values of drag and inertia coefficient, increased marine growth thickness after hurricane Juan in 1985 in GoM, resulting in 20-30% increase of hydrodynamic loads on the structure. Due to above changes in design criteria and additional loadings, there is a great challenge in documenting the structural integrity of these structures. During the reassessment, it is observed that many main structural **members and joints** show overstressing due to the upward revision of drag & inertia coefficient and marine growth parameters. In number of cases, ultimate capacity of piles of platforms are insufficient for safe transfer of the pile axial loads to the soil. In that scenario, it requires the installation of additional piles for documenting the structural integrity assessment of the platform.

2. CASE STUDY

Institute of Engineering & Ocean Technology (IEOT) carries out structural integrity assessment studies for number of fixed offshore jacket structures. Case study of one 8-legged (4 main plus 4 skirt leg) well head platform situated in water depth approximately 80 meters has been discussed in this paper. A global static in-place analysis of the platform has been carried out for 100 year return period extreme storm condition for 8 directions of wave, current and wind (with wave parameters of H_{max}=18.3m, T_{max}=14.4Sec for predominant wave direction) along with other design loads and marine growth thickness of 100mm from (+) 3.0m to (-) 30m & 50mm from (-) 30m to mud-line using SACS software as per API RP 2A (WSD).

a. View of the platform   b. Structural model

The results of soil pile interaction analysis of jacket structure shows that, all piles (4 Main & 4 Skirt) are having Factor of Safety (FOS) for pile axial load transfer more than the API-RP-2A WSD minimum required value of 1.50 (extreme condition) under compression and tension condition. UC ratio of all the piles are less than the permissible value of 1.00. Hence, jacket structure of the platform is satisfying the structural integrity requirements in terms of pile FOS and pile axial stress.

The reassessment analysis results reveals that Eight structural members and four X-brace structural joints were found to be overstressed as they are having UC ratio more than the API-RP-2A WSD specified acceptable limit value of 1.0 and hence, do not meet the structural integrity assessment criteria in design level in-place analysis. The details of members having UC ratio more than 1.00 are depicted as UC^1 shown below in table 1.

TABLE 1: MEMBER HAVING UC RATIO MORE THAN 1.00

Member	UC^1	UC^2
1300-602L	1.317	1.024
202L-0151	1.273	1.168
204L-0150	1.368	1.248
503L-0200	1.034	0.928
303L-0196	1.105	0.978
404L-0196	1.134	0.999
203L-1146	1.166	1.024
304L-0205	1.022	0.904

UC^1 : UC ratio in design level in-place analysis with 85% Environmental loading
UC^2 : UC ratio in simplified ultimate strength analysis with 100% Environmental loading

4. SIMPLIFIED ULTIMATE STRENGTH ANALYSIS

The members & joints which are not passing in design level assessment are checked in simplified ultimate strength analysis. Member and joint simplified ultimate strength check results for jacket structure with respect to API 2SIM as per Clause No.8.5.4.2 shows that, four out of eight members which were not passing in the design level analysis, are still having UC ratio more than the API-RP-2A WSD specified acceptable limit value of 1.00, hence, structural integrity assessment could not be documented even in simplified ultimate strength analysis. The details of members checked in simplified ultimate strength analysis are depicted as UC^2 shown below in table 1.

5. NON-LINEAR ULTIMATE STRENGTH ANALYSIS

The non-linear ultimate strength analysis has been carried out, as per API R 2SIM, Clause 8.5 (Ref. 2) and it is found that minimum Reserved strength Ratio (RSR) of jacket structure is 1.15 which is less than the target RSR value of 1.323 (as per DNV OS C101). Hence, structural integrity of jacket structure could not be documented even in non-linear ultimate strength analysis.

5.1 Mitigation measures

To document the structural integrity of jacket structure, there is need of load reduction by identifying and removing some redundant facilities. In view of above, redundant facilities comprising of one 12" riser and Fire water pump caisson were identified for removal.

The re-analysis was carried out considering removal of 12" riser and Fire water pump caisson and it was found that minimum Reserved strength Ratio (RSR) of the jacket structure is 1.20 which is still less than the target RSR value of 1.323 (DNV OS C101). Hence, structural integrity of jacket structure could not be documented even after load reduction in non-linear ultimate strength analysis.

5.2 Conductor as pile:

There are 12 numbers of 30" diameter conductors installed in the well bay area in northern portion of the platform. Wall thickness of the conductors are one inch. These conductors are driven to a depth of 70m below seabed. Reanalysis has been carried out considering these conductors as piles in non-linear ultimate strength analysis in order to provide additional lateral foundation capacity to the structural system (As per API RP2 SIM, Clause No. 8.5.4.4.4). The analysis results reveals that buckling of one of the X-brace member (610mm dia & 16mm wall thickness) occurs before reaching the target RSR 1.323. Therefore, there is a need for underwater strengthening of this critical member to meet the target RSR for documenting the structural integrity. Deflected shape of this members is depicted in Figure 2 below.

5.3 Strengthening of member and conductor as pile:

The re-analysis has been carried out considering conductor as pile and grouting chord member of X-brace joint. The equivalent thickness have been modeled for analysis. The non-linear ultimate strength analysis reveals that minimum Reserved Strength Ratio (RSR) of the jacket structure is 1.367 which is more than the target RSR value of 1.323. Hence, structural integrity of jacket structure could be documented in non-linear ultimate strength analysis with strengthening (infill cement grout which adds to 0.64 tons/m length of the member) of one X-brace member and removal of identified redundant facilities.

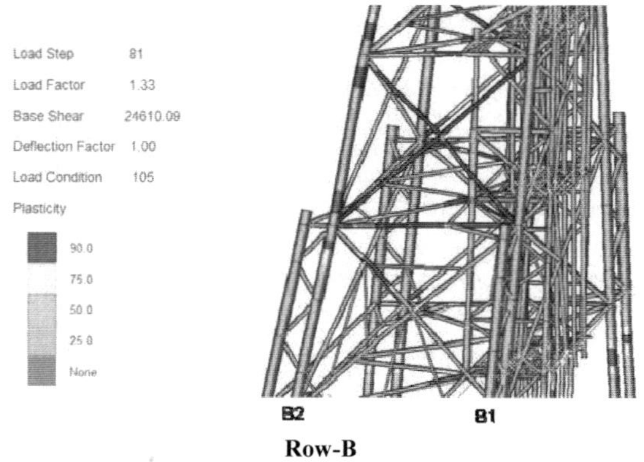

FIGURE 2: DEFLECTED SHAPE OF THE MEMBER IN NON-LINEAR ULTIMATE STRENGTH ANALYSIS

6. RESULTS AND DISCUSSION

The reassessment analysis results reveals that the jacket structure of the platforms was not meeting the structural integrity assessment criteria in design level in-place analysis with overstressing of members and joints. Structural integrity of the jacket structure could be documented in non-linear ultimate strength analysis only by strengthening (infill cement grout) of X-brace member and removal of 12" riser and Fire water pump caisson.

7. CONCLUSIONS

ONGC has been operating in the western offshore since 1976. Many platforms installed since then have outlived their design lives, however, are still required to stay under production due to advancement in the production technology. Thus, life extension needs to be carried out for them. For some of the platforms, during life extension studies, it is found that the structure does not meet the present day re-assessment criteria due to over utilization of some structural components. This becomes more important with respect to overstress in piles, which in general cases would ask for installation of additional ones for extended use of the platforms. IEOT, for such cases has recommended, as in the present study, to remove all the redundant facilities and strengthen the platform for documenting its structural integrity. This will result in substantial benefit to the organization in terms of continued production and revenue generation from the facility.

8. ACKNOWLEDGEMENT

The views expressed in the paper are those of authors only. We acknowledge the support and resources provided by ONGC required for carrying out this study. We are immensely grateful for the continuous motivation, generous support and encouragement received from Shri Dinesh Kumar, ED- Head of Institute, IEOT.

9. REFERENCES

i. API – "Planning, Designing and Constructing Fixed Offshore Platforms – Working Stress Design"; API RP 2A-WSD.

ii. Structural Integrity Management of Fixed Offshore Structures, API Recommended practice 2SIM.

iii. Re-qualification methodology for fixed offshore platforms in west coast of India" DOC IEOT-LIFE-001, REV-4, IEOT, ONGC, 2012

iv. Design of Offshore Steel Structures, General (LRFD Method), DNVGL-OS-C101.

Proceedings of the 6th International Conference on Ship and Offshore Technology
ICSOT 2019
November 7-8, 2019, IIT Kharagpur, India

ICSOT INDIA 2019-03

SPECTRAL FATIGUE ANALYSIS OF SHIP
PART 1: USING BEAM THEORY BASED STRUCTURAL RESPONSES

Amresh Negi
Indian Register of Shipping
Mumbai, Maharashtra, 400072
India
amresh.negi@irclass.org

Yogendra Parihar
Indian Register of Shipping
Mumbai, Maharashtra, 400072
India
yogendra.singh@irclass.org

Suhas Vhanmane
Indian Register of Shipping
Mumbai, Maharashtra, 400072
India
suhas.vhanmane@irclass.org

ABSTRACT

The Spectral Fatigue Analysis (SFA) is the comprehensive fatigue life assessment method for vessels. This analysis is to be carried out for the many critical locations of the ship structure. The spectral fatigue analysis is performed through the process of hydrodynamic response analysis, global and local structural analysis, and eventually prediction of fatigue damage using the long-term distribution models. To perform these numerically intense stages, enormous amount of computational resources are required. Considering the computational efficiency, it is not worthwhile to go for such analysis for the entire ship. Therefore, individual stages can be simplified to cut down the analysis time without upsetting the physics of the problem. This type of approach by simplifying the intermediate stages can serve as some sort of preliminary level of SFA to limit the number of cases for the detailed SFA. The present paper discussed these types of simplification in SFA though application of methods using a sample ship problem.

INTRODUCTION

Last two decades, fatigue damage assessment for ship structure gained a lot of interest in the shipping industry and special attention has been paid towards this failure mode. Being a compulsory requirement from the ship stakeholders, many classification societies published guidelines to carry out the detailed fatigue analysis for variety of marine structures. Shipping industry developed its concern towards accepting the fatigue assessment of critical structural elements after many bulk carriers and lives has been lost [1]. Side shell cracks were observed and found responsible for these accidents. Not only bulk carriers but other ship accidents were also recorded with the cracks in the hull structure within the short period of commencement in service. Fatigue was considered to be an important contributor for these structural damages. It has been recognized that even though fatigue damage does not result complete structural failure but the estimated cost of repair and consequences to marine pollution are high. Apart from past bad experience with ship structural failure and damages, there were other reasons which contributed the inclusion of fatigue assessment as new criteria to be considered in design stage. Some of these reasons are mentioned here as:

- Optimize hull structure to improve the strength-to-weight ratio by introducing new material such as Aluminium and high tensile steel.
- Rise in number of ageing ships with lack of maintenance.
- Growing concern towards the safety of ship, human and environment.

In general, there are two approaches which are used for performing the fatigue assessment, namely S-N approach and fracture-mechanics approach. Fracture-mechanics approach may be useful in evaluating the crack growth after a crack is spotted. This approach is used in developing the plans for inspection and repair. In practice, the S-N approach is widely used for the fatigue assessment and design. Experimentally accomplished S-N curves are used for fatigue strength characterization. The S-N curve based approach consists of three methods namely simplified fatigue method, deterministic fatigue method and spectral fatigue method. In simplified fatigue life assessment method, the dominant loads which determine the stress range for the structural locations are calculated by empirical formulas provided by various ship classification societies. This method involves the long-term distribution of stress ranges to be characterized by Weibull distribution. Being a simple method to apply, this does not account for the specific ship details and operating conditions. In the deterministic method, a sea state is simply characterized using a deterministic wave height and period. In contrast to spectral method, deterministic method does not consider the spectral energy corresponding to sea state. This method is applicable for special marine structures and specific operating conditions [2].

The spectral-based method is a frequency domain assessment method which is complex and numerically intensive technique. This method relies on the assumption of linearity between wave-induced loads with respect to waves and presumes linear relationship between structural responses and the wave-induced loads. The spectral based fatigue assessments of ship predict the fatigue life; therefore this method is referred as direct method. Fatigue assessment for ship structures are typically conducted using direct calculation procedures to compute fatigue loads. On contrary to conventional-rule based design approach, the direct calculation approach includes the structural and operational details pertaining to each individual vessel. Time domain method along with rain flow counting technique can also be employed for the fatigue assessment of the structures for which the non-linear responses are important and need to be considered. The applicability of time domain method is limited to specific offshore structures [2].

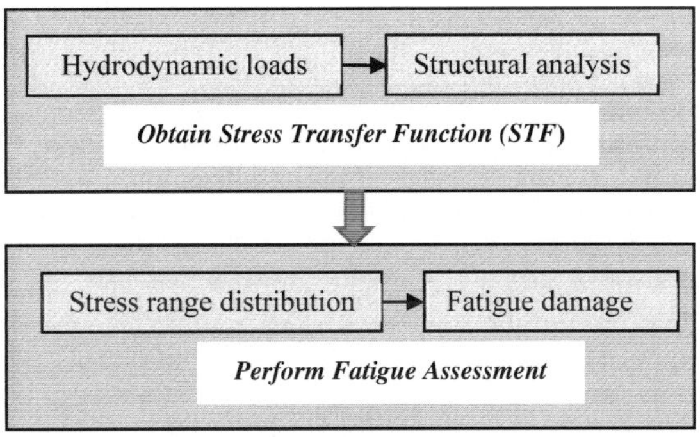

Figure 1. SCHEMATIC DIAGRAM FOR DIRECT FATIGUE ASSESSMENT

With reference to rule based simplified and direct calculation approaches, DNV-GL [3] suggested different possible analysis combinations using both approaches where intermediate results of one approach can be interchanged with other approach. Therefore, the fatigue analysis may be performed based on a combination of simplified and refined techniques such as spectral analysis. Determination of loads for the representative loading condition serves the initial step for both the approaches. Combination of stresses resulting from the action of global and local loads is to be performed according to each class society criteria and with consideration given to the probability level.

Rule based simplified approach consist of the evaluation of loads based on individual classification society's rules or methodology based on Common Structural Rule [4]. However, direct calculation approach requires the numerical computation of loads based on the operational parameters and service route of the ship. Numerically evaluate stresses based on the actual ship operating conditions can also be used for the simplified approach rather than using the rule based empirical formulations. This can be achieved by evaluating Weibull parameters by fitting Weibull distribution to long-term stress range. In this manner, actual ship and operational parameters can be taken into account for classification society's simplified approach. This approach is also included in present paper to predict the damage for the sample ship in addition to SFA approach.

FATIGUE ASSESSMENT METHODS

Spectral fatigue analysis primarily consists of four different stages: Computation of hydrodynamic loads, structural analysis, long term and short term stress distribution and damage calculation as shown in Figure 1. The fundamental task of a spectral fatigue analysis is the determination of the stress range transfer functions or Stress Transfer Functions (STF), which express the stress response of a structural location for unit wave amplitude of specified wave frequency and heading. Once STFs are known, remaining task is to perform statistical analysis.

The most comprehensive method to get structural responses i.e. STF, is to perform Finite Element (FE) with the direct application of hydrodynamic loads. Where, loads are to be computed using 3D panel method. This requires enormous computation resources and time. However, In available literature, authors try to simplify this step by the application of global loads calculated using 2D strip theory for FE structural application [5] or even simply using beam theory [6]. Such a simplification may serve an initial level of SFA to sort out the number of critical location for full 3D hydro-structural analysis to get the STF or predict the critical locations which require immediate attention during ship structural survey.

In this paper work, spectral fatigue analysis of a bulk carrier has been conducted for the butt- welded plate joints at deck and side shell locations at mid-ship (*0.5L*). 2D strip theory and close–form semi analytical formulation for the evaluation of global loads has been used. Stress range transfer functions have been obtained in simpler manner using beam theory. However, steps and methods to perform statistical analysis based spectral fatigue assessment have been explained elaborately.

Simplified method has also been applied for which shape and scale parameter were obtained numerically by fitting the Weibull distribution on long-term stress range. Finally, the results were discussed with respect to the different methods used in present work and conclusions were drawn.

Table 1. SUMMARY OF THE FATIGUE ASSESSMENT APPROACHES

Method – IDs	Load Evaluation Method	Fatigue Damage Approach
FAM-1	Semi analytical formulation	Closed form approach based on long-term response
FAM-2	2D Strip theory	Closed form approach based on long-term response
FAM-3	Semi analytical formulation	*Spectral approach Based on short term response
FAM-4	2D Strip theory	*Spectral approach Based on short term response
FAM-5	3D Panel method	*Spectral approach Based on short term response

*long term stress range distribution is defined through a short-term Rayleigh distribution within each short term sea state

Table 1 summarizes the fatigue damage prediction approaches which have been followed in the present papers (Part-1 and Part-2). These approaches are the combination of the different load evaluation and damage prediction methods. Hence forth the method IDs are used when referring the various methods as mentioned in Table 1.

Present paper focused on FAM-1 to FAM-4 fatigue assessment methods using beam theory based structural responses i.e. STF evaluation. FAM-5 consists of the direct application of hydrodynamic loads (pressure and motions) to FE model of ship. Complete description of the process has been provided in Part-2 of the paper.

STRESS RANGE TRANSFER FUNCTION

The primary step to evaluate the stress transfer function is the computation of hydrodynamic loads using potential theory (2D strip or 3D panel method) based seakeeping programs or any other alternate method such as semi-analytical closed form formulation for vertical bending moment. Once the wave induced loads and ship motions are obtained, the next step is to perform structural analysis either direct application of loads on FE model (explained in detail in Part-2) or using beam theory.

Hydrodynamic Load

Hydrodynamic loads need to be calculated for each representative loading condition for the ship. During voyages, vessel encounters the ocean waves from different directions and as result it undergoes through so called wave induced loads. If ship considered as flexible beam subjected to random sea environment, which bends the ship hull girder upward direction (hogging) and downward direction (sagging) depends on the position of the wave crests along the ship hull. Repetitive nature of these loads makes structure element (Plates and stiffeners) fail in fatigue mode. In case of bulk carrier the deck and side stake plates predominantly subjected to global loads under the normal operating environmental conditions. Here, following two methods have been used to obtain load transfer functions or load RAOs. Global loads were computed using Strip theory and semi analytical formulations in the present part of the paper.

VBM and HBM Using Strip Theory: In 2D strip theory calculations, the wave loads on a ship are found by integrating the two dimensional loads on the cross sections of an un-restrained ship over the ship length [7]. The dynamic loads (vertical and horizontal bending moments; VBM and HBM) at any section is the difference between the inertia force and the sum of the external forces acting on the portion of the hull. If the external forces are separated as static restoring force/moment R_j, the exciting force/moment E_j, and hydrodynamic force/moment due to body motion D_j, we find the load Eqn. (1)

$$V_j = I_j - R_j - E_j - D_j \tag{1}$$

Where, j = load index (j = 5 for VBM and 6 for HBM)

The inertia is the mass times the acceleration. Here, the inertia force, I_j is expressed in terms of the sectional inertia force. Hydrostatic moments are linear and computed by considering the actual variation of the individual sectional draft and thus accounting for the vessel motions. Since there is no restoring force/moment in horizontal plane, therefore $R_6 = 0$. For excitation forces, Froude-krylov and diffraction moments need to be evaluated. The hydrodynamic moments are caused due to the body motion. So, the D_j term in Eqn. (1) consists of sectional added mass and damping. All the terms of the dynamic load equation can be computed using numerical seakeeping program. While computing dynamic loads, a critical test for consistent treatment of forces and moments is to be conducted which intended all sea loads must be equal to zero at the aft and forward of the ship. This condition needs to be satisfied through careful attention to several details such as hydrostatic balancing of forces and moments.

VBM using Semi Analytical Close-From Formulation: The most important design parameter in assessing the ship strength is the vertical bending moment. This load becomes most important at design level to estimate the section modulus of the ship. Prevailing practice to determine the wave bending moment has been remained the use of formulas issued by the classification societies. Based on the first principle and with simplifications, semi-analytical approach has been used to derive frequency response functions for the wave induced vertical bending moments for mono-hull ships [8].

Input information require for the closed-form expression is restricted to the main dimensions: length, breadth, draught, block coefficient and water plane area together with speed and heading. The formula, Eqn. (2) makes it simple to obtain quick estimate of the wave-induced vertical bending and used as an alternate to numerical computation of VBM transfer function.

$$\frac{\Phi_M}{\rho g B L^2} = \kappa \frac{1-kT}{(k_e L)^2}\left[1-\cos\left(\frac{k_e L}{2}\right)-\frac{k_e L}{4}\sin\left(\frac{k_e L}{2}\right)\right]F_V(F_n)F_C(C_b) \tag{2}$$

$$\kappa = \exp(-k_e T) \qquad k_e = |k\cos\theta|$$

Where, V is the forward speed, θ is the heading angle (180° corresponding to head sea), B and T are the breadth and draught, k is the wave number, ω is the wave frequency ($\omega^2 = kg$). $F_C(C_b)$, $F_V(F_n)$ are the correction factors for the block coefficient ($C_b \geq 0.6$) and speed ($F_n < 0.3$) respectively. The details are provided by Jensen [8].

Structural responses or load Response Amplitude Operators (RAOs) need to be obtained for recommended range of wave frequencies and all directional wave headings for specified average service speed. Semi analytical expression (close form) results in VBM load transfer functions. Whereas, other set of loads i.e. VBM and HBM loads and phases were numerically computed using strip theory.

Structural Response

Stress transfer function is found using the application of beam theory for each set of the loads obtained using the two different approaches as motioned in previous section. Vertical and Horizontal bending moment RAOs ($RAO_{M,V}$ and $RAO_{M,H}$) are converted into stress transfer function (Vertical and horizontal bending stress RAO) as:

$$RAO_{\sigma,h} = \frac{y}{I_{ZZ}} RAO_{M,H} \tag{3}$$

$$RAO_{\sigma,v} = \frac{z - z_0}{I_{YY}} RAO_{M,V} \tag{4}$$

Where, z is vertical distance of the structural part from the base line in meters. y is the horizontal distance of the structural part from the center line in meters. z_0 is the distance of the neutral axis from the base line. I_{YY}, I_{ZZ} are ship section modulus with respect to neutral axis and the center line axis of the vessel respectively in m^4. The combined transfer function can be derived as:

$$RAO_\sigma = \begin{bmatrix} (RAO_{\sigma,v})^2 + (RAO_{\sigma,h})^2 + \\ 2.\, RAO_{\sigma,v}.\, RAO_{\sigma,h}.\, \cos(\varepsilon_v - \varepsilon_h) \end{bmatrix}^{1/2} \tag{5}$$

Where ε_v and ε_h denotes phase of the stress process due to vertical and horizontal hull girder bending respectively. Stress transfer function can be obtained

$$H_\sigma(\omega|\theta) = RAO_\sigma \tag{6}$$

Semi analytical close-form load evolution approach is used to determine vertical bending moment only. Therefore, STFs calculation contains the effect of VBM.

FATIGUE DAMAGE ASSESSMENT

Operating Conditions

For fatigue damage assessment, it is require having detailed information about ship operating condition which includes:

1. Percentage time spent in each wave environment if ship is operating on specific route (p_{ij}).
2. Fraction of time spent in each loading condition (p_l).
3. Probability of encountering for each wave heading (p_k).
4. Vessel's average speeds and time spend in each average speed when sailing in a particular sea state.

In present work, Scatter diagram for the worldwide trade excerpt are taken from the DNVGL guideline [3]. Standard loading conditions for the SFA of bulk carrier are defined as homogeneous, alternate, normal ballast and heavy ballast as[9].

The fractions of time spend in each loading conditions of bulk carrier are taken as per rule [9]. Usually, an equal probability of ship heading with respect to direction of wave is considered and followed in the present paper. The vessel average speed is taken as 75% of the service speed [10].

Wave Environment

Ocean waves are considered to be main source of fatigue damage. The wave data usually available in the form of scatter diagram for various regions of the entire world ocean. Wave scatter diagram represents the standard environmental conditions. This contains the probability of occurrence of different sea states defined with significant wave heights (H_S) and zero crossing periods (T_Z). For each combination of H_S and T_Z, the probability of occurrence is found by dividing the observation for a sea state with total number of observations. The Pierson-Moskowitz (PM) spectrum was used to describe the short-term sea states.

$$S_\xi(\omega|H_S,T_Z,\theta)=\frac{H_S^2}{4\pi}\left(\frac{2\pi}{T_Z}\right)^4\omega^{-5}exp\left(-\frac{1}{\pi}\left(\frac{2\pi}{T_Z}\right)^4\omega^{-4}\right) \quad (7)$$

Spectral Moments

The response spectrum of the ship based on the linear model is obtained directly from the wave spectrum as defined using Eqn. (7) and stress transfer function as obtained using Eqn. (6). Stress response spectrum can be obtained as

$$S_\sigma(\omega|H_S,T_Z,\theta)=|H_\sigma(\omega|\theta)|^2.S_\xi(\omega|H_S,T_Z) \quad (8)$$

3D irregular sea way can be modeled using the spreading function. Cosine-squared spreading is assumed from +90 to – 90 degrees on either side of the selected dominant wave heading. Spectral moments for each short-term sea states can be computed using Eqn. (9):

$$m_n=\int_0^\infty\int_{\theta'=\theta-90}^{\theta'=\theta+90}\left(\frac{2}{\pi}\right)cos^2\theta'.(\omega_e^n S_\sigma).d\theta'.d\omega \quad (9)$$

$$\omega_e=\omega-\frac{V\omega^2}{g}cos\theta'$$

Where θ' is the spreading angle between a wave component and the dominant wave direction. Short-term and long-term stress distributions in the form of response spectrum are to be obtained. This involves certain details and calculations for number of investigated load cases in a lifetime. The following section provides the theoretical background for the same.

FATIGUE DAMAGE ASSESSMENT

S-N curve gives a relationship between the applied stress amplitude (S) and number of cycle (N) to failure at that stress amplitude. In general, it illustrates the material or structural element capacity to fatigue failure at constant stress. The variable amplitude cyclic load are divided into block of stresses using the Palmgren-Miner summation rule. It is assumed that the total cumulative damage of a structural element is linear summation of the damage in each stress block, and can be given by following Eqn. (10)

$$D=\sum_{i=1}^{n_t}\frac{n_i}{N_i} \quad (10)$$

Where, n_i is the number of cycles of constant amplitude stress ranges and can be expressed in terms of a stress distribution, which may be fitted into probability distributions function, N_i is the total number of cycles to failure under a constant amplitude stress range and can be determined using S-N curve characteristics. n_t represents the total number of stress blocks.

Stress range corresponding to 10^{-2} probability level has been considered for the mean stress factor calculation. Appropriate factors for correction of mean stress effect, thickness effect and material factor are taken in analysis [9]. A factor of 0.85 is considered to account for the exclusion of harbor operations.

In present paper, the nominal stress approach is used to determine the fatigue damage of all transverse butt-welded joints. The S-N curve 'FAT80' was selected for the assessment, which considers an axial misalignment of 10% in plate thicknesses [12].

Closed Form approach based on long-term response

The probability density function of peak values of the stress range response for a ship lifetime $[f(\Delta\sigma)]_L$ can be presented as a weighted sum of the various short term probability density functions [13].

$$\frac{1}{n_L}=[f(\Delta\sigma)]_L=\frac{\sum_i\sum_j\sum_k\sum_l\bar{n}\,p_{ij}p_kp_l.[f(\Delta\sigma)]_{ijkl}}{\sum_i\sum_j\sum_k\sum_l\bar{n}\,p_{ij}p_kp_l} \quad (11)$$

Where, $[f(\Delta\sigma)]_{ijkl}$ is the probability density function for amplitudes of the short-term responses, n_L is the total number of responses expected in the lifetime of the ship.

$$[f(\Delta\sigma)]_{ijkl}=1-[F(\Delta\sigma)]_{ijkl}=exp\left(-\frac{\Delta\sigma^2}{8m_{0,ijkl}}\right) \quad (12)$$

$$\bar{n}=\frac{1}{2\pi}\sqrt{\frac{m_{2,ijkl}}{m_{0,ijkl}}} \quad (13)$$

Where i, j, k, l represents the index numbering for significant wave height, average zero crossing period, wave heading and loading condition.

The long term stress range for fatigue assessment is approximated using a two parameter Weibull distribution. Weibull shape (α) and scale parameters (k) are evaluated for the butt welded joint location.

$$F(\Delta\sigma)=1-\exp\left(\frac{\Delta\sigma}{k}\right)^\alpha \quad (14)$$

The least square method is used for fitting of the Weibull distribution to the sum of Rayleigh distributions for a number of probability levels of exceedance. As a result, shape and scale parameters can be determined straight forward for each structural location.

When the long-term stress range distribution is defined applying Weibull distributions, and a two-slope S-N curve is used, the fatigue damage for single loading condition is given by [3]

$$D=\nu_0T_d\left[\frac{q^{m_1}}{\bar{a}_1}\Gamma\left(1+\frac{m_1}{h};\left(\frac{S_1}{q}\right)^h\right)+\frac{q^{m_2}}{\bar{a}_2}\gamma\left(1+\frac{m_2}{h};\left(\frac{S_1}{q}\right)^h\right)\right] \quad (15)$$

Where,

ν_0 = long-term average zero-up-crossing-frequency (Hz)

T_d = Design life in seconds.

S_1 = Stress range for which change of slope of S-N curve occur

$\bar{a}_1$, m_1 = S-N fatigue parameters for N < 10^7 cycles

$\bar{a}_2$, m_2 = S-N fatigue parameters for N > 10^7 cycles

γ = Incomplete Gamma function

Γ = Complementary Incomplete Gamma function

Spectral Approach Based on Short Term Response

Equation for the fatigue damage in specific sea state the stress range is normally expressed in terms of probability density functions for different short-term intervals corresponding to the individual cells of the wave scatter diagram. Linear addition of short term damages sustained over all the sea states gives the total damage for the structure element. Total fatigue damage accumulated over operational service life ($T_D = 25$ years) can be estimated by accounting for all sea states encountered with the different wave directions and represented loading conditions.

When the long term stress range distribution is defined through a short term Rayleigh distribution within each short term sea state for the each different loading conditions, and a two-slope S-N curve is used, the fatigue damage expression is given as [3]

$$D = \nu_0 T_d \sum_{i=1,j=1}^{\substack{\text{All sea states} \\ \text{All headings}}} r_{ij}\,[A + B] \tag{16}$$

$$A = \frac{\left(2\sqrt{2m_{0ij}}\right)^{m_1}}{\bar{a}_1} \Gamma\left(1 + \frac{m_1}{2}; \left(\frac{S_0}{2\sqrt{2m_{0ij}}}\right)^2\right)$$

$$B = \frac{\left(2\sqrt{2m_{0ij}}\right)^{m_2}}{\bar{a}_2} \gamma\left(1 + \frac{m_2}{2}; \left(\frac{S_0}{2\sqrt{2m_{0ij}}}\right)^2\right)$$

Where,

r_{ij} = the relative number of stress cycles in short-term condition *i, j*

S_0 =Stress range in *S-N* curve, where the change of slope occurs

TOTAL FATIGUE DAMAGE

Total fatigue damage is taken as a sum of damage occurred in a particular loading condition considering the fraction of time spent in each of the loading conditions. Let us assume that D_1, D_2, D_3 and D_4 are the damages occurred in homogeneous, alternate, normal ballast and heavy ballast loading condition respectively. Therefore, combined fatigue damage is represented by Eqn. (17) for the case of bulk carrier having length more than 200 m. [9].

$$D = 0.25\,D_1 + 0.25\,D_2 + 0.2\,D_3 + 0.3\,D_4 \tag{17}$$

NUMERICAL COMPUTATION

In present study, fatigue damage assessment of transverse butt-welded plates in the two deck locations (DK1 and DK2) and side shell (SS1) structure of a bulk carrier have been carried out. Ship particulars are shown in

Table 2.

Figure 2 shows the representative mid-ship section indicating the butt welded joint locations. Load transfer functions for the mid-ship section are computed using strip theory (VBM and HBM amplitudes and phases) and semi analytical VBM formulation. Seakeeping analysis using 2D strip theory requires three main inputs namely geometry in 2D sectional format, mass distribution along the ship hull along with parameters related to loading condition and wave definition. The load transfer function is calculated for the following set of parameters:

- Frequency: λ/L = 0.2 ~ 5.0
- Wave headings: 0~330 (step of 30 deg.)
- Speed profile (75% of the service speed)

Table 2. Ship particulars

Ship type	Bulk carrier
Length overall [m], L_{oa}	287.50
LBP [m], L_{BP}	279.00
Breadth (moulded) [m], *B*	45.00
Depth (moulded) [m], *D*	24.10
Scantling Draught [m], T_{sc}	18.49
Max Service speed [knots], V_s	14.60

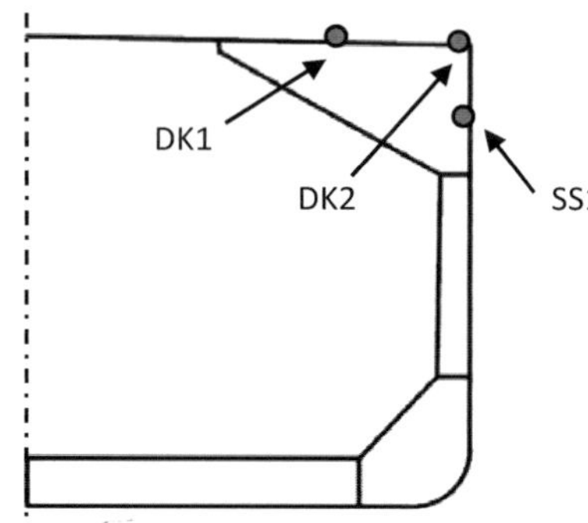

Figure 2. REPRESENTATIVE MID-SHIP SECTION SHOWING THE BUTT-WELDED PLATE JOINTS WITH IDS (DK1, DK2 AND SS1)

Figure 3 and Figure 4 shows the VBM and HBM loads transfer function obtained for Alternate loading condition of the ship. Non-dimensional load RAOs has been plotted against the wave length to ship length ratio. It has been observed that for this loading condition, following sea (0 deg.) contributed to the maximum value of load RAO. Stern quartering seas (30 and 60 deg.) provide the larger values of RAO then the head sea. The HBM for the head (180 deg.) and following sea are almost zero. For the wave heading 60 deg. and 120 deg. ship experiences the higher values of HBM than the other headings. Similarly, load RAOs have been computed for remaining other three loading conditions.

Figure 5 shows the vertical bending moment RAO obtained for the alternate loading condition using close-form semi analytical method. This VBM formulation suffers the limitation due to assumption of ship's aft and forward symmetry. Therefore, it predicts the same value for a pair of equivalent headings when wave encounters from stern and head directions.

Sectional properties are determined for the plane of butt-welded joints. Beam theory has been applied to obtain the STFs for the three specified butt-welded locations using Eqn. (3) and Eqn. (4). Figure 6 to Figure 8 show the STF for the all the three mid-ship locations based on the strip theory load transfer functions for various headings. The stress transfer functions have been calculated for each load components. The structural responses have been combined using the respective phases (Eqn. (5)). All location experienced maximum stress range transfer function for the following sea condition (0^0 wave heading). Location SS3 experienced overall lesser values of stress transfer function compare to locations, DK1 and DK2. In same fashion, the STFs have also been calculated using the beam theory and VBM load transfer function results obtained from semi analytical close-form formula.

Spectral analysis is performed for the each sea states encountered by the ship described by the World wide scatter diagram [3]. Approaches followed for fatigue assessment are Weibull long term stress distribution (Close Form) and short term Rayleigh distribution (spectral) as described above sections. Table 1 is summarizing the different fatigue assessment approaches used in this paper except FAM-5.

For each individual methods and structural locations, calculated fatigue damage is shown in Figure 9 to Figure 12. Fatigue damage is shown for all four loading conditions. Combined damage is summation of damages in corresponding loading conditions as per Eqn. (17). A large scatter in fatigue damage values can be noticed for all the loading conditions which are quite obvious primarily due to different hydrodynamic loads and mean stress effect consideration. FAM-1 and FAM-3 (Figure 9 to

Figure 11) predict the maximum fatigue damage for normal ballast loading condition as compare to other load cases. This is mainly due to larger VBM load transfer function values obtained using semi analytical formulation.

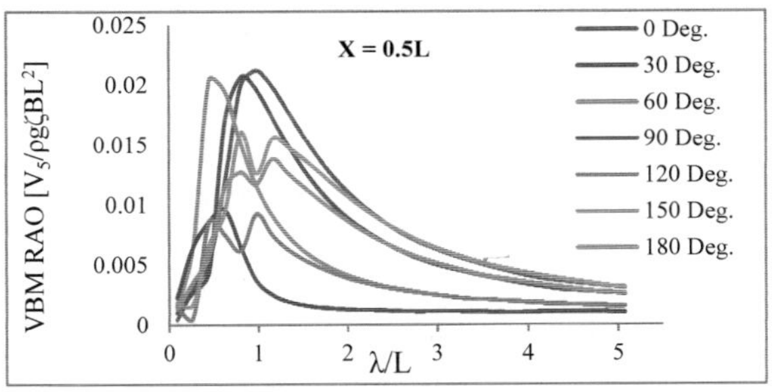

Figure 3. VBM RAO (2D STRIP THEORY) FOR ALTERNATE LOADING

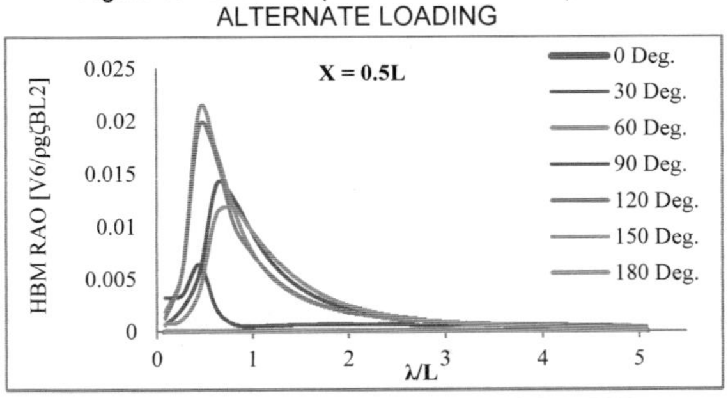

Figure 4. HBM RAO (2D STRIP THEORY) FOR ALTERNATE LOADING

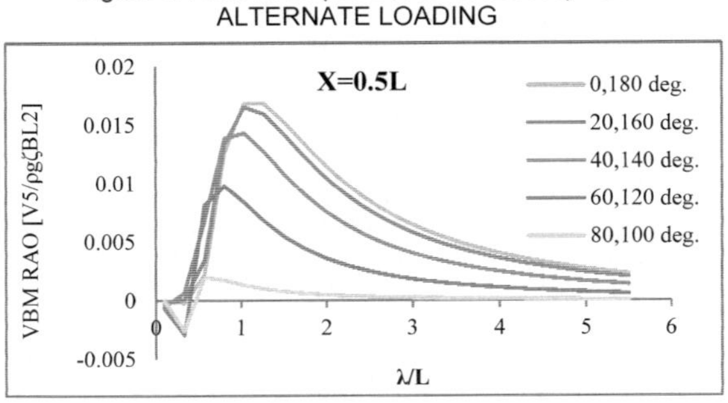

Figure 5. VBM RAO (SEMI ANALYTICAL) FOR ALTERNATE LOADING

Despite scattering of damage values for each individual loading condition, the final combined damage values are comparable at some level, refer to Figure 13. In general, the closed form approach produces the larger fatigue damage than the spectral approach when comparing the load evaluation method-wise (FAM-1 and FAM-3) and (FAM-2 and FAM-4). For FAM-2 and FAM-4 strip theory was used for the estimation of loads. Therefore, influence of the HBM can be seen for the location DK2 where the fatigue damage is more than for location DK1 and SS1 using combined stress range due to VBM and HBM. This sort of effect is completely missing across all the results (FAM-1 and FAM-3) which used close-form semi analytical formulation to consider the load effect due to *VBM* only.

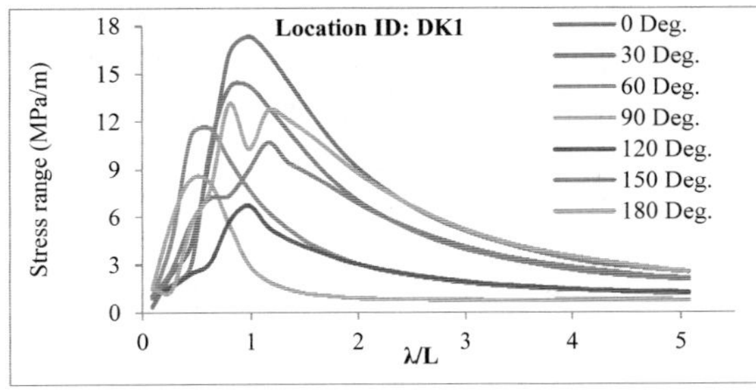

Figure 6. STFS FOR LOCATION DK1

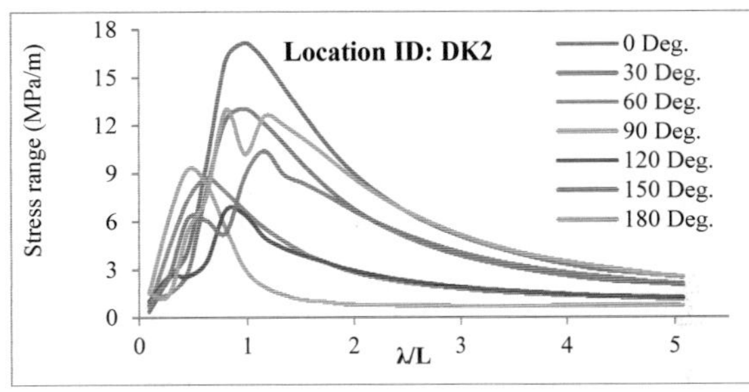

Figure 7. STFS FOR LOCATION DK2

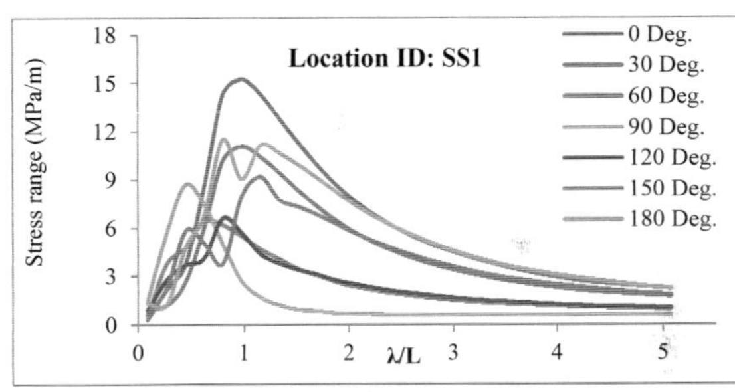

Figure 8. STFS FOR LOCATION SS1

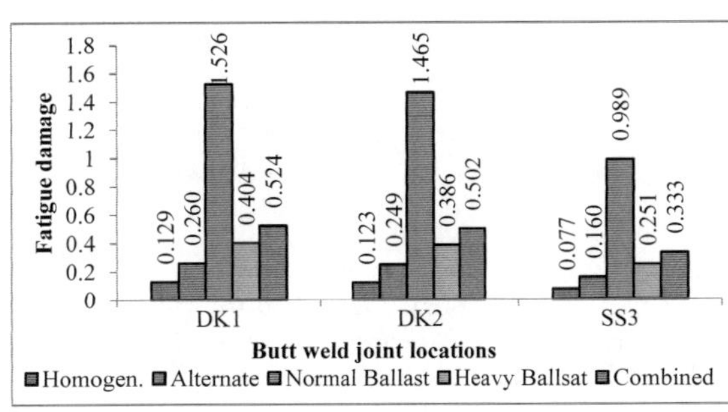

Figure 9. FATIGUE DAMAGE USING FAM-1

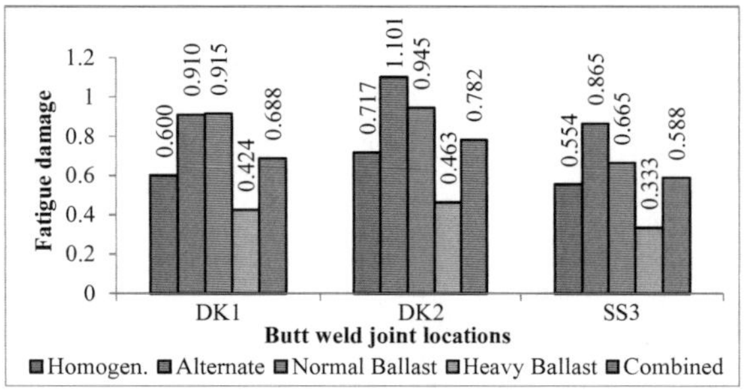

Figure 10. FATIGUE DAMAGE USING FAM-2

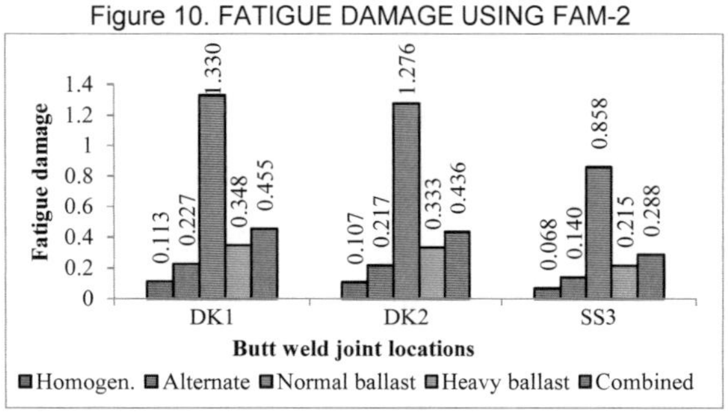

Figure 11. FATIGUE DAMAGE USING FAM-3

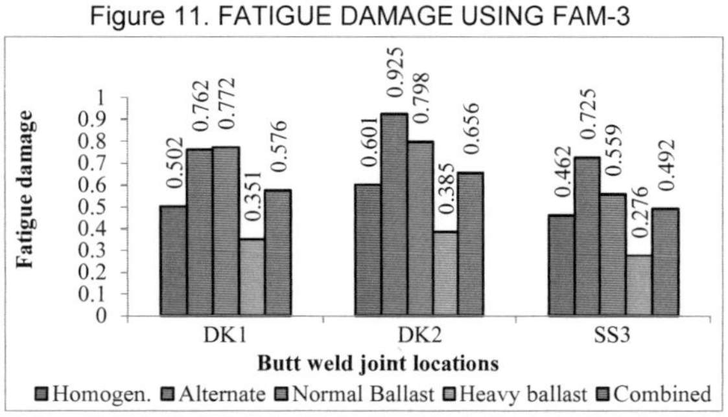

Figure 12. FATIGUE DAMAGE USING FAM-4

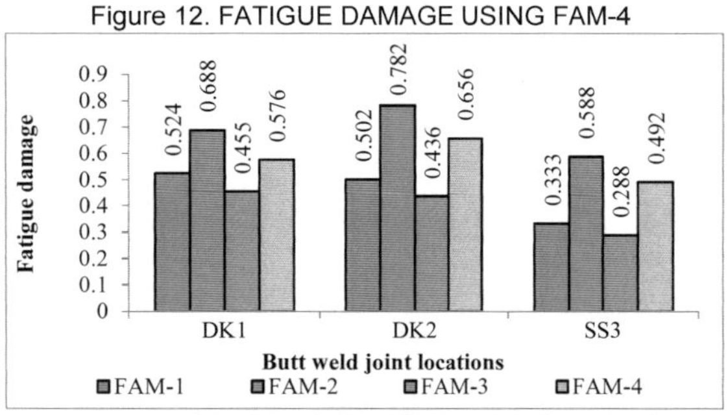

Figure 13. COMBINED FATIGUE DAMAGE USING DIFFERENT METHOD

CONCLUSION

In this paper, the fatigue analysis of butt-welded joints in the deck and side shell of a bulk carrier has been carried out using four different method of the fatigue assessment. These methods comprised the combination of two different loads evaluation methods and two different fatigue assessment approaches. Actual ship operational and environmental details have been considered to predict fatigue damage using both the fatigue assessment methods. However, spectral based fatigue assessment approach the simplifications were made in obtaining the structural responses.

All methods were demonstrated using a sample ship problem. The fatigue damage was obtained for the three different butt-welded location of a bulk carrier. For fatigue analysis using direct methods, evaluation of STFs is one of the elementary requirements. STFs have been obtained for three different locations of the ship using beam theory. All in all, the structural responses have been obtained in simplistic manner.

Due to its massive structural size, performing direct FE based structural analysis for each structure element is time consuming process. However, entire ship structural elements can be analyzed for the fatigue failure using the methodologies explained in this paper which may provide an initial level of investigation for the fatigue failure. The outcome of this initial fatigue assessment can reduce the number of critical locations to perform a comprehensive structural analysis such as using FE methods. Fatigue predictions based on FE methods cannot be ignored for the more realistic results and has been discussed in our paper part-2. The fatigue assessment for the three butt-welded joints of bulk carrier shows that

- Large variation of fatigue damage for individual loading condition is pertaining to the differences in the load response transfer function level and determination of mean stress effect.
- For the common load evaluation method, closed form approach based on long-term response predicts higher fatigue damage than spectral approach based on short term response for all the three butt plate joints.
- Notable influence on combined fatigue damage can be observed using HBM.
- Though in some of the loading the fatigue damage is more than 1.0, but the combined fatigue damage compensated due to fraction of time spent by ship in each loading conditions.

ACKNOWLEDGEMENT

Authors would like to thank Dr. Shivaji Ganesan for his useful insights on probabilistic part of the fatigue damage formulations.

REFERENCES

[1] IMO, "MSC 65/INF - Bulk carrier safety - Report of the correspondence group on the safety of ships carrying solid bulk cargoes," 1995.

[2] ABS, "Guide for fatigue assessment of offshore structures," 2018.

[3] DNVGL, "DNVGL-CG-0129 - Fatigue assessment of ship structures," 2015.

[4] IACS, "Common structural rules for bulk carriers and oil tankers," London, UK, 2018.

[5] C. G. Soares, Y. Garbatov, and H. Von Selle, "Fatigue Damage Assessment Of Ship Structures Based On The Long-Term Distribution Of Local Stresses," *Int. shipbiulding Progr.*, vol. 50, no. 1 & 2, pp. 35–55, 2003.

[6] A. Negi and S. S. Dhavalikar, "Spectral Fatigue Analysis of a Transportation Barge," in *International Offshore and Polar Engineering Conference*, 2011, vol. 8, pp. 936–942.

[7] B. N. Salvesen, E. O. Tuck, and O. Faltinsen, "Ship Motions and Sea loads," in *Society of Naval Architects and Marine Engineers*, 1970, vol. 6, no. 6, pp. 250–287.

[8] J. J. Jensen and A. E. Mansour, "Estimation of Ship Long-term Wave-induced Bending Moment using Closed-Form Expressions," *International Journal of Maritime Engineering*. pp. 41–55, 2002.

[9] IRS, "Rules for bulk carriers and oil tankers," Mumbai, 2019.

[10] IRS, "Guidelines on Spectral Ship Assessment for Ship Structures," Mumbai, 2018.
[11] J. Fukuda, "Theoretical Determination Of Design Wave Bending Moments," *Japan Shipbuild. Mar. Eng.*, vol. 2, no. 3, 1967.
[12] IACS, "Recommendation No 56 - Fatigue Assessment of Ship," 1999.
[13] Guedes Soares C. et al., "Fatigue damage assessment of ship structures based on the long-term distribution of local stresses", International Shipbuilding Progress 50, 35-55, 2003.

Proceedings of the 6th International Conference on Ship and Offshore Technology
ICSOT 2019
November 7-8, 2019, IIT Kharagpur, India

ICSOT INDIA 2019-04

ASSESSMENT OF PURE LOSS OF STABILITY FAILURE MODE FOR 2ND GENERATION INTACT STABILITY

Amresh Negi
Indian Register of Shipping
Mumbai, Maharashtra, 400072
India
amresh.negi@irclass.org

Shivaji Ganesan T
Indian Register of Shipping
Mumbai, Maharashtra, 400072
India
shivaji.ganesan@irclass.org

ABSTRACT

One of the most important element in the ship design is the transverse intact stability. Ships compulsorily comply the 2008 IS code ensuring the safety of ship based on the ship stability assessment in calm water. When ship encounters harsh weather conditions it is exposed to large variations of stability in waves. Extremely steep waves in following seas along with small ship stability due to large reduction of the righting lever in a wave crest at amidship location can lead to sudden capsize of ship. This phenomenon occurs due to ship's dynamic instabilities in longitudinal waves and termed as pure loss of stability. This paper presents the sample calculations of ship's vulnerability to such phenomenon, based on in-house developed computer program. Numerical algorithm to calculate the ship stability in waves and still water which has been developed is discussed elaborately. The calculations concerning the assessment of the vulnerability to Level-1 and Level-2 criteria of pure loss of stability of a cargo ship has been explained using [1] procedure.

INTRODUCTION

The 2008 IS Code [2], can be traced to the pioneering work of many researchers and developments over more than a century. This stability code is the revised version of first generation of intact stability code (FGISC) originally codified at IMO in 1993. The 2008 IS Code provides general safety measures against capsizing based on the metacentric height (*GM*) and properties of the righting lever (*GZ*) curve. 2008 IS Code is based on a traditional empirical/statistical approach, with the exception of the weather criterion. The revised version of FGISC does not explicitly address the situations which are considered dangerous from the ship stability point of view. Therefore, a preamble was added in the 2008 IS Code in support of the long-term future development of stability code considering the ship hydrodynamic aspects and stability analysis in a seaway.

Therefore, as a part of further development, a separate project was taken up by IMO Sub-Committee on Ship Design and Construction (SDC) to formulate the stability assessment based on the scientific principles and termed as the development of second generation of intact stability code (SGISC). At present, the development and validation work of SGISC is going on at IMO/SDC. In SGISC work progress, following five scenarios or stability failure modes have been identified as dangerous situations:

- Parametric Rolling (PL) in following and head seas
- Loss of Stability (PLS) in the wave crest
- Broaching to and surf-riding (B/SR)
- Dead Ship (DS) condition in beam sea
- Excessive Accelerations (EA) when rolling in beam sea

The above listed five stability failure modes are belonging to three fundamental categories of problems:

- Restoring arm variation (PLS, PR),
- Ship maneuvering problem (B/SR). and
- Stability under dead ship condition (DS).

Problem of lateral accelerations (EA) has been identified and added lately, mainly concern with the safety of crew members in working areas.

Each of the failure modes consist of three levels (*Level-1*, *Level-2* and *Level-3*) of stability assessment. *Level-3* stability assessment is based on the direct assessment of the stability failure mode through numerical computations. The complete procedure to carryout *Level-3* is still under discussion and yet to be finalize at IMO. However, calculation procedures for *Level-1* and *Level-2* of all the identified failure modes are almost complete.

Level-1 formulation is based on simple calculations and required hull geometry and forward speed inputs. *Level-1* produce conservative results compared to *Level-2* criteria. *Level-2* formulation is based on simplified physics following probabilistic approach which is lesser conservative than *Level-1*. When a ship failed to meet both *Level-1* and *Level-2* vulnerability criteria of any failure mode, then *Level-3* or Direct Assessment (*DA*) shall be performed. If a ship fails to comply with all the three levels, the ship could be allowed to navigate with Operational Guidelines (*OG*) based on the direct stability assessment procedures or Operational Limitations (*OL*) based on the *Level 2* vulnerability criteria. This is referred as multi-tier approach in 2nd generation of intact stability.

Pure Loss of Stability

This paper work is focused on the loss of stability which is related with restoring arm variation problem in waves. The restoring moment increases in wave trough and become smaller in wave crest. This phenomenon of reduced stability in waves is usually known as "pure loss of stability''. This mode of stability failure is not new but known since [3]. Over the period, number of researchers had worked upon the problem of restoring arm variation in waves. The first calculations were performed by Paulling [4]. To gain the physical insight, series of experimental studies were conducted in between 1970 to 1980 (summary available from [5]). Complete descriptions of ship capsize failure modes are provided in the ITTC report [6]. However, In the 48th session of the IMO SLF Sub-Committee [7] it has been decided to include restoring arm variation problem as one of the three distinct physical phenomena responsible for stability failure as mentioned in previous section of this paper.

Level-1 of PLS vulnerability assessment focus on the evaluation of minimum *GM* (GM_{min}), which is to be calculated for wave length equal to ship length and having a specific wave steepness coefficient S_W. This wave steepness is to be derived based on the environmental data provided for the area of ship operation. A ship is qualified as vulnerable to pure loss of stability at *Level-1* when calculated GM_{min} is less than the standard value defined at this level.

Level-2 vulnerability to pure loss of stability parameters are related to the shape of righting levers curves which are calculated for various ship positions relative to the wave and for wave steepness parameters. Ship requires to be balanced in trim and sinkage before the evaluation of righting arm curves for each ship position and wave condition. The *Level-2* criterion is based on the following two *GZ* curve characteristics which have been considered in the *Level-2* criteria formulation.

- Minimum angle of vanishing stability
- Maximum angle of static heel under the action of specified heeling lever.

These parameters are evaluated for each bin of the scatter diagram. Depending on whether ship failed or passed the criteria, logical indices either 1 or 0 are assigned. The evaluation of logical indices is based on ship's large angle stability calculation in each wave condition and criteria mentioned for the *GZ* parameters. For both the *GZ* curve parameters based criteria, weighted criteria has to be determined by summing up the weightage factors provided for each bin of scatter diagram considering the indices. Sea states having zero weightage factors are to be excluded from the computation. Eventually, two separate values corresponding to weighted criteria have been obtained. The maximum value of the two is compare against the standard value defined at this level. This paper presents the *Level-1* and *Level-2* computations for PLS vulnerability criteria using in-house developed numerical program and the same has been discussed in detail.

PHENOMENON - PURE LOSS OF STABILITY

As discussed, according to the International Stability Code 2008 [2] the two basic parameters which are required in assessing intact stability are the righting lever (*GZ*) and the initial metacentric height (*GM*). These parameters are strictly related to loading condition of the vessel and particularly sensitive to the vertical position of center of gravity. Ship hydrostatic parameters required to evaluate the transverse stability are usually calculated in calm water. Accordingly, the underwater volume of the ship remains constant and does not vary as ship moves in calm water (assumption valid only for hydrostatic calculation). The Water-Plane Area (WPA) of ship can be represented by an un-deformed plane parallel to still water.

However, in practical scenario, moving ship encounters the sea waves which alter its equilibrium position and therefore require balancing in trim and sinkage beforehand the computation of stability parameters. Ship in presence of waves, alternately finds itself on the wave crest and trough, and some transitory state when the wave passes along the ship hull. In such situations, the underwater geometry of the hull changes instantaneously [8]. Consequently, ship's water plane area starts deforming subsequent to the wave profile. However, not only wave profile but geometry of the ship is also responsible for the change in WPA. Deformed WPA leads to change in moment of inertia of water plane, shift in the position of center of buoyancy and other parameters associated with the underwater geometry of hull. All in all, ship movement in waves, significantly affects the *GM* and *GZ* values at various heel angles.

The most unfavorable situation occurs during the ship-hogging condition (i.e. wave crest located at midship) for which the wave length (λ) is comparable to ship length (L), i.e. $L = \lambda$. In this situation, ship has decreased water plane area and causing significant reduction in *GZ* and *GM*. Thus, leading to phenomenon termed as pure loss of stability [8]. Extremely steep waves in following seas along with small ship stability due to large reduction of the righting lever in a wave crest can lead to sudden capsize of ship. The most adverse conditions to cause the occurrence of this phenomenon can be listed as

- Direction of wave propagation (following waves).
- Ship speed equal to wave speed.
- Wave length equals to ship's length.
- Wave height/s which needs to be derived from the actual wave environment.
- Froude no. corresponding to service speed is greater than equals to 0.24.

RESTORING ARM VARIATION IN WAVES

The underwater portion of the hull continuously changes as ship sail though longitudinal waves and experiences a completely different shape of the underwater volume as compared with the ship in still water. The change in the underwater hull is pertaining to hull geometry itself and the relative speed of ship and encounter waves. The change in the submerge part of hull becomes significantly important if ship length is approximately equals to the wave length.

Let us consider a situation when trough of a wave is located amidships and the wave length is same as ship length. One may notice that in this condition the wave elevation is higher towards the ship aft and bow part. Refer figure 1. (a), the underwater geometry is shown with green color. For most of the cargo ship, the sections are wider at forward part due to bow flare to avoid green water loading. The upper overhung stern part provides the space requirement for the steering gear. However, the midship part remains almost the wall-sided. Therefore, small change in the WPA can be observed around amidships region compare to the aft and bow, see Figure 1. (b). In present scenario of wave trough at mid ship, there would be an increase in water plane area due to the wave crest location at forward perpendicular (FP) or aft perpendicular (AP). As a result, there would be an overall increase in WPA and ship stability increases sharply compare to still water case (see GZ curve in figure 1(c)).

Situation changes significantly when wave crest is located near amidships. Wave elevation at bow and aft sections reduced drastically (refer Figure 1 (a)). Cargo ships are designed with narrow bulbous bow and underwater aft sections with due consideration to the resistance and propulsion efficiency respectively. Ships with higher Froude number and lower prismatic coefficients have very fine underwater bow sections. When wave trough is located at AP and FP, water plane reduces severely (see figure 2. (b)). Little change in the water plane area around amidships can be observed due to wall-sided sections. Overall, there is reduction in WPA mainly from aft and forward sections. Consequently, ship stability is reduced drastically compare to still water (see GZ curve in Figure 2. (c)).

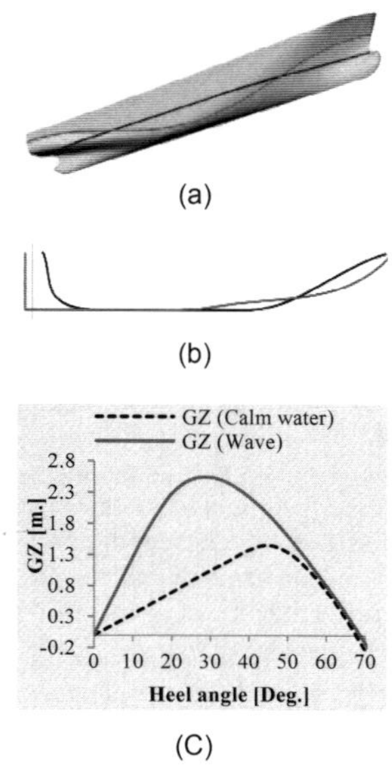

Figure 1. CHANGE IN THE HULL GEOMETRY WHEN WAVE TROUGH AMIDSHIP (A) 3D VIEW OF SHIP HULL (B) DEFORMED WPA COMPARISON WITH THE STILL WATER WPA (C) CHANGE IN RESTORING ARM.

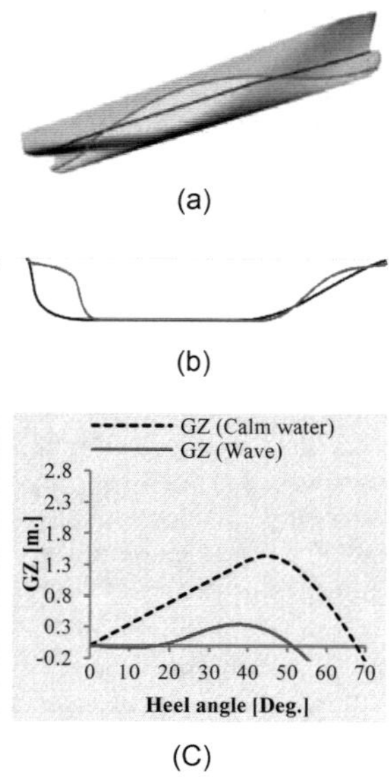

Figure 2. CHANGE IN THE HULL GEOMETRY WHEN WAVE CREST AMIDSHIP (A) 3D VIEW OF HULL (B) DEFORMED WPA COMPARISON WITH THE STILL WATER WPA (C) CHANGE IN RESTORING ARM.

As discussed, figure 1 and figure 2 clearly depict the influence of amidship positioned wave trough/crest on ship stability. These are the two extreme situations where the ship experienced the maximum and minimum stability change. But, in reality ship stability undergoes a continuous change as the wave passes the ship. The righting moment of the ship varies in time with the passing wave. On that account, ship stability is to be evaluated for various wave crest locations considering the actual operating environment. Therefore, *Level-1* and *Level-2* formulations of PLS vulnerability criteria consist of stability calculations for various waves crest locations in quasi-static manner.

DETERMINING WAVE STEEPNESS AND WAVE CASES

Level-1 vulnerability calculations are carried out using a conservative value for the wave steepness. On the basis for unrestricted service IACS Rec.34 scatter diagram is selected. A reference significant wave height ($H_{1/3,ref,i}$; *j=1~N, N* being the number of periods in the scatter diagram) is obtained for each spectral period ($T_{z,i}$) of the wave scatter diagram. A set of 'N' wave cases obtained for pure loss of stability (H_i and λ_i) among which maximum wave steepness factor is selected for the *Level-1* calculations [8]. Refer to Figure 3 for the complete description of the process.

Level-2 vulnerability calculations require estimating the effective wave heights (H_i) based on assumed sea spectrum for each sea state of scatter diagram. The maximum effective wave height (H_m) is selected from the scatter of H_i. The effective wave height is the function of significant wave height ($H_{1/3}$), representative spectral period (T_Z), and the ship length (L). The estimated effective wave height should not exceed the 10% of the ship length. Thus obtained wave height needs to be divided equally into ten wave heights with an additional calm water condition (See figure 3). Following formulation has been used to find the effective wave heights [8].

$$H_i(L) = 5.9725\sqrt{m_0} \tag{1}$$

Spectral moment *'m₀' is* determined as:

$$m_0 = \int_{0.01\omega_L}^{\omega_L} \left\{ \frac{\frac{\omega^2 L}{g} sin\left(\frac{\omega^2 L}{2g}\right)}{\pi^2 - \left(\frac{\omega^2 L}{2g}\right)^2} \right\}^2 A\omega^{-5} exp(-B\omega^{-4}) d\omega \; +$$
$$\int_{\omega_L}^{3\omega_L} \left\{ \frac{\frac{\omega^2 L}{g} sin\left(\frac{\omega^2 L}{2g}\right)}{\pi^2 - \left(\frac{\omega^2 L}{2g}\right)^2} \right\}^2 A\omega^{-5} exp(-B\omega^{-4}) d\omega$$

Where,

$$A = 173 H_s^2 T_{01}^{-4} \qquad B = 691 T_{01}^{-4}$$
$$g = 9.81\, m/sec^2 \qquad T_{01} = 1.086 T_Z$$
$$\omega_L = \sqrt{\frac{2g\pi}{L}}$$

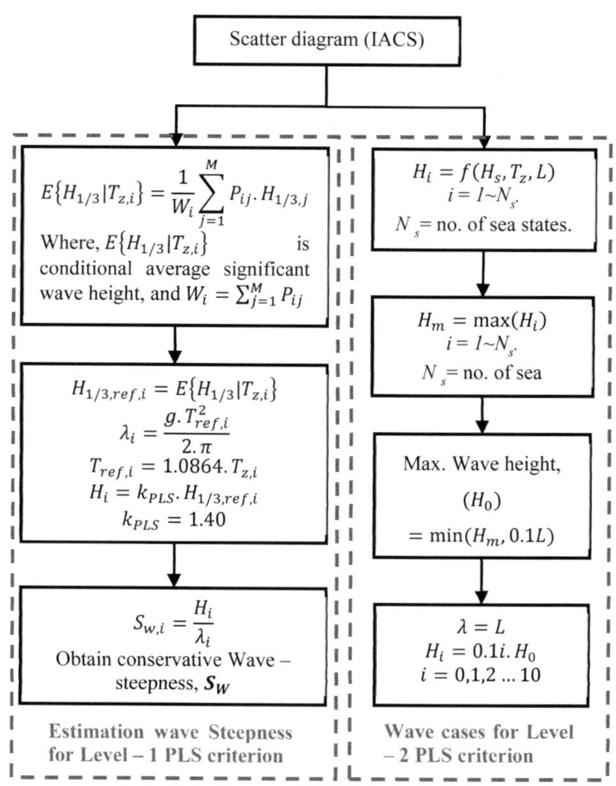

Figure 3. ESTIMATION OF CONSERVATIVE WAVE STEEPNESS AND WAVE CASES FOR LEVEL-1 AND LEVEL-2 PLS CRITERIA

BALANCING SHIP IN WAVES

Belenky [8] clearly show the importance of balancing in both trim and sinkage when performing the stability calculations. Inclusion of trim and sinkage, results in a significant difference for the moment of inertia of the water-plane, BM, KB. As a result, there is significant influence on initial stability in waves. But, the change is not limited to *GM* only; *GZ* curve also experiences the severe change which may lead to ship capsize.

Hydrostatic equilibrium computations involve an iteration method (Newton-Raphson method). These complex calculations involve iteration towards attaining zero force and moments with two variables namely sinkage and trim. The iteration performed using Newton Raphson method for two coupled equations. For a given position, the following equations are used to arrive at an estimate of the required change in sinkage and trim [9]

$$F_Z + \frac{\delta F_Z}{\delta draft}\Delta draft + \frac{\delta F_Z}{\delta trim}\Delta trim = 0 \quad (2)$$

$$M_y + \frac{\delta M_y}{\delta draft}\Delta draft + \frac{\delta M_y}{\delta trim}\Delta trim = 0 \quad (3)$$

Where,

F_z = deficit in displacement
M_y = deficit in moment

LEVEL 1 VULNERABILITY CRITERIA FOR PURE LOSS OF STABILITY

The condition [11] for a ship not to be vulnerable to the pure loss of stability failure mode if:

$$GM_{min} > R_{pla}(0.05m) \quad (4)$$

Few hull forms are more susceptible to the pure loss of stability phenomenon due to considerable change of geometry in the fore and aft sections. Additionally, ships with smaller *GM* value in calm water condition may experience a loll angle on wave crest and result in large angle of heel. Therefore, some of geometric features connected with overall hull shape are considered for the development of *Level-1* criterion. The details background of the *Level-1* formulation can be found in [8].

GM_{min} can be determine following one of the methods as mention below depend on the Vertical Prismatic Coefficient (VPC), a condition to measure the vertical "wall-sidedness". In this work, the first method is termed as *method-1* and second method is termed as *method-2* of *Level-1*[10].

Method-1 Formulation

$$GM_{min} = KB + \frac{I_L}{V} - KG \; Only \; if \frac{V_D - V}{A_W(D-d)} \geq 1.0$$

Mehtod-2 Formulation

$$GM_{min} = \begin{cases} [Minimum\ value\ of\ GM, consdering \\ the\ ship\ to\ be\ balanced\ in\ sinkage\ and\ trim \\ on\ a\ series\ of\ waves\ with\ the\ following \\ characteristics: \\ Sw = 0.0334\ (Based\ on\ IACS\ Ch. 34) \\ Wave\ length\ \lambda = L;\ Wave\ height\ h = L.S_W \\ Wave\ crest\ is\ to\ be\ centred\ amidship, and\ at \\ 0.1L, 0.2L, 0.3L\ \ 0.4L\ and\ 0.5L\ forward \\ and\ 0.1L, 0.2L, 0.3L\ and\ 0.4L\ aft\ thereof.] \end{cases}$$

Where,

KB = height of vertical centre of buoyancy corresponding to the loading condition under consideration (m).

KG = height of vertical center of gravity corresponding to the loading condition under consideration (m).

I_L = moment of inertia of the waterplane at the draft d_L (m^4).

V = volume of the displacement corresponding to the loading condition under consideration (m^3).

D = draft corresponding to the loading condition under consideration (*m*).

δd_L = defined as $Min\left(d - 0.25d_{full}, \frac{L.S_W}{2}\right)(m)$ and, (d-$0.25d_{full}$) should not be taken less than zero.

S_w = wave steepness (0.0334 based on IACS Rec. 34 scatter diagram [12].).

D = moulded depth at side to weather deck (*m*).

V_D = moulded depth at side to the weather deck (m^3).

A_w = waterplane area at the draft equal to d (m^2).

d_{full} = draft corresponding to the fully loaded departure condition (*m*).

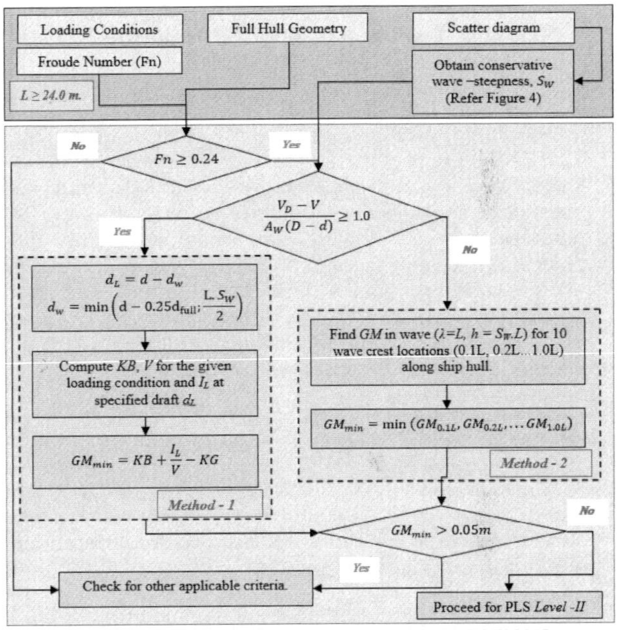

Figure 4. FLOW CHART FOR LEVEL-1 COMPUTATIONS

The proposed IMO method of evaluating the ship vulnerability to pure loss of stability applies to ships length ≥ 24 and Froude number, $F_n \geq 0.24$. Figure 4 shows the flow chart for the *Level-1* computations. The inputs can be grouped as

1. Operational conditions: which consist of ship loading conditions, ship service speed etc,
2. Hull form: includes the information related to hull geometry format.
3. Environmental conditions: contains the environmental data based on ship area of operation.

Based on the Vertical prismatic coefficient condition, *method-1* or *method-2* can be followed for *Level-1* assessment. *Method-1* involves the calculation of I_L at lower draft d_L. Lower draft formulation accounts for the wave and the max. ship draft. *Method-1* provides much simpler formulation to determine GM_{min}. *Method-2* involves more computational efforts than method-1. In *method-2*, *GM* values are calculated for a 10 wave crest locations along the ship hull for wave height $h = L.S_w$ and wave length $\lambda = L$. Minimum *GM* value among all cases is taken as GM_{min}. Both the methods require a common environmental parameter i.e. conservative wave-steepness, S_w to proceed with the evaluation of *GM* value. Refer to flow chart provided in figure 4 for the description of calculations to be performed for *Level-1* assessment.

LEVEL 2 VULNERABILITY CRITERIA FOR PURE LOSS OF STABILITY

Ship is considered not to be vulnerable to the pure loss of stability failure mode in *Level-2*, if the maximum value of weighted criteria CR_1 *and* CR_2 derived based on the righting lever curve is less than the standard value as shown by eq. 5 [1].

$$Max(CR_1, CR_2) < 0.06 \tag{5}$$

Where,

$$CR_1 = \sum_{i=1}^{N_S} W_i C1_i = Weighted\ criterion\ 1 \tag{6}$$

$$CR_2 = \sum_{i=1}^{N_S} W_i C2_i = Weighted\ criterion\ 2 \tag{7}$$

Where, N_s = total no. of sea states.

Pure loss of stability is considered to be a single wave event. Level 2 Vulnerability check is to be done for a set of regular waves (wave length = ship length) systematically covering the entire range of possible values of wave steepness. This level of assessment is pertaining to the calculation of two indices: CR_1 and CR_2, which are related to the angle of vanishing stability and the angle of static heel under action of a heeling lever specified by a formula for R_{PL3} (eq. 10).

Criterion 1 is based on the calculation of 'angle of vanishing stability' ϕ_V, as provided in following formula.

$$C1_i = \begin{cases} 1 & \phi_V < 30\ Deg. \\ 0 & Otherwise \end{cases} \tag{8}$$

Criterion 2 is based on the calculation of 'angle of heel' ϕ_S, under the action of heeling arm, R_{PL3} as shown by following formula. Where 15 deg. needs to be consider for passenger ship and 25 deg. for other ships.

$$C2_i = \begin{cases} 1 & \phi_S < 15\ or\ 25\ Deg. \\ 0 & Otherwise \end{cases} \tag{9}$$

$$R_{PL3} = 8(H_i/\lambda_i)dF_n^2 \tag{10}$$

Where:
H – Significant wave height.
λ – Wave length.
d – Draft corresponding to the loading condition.
F_n – Froude number corresponding to ship's service Speed.

Figure 5 shows the flow chart for the *Level-2* computations. The entire *Level-2* PLS analysis can be divided into three stages as:

1. *Stage-I*: Estimation of maximum wave height using the environmental conditions to determine the wave cases for the stability computation.
2. *Stage-II*: Restoring arm curve computations for each of the wave case and crest location.
3. *Stage-III*: Determination of indices for the weighted criterion (refer eq. 6 and eq. 7) calculation.

Three sets of input are required to perform *Level-2* computation similar to *Level-1* assessment. *Level-2* computations include the contribution from each sea state conversely to *Level-1*, where the computations were carried out considering the single conservative wave steepness.

The first and second levels of vulnerability check for PLS have been discussed here in detail and most of the ships are expected to pass these two levels. However, a ship failed to comply with *Level-1* and *Level-2*; then *Level-3* or direct stability assessment is required to perform. The direct computation for PLS is based on criteria of critical time period spend by ship on wave crest which require to be solved using a time domain numerical technique and is not covered in the present paper.

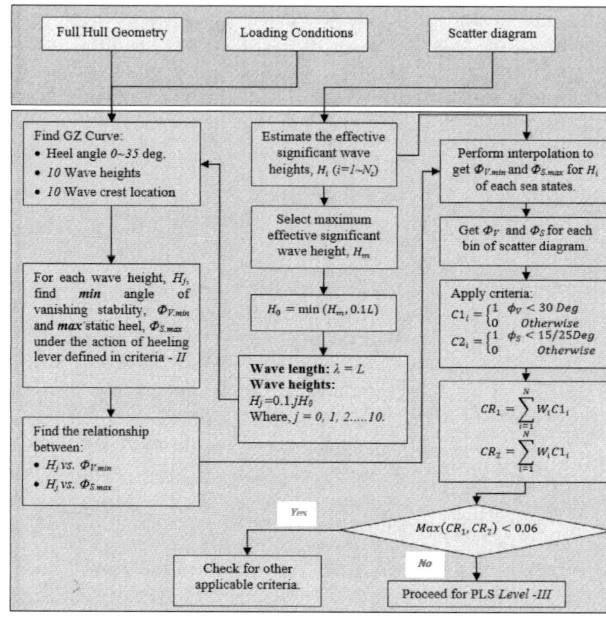

Figure 5. FLOW CHART FOR LEVEL-1 COMPUTATIONS

SAMPLE SHIP PROBLEM

For the verification of calculations, C11 class Post Panamax Container ship has been used. Total 48 sections were used for the stability computations. Ship data and loading condition are given in table 1. Stability assessment was performed using in-house developed program which has been tested for a variety of hull forms. For the present ship, results were compared with the MAXSURF. It is a standard software package which consists of stability module used for ship stability calculation. Computations were obtained using free trim and sinkage option for blanching the ship in waves and calm water. Level-1 and level -2 check calculations were verified with the results provided in IMO documents [13] and [14].

Table 1 SHIP DATA AND LOADING CONDITION

Length: L_{PP}	262.0 m
Breadth: *B*	40.0 m
Depth: *D*	24.45 m
Draft: *d*	12.0 m
Long. Centre of gravity: *LCG*	125.52 m
Vert. Centre of gravity: *KG*	18.9 m
Service speed: V_S	12.86 m/sec.

For the example ship, L = 262m ≥ 24m. Froude no (F_n) corresponding to the service speed of 12.86 m/s (F_n = 0.254 ≥ 0.24). Therefore, this ship qualifies for the PLS vulnerability assessment.

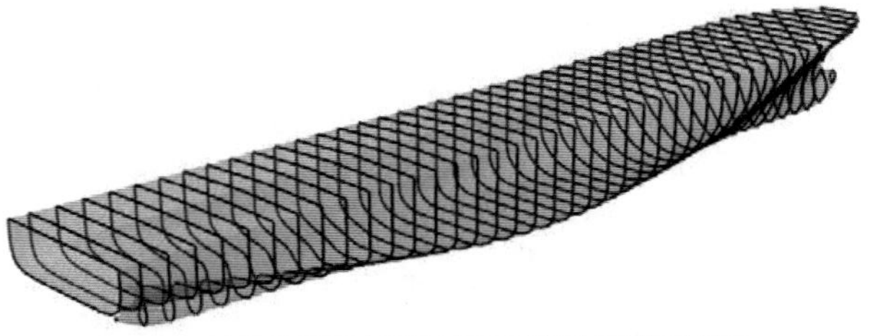

Figure 6. SECTIONS USED FOR THE STABILITY ASSESSMENT

Level-1 Check

Level-1 vulnerability assessment consists of two different methods as already discussed in this paper. VPC value is required to determine for the application of compatible methods. This condition (refer table 3) suggests the application of *method-1* for the present problem. Although, we have tried both methods to observe consistency in results predicted by these two methods.

Following the process given in figure 3 for the estimation of wave steepness for *Level-1* PLS criterion. Total 16 wave steepness have been obtained for IACS Scatter diagram (refer table 2). Conservative wave steepness factor for IACS Rec-34 [12] scatter diagram is estimated as $S_w = 0.0334$.

Table 2. WAVE CASES: OBTAIN CONSERVATIVE WAVE STEEPNESS

Case No.	Weight W	Wave length [m]	Wave height [m]	Wave steepness
1	1.30E-05	22.574	0.7	0.031
2	1.65E-03	37.316	0.99	0.0265
3	2.09E-02	55.743	1.715	0.0308
4	9.28E-02	77.857	2.589	0.0333
5	1.99E-01	103.655	3.464	**0.0334**
6	2.49E-01	133.139	4.41	0.0331
7	2.09E-01	166.309	5.393	0.0324
8	1.29E-01	203.164	6.351	0.0313
9	6.25E-02	243.705	7.25	0.0297
10	2.48E-02	287.931	8.08	0.0281
11	8.37E-03	335.843	8.841	0.0263
12	2.47E-03	387.44	9.539	0.0246
13	6.58E-04	442.723	10.194	0.023
14	1.58E-04	501.691	10.739	0.0214
15	3.40E-05	564.345	11.241	0.0199
16	7.00E-06	630.684	11.9	0.0189

Table 3. PLS LEVEL-1 CONDITION CHECK

Parameters	IRS Result	IMO/SDC
V_D	190,632 m^3	190,580 m^3
V	71,562 m^3	71,559 m^3
A_W	8,300.0 m^2	8,292.6 m^2
$\frac{V_D - V}{A_W(D-d)}$	1.152	1.153

For method-1, lowest draft, d_L is to be considered for the calculation of moment of inertia of WPA (I_L). The formulation of d_L is based on the hull geometry and wave parameter. Table 4 contains all the required parameters to calculate the minimum GM requirement according to *method-1*. I_L being associated with the lowest draft therefore can be calculated using the hydrostatic table of the ship. In this work, in-house developed program is used to compute all required stability parameters. Minimum GM was calculated as -2.43 m. This value was compared with the sample results provided by [13].

Table 4. PLS LEVEL-1 METHOD-1 RESULTS COMPARISON

Parameters	IRS Result	IMO/SDC
KB	6.837 m	6.847 m
d_L	7.625 m	7.625 m
I_L	689,622 m^4	688,080 m^4
$[GM_{min} = KB + \frac{I_L}{V} - KG]$	-2.43 m	-2.44 m

Wave conditions for the application of method-2 has been obtained as wave height (h) = 8.571m and wave length (λ) = 262 m. GM values were estimated for a single regular wave (defined by parameters: h, λ) by placing the wave crest at AP and 0.1L, 0.2L ~0.9L along the ship hull (refer to table 5). The minimum value of the GM in waves is attained for position when the wave crest is located around mid-ship and is equal to -0.566 m. The results were compared with IMO document [13]. The calculated value of minimum GM was found to be less than 0.05 m. Consequently, the ship is assessed as vulnerable to pure loss of stability by the *Level-1* criterion using both *method-1* and *method-2*. This indicates to perform *Level-2* PLS vulnerability assessment.

Table 5. PLS LEVEL-1 METHOD 2 RESULTS COMPARISON

Wave Conditions:
$h = L.S_w = 262 \times 0.0334 = 8.751$ m.
$\lambda = 262$ m.

Wave location	GM in Waves	
	IRS	IMO/SDC
0	2.490	2.426
1	0.778	0.768
2	-0.219	-0.242
3	**-0.566**	**-0.587**
4	-0.403	-0.414
5	0.589	0.593
6	2.680	2.857
7	3.388	3.689
8	3.866	4.168
9	3.806	3.976

Level-2 Check

As discussed, this level consists of three stages of computations. In stage one, maximum wave height is to be determined based on the provided environmental data for the ship operation or ship geometry particulars. Thus obtained maximum wave height is required to create wave cases to perform next stage calculations. Stage two is related with the GZ curve computation in waves for ten wave crest location along the ship hull. Stage three includes the evaluation of logical indices based on the GZ curve shape. These indices are then used to calculate the weighted criteria based on the weightage factors corresponding to each sea state of scatter diagram. In this section, sample ship has been tested for the level -2 vulnerability assessment following these three stages.

Stage-I. Determination of Max. Wave Height: The long-term characterization of the standard environmental conditions is given by a wave scatter diagram. The wave scatter diagram contains the probability of occurrence of different sea states defined with significant wave heights (H_S) and zero crossing periods (T_Z). For each combination of H_S and T_Z, W_i

(H_S, T_Z) is obtained as the value reported for the probability of occurrence divided by total number of observations for a particular sea state. As shown in figure 7, each intersection in H_S-T_Z plane represents the single sea state and the stems correspond to probability of occurrence. Only non-zero-weighted sea states have been considered for the analysis.

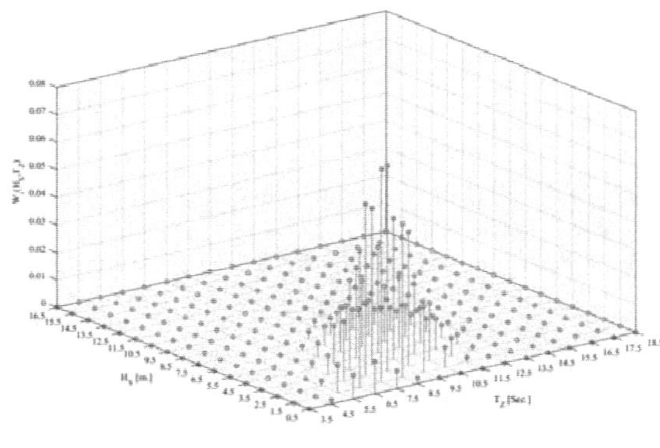

Figure 7. PROBABILITY OF OCCURRENCE (W_i) FOR EACH SEA STATE OF IACS REC. 34 SCATTER DIAGRAM

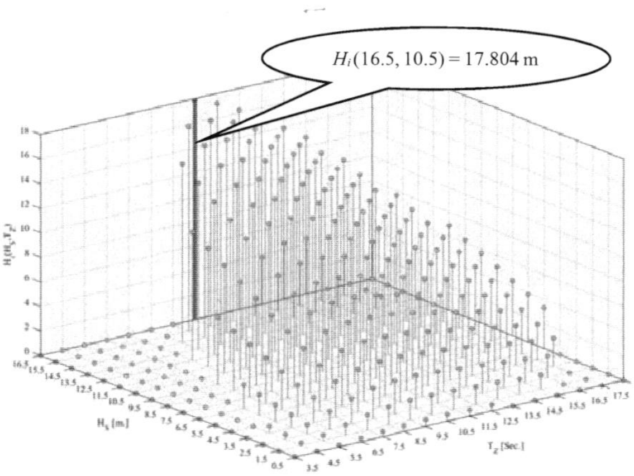

Figure 8. EFFECTIVE WAVE HEIGHTS (H_i) PLOT FOR C11 SHIP CONSIDERING IACS SCATTER DIAGRAM

Firstly, the 3% largest effective significant wave height (H_i, $i = 1 \sim N$, N = no. of sea states) needs to be calculated by filtering ocean waves within ship length for individual sea state of the IACS scatter diagram [12] using eqn. 1.

Figure 8 shows the 3-D representation of the calculated effective wave heights for all the sea states of the scatter diagram. Every single stem imply H_i corresponding to a pair of H_S and T_Z. For each H_i, λ_i is taken as equal to length of the ship = 262 m. The maximum effective wave height H_m = 17.803m (< 0.1*L) was estimated for H_S = 16.5 m and T_Z = 10.5 Sec. Therefore, maximum wave height $H_0 = H_m$ =17.803 m was selected and used to determine wave cases for the computation the *GZ* curves. In case of ($H_m > 0.1L$), then ship length would have been used to determine maximum effective wave height (0.1*L = 26.2 m).

Stage-II. Computation of GZ Curves: A computer program has been developed to obtain the hydrostatic parameters and *GZ* curves. For each wave height and wave crest location, the ship has been balanced in trim and sinkage. The program requires hull geometry in sectional-offset format. In the present example of container ship, total 48 sections were considered for stability assessment (refer to figure 6). In general, the accuracy of *GZ* computation highly depends on the definition of hull geometry. A sufficient number of sections need to be created to compute the stability in waves. Closely spaced sectional offsets at aft and forward part are used to capture the hull geometry variations. Sections need to be extended till deck to consider the variations in deck due to camber which influence the stability at large heel angles. An adequate number of offset points are required to create ship sectional curves so as to represent approximated hull shape close to the original.

For calculating the restoring moment in waves, the following wavelength and wave heights were used:

Length, $\lambda = L = 262.0\ m.$
Wave height, $H_j = 0.1\,jH_0$,
$j = 0, 1, 2,...,10; j = 0$ is corresponding to cam water condition.

The maximum effective wave height H_0 = 17.803 was used to generate 11 effective wave heights ranging from *0.0* to 17.803 m. having wave length 262.0 m. *GZ* curves then obtained from the in-house developed program were compared with the MAXSURF results for a single wave height of 17.803 m and 11 wave crest locations from AP to FP as shown in figure 9. Calm water *GZ* curve also compare and labeled as IRS-0 and MAXSURF-0. However, the results for wave crest location A.P., 0.1L...were labeled as IRS-AP, IRS-0.1L and MAXSURF-AP, MAXSURF-0.1L. Both results were found to be in good agreement for all the conditions. We can notice that when the wave crest is at AP and FP (figure 9(a) and 9(d)), resorting arm increases for all heel angles compare to *GZ* values for calm water. However, due to high *GM*, ship becomes stiffer. On the other hand, ship suffers drastic reduction in restoring arm when the wave crest is around midship (refer to figure 9(b) and 9(c)).

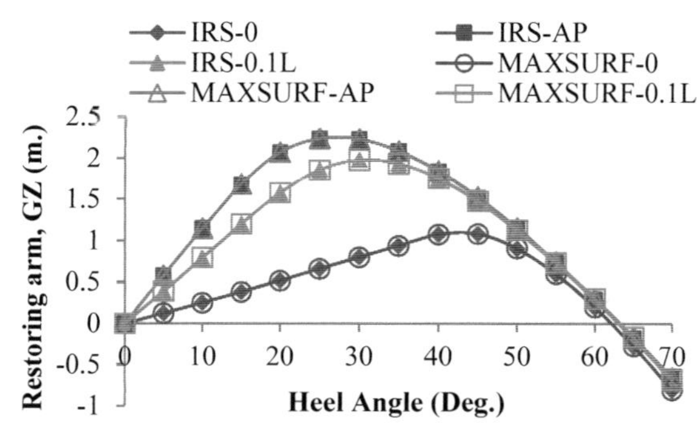

(a)

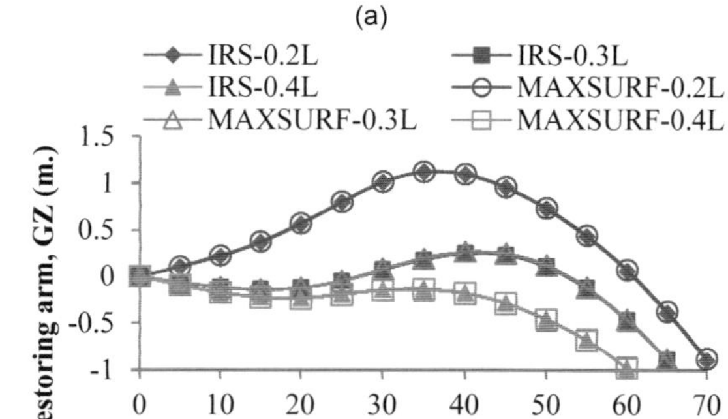

(b)

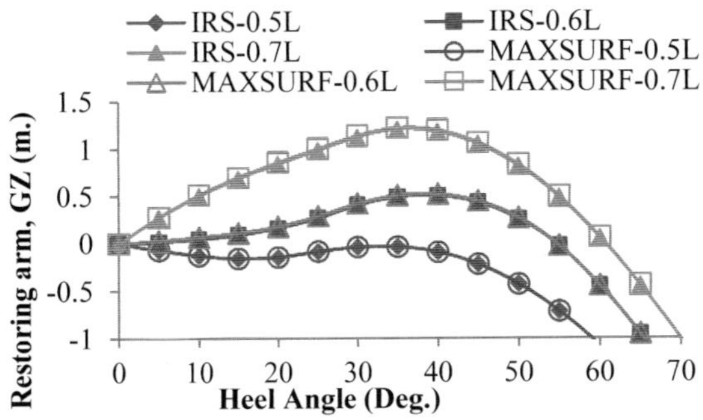

(c)

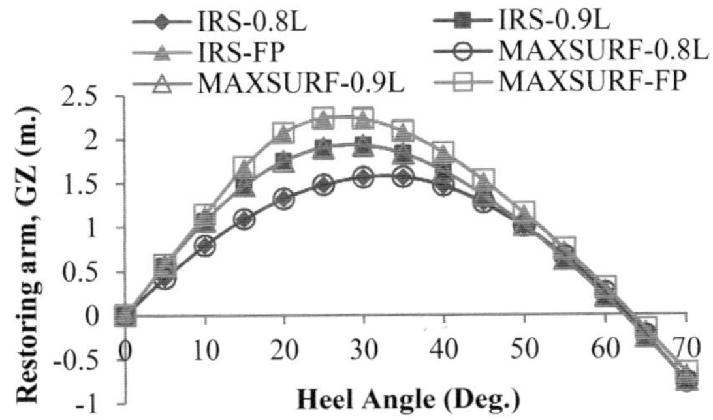

(d)

Figure 9. (a, b, c, AND d) GZ CURVES COMPARISON FOR CALM WATER AND DIFFERENT WAVE CREST LOCATION FOR MAX. WAVE HEIGHT 17.803 M.

For the present ship, *GZ* curves are computed for 0~35 deg. in a step of 5 deg. heel angle. In a view of the heel angle requirement for the *Level-2* applicable criteria, the heel angle is limited up to 35 deg. A heeling lever (R_{PL3}) is obtained for each wave case based on the Froude number, wave steepness and the draft (See Eqn. 10). Wave height 0.0 m represents the calm water condition, therefore a single *GZ* curve obtained. However, for each wave case 10 *GZ* curves were computed for 10 different wave crest location along the ship length. The angle of vanishing stability (Φ_V) and angle of stable equilibrium (Φ_S) under the action of heeling arm (R_{PL3}) are calculated for each set of *GZ* curve.

The minimum angle of vanishing stability, $\Phi_{V.min}$ and maximum static heel, $\Phi_{S.max}$ are obtained for each wave and plotted in figures 10 and 11. However, some differences can be observed between IRS results and [14] particularly obtaining $\Phi_{V.min}$ for different wave situations in figure 10. These differences may be due to the inconsistencies in the geometry input used for the computation of *GZ* curves or convergence criteria set for the balancing of trim and sinkage. The effect of this difference in the final result was found negligible.

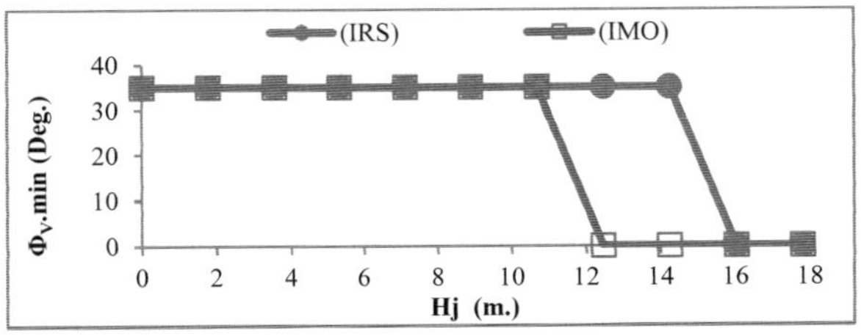

Figure 10. $\Phi_{V.min}$ vs. H_j

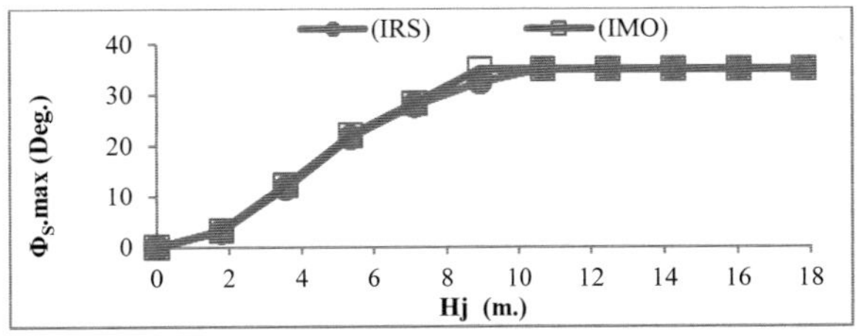

Figure 11. $\Phi_{S.max}$ vs. H_j

Stage-III. Determination of Indices for the Weighted Criterion: In stage three, using the plots in figures 10 and 11, a linear interpolation is performed to get $\Phi_{V.min}$ and $\Phi_{S.max}$ for non-zero-weighted H_i of each sea states of scatter diagram. Therefore, we have a pair of values ($\Phi_{V.min}$ and $\Phi_{S.max}$) corresponding to each sea state. Now, we apply the criteria $C1_i$ and $C2_i$ for minimum angle of vanishing stability and maximum angle of heel. As a result, we get a set of logical indices (either 1 or 0) related to two criteria for each bin of scatter diagram as shown table 6 and table 7. Weighted criteria CR_1 and CR_2 has been calculated as summing of their respective contributions from the weighted factors as specified for the each sea state of scatter diagram. The maximum value of CR_1 and CR_2 has been considered and compared with the standard value for the *Level-2* vulnerability assessment (Refer table 8).

Table 6. LOGICAL INDICES FOR CRITERIA, $C1_i$

Hs/Tz	3.5	4.5	5.5	6.5	7.5	8.5	9.5	10.5	11.5	12.5	13.5	14.5	15.5	16.5	17.5	18.5
0.5	0	0	0	0	0	0	0	0	0	0	1	1	1	1	1	1
1.5	1	0	0	0	0	0	0	0	0	0	0	0	0	1	1	1
2.5	1	0	0	0	0	0	0	0	0	0	0	0	0	0	1	1
3.5	1	0	0	0	0	0	0	0	0	0	0	0	0	0	0	1
4.5	1	1	0	0	0	0	0	0	0	0	0	0	0	0	0	1
5.5	1	1	0	0	0	0	0	0	0	0	0	0	0	0	0	0
6.5	1	1	0	0	0	0	0	0	0	0	0	0	0	0	0	0
7.5	1	1	1	0	0	0	0	0	0	0	0	0	0	0	0	0
8.5	1	1	1	0	0	0	0	0	0	0	0	0	0	0	0	0
9.5	1	1	1	0	0	0	0	0	0	0	0	0	0	0	0	0
10.5	1	1	1	1	0	0	0	0	0	0	0	0	0	0	0	0
11.5	1	1	1	1	0	0	0	0	0	0	0	0	0	0	0	0
12.5	1	1	1	1	0	0	0	0	0	0	0	0	0	0	0	1
13.5	1	1	1	1	1	0	1	1	0	0	0	0	0	0	0	1
14.5	1	1	1	1	1	1	1	1	1	0	0	0	0	0	1	1
15.5	1	1	1	1	1	1	1	1	1	1	0	0	0	0	1	1
16.5	1	1	1	1	1	1	1	1	1	1	1	0	0	1	1	1

Table 7. LOGICAL INDICES FOR CRITERIA, $C2_i$

Hs/Tz	3.5	4.5	5.5	6.5	7.5	8.5	9.5	10.5	11.5	12.5	13.5	14.5	15.5	16.5	17.5	18.5
0.5	0	0	0	0	0	0	0	0	0	0	0	0	0	0	0	0
1.5	0	0	0	0	0	0	0	0	0	0	0	0	0	0	0	0
2.5	0	0	0	0	0	0	0	0	0	0	0	0	0	0	0	0
3.5	0	0	0	0	0	0	0	0	0	0	0	0	0	0	0	0
4.5	0	0	0	0	0	0	0	0	0	0	0	0	0	0	0	0
5.5	0	0	0	0	0	0	0	0	0	0	0	0	0	0	0	0
6.5	0	0	0	0	0	1	1	1	1	1	0	0	0	0	0	0
7.5	0	0	0	0	0	1	1	1	1	1	1	1	0	0	0	0
8.5	0	0	0	0	1	1	1	1	1	1	1	1	1	0	0	0
9.5	0	0	0	0	1	1	1	1	1	1	1	1	1	1	0	0
10.5	0	0	0	0	1	1	1	1	1	1	1	1	1	1	1	0
11.5	0	0	0	0	1	1	1	1	1	1	1	1	1	1	1	1
12.5	0	0	0	0	1	1	1	1	1	1	1	1	1	1	1	0
13.5	0	0	0	0	0	1	1	1	1	1	1	1	1	1	1	0
14.5	0	0	0	0	0	1	1	1	1	1	1	1	1	1	0	0
15.5	0	0	0	0	0	0	1	1	1	1	1	1	1	1	0	0
16.5	0	0	0	0	0	0	0	1	1	1	1	1	1	0	0	0

Table 8. WEIGHTED CRITERIA CR_1 AND CR_2

Criteria	IRS	IMO/SDC
CR_1	0.000109	0.002814
CR_2	0.093690	0.094216
$Max(CR_1, CR_2)$	0.093690	0.094216
$Max(CR_1, CR_2) > 0.06$	FAIL	FAIL

Table 9. SUMMARY FOR LEVEL-1 AND LEVEL-2

PLS	IRS	IMO/SDC
Level-1 Method-1	FAIL	FAIL
Level-1 Method-2	FAIL	FAIL
Level-2	FAIL	FAIL

In summary (table 9), the present ship C11 is assessed vulnerable to pure loss of stability in *Level-1* and *Level-2* for the considered loading condition. Therefore, the ship qualifies for the *Level-3* vulnerability assessment of pure loss of stability for the specified loading condition. In same manner, all other loading conditions need to be checked for pure loss of stability failure mode.

CONCLUSION

This paper presents the vulnerability assessment of pure loss of stability for container ship based on *IMO* document [1] procedure. The following points were concluded from present work.

- An algorithm has been presented and implemented using computer program to carry out the *Level-1* and *Level-2* calculations.
- *Level-1* vulnerability assessment consist of two different methods to determine minimum GM and termed as *method-1* and *method-2*.
- '*Level-1 method-1*' calculations can be performed with the help of calculator or excel sheets provided the hydrostatic values are available.
- '*Level-1 method-2*' and *level-2* assessment necessarily required a computer program.
- For *level-2* vulnerability check, the most time consuming part remains the computation of *GZ* curve in waves for the balanced trim and sinkage scenario considering various wave conditions and crest locations.
- The present computer program for *GZ* computation is validated for other hull geometries. However, in this work, the validation of *GZ* curve with MAXSURF has been presented and found in good agreement.
- The results obtained from the in-house developed computer program for *Level-1* and *Level-2* vulnerability assessment were compared and found in consistent with the results reported in the IMO/SDC documents [13] and [14].

ACKNOWLEDGEMENT

Authors would like to thank Mr. N. Girish (HOD - RIC) and Dr. Suhas V. In-charge of Strategic research group at Indian Register of Shipping for their continual support for this research work. Also would like to thank Prof. Naoya Umeda of Dept. of Naval Architecture and Ocean Engineering, Osaka University, for allowing us the access to the necessary document from IMO working group website for this work.

REFERENCES

[1] IMO Document SDC 2/WP.4. 19 February 2015, "finalization of second generation intact stability criteria" -Report of the working group (Part 1).

[2] IMO, 2008, International code on intact stability, (2008 IS CODE)

[3] Pollard, J. and A. Dudebout, 1892, "Theorie du Navire," *Vol. 3, Paris*

[4] Paulling, J. R., 1961, "The Transverse Stability of a Ship in a Longitudinal Seaway," J. Ship Research, Vol. 4, No. 4, pp. 37-49.

[5] Kobylinski, L. K. & S. Kastner, 2003, "Stability and Safety of Ships: Regulation and Operation", Vol. 9 of Eng. Book Series, Elsevier, Amsterdam.

[6] ITTC (1999) "Final Report and Recommendations to the 22nd ITTC", *Seoul and Shanghai, Vol. II.*

[7] IMO SLF 48/4/12, 2005, "On the Development of Performance-Based Criteria for Ship Stability in Longitudinal Waves," *Submitted by Italy, IMO, London*.

[8] IMO SDC 3/WP.5, 2016, "Finalization of second generation intact stability criteria", Annex-3.

[9] Belenky V., Bassler C., Spyrou K. (2011) "Development of Second Generation Intact Stability Criteria", *Naval Surface Warfare Center Carderock Division- Hydromechanics department report.*

[10] Barltrop N.D.P., 1998, "Floating Structures: a guide for design and analysis", *Volume-1 CMPT.*

[11] IMO SDC 2/WP.4, 2015 "Development of second generation intact stability criteria", Annex - 1

[12] IACS Rec.34, 2000 "Standard Wave Data No.34 – IACS", www.iacs.org.uk/download/1978.

[13] IMO SDC5/INF.4, 2017 "Finalization of second generation intact stability criteria" *Report submitted by France, Annex-13.*

[14] IMO SDC4/5/1/Add.5, 2016 "Finalization of second generation intact stability criteria" *Report submitted by Japan, Annex-6.*

Proceedings of the 6th International Conference on Ship and Offshore Technology
ICSOT 2019
November 7-8, 2019, IIT Kharagpur, India

ICSOT INDIA 2019-05

RADAR CROSS SECTION AND SUPERSTRUCTURE GEOMETRY - A STUDY ON AUGMENTATION OF STEALTH IN MODERN WARSHIPS

Ananth Kishore
Dept of Applied Mechanics
IIT Delhi
New Delhi -110016
India
amx185503@am.iitd.ac.in

K Vignesh Kumar
Dept of Applied Mechanics
IIT Delhi
New Delhi -110016
India
kvkumar@am.iitd.ac.in

ABSTRACT

Stealth technology has steadily gained prominence in the design philosophies of modern warships. As radar detection capabilities and weapons technology have exhibited significant advancements in recent times, it necessitates the incorporation of stealth or 'low-observable' technology into modern surface combatants. While warship designers employ multiple aspects of signature suppression such as visual, acoustic, infra-red, etc., reduction of the Radar Cross-Section (RCS) is an effort to avoid detection, identification, and offers better survivability against radar-based homing anti-ship missiles. This study focusses on the reduction of RCS of typical modern warship superstructure.

Firstly, this paper evaluates a few canonical geometries to study the primary effect of shaping, on RCS and also validate the computational tools employed. Subsequently, RCS of generic, conventional warship shapes are evaluated and compared against the ONR Tumblehome model, (an open-access stealth ship model available for fundamental research). The study investigates the change in stealth characteristics of the ship achieved by imparting parametric variations to its superstructure geometry. Finally, the study evaluates some innovative superstructure shapes to reduce RCS while not compromising on valuable internal volume.

NOMENCLATURE

A	Area of Cross Section
σ	Radar cross-section
dbsm	Decibels per square meter
ε_o	Free space permittivity
μ_o	Free space permeability
$\vec{E}$	Electric field
$\vec{B}$	Magnetic field

INTRODUCTION

Modern military warfare has always been connected with the scheme of understanding the enemy position in advance. In the older days, ships could only be detected through the naked eye. However, with the advancements in military technology, the process of detection and the parameters involved underwent a radical change. After the 1930s, radar technology was introduced in military warfare. With the help of radar, ships could be detected, tracked, and classified from a far distance. A modern-day radar consists of a transmitter, a receiver, a duplexer, and an antenna. Radar detects a target by measuring the time required for a transmitted wave signal to get to the target and return. The radar captures the details about the target in terms of Radar Cross-section. Radar cross-section is effectively an area that may or may not be related to the physical area of the target. It is a measure of how big or how bright the target appears to the radar. A target's RCS is represented as the product of three factors [1]:

$$\sigma = \text{Projected cross-section x Reflectivity x Directivity} \quad (1)$$

RCS of a target is defined as the projected area of a metal sphere that would scatter the same power in the same direction as the target. Detection of a ship with lesser RCS is a mighty task, and it can be easily misunderstood for a much smaller vessel. This brings the element of surprise as these ships will be detected late. Thus, radar signature or RCS reduction is a critical factor in the design of a stealth ship.

The unit of area or cross-section is given in square meters. Due to the high contrast in the RCS pattern from one aspect angle to another, it is convenient to display the RCS in a logarithmic form. Therefore, RCS is usually expressed in decibels per square meter.

$$\text{RCS (dBsm)} = 10 \log (\sigma \text{ in m}^2 / \text{ref value of } 1\text{m}^2) \quad (2)$$

The radar transmits energy in the form of electromagnetic waves in all directions. Energy from the incident field gets scattered in all directions by the target, and the key to RCS reduction lies in deflecting the energy to the direction where the threat is not present. The scattered energy, especially from a large target such as a ship or an aircraft, obeys the basic principle of optics. A reflection is a process by which waves bounce off surfaces in a mirror-fashioned way explained by Snell's law, which states that the angle of a reflected ray equals the angle of the incident ray.

RCS Prediction Techniques

Methods used to determine the RCS of a naval ship or other complex structures are:

a. Scaled measurements

b. Computational prediction techniques

Scaled measurements involve both full-scaled and low scaled measurements. Full-scale measurements offer more accuracy as the data is taken in the open sea that resembles its operational environment. However, it has its limitations, which include high cost, the requirement of the ship to be at sea, and the inability to differentiate between RCS characteristics of individual parts of the ship. Furthermore, the inability to incorporate many changes in a fully built ship at this stage limits the relevance of the effort in measuring and attempting to reduce the RCS.

Scaled measurements involve the RCS prediction using scaled models of the full-scale ship, which is carried out using high-frequency emitters in closed conditions. The models are usually made of brass. This is less expensive when compared to a full-scale measurement. But the method is laborious and tedious.

Finally, the RCS of an object can be determined using computational prediction techniques. This is a comprehensive technique that is both less expensive and less time-consuming. The scaled methods are confined to relatively small and simple objects in the Rayleigh region, in which the size of the target is smaller than the wavelength of the illuminating wave and Resonance regions in which the wavelength of the wave and the sizes of the target are comparable [2]. High-frequency methods are used for complex objects in the Optics region, also called high-frequency region. This is the region that deals with RCS prediction of aircraft and ships. For computerized methods, a far-field approximation is usually considered for allowing the incident wave to be represented as a plane wave.

NUMERICAL TOOL USED IN PRESENT STUDY

Ansys® High-Frequency Structural Simulator (HFSS) is a full-wave 3D electromagnetic high-frequency structure simulator. HFSS is based on FEM simulation technology. It is one of several tools used for antenna design, radar, and the design of complex RF problems. Maxwell equations [3] are solved to find the RCS using the finite element method. Maxwell's equations describe how electric currents and electric charges create magnetic and electric fields. Further, they describe how a magnetic field can generate an electric field and vice versa. The four general Maxwell equations are given as:

$$\nabla . \vec{E} = \frac{\rho}{\varepsilon_0}$$

$$\nabla . \vec{B} = 0$$

$$\nabla \times \vec{B} = -\frac{d\vec{B}}{dt}$$

$$\nabla \times \vec{E} = \mu_0 \vec{J} + \frac{1}{c^2}\frac{d\vec{E}}{dt} \qquad (4)$$

SCOPE OF PRESENT STUDY

RCS reduction techniques involve the shaping of above waterline geometry, active cancellation of the incoming wave, and usage of radar-absorbent materials. This study deals with the shaping of the superstructure to optimize the RCS values for a stealth ship. The study was undertaken in a sequential manner commencing with simple canonical structures and maturing to complex shapes and eventually to stealth warship geometries. This was done to understand the fundamental workings of numerical tools and how it handles the physics behind stealth RCS signature calculations and after that, exploit it to derive design inputs for future stealth ships.

RCS of a Sphere

The plane wave scattering from a sphere is a classic problem and is often used as a reference because of its symmetry. i.e., the scattering from a sphere is uniform in all directions, so it gives a constant value of RCS for all the aspect angles. For a sphere of radius, 'a' the incident radiation is uniformly scattered in all directions, and the RCS is defined as same as that of the projected area,
Therefore, RCS for a sphere [2]

$$\sigma = \pi(r)^2 \qquad (2)$$

This equation shows that a sphere of radius 0.565 m gives a value of RCS 1m^2. Thus, a sphere of radius 0.565 m was modeled in ANSYS HFSS to validate the results from the software, and the result obtained is given in Figure 1.

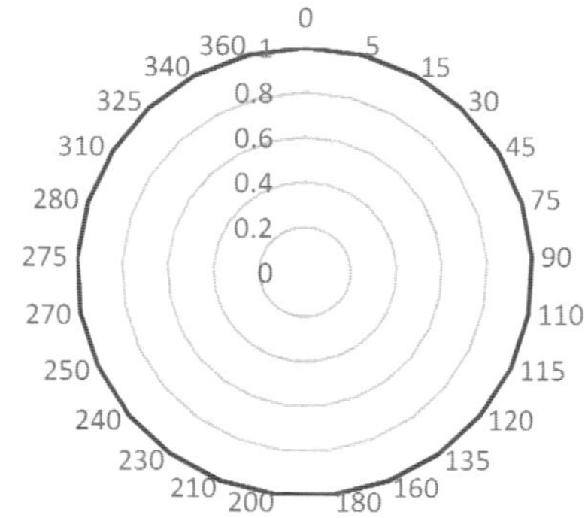

Figure 1. RCS OF SPHERE

As shown in the above pic, the RCS of the sphere calculated from the software was approximately 1m^2

RCS of a Cube

Another classic geometry is a cube. A cube has a different projected area at different angles. For a flat plate, RCS is given by [2].

$$\sigma = (4\,\pi\, h^2\, w^2)/\,\lambda^2 \qquad (3)$$

Where h and w are the sides of the plate, and λ is the wavelength. This equation shows that at the broadsides, a cube of 1m, at 1 GHz, gives a value of σ = 140m^2.
Now, a cube of 1m was modelled in the ANSYS HFSS, and RCS was estimated. The value at the broadside came out to be 21.45 dbsm= 139.6 m^2, which is almost equal to the actual value of 140 m^2. The modeled cube and the plot of RCS along with the cube orientation is shown in Figure 2.

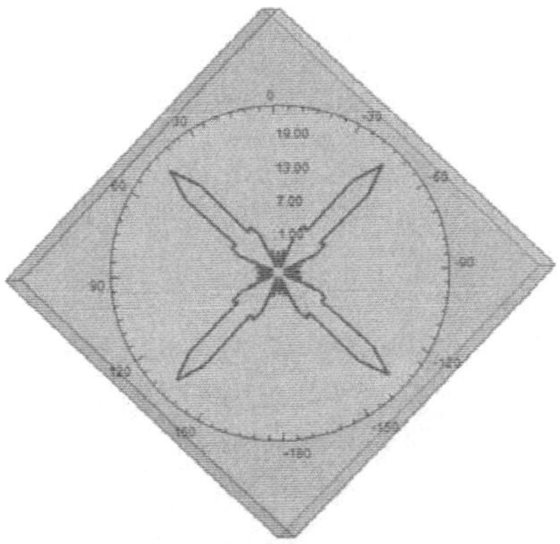

Figure 2. RCS OF CUBE

Parametric changes to the cube

A parametric transformation is now carried out for a cube in which two vertical faces of the cube were inclined to investigate the variation of RCS with respect to the inclination angles. The taper angle θ is shown in Figure 3.

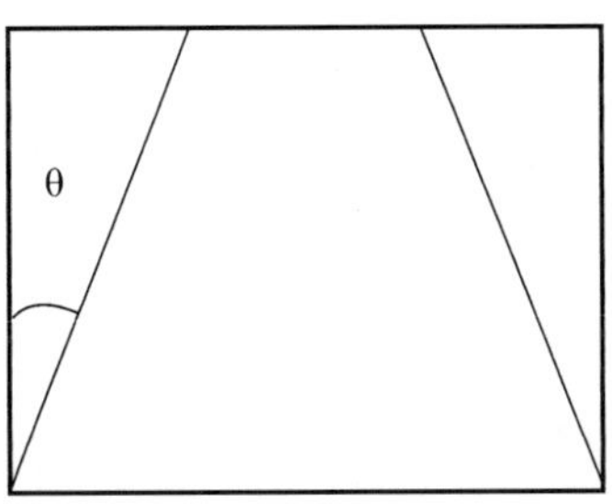

Figure 3. INCLINING SIDES OF CUBE

The taper angle is varied from 5 degrees to 30 degrees, and the corresponding geometries were developed in the software. The shapes and values of RCS are given in Table 1.

Table 1: VARIATION OF RCS WITH CHANGE IN TAPER ANGLES

Taper angle	Shape	RCS value in dB
5	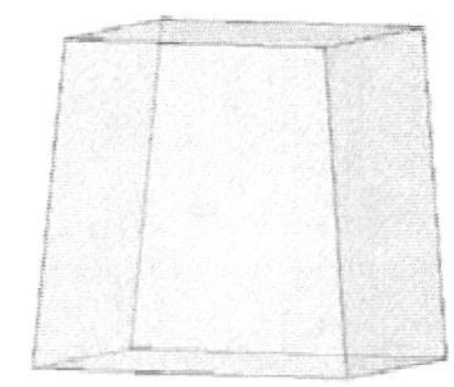	34.41
15	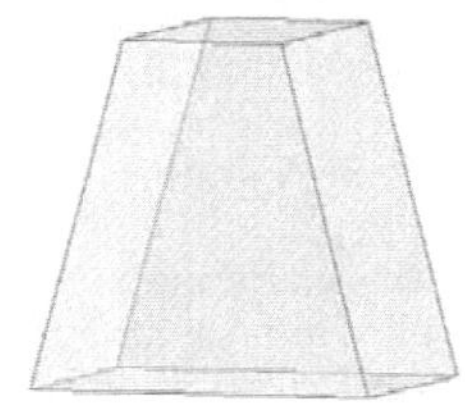	32.56
30	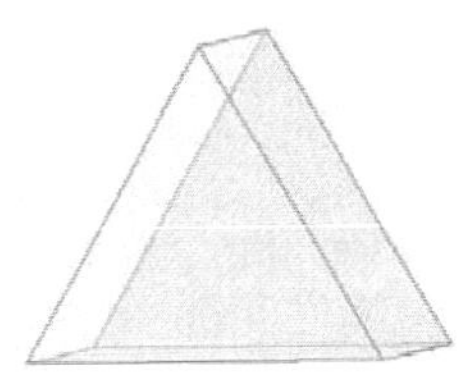	28.16

It was observed that with increasing taper angles, the peak values of RCS decreased. The variation in peak values of RCS of the tapered cube at various angles is given in Figure 4.

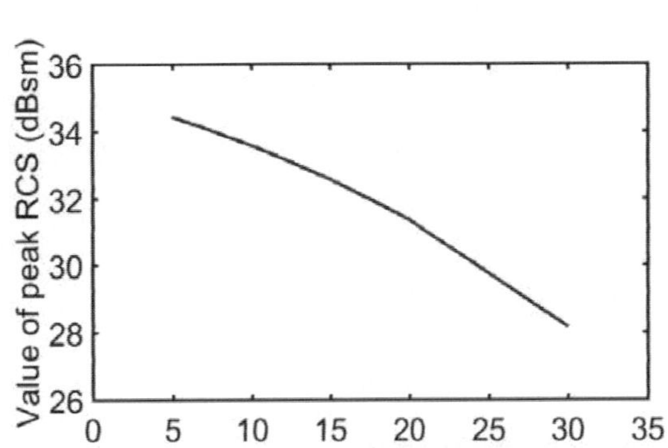

Figure 4. VARIATION IN PEAK VALUES OF RCS OF TAPERED CUBE

Now in the next section, the same idea of varying the sides of the cube is applied to the superstructure of ship geometries to study the variation in RCS.

ONR TUMBLEHOME

The ONR Tumblehome model is a preliminary design of a modern surface combatant, which is publicly accessible for fundamental research [4]. The dimensions of the vessel are:

LWL(m)	154.0
BWL(m)	18.78
D (m)	14.5
T (m)	5.494
Δ	8507 ton

Figure 5. ONR TUMBLEHOME MODEL

The ONR tumblehome model was scaled down to 1:100 for ease of simulation and reducing computation time. The ONRT superstructure has superstructure sides with an inclination of 15 degrees from the vertical. Simulations were carried out varying this inclination angle from 5 degrees to 30 degrees. The variation in peak values of RCS of ONR at various angles is given in Figure 6.

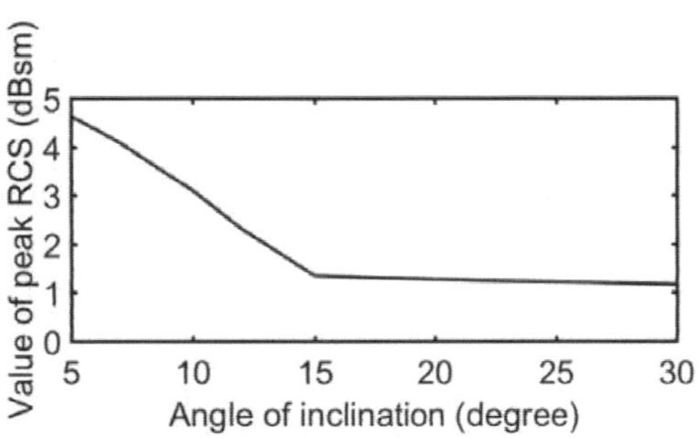

Figure 6. VARIATION IN PEAK VALUES OF RCS OF ONRT

It is observed that the peak RCS is reducing with increasing taper angles as expected. However, the rate of reduction is steep till about 15 degrees, beyond which the RCS change is not appreciable. Since the inclination of superstructure sides in a ship is associated with a reduction in usable internal volume as well, it is opined that the region for best tradeoff is between 5 – 15 degrees, with any further inclination giving a marginal reduction in RCS. This observation also correlates well with the designed inclination of the ONR superstructure being 15 degrees.

ONR SUPERSTRUCTURE WITH 'X' FORM

In the next stage of the study, a variation was imparted to the ONRT geometry. Keeping the hull geometry unchanged, the superstructure was morphed into an X form, while maintaining the volume of superstructure constant. This was done to explore if an X form accorded a wider range of angles and a relatively better reduction in RCS. For the same volume, the X form is expected to be better than a tapered cuboid in terms of internal space utilization as both the base and top are broader than the waist enabling better internal disposition of components. The X form, named ONRTX, is depicted in Figure 7, and the angle definition is included in Figure 8.

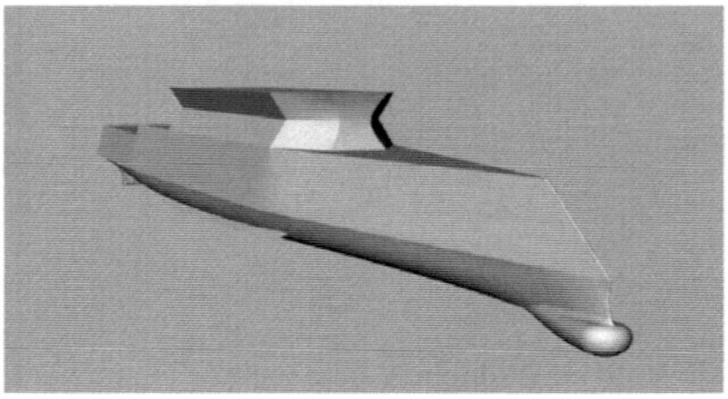

Figure 7. ONRTX MODEL

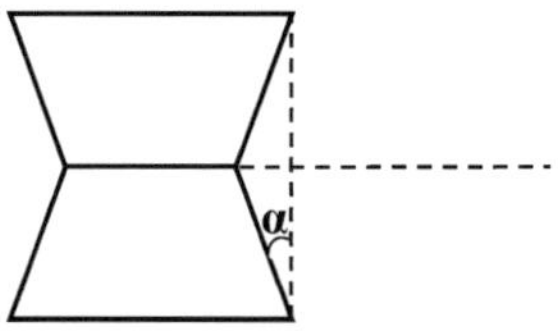

Figure 8. Taper Angle for X hull form

RCS simulations were carried out across a range of angles (ά), and the peak RCS variations are given in Figure 9. It is observed that the slope of the graph (rate of RCS reduction with increasing angles of inclination) showed almost a linear trend for the angles from 5 to 12 degrees.

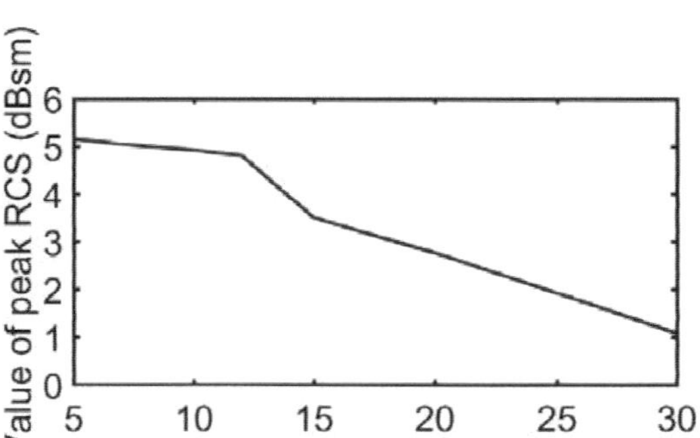

Figure 9. VARIATION IN PEAK VALUES OF RCS OF X HULL FORM

To investigate the form further, a study was conducted to see how the X shape performed at larger angles. The RCS of the X hull form was further developed for angles up to 35 degrees, and it was observed that at angles greater than 30 degrees, the X hull form performed better and showed less RCS than the ONRT form at corresponding angles of inclination. The comparison of both the hull forms is shown in Figure 10.

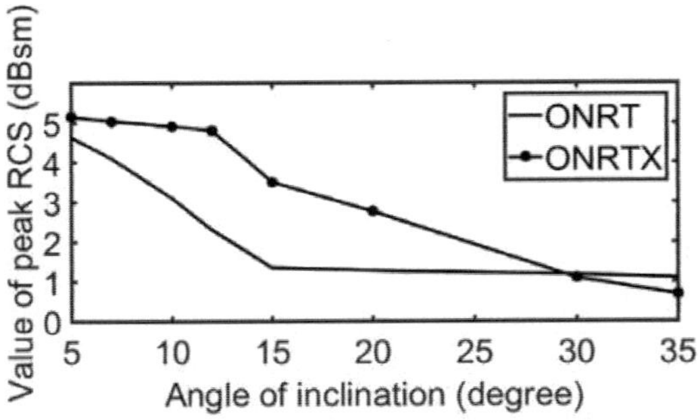

Figure 10. COMPARISON OF X HULL FORM AND INCLINED HULL AT LARGE ANGLES

CONCLUSION

The RCS of basic shapes has been computed using high frequency simulation tool. The reduction in RCS with changes in inclination angles has been studied. Tumblehome superstructure has been compared with a novel X form superstructure for better RCS performance, and it has emerged that the latter presents a reduction of about 5-15% RCS at large angles (beyond 30 degrees) for the same internal volume.

Modern-day warship ship designs involve the inclusion of stealth at a much later stage in the design spiral. The incorporation of stealth at this stage leaves the naval architect with many constraints. The application of these results in the early design spiral of a warship helps to incorporate stealth as a key factor. Thus, naval architects can incorporate stealth in the early design stage itself. This study demonstrates that incorporating efforts for RCS reduction at the initial design stage of warships without compromising the usable internal volume is possible and can significantly reduce RCS and in turn, provide a tactical advantage in modern-day warfare. Further modifications in terms of variations in larger angles or combining superstructure inclination with above water hull geometry modifications may be explored using a similar methodology for stealth optimized design of warships.

REFERENCES

[1] Eugene F. Knott, Merrill I. Skolnik,1990. "*The radar handbook*," *McGraw-Hill.* Houston, USA.

[2] Naval Air Warfare Weapon Centre 2013. "*Electronic warfare and radar systems engineering handbook*". Naval air warfare center weapons division, Point mugu, California ,USA

[3] Clerk-Maxwell, J.,1861, *"On Physical Lines of Force"*, Philosophical Magazine, Volume XXI, Fourth Series, London.

[4] Sanada, Y., Tanimoto, K., Takagi, K., Toda,Y., Stern, F., 2013, "*Trajectories and Local Flow Field Measurements around ONR Tumblehome in Maneuvering Motion*", Ocean Engineering, Vol. 72.

Proceedings of the 6th International Conference on Ship and Offshore Technology
ICSOT 2019
November 7-8, 2019, IIT Kharagpur, India

ICSOT INDIA 2019-06

ANALYSIS OF FORCES AND FLOW FEATURES OF A WAVY CYLINDER AT LOW REYNOLDS NUMBER

Arijit Pradhan
IIT(ISM) Dhanbad
Dhanbad, Jharkhand, 826004
India
arijit575@gmail.com

Ravi Chaithanya Mysa
Singapore University of
Technology and Design
Singapore, 487372
Singapore
ravichaithanya.mysa@gmail.com

Subhankar Sen
IIT(ISM) Dhanbad
Dhanbad, Jharkhand, 826004
India
subhankars@mail.com

ABSTRACT
The attempt to use wavy cylinders in the offshore industry is bio-inspired from the vibrissa of the harbor seal vibrissa. The ability to detect the prey in worse conditions by the seal vibrissa is due to the sensitivity of the wavy cylinder to the wakes of the prey. A clear understanding of the flow physics of the stationary wavy cylinder and its vortex-induced vibrations in a uniform flow, as well as the non-uniform flow, has to be studied in detail for venturing into the applications in off-shore industry. In the present numerical study, the unsteady flow physics and the force variation due to the wavy nature of the stationary cylinder is studied in detail. The Reynolds number considered for the study is 100. The geometry of the cylinder is parametrized by the wavelength λ, wave amplitude a, and mean diameter, D_m of the cylinder sinusoidal curve. The Results are presented for the non-dimensional parameter λ/D_m varying between 3.5 and 9. The wave amplitude along the span of the cylinder for the numerical investigation has been fixed to 0.1. A detailed analysis based on the pressure variation along the span of the wavy cylinder and it's near flow features is done and an explanation is provided for the source of force in the paper.

INTRODUCTION
Vortex-induced vibration (VIV) of bluff bodies finds application in off-shore, tall buildings, power lines, bridges, chimneys, and energy conversion equipment. The vortex shedding from these vortices generates oscillatory forces on the bluff body. The control of vortex shedding for suppression of VIV and extracting energy is a challenging issue. Lam and Lin [1] investigated a series of wavy cylinder with different combinations of wave amplitude and wavelength to study the effect of drag force over the cylinder surface at low Reynolds number. It has been shown that the free shear layers from the wavy cylinder are difficult to roll up to vortex and it causes an increase in wake formation length of the wavy cylinder. Yoon et. al [2] have studied the reduction in drag coefficient and lift coefficient on an asymmetric wavy cylinder at Reynolds number 3000. They have shown that the length of vortex formation for an asymmetric wavy cylinder is longer than that of a symmetric wavy cylinder. And there is a significant reduction in drag and lift coefficient than the symmetric cylinder. Darekar and Sherwin [3] have numerically investigated flow past square wavy cylinder. The introduced waviness at Reynolds number 100 results in stabilization of near wake compared to fully developed shedding in a straight square cylinder. It has been found out that an increase in wave amplitude results in the emergence of hairpin vortices from the near wake region. The introduction of waviness at a wavelength close to the mode A wavelength leads to the suppression of the Karmann street at a minimal waviness amplitude. Ahmed et. al [4] have experimentally studied the transverse flow over a set of wavy cylinder with different axial wavelengths. The Presence of spanwise pressure gradients results in three-dimensional separation lines and the formation of streamwise trailing vortex structure near the geometric nodes. They have shown that there is a larger sectional drag coefficient at the geometric nodes than at the geometric saddles.

Lam et. al [5] have done numerous experiments to study the near wake of a wavy cylinder. The crossflow mechanism around the wavy cylinder reveals the average vortex formation length behind a wavy cylinder is longer than that of a circular cylinder, which effectively reduces the drag and suppresses the vibration. The free shear layers shed from the points near the saddles extend along the spanwise direction, while the shear layer near the nodes contract. Lam et. al [6] conducted wind tunnel experiments to study the effects of the surface waviness of wavy(varicose) cylinders. The experiments are conducted with Reynolds number between $2.0\text{x}10^4$ and $5.0\text{x}10^4$ and free stream turbulence intensity less than 0.2%. It has been concluded from the experiment that drag reduction of up to 20% can be achieved. The spectral analysis of both hot-wire and load cell signal shows the Strouhal number of a wavy cylinder and corresponding circular cylinder remains the same. Vibrations of a thin and flexible cylinder under nominally constant towing conditions are studied by Dowling [7]. They have shown that the displacement near to critical points depends on normal and tangential drag coefficient C_N and C_T. Kim and Yoon [8] have studied the effect of angle of attack on harbor seal vibrissa shaped cylinder on flow characteristics at Reynolds number 500. They have shown that there exist longer shear layers for the said cylinder than elliptic cylinder at smaller angle of attack. But, with increase in angle of attack, the flow contours become similar to that of an elliptic cylinder.

In the present work, numerical simulation has been carried out to study the effect of waviness on the net fluid force exerted on the cylinder. We are intending to study the detailed analysis on the origin of forces in this work. In later part of the work, the vortex-induced vibration (VIV) pattern and its cause, has been analyzed for a wavy cylinder. A series of different values of non-dimensional parameter, λ/D_m has been taken into account. The ratio of wave amplitude to mean cylinder diameter, a/D_m is fixed at 0.1. The whole study has been carried out at Reynolds number Re, 100. In the next section, the mathematical concept behind the numerical simulation has been explained.

MATHEMATICAL MODELING

The governing Navier-Stokes equation for the unsteady incompressible flow in the Arbitrary Lagrangian Eulerian (ALE) approach is described by:

$$\nabla . U = 0 \tag{1}$$

$$\frac{\partial U}{\partial t} + \nabla .((U - Ug)U) = -\nabla p + \nabla .(\nu \nabla U) \tag{2}$$

Where **U**, p, $\mathbf{U_g}$ and ν represent the fluid velocity, pressure, velocity of the moving mesh and kinematic viscosity of the fluid respectively. The momentum equation is discretized using an implicit one step Euler discretization in time,

$$a_p U_P^{n+1} = \sum_f a_f U_f^{n+1} = -\nabla p^n + \frac{U^n}{dt} \tag{3}$$

The above equations are solved over multiple outer and inner corrector steps for forwarding the solution over one time step in the openFoam solver *pimpleDymFoam* [5]. For each outer corrector, the non-linear term $(\mathbf{U}\text{-}\mathbf{U_g})\mathbf{U}$ is linearized such that the term $(\mathbf{U}\text{-}\mathbf{U_g})$ appears as relative face flux at a previous outer corrector. The term a_p and a_f contain the coefficients of discretization, which also includes the relative face flux.

The inner correctors follow the Pressure Implicit with Splitting of Operators (PISO) algorithm. This can be achieved by casting the momentum equation in such a way that it enforces a zero discrete at the following inner correctors as follows:

$$U_f^{n+1} = \frac{H(U^{n+1})}{a_{p,f}} - \frac{\nabla p_f^{n+1}}{a_{p,f}} \tag{4}$$

Where,

$$H = \sum_f a_f U_f^{n+1} + \frac{U^n}{dt} \tag{5}$$

The Poisson equation for pressure can be obtained by taking the discrete divergence of equation - (4).

$$\sum_f \frac{\nabla p_f}{a_{p,f}}.S = \sum_f \frac{H}{a_{p,f}}.S \tag{6}$$

Where **S** is the area vector of the given face. *pimpleDymFoam* is developed by merging the PISO and SIMPLE algorithms for a dynamic mesh case. The pressure is relaxed and end of each inner corrector step in the PISO algorithm, which gives rise to the SIMPLE algorithm. The number of outer correctors are determined through a residual criteria, whereas the number of inner correctors is fixed to 2-3 for this work.

The different non-dimensional parameters used in this study are mass ratio m^*, Reynolds number *Re*, reduced velocity U_r, and damping ratio ς are defined as follows:

$$m^* = \frac{m}{m_f}, \quad Re = \frac{U_\infty D_m}{\nu}, \quad U_r = \frac{U_\infty}{f_N D_m}, \quad \zeta = \frac{c}{\sqrt{km}} \tag{7}$$

Where *m* is mass of the bluff body, m_f is the mass of the fluid displaced by the bluff body, *c* is the damping coefficient, and *k* being the stiffness constant for an equivalent spring-mass-damper system. U_∞ and D_m are the free stream speed and mean diameter of the cylinder respectively. The natural frequency of the bluff body is defined as $f_N = (1/2\pi)\sqrt{(k/m)}$. The drag coefficient C_D is defined as:

$$C_D = \frac{1}{\frac{1}{2}\rho U^2 D}\int_S \left(\left(-p\mathbf{I} + \mu\left(\nabla \mathbf{U} + (\nabla \mathbf{U})^{\mathbf{T}}\right)\right).\mathbf{n}\right).\mathbf{n_x}dS \tag{8}$$

$$cd_{avg} = \frac{\int_0^T cd(t)dt}{\int_0^T dt} \tag{9}$$

Where **S** is the surface area of the wavy cylinder, μ and ρ being the dynamic viscosity and density of fluid respectively, **n** is the unit normal vector to the plane and **I** is the identity matrix. The lift coefficient due to pressure force C_{Lp} and lift coefficient due to viscous force $C_{L\mu}$ are defined as:

$$C_L = \frac{1}{\frac{1}{2}\rho U^2 D}\int_S \left(\left(-p\mathbf{I} + \mu\left(\nabla \mathbf{U} + (\nabla \mathbf{U})^{\mathbf{T}}\right)\right).\mathbf{n}\right).\mathbf{n_y}dS \tag{10}$$

$$Cl_{rms} = \sqrt{\frac{\int_0^T [cl(t)]^2 dt}{\int_0^T dt}} \tag{11}$$

$$C_{Lp} = \frac{1}{\frac{1}{2}\rho U^2 D}\int_S \left(\left(-p\mathbf{I}.\mathbf{n}\right).\mathbf{n_y}dS\right) \tag{12}$$

$$C_{L\mu} = \frac{1}{\frac{1}{2}\rho U^2 D}\int_S \mu\left(\left(\nabla \mathbf{U} + (\nabla \mathbf{U})^{\mathbf{T}}\right).\mathbf{n}\right).\mathbf{n_y}dS \tag{13}$$

PROBLEM DESCRIPTION

A series of wavy cylinders with different combination of parameters has been studied in this work. The sinusoidal function for generating the wavy surface of the cylinder is being defined as follows:

$$D = D_m + 2a\cos(\frac{2\pi z}{\lambda}) \quad (14)$$

Figure 1(a) shows the geometric parameters of the wavy cylinder. Figure 1(b) depicts the schematic diagram of the flow domain with different boundary conditions. The base of the wavy cylinder is circular. The wave amplitude is denoted by 'a' and the wave length is denoted by λ.

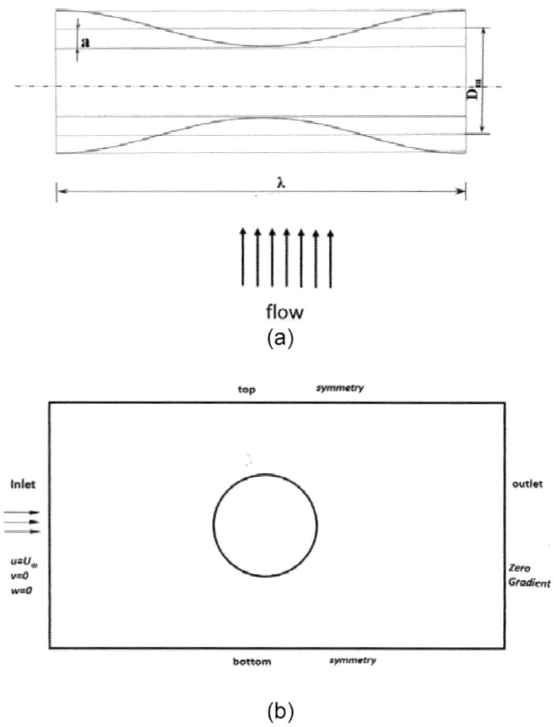

Figure1: SCHEMATIC REPRESENTATION (a). GEOMETRIC PARAMEMTERS OF WAVY CYLINDER (b). COMPUTATIONAL DOMAIN

VERIFICATION AND MESH CONVERGENCE

The mesh convergence study has been carried out for two test cases with λ/D_m=3.5. The results for the comparison has been presented in Table-1. The mesh with size near to 700,000 gives converged results. Thus, this grid size has been chosen for further computation purpose.

Table 1. MESH CONVERGENCE STUDY

Case	Number of Cells	Cd_{avg}	%Change in Cd_{avg}
1	1381527	1.3395	0.1621
2	727982	1.3416	

The validity of the formulated numerical method is need to be checked before further discussion. This section presents the mesh convergence test and validity of the computational data with reference to the work of Lam et. al [1]. The average drag coefficient, Cd_{avg} has been compared for different values of λ/D_m. Table 2 shows the comparison of Cd_{avg} with the results from the work of Lam et. al [1]. From the table it has been observed that the % change is very small, thus validates our numerical method.

Table 2. VALIDATION OF THE NUMERICAL METHOD

λ/D_m	$Cd_{avg,\ Reference}$	$Cd_{avg,\ computation}$	%Change in Cd_{avg}
3.5	1.3297	1.3188	0.8139
4.0	1.2950	1.2797	1.1816
5.4	1.1599	1.1489	0.9522
5.8	1.1411	1.1411	0.4085
9.0	1.1391	1.1391	8.0214

RESULTS AND DISCUSSION

In this text the behavior of wavy cylinder has been studied with carrying several numerical computations at Re = 100. To study the effects of forces due to the curvature of wavy cylinder, the time dependency of coefficient of drag and lift need to be considered for different values of λ/D_m. For the

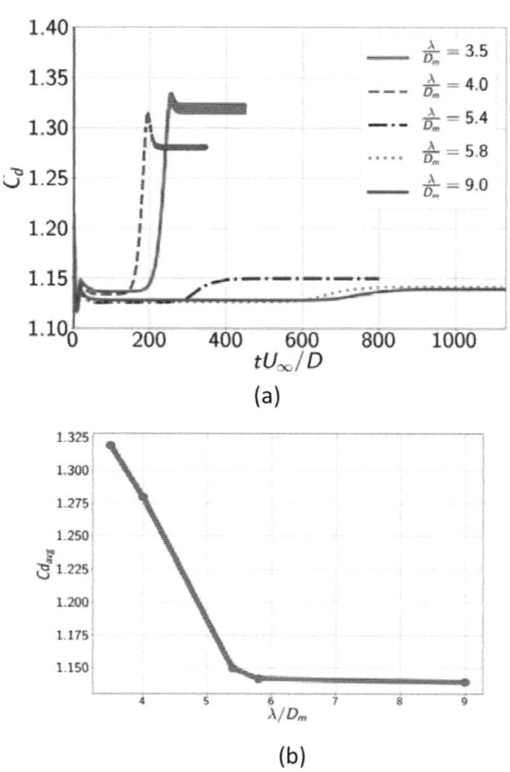

Figure 2: (a). TIME HISTORY VARIATION OF DRAG COEFFICIENT (b) VARIATION OF MEAN DRAG COEFFICIENT WITH DIFFERENT VALUES OF λ/D_m.

current study, the value of a/D_m has been fixed at 0.1. Figure 2 shows the variation of drag coefficient, C_d with time and the geometrical parameter, λ/D_m. As the value of λ/D_m increases, the value of Cd_{avg}, decreases. The graph asymptotically approaches 0.12 as we increases the value of λ/D_m. For circular cylinder the value of drag coefficient is near to 1.33. Thus adding waviness to a circular cylinder reduces the drag force.

Figure 3 shows the variation of lift coefficient, C_l. The RMS value of lift coefficient varies in a similar fashion as Cd varies with λ/D_m. As the value of λ/D_m increases, the root mean square value of lift coefficient decreases. Figure 2 - 3 shows that the waviness decreases the net fluid force exerted on the bluff body.

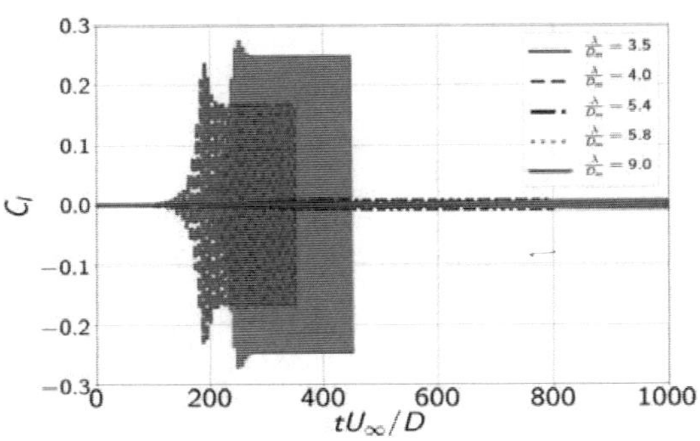

(a)

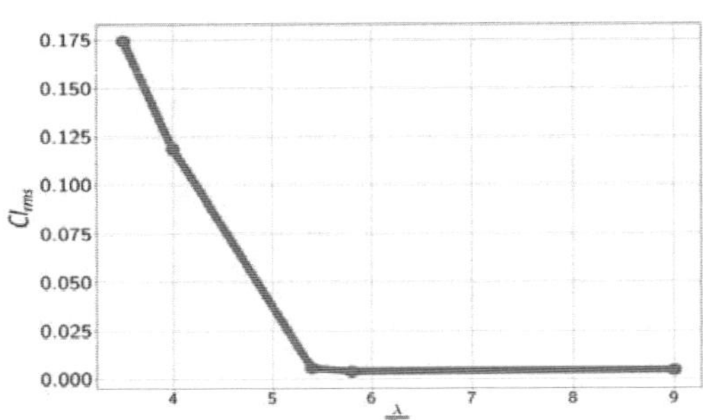

(b)

Figure 3: (a) TIME HISTORY VARIATION OF LIFT COEFFICIENT (b). VARIATION OF LIFT COEFFICIENT WITH DIFFERENT VALUES OF λ/D_m

Figure 4 shows the vorticity contours of wavy cylinder at three different sections for different values of λ/D_m. For the purpose of analysis, nodal plane $z/\lambda=0$, saddle plane $z/\lambda=0.5$ and the mid plane of these two cross sections $z/\lambda=0.25$ has been chosen. Here it's worth mentioning that the wavy cylinder is symmetrical about the mid plane, i.e., $z/\lambda=0.5$. Thus the observed parameters also do behave symmetrically with respect to the mid plane. As the value of λ/D_m increases, it is being observed that the vortex sheet do not roll to form vortex, which causes suppression of vortex shedding near to the cylinder. For $\lambda/D_m=9.0$, the vortex formation is least, whereas for $\lambda/D_m=3.5$, vortex shedding do occur. This facilitates the previous observation that with increase in λ/D_m, there is a reduction in the net fluid forces over the cylinder. Now, for each λ/D_m, there is a difference in vorticity contours for the three shown planes. Thus instead of two-dimensional vortices as in the case of a circular cylinder, here the vortex contour pattern is three dimensional in nature. Due to the three dimensional nature of the vortex formation pattern, the vortex

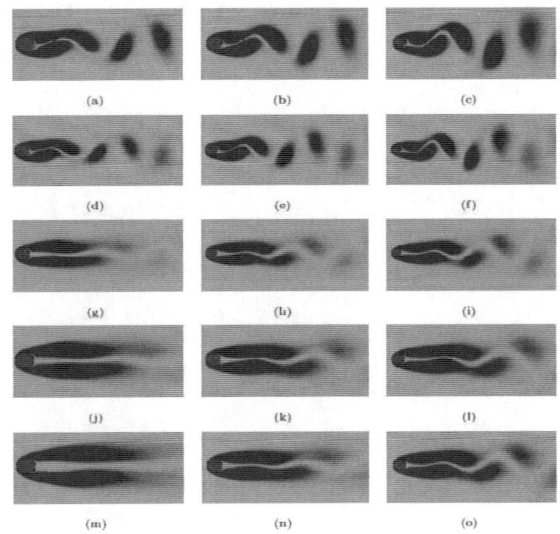

Figure 4. VORTICITY CONTOURS AT THE INSTANT OF TIME WHEN C_{lp} is MAXIMUM. 1ST COLOUMN $z/\lambda=0$, 2nd COLOUMN $z/\lambda=0.25$, 3rd COLOUMN $z/\lambda=0.5$ (a)-(c) $\lambda/D_m=3.5$, (d)-(f) $\lambda/D_m=4.0$, (g)-(i) $\lambda/D_m=5.4$, (j)-(l) $\lambda/D_m=5.8$, (m)-(o) $\lambda/D_m=9.0$

sheet finds it difficult to roll into a full vortex. As the vortex sheet pattern on each plane is different, thus it does not roll to form vortex in the near wake region. Thus the fluctuating forces are absent or very low in case of wavy cylinder at higher value of λ/D_m.

PRESSURE DISTRIBUTION

To understand the nature of fluid forces around the periphery of the cylinder, we will now look at the coefficient of pressure distribution, C_p on the cylinder. The coefficient of pressure, C_p is mathematically defined as $C_p = 2p/\rho U_\infty^2$. Figure 5 shows three graphs showing the variation of C_p with respect to θ measured anticlockwise. The pressure coefficient variation with respect to theta on $z/\lambda=0$ plane gets superimposed on that of $z/\lambda=1$ plane. And similarly the pressure variation with respect to theta on $z/\lambda=0.25$ plane and $z/\lambda=0.75$ plane overlap each other. Thus it's graphically verifies our previous observation that the properties are symmetric about the mid plane i.e., $z/\lambda=0.5$ plane.

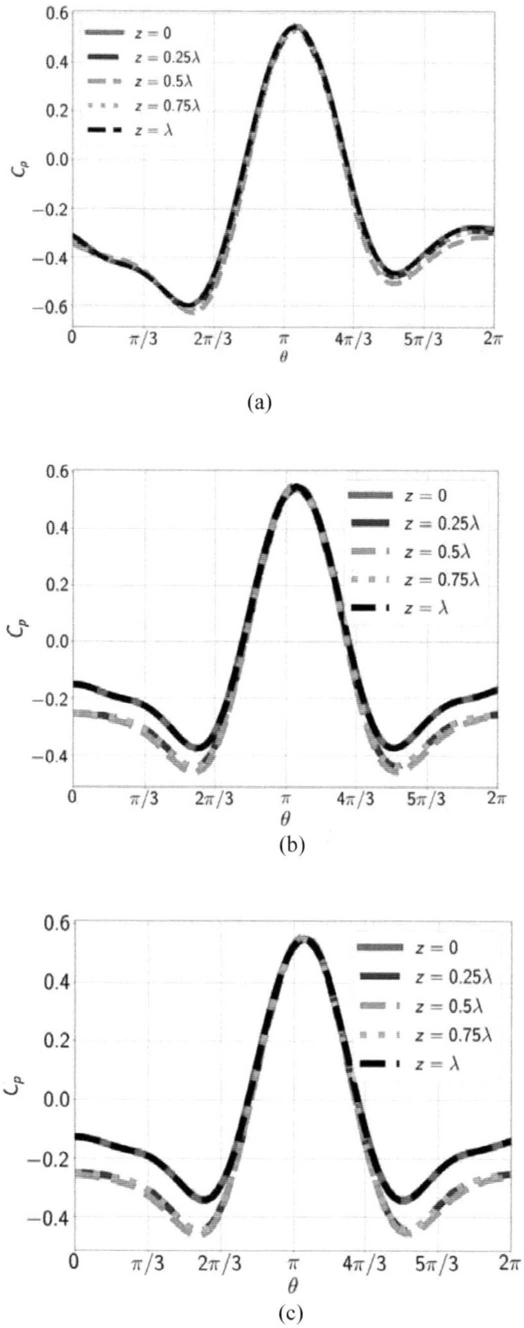

Figure 5: C_p-θ VARIATION ACROSS 5 PLANES i.e., z/λ=0, 0.25, 0.5, 0.75, 1.0 FOR λ/D_m= (a).3.5 (b). 5.8 (c). 9.0

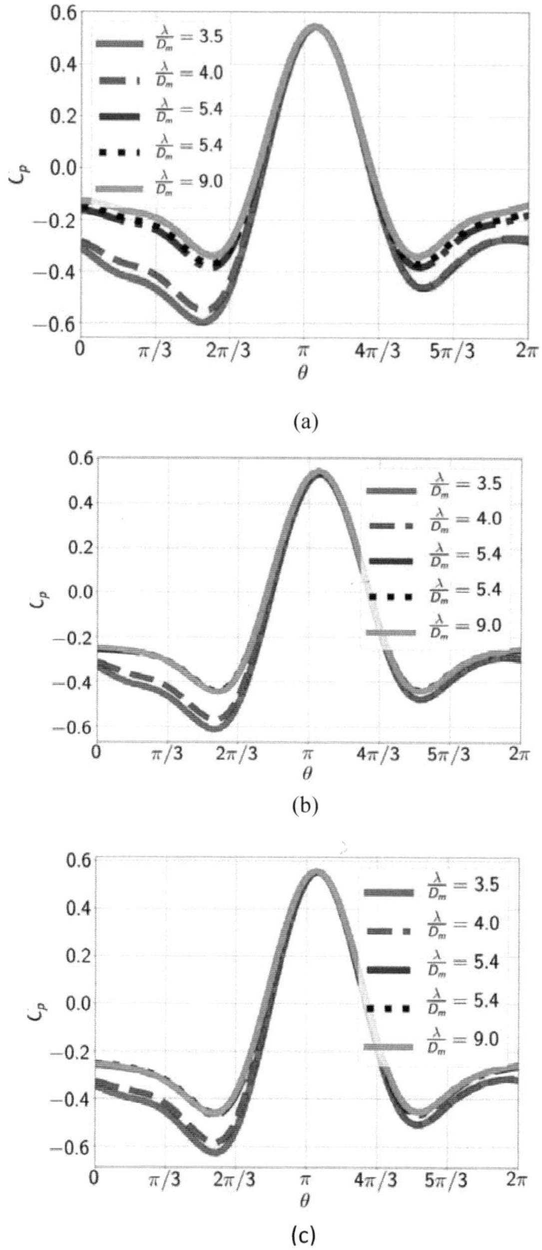

Figure 6: C_p-θ VARIATION AT THE INSTANT OF $C_{l,pmax}$ FOR DIFFERENT VALUES OF λ/D_m FOR z/λ = (a). 0 (b). 0.25 (c). 0.5

At the instant when the lift force, $C_{l,p}$ is maximum, then the suction pressure on top part of the cylinder, $0 \leq \theta \leq \pi$ is more than the bottom part of the cylinder. Thus the top region is designated as the high suction pressure region, where as the bottom portion is called as the low suction pressure region. This designations get reversed if we will talk about the C_p-θ plot at the instant when $C_{l,p}$ is minimum. This is depicted in figure 6. Here, the stagnation point lies on the low suction region, irrespective of the value of λ/D_m The stagnation point for a given normalized plane remain constant for different values of the parameter λ/D_m. The suction pressure is maximum for λ/D_m =3.5. Since for λ/D_m=3.5, rolling of vortex occurs, which causes maximum suction pressure for lower value of λ/D_m. As λ/D_m increases, the value of suction pressure decreases, which is coherent with our previous observation. For z/λ=0 plane the variation in suction pressure is more compared to the saddle plane i.e, z/λ=0.5.

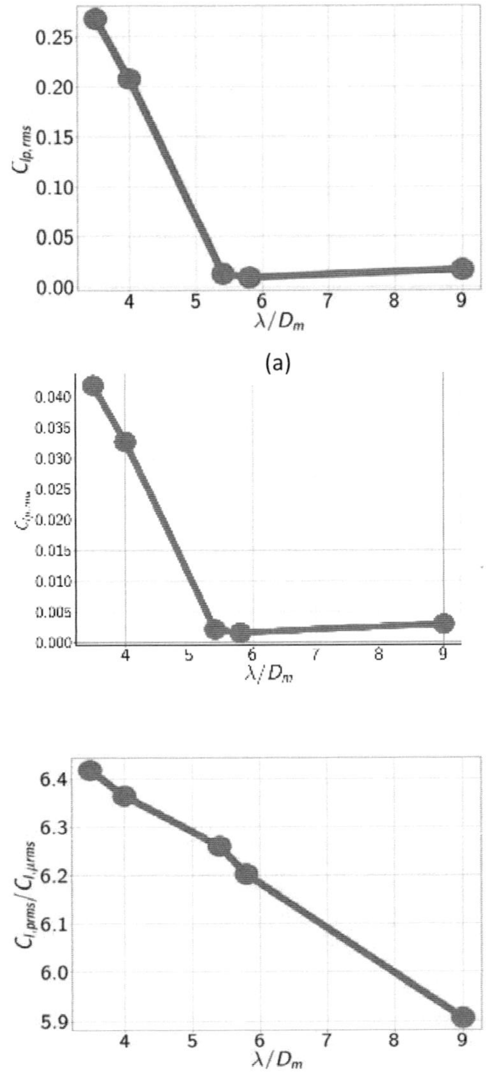

Figure 7: (a). VARIATION OF C_{lp} FOR DIFFERENT VALUES OF λ/D_m (b). VARIATION OF $C_{l\mu}$ FOR DIFFERENT VALUES OF λ/D_m (c). VARIATION OF $C_{lp,\,rms}/C_{l\mu,\,rms}$ FOR DIFFERENT VALUES OF λ/D_m

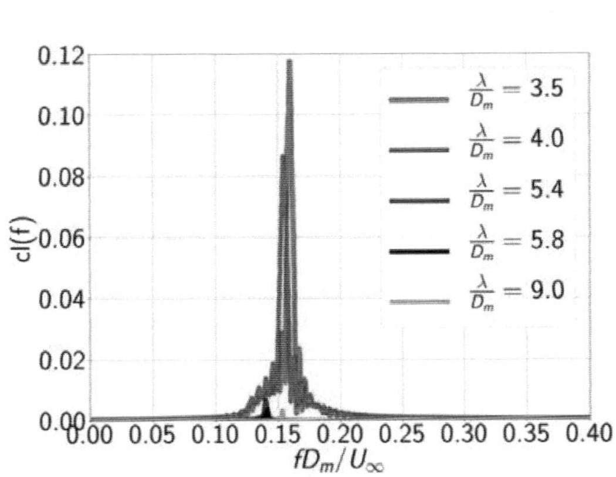

(a)

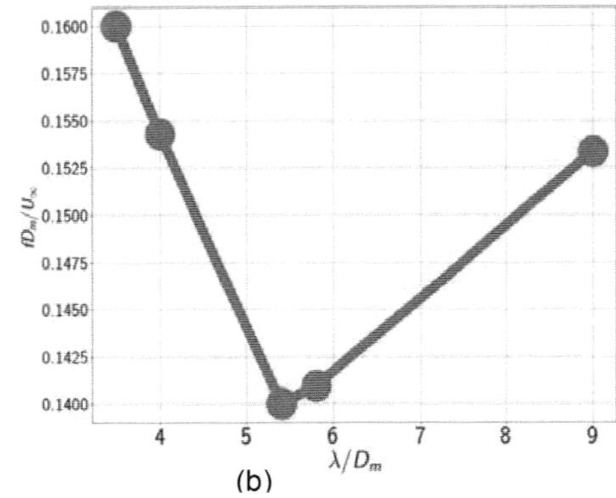

(b)

Figure 8: (a) VARIATION OF LIFT COEFFICIENT IN FREQUENCY DOMAIN (b). VARIATION OF FREQUENCY RATIO WITH λ/D_m

For clear understanding of the effect of transverse force on the cylinder, the lift force coefficient has been decomposed into two parts. The two parts are lift force due to pressure, $C_{l,p}$ and lift force due to viscosity, $C_{l,\mu}$. The root mean square value of lift coefficient due to pressure and viscosity decrease steadily with λ/D_m, conforming to our previous observation. Figure 7 (a)-(b) shows this behavior. Figure 7(c), shows that the effect of $C_{l,p}$ per unit $C_{l,\mu}$ decreases with λ/D_m. The effect of $C_{l,\mu}$ component increases with λ/D_m.

Figure 8 shows the contribution of different frequency components in lift coefficient. The maximum value of normalized frequency for λ/D_m=3.5 to 9.0 is 0.16, which is very low compared to circular cylinder. Figure 8(b) shows that the frequency component decreases with λ/D_m upto λ/D_m =5.4, and then it increases.

VIV OF WAVY CYLINDER

In this section, we will briefly discuss about the VIV of the wavy cylinder. The results of wavy cylinder with geometrical parameter, λ/D_m=5.8 at Re=100 and m^*=10 is being compared with the results from the work of Leontini et. al [10] for a circular cylinder at Re=200 and m^*=10. The behavior of circular cylinder at Re=200 will not differ much from that of Re=100. Thus the works of Leontini et. al[10] will serve our purpose. Figure 9(a) shows the variation of frequency ratio, f/f_N with respect to U_r. The lock-in region for a wavy cylinder

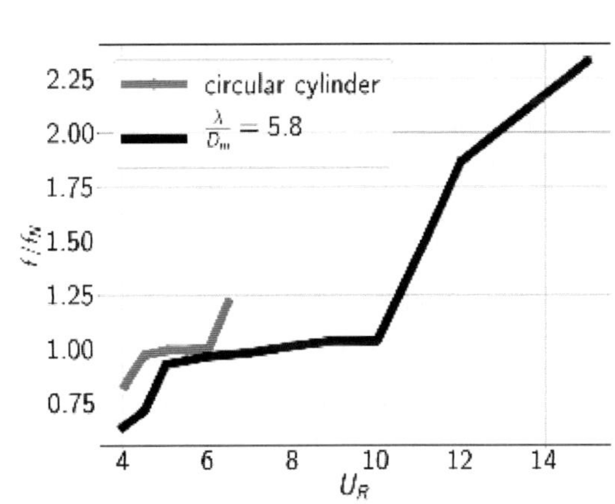

(a)

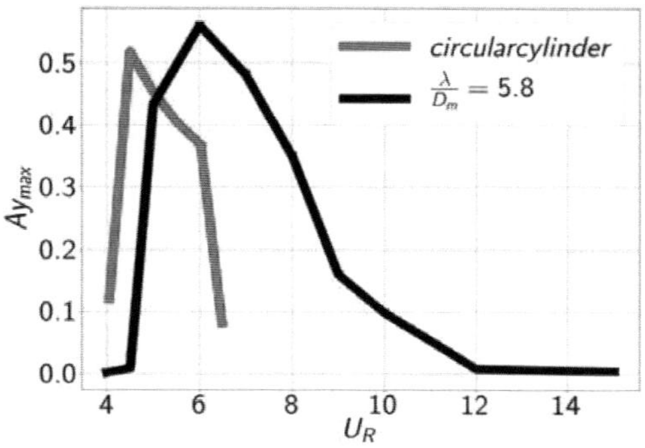

(b)

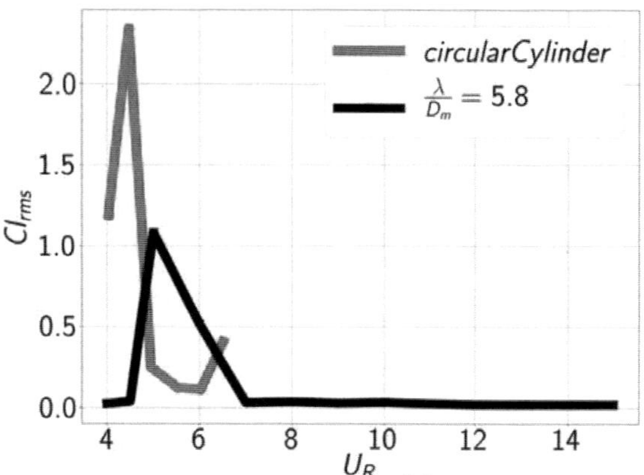

(c)

Figure 9: COMPARISON OF PROPERTIES OF WAVY CYLINDER WITH λ/D_m =5.8, Re=100, m^*=10 WITH CIRCULAR CYLINDER DATA FROM LEONTINI et. al HAVING Re=200, m^*=10 (a). VARIATION OF FREQUENCY RATIO f/f_N WITH U_r (b). VARIATION OF $A_{y,max}$ WITH U_r (c). VARIATION OF $C_{l,rms}$ WITH U_r

with λ/D_m =5.8 will be between U_r=4.5 to U_r=10.0. But, for circular cylinder, it will lie between U_r=4.5 to U_r=6.0. The lock-in region for a wavy cylinder is larger than that of the circular cylinder. Figure 9(b) shows the variation of $A_{y,max}$ with respect to U_r. $A_{y,max}$ is calculated by using the formula $A_{y,max}=\sqrt{2}A_{y,rms}$. Here, the maximum value of amplitude, $A_{y,max}$ for both circular cylinder as well as wavy cylinder is near to 0.5. For both circular cylinder and the wavy cylinder, the amplitude curve follows the same pattern i.e., increasing up to a particular point and then decreasing steadily. The critical point for the wavy cylinder is at U_r=6, whereas for circular cylinder, it is Ur=4.5. Figure 9(c) shows the variation of Cl_{rms} with respect to U_r. For circular cylinder the root mean square value of lift coefficient is generally higher than that of a wavy cylinder with λ/D_m =5.8. The maximum value of Cl_{rms} do occur at Ur=4.5 for the circular cylinder, but for the wavy cylinder, it do occur at U_r=5.0.

CONCLUSION

In the present work the flow features of a wavy cylinder at Reynolds number, Re=100 has been studied for stationary as well as VIV cases. A series of wavy cylinder has been taken for numerical simulation with λ/D_m from 3.5 to 9.0. The ratio of wave amplitude to the mean diameter of the wavy cylinder, a/D_m is 0.1. The VIV of wavy cylinder has been compared with the works of Leontini et. al [10]. It has been observed that with increasing λ/D_m, the net fluid force on the cylinder decreases. The three-dimensional nature of the vortex formation pattern makes it difficult for the vortex to roll. So, with increase in the value of λ/D_m, the tendency of vortex to roll decreases and so is the vortex shedding. The pressure distribution around the cylinder periphery depicts the same observation. It has been observed that for a wavy cylinder with λ/D_m =5.8, the lock-in region is larger compared to a circular cylinder. The transverse force on a wavy cylinder is significantly lower than the circular cylinder.

REFERENCES

[1] K. Lam, Y. F. Lin, Drag force control of flow over wavy cylinders at low reynolds number Journal of Mechanical Science and Technology 21 (9) (2007) 1331. Doi: 10.1007/BF03177417 https://doi.org/10.1007/BF03177417

[2] H. S. Yoon, H. Shin, H. Kim, Asymmetric disturbance effect on the flow over a wavy cylinder at a subcritical Reynolds number, Physics of Fluids 29 (9) (2017) 095102 doi:10.1063/1. https://doi.org/10.1063/1.5001968

[3] R. M. DAREKAR, S.J. SHERWIN, Flow past a square-section cylinder with a wavy stagnation face, Journal of Fluid Mechanics 426 (2001) 263-295, doi:10.1017/S0022112000002299

[4] A. Ahmed, B. Bays-Muchmore, Transverse flow over a wavy cylinder, Physics of Fluids A: Fluid Dynamics 4 (9) (1992) 1959-1967, doi: 10.1063/1.858365, https://doi.org/10.1063/1.858365

[5] K. Lam, F. H. Wang, R. M. C. So, Three-dimensional nature of vortices in the near wake of a wavy cylinder, Journal of Fluids and Structures, 19 (2004) 815-833, doi: 10.1016/j.jfluidstructs.2004.04.004, http://adsabs.harvard.edu/abs/2004JFS....19..815L

[6] K. Lam, F.Wang, J. Li, R. So, Experimental investigation of the mean and fluctuating forces of wavy (varicose) cylinders in a cross-flow, Journal of Fluids and Structures, 19 (3) (2004) 321-334 doi: https://doi.org/10.1016/j.jfluidstructs.2003.12.010 http://www.sciencedirect.com/science/article/pii/S0889974604000271

[7] A. P. Dowling, The dynamics of towed flexible cylinders Part 1. Neutrally buoyant elements, Journal of Fluid Mechanics 187 (1988) 507-532. doi: 10.1017/S0022112088000540

[8] H. Kim, H. S. Yoon, Effect of the orientation of the harbor seal vibrissa based biomimetic cylinder on hydrodynamic forces and vortex induced frequency, AIP Advances 7(10) (2017) 105015. doi: 10.1063/1.5008658 https://doi.org/10.1063/1.5008658

[9] H. Jasak, A. Jemcov, Z.Tukovic, OpenFOAM: A C++ library for complex physics simulations

[10] J.S. Leontini, M.C. Thompson, K. Hourigan, The beginning of branching behaviour of vortex-induced vibration during two-dimensional flow, Journal of Fluids and Structures. doi:https://doi.org/10.1016/j.jfluidstructs.2006.04.003 http://www.sciencedirect.com/science/article/pii/S088997460600051X

Proceedings of the 6th International Conference on Ship and Offshore Technology
ICSOT 2019
November 7-8, 2019, IIT Kharagpur, India

ICSOT INDIA 2019-07

DYNAMIC BEHAVIOR OF FUNCTIONALLY GRADED CNT REINFORCED COMPOSITE PLATES: AN ISOGEOMETRIC FINITE ELEMENT APPROACH

Ashish K. Singh
Department of Applied
Mechanics, MNNIT Allahabad
Prayagraj, U.P, 211004
India
aashish010389@gmail.com

A. Bhar
Department of Applied
Mechanics, MNNIT Allahabad
Prayagraj, U.P, 211004
India
anindyab@mnnit.ac.in

ABSTRACT
This paper presents free and forced vibration behavior of Functionally Graded CNT Reinforced Composite (FGCNTRC) plates. For this purpose, the recently developed isogeometric finite element method based on the use of NURBS to interpolate the geometry as well as the field variables, is adopted. Relevant formulation and a computer program for implementation of the same is developed using a semi-refined higher order shear deformation theory, introduced by the second author. Different types of distribution of the CNTs through the plate thickness, making the plate functionally graded through the thickness, are considered.

After successful free vibration analysis to study the natural frequencies of such plates, forced vibration analysis is performed for different types of excitation forces with sinusoidal, step and triangular variation in time domain. The Newmark-Beta method is used to perform the required numerical time integration. The cases of free vibration post forced vibration up to the desired time of application of force, are also considered here. Parametric studies are carried out to investigate the effect of variation of volume fraction / distribution of CNT, boundary conditions, plate aspect ratios and other parameters, on the natural frequency and transient response in terms of central deflection of the plate.

NOMENCLATURE

IGA	Isogeometric analysis
FG-CNTRC	Functionally graded carbon nanotube reinforced composite
NURBS	Non-uniform rational B-spline basis functions
CNTs	Carbon nanotubes
E_{11}^{CNT}, E_{22}^{CNT}	Young's moduli of CNTRC
G_{12}^{CNT}	Shear Modulus of CNTRC
E^m	Young's Modulus of isotropic matrix
G^m	Shear Modulus of isotropic matrix
V_{CNT}	Volume fraction of CNT
V_m	Volume fraction of Matrix
U, v, w	Membrane displacement
θ_x	Rotation around Y-axis
θ_y	Rotation around X-axis
θ_x^*, θ_y^*	Higher order terms
t_1, t_2	Time limits
T	Kinetic energy
U	Internal potential energy
V	Potential energy due to external force

INTRODUCTION
Isogeometric analysis has been recently introduced by Hughes et. al. [1] with possibly integrating the gap between design and analysis. Isogeometric analysis uses non uniform rational B-splines (NURBS) basis functions for constructing exact geometrical model. Basis functions used in isogeometric analysis is capable for both interpolating the field variables and constructing the geometrical model.

A lot of research has been published on IGA in the field of plate structures. Static, dynamic and buckling analysis of functionally graded plates using higher order shear deformation theory (HSDT) using IGA has been studied by Tran et. al. [2]. Phung-van et. al. [3] studied CNT reinforced composite plate based on HSDT using IGA. They used 5 degree of freedom which are associated at each control points. But in this study we use HSDT of 7 degree of freedom to achieve more accurate results using IGA. The analysis of functionally graded plates has been studied by Valizadeh et. al. [4] using NURBS based IGA. Similarly the study of static and free vibration analysis of functionally graded sandwich plates has been studied by Thai et. al. [5]. Shojaee et. al. [6] used classical plate theory (CPT) for free vibration and buckling analysis of laminated composite plates containing hole of complex shape. They used NURBS basis functions for exactly representing this type of complex shape.

Carbon nanotube has been widely studied since 1991. The material properties are widely amazing; yield strength of CNT are greater than that of high strength steels; excellent thermal conductivity and electrical conductivity [7-8]. Because of unique geometrical, mechanical and electrical properties of CNT reinforced composite plates, they are used in aerospace, aeronautics and marine industries. Shen et. al. [9] has introduced the concept of distribution of CNT reinforced composite along the thickness direction of plate in which the material properties are obtained by the molecular dynamics and rule of mixtures. After that Zhu et. al. [10] has done the static and free vibration using first order shear deformation theory (FSDT) of functionally graded CNT reinforced composite plates which are having four types of reinforcement distribution in which one is uniformly distributed and three are functionally graded distributions.

In this paper, free vibration and forced vibration behavior of CNT reinforced composite plates are investigated using IGA based on semi refined HSDT (SR-HSDT) [11]. CNT are considered to be dispersed into matrix in four different types of volume fraction distribution varying across the thickness as per Van et. al.[3]. Effect of volume fraction of CNT, boundary conditions, plate aspect ratio and other parameters has been investigated after due validation.

CARBON NANOTUBE REINFORCED COMPOSITES

The analysis performed on functionally graded CNTRC plate having length a, width b and height h as shown in figure1using IGA formulation based on HSDT. Four types of volume fraction distribution of CNT i.e. uniformly distributed (UD) CNTRC, functionally graded V (FG-V) type distribution, functionally graded X (FG-X) type distribution and functionally graded O (FG-O) as shown in figure1. The effective modulus of elasticity, employing the extended rule of mixture, is expressed as.

$$E_{11} = \eta_1 V_{CNT} E_{11}^{CNT} + V_m E^m \quad (1)$$

$$\frac{\eta_2}{E_{22}} = \frac{V_{\mathrm{CNT}}}{E_{22}} + \frac{V_{\mathrm{m}}}{E^{\mathrm{m}}} \quad (2)$$

$$\frac{\eta_3}{G_{12}} = \frac{V_{\mathrm{CNT}}}{G_{12}^{CNT}} + \frac{V_{\mathrm{m}}}{G^{\mathrm{m}}} \quad (3)$$

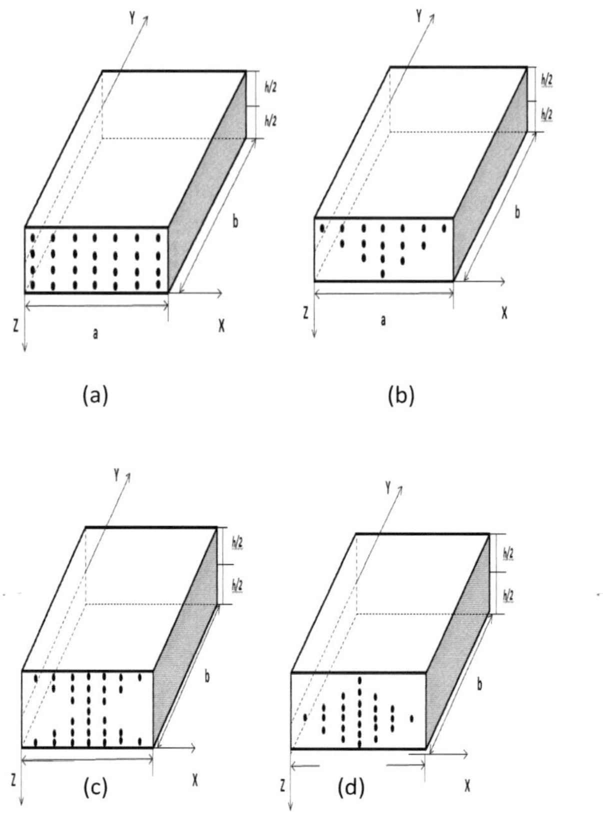

Figure 1. CONFIGURATION OF CARBON NANOTUBE REINFORCED COMPOSITE PLATES. (a) UD CNTRC PLATE (b) FG-V CNTRC PLATE (c) FG-X CNTRC PLATE (d) FG-O CNTRC PLATE.

Mathematically four types of distributions of CNT along the thickness directions can be written as:

$$\begin{cases} V_{CNT} = V_{CNT}^{*} & (UD\ CNTRC) \\ V_{CNT}(z) = \left(1 + \dfrac{2z}{h}\right) V_{CNT}^{*} & (FG\text{-}V\ CNTRC) \\ V_{CNT}(z) = 2\left(1 - \dfrac{2|z|}{h}\right) V_{CNT}^{*} & (FG\text{-}O\ CNTRC) \\ V_{CNT}(z) = 2\left(\dfrac{2|z|}{h}\right) V_{CNT}^{*} & (FG\text{-}X\ CNTRC) \end{cases} \quad (4)$$

GOVERNING EQUATIONS OF CNT-REINFORCED COMPOSITE PLATES:

Displacement field: According to the higher order shear deformation theory the displacement field of CNTRC plate can be expressed as (Bhar et al. [11]);

$$\left.\begin{aligned} U(x,y,z,t) &= u(x,y,t) + z\theta_x(x,y,t) + z^3\theta_x^*(x,y,t) \\ V(x,y,z,t) &= v(x,y,t) + z\theta_y(x,y,t) + z^3\theta_y^*(x,y,t) \\ W(x,y,z,t) &= w(x,y,t) \end{aligned}\right\} \quad (5)$$

The nonzero strain components derived from the displacement assumptions, are

$$\{\varepsilon\}_k^T = \{\varepsilon_{xx}\ \varepsilon_{yy}\ \gamma_{xy} \mid \gamma_{xz}\ \gamma_{yz}\}_k = \left\{\{\varepsilon_p\}_k^T \mid \{\varepsilon_t\}_k^T\right\} \quad (6)$$

The components of the spatial stress vector $\{\sigma\}_k$, energy conjugate to the strain vector $\{\varepsilon\}_k$, are

$$\{\sigma\}_k^T = \left\{\sigma_{xx}\ \sigma_{yy}\ \tau_{xy} \mid \tau_{xz}\ \tau_{yz}\right\}_k = \left\{\{\sigma_p\}_k^T \mid \{\sigma_t\}_k^T\right\} \quad (7)$$

The laminate constitutive relations are

$$\{\bar{\sigma}\} = \begin{Bmatrix} \{\bar{\sigma}_P\} \\ \{\bar{\sigma}_t\} \end{Bmatrix} = \begin{bmatrix} [\bar{D}_P] & [0] \\ [0] & [\bar{D}_t] \end{bmatrix} \begin{Bmatrix} \{\bar{\varepsilon}_P\} \\ \{\bar{\varepsilon}_t\} \end{Bmatrix} = [\bar{D}]\{\bar{\varepsilon}\} \quad (8)$$

NURBS BASED ISOGEOMETRIC FORMULATION OF PLATE

Using NURBS basis function the field variables at all control points can be expressed as:

$$u = \begin{Bmatrix} u \\ v \\ w \\ \theta_x \\ \theta_y \\ \theta_x^* \\ \theta_y^* \end{Bmatrix} = \sum_{I=1}^{n\times m} N_I(\xi,\eta) \begin{Bmatrix} u_I \\ v_I \\ w_I \\ \theta_{xI} \\ \theta_{yI} \\ \theta_{xI}^* \\ \theta_{yI}^* \end{Bmatrix} = \sum_{I=1}^{n\times m} N_I d_I \quad (9)$$

The governing equation of motion for a conservative system in dynamic equilibrium using Hamilton's principle can be expressed as:

$$\int_{t_1}^{t_2} (\delta T - \delta U - \delta V)\, dt = 0 \quad (10)$$

Putting the expressions for kinetic energy, internal potential energy and potential energy due to external forces, followed by subsequent rearrangement, the relation becomes

$$\left\{[M]\{\ddot{d}\} - [K]\{d\} + \{F\}\right\} = \{0\} \quad (11)$$

Above equation is the basic governing equation to solve the general dynamic problems. For the special case of free vibration using simultaneous iteration method [12], the equation becomes

$$[K]\{d\} - [M]\{\ddot{d}\} = \{0\} \quad (12)$$

For transient analysis the discretized system of equation for CNTRC plate is

$$[M]\{\ddot{d}\} + [K]\{d\} = F(t) \quad (13)$$

The well-known New-mark beta method is used here to solve these transient (time dependent) problems.

FREE VIBRATION ANALYSIS

In this section, analysis is being done by NURBS based isogeometric analysis on bare plates (UD, FG-V, FG-O, and FG-X) to ensure the validity of the developed formulation into computer program using FORTRAN. The materials properties of matrix are assumed as [3] Poison's ratio (v^m) =0.34 and Elastic modulus (E^m)=2.1 GPa. The values of elastic modulus, shear modulus and poison's ratio of CNT are: $E_{11}^{CNT} = 5.6466\times10^3 Gpa$, $E_{22}^{CNT} = 7.08\times10^3 Gpa$, $G_{12}^{CNT} = 1.9445\times10^3 Gpa$, and $v_{12}^{CNT} = 0.175$.

The CNT efficiency parameter for the case of V^*_{CNT}=0.11, η_1=0.149 and η_2=0.934, for the case of V^*_{CNT}=0.14, η_1=0.150 and η_2=0.941 and for the case of V^*_{CNT}=0.17, η_1=0.149 and η_2=1.381.Where η_1 and η_2 are the CNTs efficiency parameters. In this paper there are two types of boundary conditions are used simply supported (SSSS) and clamped (CCCC) boundary conditions. In simply supported boundary condition all edges are simply supported and in clamped all edges are clamped. The results are obtained in terms of non-dimensional natural frequency $\bar{\omega} = (\omega a^2 / h)\sqrt{\rho^m / E^m}$ using NURBS based isogeometric analysis for obtaining Eigen solution.

As evident from Table-1, the present results in terms of non-dimensional fundamental natural frequency are found to be in close conformity with those from published literature, for rectangular CNTRC plates with clamped boundary condition.

Table1. NON-DIMENSIONAL FREQUENCY PARAMETERS of CNTRC PLATE WITH CLAMPED BOUNDARY CONDITION

			UD		FG-V		FG-O		FG-X	
V^*_{CNT}	**h/a**	**Mode**	**Van et al. [2015]**	**Present (IGA)**	**Van et al. [2015]**	**Present (IGA)**	**Van et al. [2015]**	**Present (IGA)**	**Van et al. [2015]**	**Present (IGA)**
0.11	**0.1**	**1**	19.473	18.120	18.811	17.676	18.198	17.145	19.995	18.604
		2	24.659	23.503	24.221	23.238	23.810	22.652	25.023	24.077
		3	35.237	33.925	35.025	33.859	34.821	33.103	35.425	34.759
		4	37.294	35.080	37.047	34.329	36.177	33.229	37.339	35.371
		5	38.033	37.385	37.334	37.510	37.339	36.673	38.835	37.508
		6	41.073	38.403	40.264	37.839	39.554	37.500	41.734	38.803
	0.05	**1**	30.391	29.157	27.709	27.026	25.592	25.201	32.901	31.245
		2	34.828	33.595	32.651	32.107	30.931	30.367	36.966	35.810
		3	45.827	45.064	44.377	44.348	43.202	42.643	47.379	47.349
		4	64.011	60.128	61.232	57.065	57.724	54.206	65.023	63.219
		5	65.386	63.011	63.176	60.307	60.742	57.546	68.983	65.659
		6	67.826	63.025	63.996	63.037	62.455	61.171	71.200	66.051
	0.02	**1**	40.332	40.267	34.413	34.782	30.452	31.002	47.238	46.686

		2	44.438	44.979	39.273	40.287	35.843	36.792	50.746	50.983
		3	55.894	57.298	52.074	53.940	49.533	50.739	60.969	62.634
		4	78.249	79.162	75.791	77.077	74.080	73.894	81.883	84.016
		5	101.566	98.709	87.974	86.697	78.417	77.984	116.195	111.549
		6	103.738	101.438	90.649	89.982	81.422	81.524	117.486	113.980
0.14	**0.1**	**1**	20.112	18.569	19.516	18.199	18.962	17.682	20.566	19.064
		2	25.311	23.987	24.910	23.804	24.540	23.147	25.622	24.677
		3	35.997	34.514	35.798	34.521	35.628	33.635	36.141	35.502
		4	37.954	35.501	37.988	34.894	37.457	34.173	37.995	36.313
		5	39.154	38.034	38.248	38.225	37.995	37.621	39.873	38.231
		6	42.185	38.857	41.434	38.425	40.798	38.208	42.769	39.768
	0.05	**1**	32.268	30.492	29.627	28.703	27.517	26.880	34.639	32.753
		2	36.585	35.286	34.401	33.664	32.661	31.824	38.623	37.346
		3	47.472	46.802	45.977	45.827	44.770	43.859	48.967	49.035
		4	65.710	62.818	64.566	59.928	61.233	57.204	66.682	65.713
		5	68.494	65.061	64.816	63.066	64.075	60.361	71.812	67.695
		6	70.862	65.652	67.218	64.639	64.118	62.334	73.977	68.545
	0.02	**1**	44.511	45.450	38.013	39.747	33.678	33.916	51.893	52.079
		2	48.370	48.632	42.588	43.438	38.762	39.490	55.194	55.175
		3	66.682	65.713	54.963	56.724	52.054	52.924	65.003	66.611
		4	81.416	82.143	78.458	79.741	76.464	75.788	85.524	87.978
		5	110.865	107.024	96.503	94.560	86.363	85.381	121.074	120.094
		6	112.889	109.564	98.985	97.625	89.153	88.625	125.823	122.416
0.17	**0.1**	**1**	24.299	22.526	23.437	21.995	22.655	21.225	24.970	23.316
		2	30.805	29.212	30.243	28.977	29.732	27.968	31.261	30.324
		3	44.058	42.266	43.798	42.140	43.574	40.976	44.266	43.858
		4	46.660	42.958	46.208	42.346	45.101	41.093	46.727	44.211
		5	47.489	46.761	46.716	46.664	46.727	45.470	48.511	47.136
		6	51.306	47.212	50.259	47.130	49.378	47.098	52.131	48.639
	0.05	**1**	37.741	35.823	34.300	33.121	31.644	30.738	40.936	38.702
		2	43.336	41.740	40.563	39.667	38.419	37.189	46.048	44.727
		3	57.163	56.395	55.344	55.274	53.898	52.486	59.115	59.739
		4	79.971	74.388	76.015	70.376	71.569	66.589	81.214	78.602
		5	81.390	78.108	78.956	74.571	75.416	70.779	85.988	82.352
		6	84.469	79.235	79.526	78.918	78.100	75.515	88.769	83.297
	0.02	**1**	49.777	50.380	42.324	41.659	37.443	37.012	58.403	58.121
		2	54.982	54.696	48.517	49.062	44.308	44.555	62.844	62.403
		3	69.447	70.476	64.720	66.643	61.620	62.017	75.747	77.868
		4	97.561	98.193	94.571	96.084	92.497	90.857	102.050	105.764
		5	125.557	119.702	108.326	104.704	96.479	93.941	143.956	136.138
		6	128.316	123.252	111.750	109.031	100.327	98.444	146.263	139.425

In Table 1, it is seen that at higher h/a ratio the difference between present results and literature results are much compared to the results at lower h/a ratio because at lower h/a ratio the thickness of the plate is very less so, the deflection will be more that's why the variation is more in comparison with the higher h/a ratio.

First, we analyze the effect of aspect ratio and volume fraction of CNTs on the non-dimensional natural frequency with Clamped (CCCC) boundary condition and it can be seen that as the aspect ratio increases, the non-dimensional natural frequency decreases and increases with increase of volume fraction of carbon nanotube as shown in Table 2 and the mode of vibration changes with the change of b/a ratio. In Table 2, only first mode of vibration is shown.

Table 2. VARIATIONS OF NON-DIMENSIONAL NATURAL FREQUENCY ($\bar{\omega} = \left(\omega a^2 / h\right)\sqrt{\rho^m / E^m}$) OF CLAMPED FG-CNTRC PLATE FOR DIFFERENT ASPECT RATIOS (b/h=100)

V^*_{CNT}	Distribution type	b/a: 1	1.2	1.4	1.6	1.8	2	3
0.11	**UD**	44.325	43.001	41.978	41.081	40.234	39.410	35.335
	FG-V	37.742	36.514	35.651	34.952	34.329	33.746	30.960
	FG-O	33.413	32.246	31.467	30.866	30.354	29.890	27.779
	FG-X	52.567	50.977	49.631	48.373	47.144	45.926	39.964
0.14	**UD**	49.154	47.702	46.521	45.447	44.410	43.389	38.339
	FG-V	41.736	40.444	39.494	38.693	37.958	37.255	33.816
	FG-O	36.891	35.697	34.866	34.199	33.611	33.065	30.479
	FG-X	58.343	56.513	54.895	53.350	51.823	50.310	43.024
0.17	**UD**	53.270	51.660	50.444	49.397	48.421	47.478	42.832

FG-V	45.300	43.775	42.729	41.899	41.174	40.504	37.355
FG-O	40.064	38.629	37.684	36.967	36.366	35.828	33.432
FG-X	63.317	61.399	59.820	58.371	56.967	55.581	48.761

FORCE VIBRATION

We now examine the dynamic response of CNTRC plate with simply supported boundary condition subjected to transverse load which is sinusoidally distributed in spatial domain and varies with time as

$$q = q_0 \sin\left(\frac{\pi x}{a}\right)\sin\left(\frac{\pi y}{a}\right)F(t) \quad (14)$$

Where,

$$F(t)=\begin{cases} \textit{Step loading} & \begin{cases} 1 & 0\le t\le t_1 \\ 0 & t\ge t_1 \end{cases} \\ \textit{Triangular loading} & \begin{cases} 1-\dfrac{t}{t_1} & 0\le t\le t_1 \\ 0 & t\ge t_1 \end{cases} \\ \textit{Sine loading} & \begin{cases} \sin\left(\dfrac{\pi t}{t_1}\right) & 0\le t\le t_1 \\ 0 & t\ge t_1 \end{cases} \end{cases} \quad (15)$$

Figure 2-4 show the central deflection histories of CNTRC plate with *h/a* ratio is 0.02, volume fraction of CNT is 0.11 in which $t_1 = 0.006$s, $q_0 = 3.448$MPa for four type of CNTRS i.e. UD, FG-V, FG-O and FG-X plate under step, triangular and sine loading. It can be seen that the present results matches well with those of Van et.al. [3]. From all the graph it is again observed that FG-V and FG-O has more central deflection than the UD and FG-X and the reason is same that in FG-O and FG-V the stiffness is less than the FG-X and UD distribution. Fig.2-4 shows the two separated state one state is subjected to the transverse load (force vibration) and in the other state the plate freely oscillates (free vibration) when the load is removed.

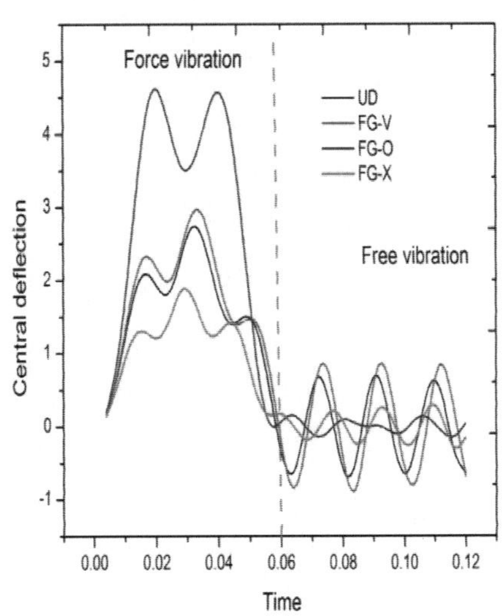

Figure 2. THE FORCE VIBRATION RESPONSE OF CNTRC PLATE HAVING VOLUME FRACTION 0.11

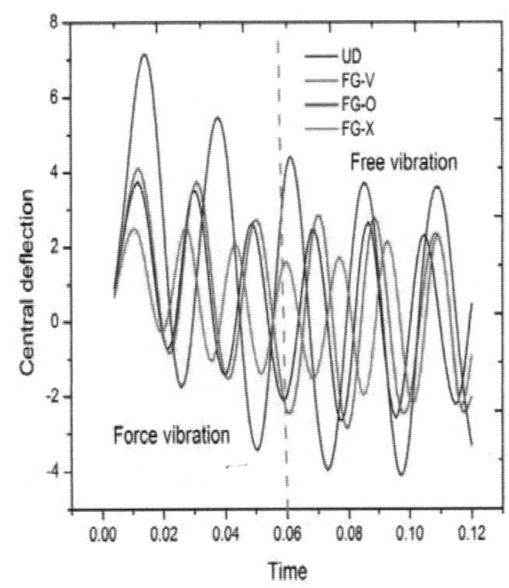

Figure 3. THE FORCE VIBRATION RESPONSE OF CNTRC PLATE HAVING VOLUME FRACTION 0.11

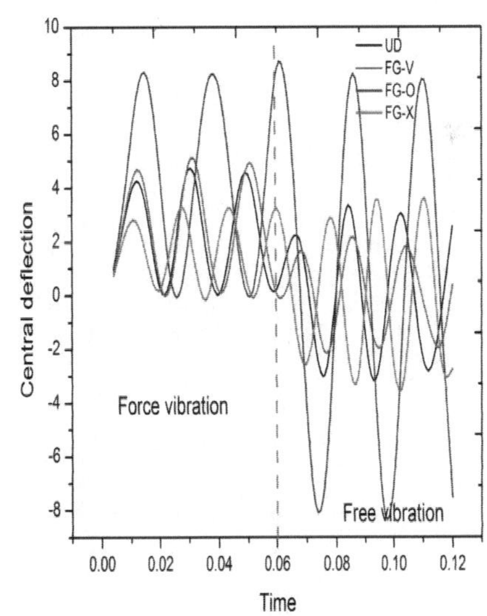

Figure 4. THE FORCE VIBRATION RESPONSE OF CNTRC PLATE HAVING VOLUME FRACTION 0.11

REFERENCES

[1] Hughes, T.J.R., Cottrell, J.A., Bazilevs, Y., 2005. "Isogeometric analysis: CAD, Finite elements, NURBS, Exact Geometry and mesh refinements". *Comput. Methods Appl. Mech. Engrg.*, **194**, October, pp. 4135–4195.

[2] Tran, Loc, V., Ferreira, A.J.M., Nguyen-Xuan, H., 2013. "Isogeometric analysis of functionally graded plates using higher-order shear deformation theory". *Composites: Part B,* **51**, March, pp. 368–383.

[3] Phung-Van, P., Abdel-Wahab, M., Liew, K.M., Bordas, S.P.A., Nguyen-Xuan, H., 2015. "Isogeometric analysis of functionally graded carbon nanotube-reinforced composite plates using higher-order shear deformation theory". *Composite Structures,* **123**, December, pp. 137–149.

[4] Valizadeh, N., Natarajan, Sundararajan, Gonzalez-Estrada Octavio, A., Rabczuk, T., Quoc Bui, T., Bordas, S.P.A., 2013. "NURBS-based finite element analysis of functionally graded plates: Staticbending, vibration, buckling and flutter". *Composite Structures,* **99**, December, pp. 309–326.

[5] Thai, Chien, H., Ferreira, A.J.M., Carrera, E., Nguyen-Xuan, H., 2013. "Isogeometric analysis of laminated composite and sandwich plates using a layerwise deformation theory". *Composite Structures,* **104**, April, pp. 196-214.

[6] Alesadi, A., Galehdari, M., Shojaee, S., 2017. "Free vibration and buckling analysis of cross-ply laminated composite plates using Carrera's unified formulation based on Isogeometricapproach". *Computers and Structures,* **183**, February, pp. 38-47.

[7] Treacy, M. M. J., Ebbesen, T. W., Gibson, J. M., 1996. "Exceptionally high Young's modulus observed for individual carbon nanotubes". *Nature,* **381**, June, pp. 678-680.

[8] Poncharal, P., Wang, Z. L., Ugarte, D., De Heer, W. A., 1999. "Electrostatic deflections and electromechanical resonances of carbon nanotubes". *Science,* **283**, March, pp. 1513-1516.

[9] Shen, H.S., 2009. "Nonlinear bending of functionally graded carbon nanotube-reinforced composite plates in thermal environments". *Compos. Struct.,* **91**, November, pp. 9-19.

[10] Zhu, P., Lei, Z. X., Liew, K. M., 2012. "Static and free vibration analyses of carbon nanotube reinforced composite plates using finite element method with first order shear deformation plate theory". *Compos. Struct.,* **94**, March, pp. 1450-1460.

[11] Bhar, A., Phoenix, S.S., Satsangi, S.K., 2010. "Finite element analysis of laminated composite stiffened plates using FSDT and HSDT: A comparative perspective". *Compos. Struct.,* **92**, August, pp. 312-321.

[12] Corr, R.B., Jennings, A., 1976. "A simultaneous iteration algorithm for symmetric eigenvalue problems". *International Journal for Numerical Methods in Engineering,* **10**, March, pp. 647-663.

Proceedings of the 6th International Conference on Ship and Offshore Technology
ICSOT 2019
November 7-8, 2019, IIT Kharagpur, India

ICSOT INDIA 2019-08

SIMPLIFIED 3D FE ANALYSIS OF FRP COMPOSITE BOLTED JOINTS WITH LAMINATE TAPERING

Bibekananda Mandal
Visiting Assistant Professor
Thapar Institute of Engineering and Technology
Patiala, Punjab, 147004
India
bibekdce@thapar.edu

Anupam Chakrabarti
Professor
Indian Institute of Technology, Roorkee
Roorkee, Uttarakhand, 247667
India
Anupam1965@yahoo.co.uk

ABSTRACT

Major savings in weight can be achieved by tapering laminates away from bolts used in a joint of a composite structure, such as ships and offshore structures. Usually, a three-dimensional (3D) finite element (FE) model is used for the analysis of laminated composite bolted joints as the stresses and strains vary in all directions due to several factors. In case of tapered laminates, it is more complicated due to high out-of-plane displacements and therefore, a 3D FE layered model can usually predict the responses accurately. To overcome this issue, a homogenization scheme is proposed for the simplified 3D FE analysis of such joints. The model can determine the basic mechanical behaviors of a laminated composite joint with significantly lesser computational effort than a layered 3D FE model. The present formulation is based on Mindlin-Reissner plate theory, which takes into account the effect of transverse shear deformation in composite laminates. The proposed model is found to be capable of finding quick and accurate estimation of the basic responses for the composite bolted joints with different tapering configurations in the laminates.

NOMENCLATURE

$[C]$	Elastic constitutive matrix
E	Young's modulus of elasticity
ε	Strain
G	Shear modulus
v	Poisson's ratio
M	Moment resultants
N	Force resultants
$[S]$	Elastic compliance matrix
S	Shear strength
X_c , X_t	Longitudinal compressive and tensile strength
Y_c, Y_t	Transverse compressive and tensile strength
Z_c , Z_t	Compressive and tensile strength in the stacking direction of lamina

INTRODUCTION

Bolted joints are widely used for joining different components of composite structures such as ships and offshore structures because of their high load transfer capability and ease of assembly. Also, bolted joints are crucial components in designing lightweight structures made of laminated composite materials. However, drilling of the plates is required for fixing bolts; which causes weak zones, and damage usually initiates from these locations. The analysis and design of laminated composites joints are complex because the stress and strains vary in three dimension and there may exist some extra failure modes compared to conventional steel joints, due to its laminated nature. For the joints with laminate tapering which become more complex due to high out-of-plane displacements and a 3D FE model is required for a numerical analysis.

There are several literatures available on load distribution, strength and failure of laminated composite bolted joints. McCarthy et al. (2005) [1] presented a 3D FE model of a single-lap, single-bolt composite joint to study the joint stiffness, out-of-plane displacement, strains etc. McCarthy and McCarthy (2005) [2] further modified the FE model to study the effect of bolt-hole clearance and the initiation of failure. McCarthy et al. (2005a) [3] developed a 3D FE progressive damage model of double-lap, multi-bolt laminated composite joints. The computational time required is very high for 3D FE analysis of FRP composite bolted joints. Homogeneous models are already popular for the sandwich structures with various core configurations Equivalent homogeneous model for the sandwich panels with rib core [4], truss-core [5, 6], corrugated core [7] and Z-core [8] are already available in literature. But, homogeneous model for laminated composite thick plate based on 3D elasticity which will be useful for FRP bolted joint, is rare in literature.

An equivalent homogeneous 3D elasticity model is developed and presented in this paper for simplified analysis of laminated composite bolted joints with significantly lesser computational effort. Fasteners (bolts/nuts/washers) used in laminated composite bolted joints are usually made of homogeneous and isotropic materials. In the present formulation the equivalent stiffness matrix is calculated for inhomogeneous, anisotropic laminated plates of composite bolted joints. The Mindlin-Reissner plate theory which takes into account transverse shear deformations through-the-thickness of plate is used in the formulation of classical laminate theory. The present method is capable of finding equivalent homogeneous model for the laminated composite joints having any type of lamination scheme [9]. For the failure analysis, at first the 3D progressive damage model is used to the layered FRP composite plate to find out equivalent strength properties of the laminate. The important mechanical behaviors of composite bolted joint, such as, joint stiffness, out-of-plane-displacements along with failure load have been calculated using the present equivalent model.

HOMOGENIZATION MODEL

If $\left[S_{ij}\right]_k$ is the elastic compliance matrix for the kth lamina of the laminated plate then, orthotropic stress-strain relationship for the lamina can be written as: $\{\sigma\}_k = \left[C_{ij}\right]_k \{\varepsilon\}_k$, where $\left[C_{ij}\right]_k$ is the elastic constitutive matrix of kth lamina with respect to the material axis system. If θ is the lamination angle of the kth lamina with respect to global x axis then with respect to global axis system $\left[C_{ij}\right]_k$ can be represented as:

$$\left[C_{xy}\right]_k = [T]^{-1} \times \left[C_{ij}\right]_k \times [T]^{-T} \tag{1}$$

where $[T]$ is the transformation matrix, the detailed derivation of which is given in [9]. Resultant force, moment and shear forces for the plate can be written as:

$$\begin{aligned} \boldsymbol{N} &= \boldsymbol{A} \times \boldsymbol{\varepsilon}^0 + \boldsymbol{B} \times \boldsymbol{\kappa}, \\ \boldsymbol{M} &= \boldsymbol{B} \times \boldsymbol{\varepsilon}^0 + \boldsymbol{D} \times \boldsymbol{\kappa} \text{ and,} \\ \boldsymbol{Q} &= \boldsymbol{A}^z \times \boldsymbol{\varepsilon}_{trans} \end{aligned} \tag{2}$$

where

$$[A] = \sum_{k=1}^{n} \left[C_{xy}\right]_k (h_k - h_{k-1}),\ [B] = \frac{1}{2}\sum_{k=1}^{n} \left[C_{xy}\right]_k \left(h^2{}_k - h^2{}_{k-1}\right),$$

$$[D] = \frac{1}{3}\sum_{k=1}^{n} \left[C_{xy}\right]_k \left(h^3{}_k - h^3{}_{k-1}\right)$$

and, $\left[A^z\right] = \kappa_s \times \sum_{k=1}^{n} \left[C_{xy}\right]_k (h_k - h_{k-1})$. κ_s denotes the shear correction factor. All the forces and moments can be expressed as: $\{\boldsymbol{N}\ \ \boldsymbol{M}\ \ \boldsymbol{Q}\}^T = \left[K_c\right]\{\boldsymbol{\varepsilon}^0\ \ \boldsymbol{\kappa}\ \ \boldsymbol{\varepsilon}_{trans}\}^T$ where, the final combined stiffness matrix $\left[K_c\right]$ for the laminate can be written as:

$$\left[K_c\right] = \begin{bmatrix} \boldsymbol{A} & \boldsymbol{B} & 0 \\ \boldsymbol{B} & \boldsymbol{D} & 0 \\ 0 & 0 & \boldsymbol{A}^z \end{bmatrix} \tag{3}$$

To find the equivalent elastic constants, equivalent elastic compliance matrix $\left[S^{Eq}\right]$ is recalculated from the stiffness matrices $[D]$ and $[Q]$ and the equivalent 4×4 elastic constitutive matrix $\left[C^{Eq}\right]$ can be written as:

$$\left[C^{Eq}\right] = [D] / z_3^{'} \tag{4}$$

where $z_3^{'} = \frac{1}{3} \times \left\{(t/2)^3 - (-t/2)^3\right\}$ i.e. $z_3^{'} = t^3/12$. Here, t is the total thickness of the laminated plate. The equivalent elastic compliance matrix $\left[S^{Eq}\right]$ from the $\left[C^{Eq}\right]$ matrix can be written as: $\left[S^{Eq}\right] = \left[C^{Eq}\right]^{-1}$. Comparing with elastic compliance matrix of orthotropic material, the elastic properties can be written as:

$$\begin{aligned} &E_1 = S_{11}{}^{-1},\ E_2 = S_{22}{}^{-1},\ E_3 = S_{33}{}^{-1}, \\ &G_{12} = S_{66}{}^{-1}, G_{13} = S_{44}{}^{-1},\ G_{23} = S_{55}{}^{-1}, \\ &\nu_{12} = -S_{21} \times E_1,\ \nu_{13} = -S_{31} \times E_1,\ \nu_{23} = -S_{32} \times E_2. \end{aligned} \tag{5}$$

FINITE ELEMENT MODELLING

The proposed homogeneous model has been validated by comparing the present results with a benchmark model of laminated composite single-bolt and single-lap joint as described in [1]. The geometry of the single-bolt and single-lap joint is shown in Figure 1. A 3D finite element half-symmetric model of the laminated composite bolted joint has been developed for this purpose using ABAQUS as shown in Figure 2. A full layered 3D model considering the actual quasi-isotropic lamination scheme has also been developed for the comparison purpose. Eight node solid brick element (C3D8R) with reduced integration was used to develop the three dimensional FE model.

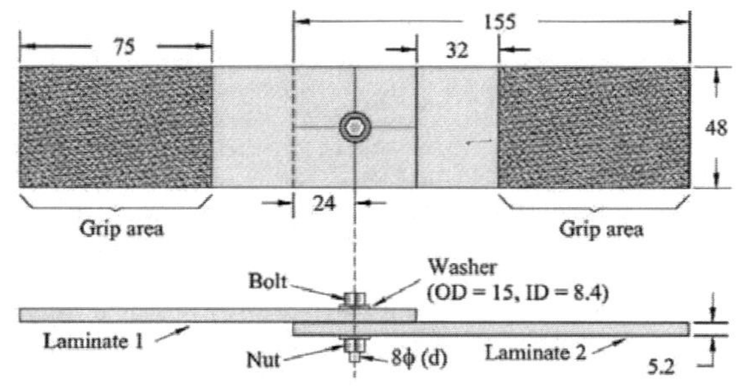

Figure 1. THE SINGLE-LAP SINGLE-BOLT COMPOSITE JOINT GEOMETRY

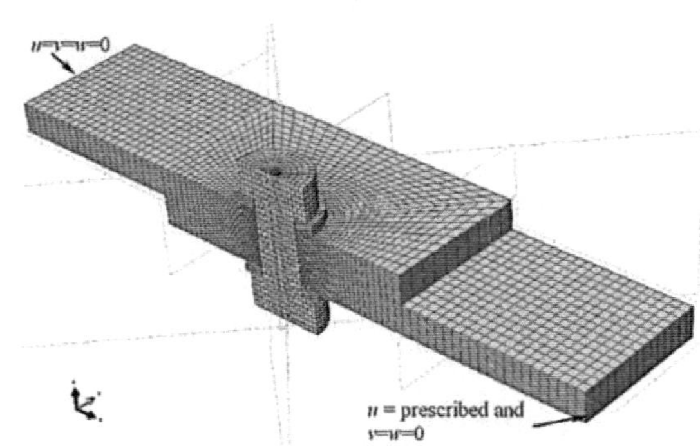

Figure 2. 3D HALF SYMMETRIC FE MODEL OF THE SINGLE-LAP SINGLE-BOLT COMPOSITE JOINT

In ABAQUS the contact between different components were defined by master-slave algorithm. Frictional coefficient of 0.7 between the plates, 0.1 for bolt to laminate interfaces and 0.3 for washer to bolt/laminate were considered. The boundary condition for the grip edge of the top plate was considered as fixed and an axial displacement (along x-axis) was applied at the bottom plate grip edge. The bolt was subjected to 0.5 N-m torque which represents a 'finger-tight' condition. A bolt pre-stress of magnitude 7.2 MPa was applied for this 0.5 N-m torque as suggested by McCarthy et al. (2005b).

A progressive material damage model has been assimilated with FE software package ABAQUS using the user-subroutine UMAT. Solution-dependent state variables have been used to store calculated failure indexes. The Hashin failure criteria [10] was considered here along with delamination failure criteria and fiber/matrix shear failure criteria [11, 12] for determining failure of a material point of the composite laminate. The correlative elastic constants (ECs) have been discounted by multiplying with a very small value if failure occurred in one direction. Homogeneous equivalent strength properties are calculated using the developed progressive damage model. For equivalent longitudinal and transverse strength, the failure load for a FE model of layered FRP composite plate has been calculated by applying axial forces from global x and y direction to calculate X_t and Y_t respectively. The equivalent shear strength has been estimated numerically using a V-notched beam FE model developed as per ASTM D5379 / D5379M-12 code. The equivalent axial strength in global z-direction was kept same as lamina strength because, for both the cases this is matrix direction.

RESULTS

In this section, the joint stiffness, out-of-plane displacement, surface strains and through the thickness radial stress distribution at bolt-hole have been calculated using the present homogeneous model as well as by using a full 3D model with layered properties. The carbon/epoxy material (HTA/6376) with unidirectional material properties (Table 1) has been used for the laminated plates as described in McCarthy et al. (2005) [1]. Lamination schemes was considered as quasi-isotropic with $[45/0/\text{-}45/90]_{5s}$ lay-ups. The laminate thickness was considered as 5.2 mm where the thickness of each ply was 0.13 mm. Titanium bolts with nominal diameter 8 mm and steel nuts together with steel washers were used. The isotropic material properties used for titanium bolt and steel washers were: E_b = 110 GPa, ν_b = 0.29 for the bolt and E_w = 210 GPa, ν_w = 0.3 for the washers.

Table 1. LAMINA PROPERTIES FOR THE HTA/6376 MATERIAL (ALL E AND G VALUES ARE IN GPA)

E_1	E_2	E_3	G_{12}	G_{13}	G_{23}	ν_{12}	ν_{13}	ν_{23}
140	10	10	5.2	5.2	3.9	0.3	0.3	0.5
X_t	X_c	Y_t	Y_c	Z_t	Z_c	S_{12}	S_{13}	S_{23}
2200	1600	70	250	50	300	120	120	50

Joint Stiffness and Out-of-Plane Displacements

For the quasi-isotropic layup, joint stiffness and out-of-plane displacements have been calculated for the single-lap single-bolt composite joint using the proposed homogeneous model. The results estimated from the present homogeneous model are compared with the results obtained from FE analysis of actual layered model along with the results reported in [1] as shown in Table 2. The plots of out-of-plane displacements of the joint are shown in Figure 3. The results match quite well and it can be concluded that the present equivalent model gives quite satisfactory results.

Table 2. JOINT STIFFNESS AND MAXIMUM OUT-OF-PLANE DISPLACEMENTS OF THE SINGLE-BOLT FRP JOINT

	Present homogeneous model	Layered full model (Present)	McCarthy et al. (2005) [1]
Joint stiffness (kN/mm)	34.8	31.6	34.6

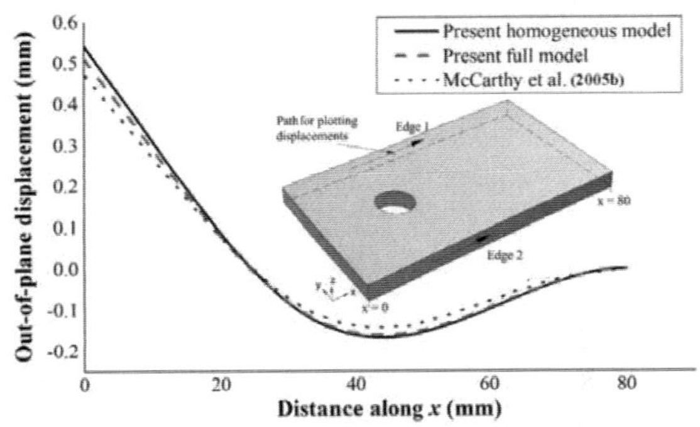

Figure 3. PLOT OF OUT-OF-PLANE DISPLACEMENTS OF THE SINGLE-LAP FRP JOINT

Failure Analysis of a Tapered Multi-Bolt FRP Joint

Failure load has been calculated for a tapered double-lap multi-bolt composite joint (Figure 4) consisting of three bolts with 36 mm center-to-center. The tapering was considered up to half of the non-overlapped portion (U_L) for the present analysis. Material properties for the carbon/epoxy composite (HTA/6376) and titanium bolts are as described in previous single-lap joint problem. The lamination scheme was considered as quasi-isotropic with $[45/0/\text{-}45/90]_{ns}$ lay-up comprising of 16 plies (n = 2) for the splice plates and 32 plies (n = 4) for the skin plate. The present FE model has been validated by modeling the same joint without tapering and compared with the available results in literature. The tapered joint model has also been validated by comparing the results from a full layered FE model. Failure loads estimated from the present analysis and corresponding failure loads as obtained from literature are shown in Table 3.

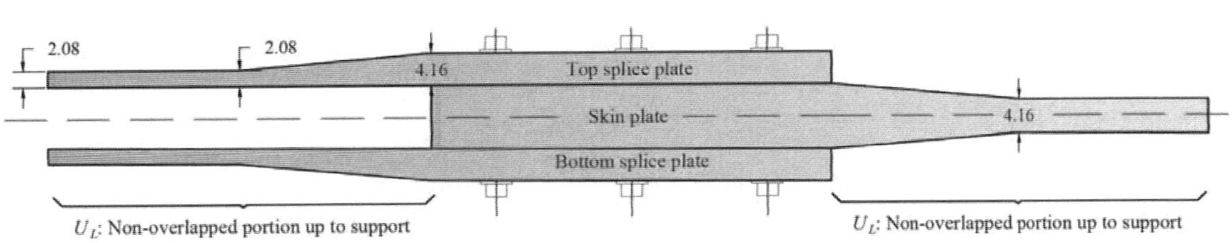

Figure 4. THE DOUBLE-LAP MULTI-BOLT COMPOSITE JOINT GEOMETRY

Table 3. FAILURE LOADS (KN) OF THE MULTI-BOLT COMPOSITE JOINT

	Failure load (kN)	Experimental [13]	McCarthy et al. (2005a) [3]	Present layered model	Present homogeneous model
No tapering	1st significant failure	50	45	43.74	-
No tapering	Failure of bolt hole 2	-	52	50.23	-
No tapering	Other failure	-	-	-	50.39
Tapered	1st significant failure	-	-	42.58	-
Tapered	Failure of bolt hole 2	-	-	49.46	-
Tapered	Other failure	-	-	-	49.64

It can be observed that, the present homogeneous model gives quite satisfactory results when compared to full layered model as well as experiment. The failure load for the tapered bolted joint estimated from the present homogeneous model is found to be 49.64 kN corresponding to the 30% drop in joint stiffness value. No significant drop in bolt load was noticed which was observed from actual layered model due to bearing failure of bolt hole. So, the homogeneous model gives quite satisfactory result in terms of basic mechanical behaviors but could not distinguish between different modes of failure. The damaged elements near bolt hole 1 of the splice plate are shown in Figure 5. The ultimate mode of failure of the joint using full layered model was found as net-section failure but, the present homogeneous model cannot predict the mode of failure clearly.

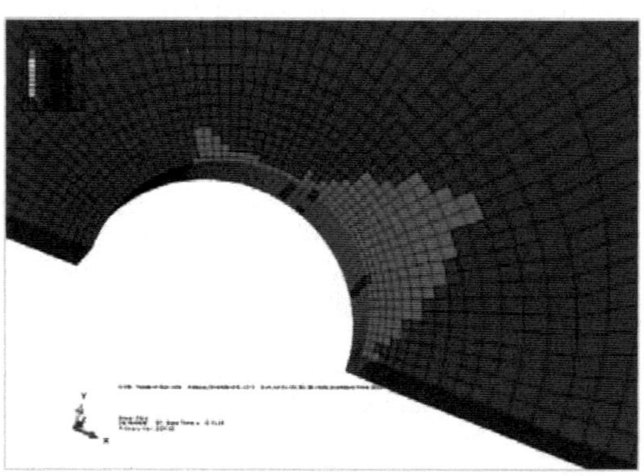

Figure 5. DAMAGED ELEMENTS NEAR BOLT HOLE 1 OF THE SPLICE PLATE

CONCLUSIONS

A homogeneous model has been presented in this paper for analyzing composite bolted joints with significantly low computational time. The present homogeneous model is based on Mindlin-Reissner plate theory which takes into account the effect of transverse shear deformation in composite laminates. 3D Progressive damage analysis of FRP composite joints with and without laminate tapering are performed using the homogeneous model in FE based software ABAQUS. Material constitutive relations and the progressive damage algorithms are integrated with ABAQUS using user-subroutine UMAT. The important mechanical behaviors of composite bolted joint, such as, joint stiffness, out-of-plane-displacement along with failure load are calculated using the present equivalent model. It has been observed that, the present homogeneous model gives quite satisfactory result in terms of basic mechanical behaviors but could not distinguish between different failure events. So, for accurate progressive damage analysis with the actual layered FE model is recommended.

REFERENCES

[1] McCarthy, M., McCarthy, C., Lawlor, V. and Stanley, W., 2005. "Three-dimensional finite element analysis of single-bolt , single-lap composite bolted joints : Part I — Model development and validation". *Compos. Struct.*, **71**(2), pp. 140–158.

[2] McCarthy, C. and McCarthy, M., 2005. "Three-dimensional finite element analysis of single-bolt, single-lap composite bolted joints: Part II—Effects of bolt-hole clearance". *Compos. Struct.*, **71**(2), pp. 159–175.

[3] McCarthy, C., McCarthy, M., and Lawlor, V., 2005. "Progressive damage analysis of multi-bolt composite joints with variable bolt-hole clearances". *Compos. Part B Eng.*, **36**(4), pp. 290–305.

[4] Aref, A., Alampalli, S., and Y. Hey, 2001. "Ritz-based static analysis method for fiber reinforced plastic rib core skew bridge superstructure". *J. Eng. Mech.*, **127**(5), pp. 450–458.

[5] Lok, T., Cheng, Q., and Heng, L., 1999. "Equivalent stiffness parameters of truss-core sandwich panel," *Proceedings of the International Offshore and Polar Engineering Conference.*

[6] Lok, T., and Cheng, Q., 2000. "Elastic stiffness properties and behavior of truss-core sandwich panel," *J. Struct. Eng.*, **126**, pp. 552–559.

[7] Buannic, N., Cartraud, P., and Quesnel, T., 2003. "Homogenization of corrugated core sandwich panels," *Compos. Struct.*, **59**(3), pp. 299–312.

[8] Fung, T., Tan, K., and Lok, T., 1995. "Elastic constants for Z-core sandwich panels". *J. Struct. Eng.*, **120**(10), pp. 3046–3055.

[9] Mandal, B., and Chakrabarti, A., 2015. "A simple homogenization scheme for 3D finite element analysis of composite bolted joints". *Compos. Struct.*, **120**, pp. 1–9.

[10] Hashin, Z. 1980. "Failure criteria for unidirectional fibre composites". *J. Appl. Mech.*, **47**(2), pp. 329–334.

[11] Naderi, M. and Maligno, A., 2012. "Fatigue life prediction of carbon/epoxy laminates by stochastic numerical simulation". *Compos. Struct.*, **94**(3), pp. 1052–1059.

[12] Lessard, L., and Shokrieh, M., 1995. "Two-dimensional modeling of composite pinned-joint failure". *J. Compos. Mater.*, vol. **29**(5), pp. 671–697.

[13] Lawlor, V., McCarthy, M., and Stanley, W., 2005. "An experimental study of bolt-hole clearance effects in double-lap, multi-bolt composite joints". *Compos. Struct.*, **71**(2), pp. 176–190.

Proceedings of the 6th International Conference on Ship and Offshore Technology
ICSOT 2019
November 7-8, 2019, IIT Kharagpur, India

ICSOT INDIA 2019-09

FULLY NON-LINEAR POTENTIAL FLOW METHOD FOR MULTIPLE FLOATING BODIES IN CONFINED WATERS

Francesco Coslovich
Chalmers University of Technology
Gothenburg, Sweden
francesco.coslovich@chalmers.se

Magnus Östberg
Flowtech International AB
Gothenburg, Sweden
magnus@flowtech.it

Leif Broberg
Flowtech International AB
Gothenburg, Sweden
broberg@flowtech.it

ABSTRACT

Prediction of ship motions and added resistance in waves using numerical methods has become an important part in the design and optimization of ships. Increasing requirements for improved performance come from several different directions, not least through new regulations and requirements for safety and energy efficiency from the International Maritime Organization (IMO), but also through economic incentives in the global transport system.

One successful type of numerical method is the 3D time-dependent fully non-linear potential flow method. The motivation behind the development of these methods is primarily to find an accurate technique for computing forces and motions in waves without the large computational effort required by its viscous counterparts. In this paper SHIPFLOW MOTIONS, one of such methods, is described along with some newly developed techniques which enable more challenging applications such as ship-to-ship interactions in constrained waters. In particular, the paper describes a flexible adaptive grid refinement technique that allows for an improvement in accuracy or computational speed given a fixed number of panels. The technique is also capable of automatically generating grids for complex geometries including multiple bodies and environment constraints such as shallow water and harbors. The paper describes results of a lightering operation, obtained using this technique.

INTRODUCTION

Hydrodynamic interaction between ships has been investigated in late years by the scientific community. When two ships are sailing close enough for one of them to feel hydrodynamic effects from the other, dangerous conditions can arise. One of the main effects of this interaction is a variation in the maneuvering characteristics of the ship. In fact, effectively maneuver and steer the ship can become challenging, due to forces that arise from an unbalanced pressure between portside and starboard caused by the asymmetrical flow. In extreme cases, this unbalance can cause attraction forces between the ships or the generation of a yaw moment that can result in a collision. When ships are sailing in shallow water and in channels, the interaction forces can be even larger, leading to more critical situations.

One condition that can be critical is the lightering operation. A lightering operation consists in transferring cargo from a ship to another while the ships are sailing. To achieve this, two ships must sail close to each other at the same speed. When the transferring is done, the two ships disengage each other and the service ship, the one that received the cargo, overtakes the other one. As reported in [1], an even worse situation which can lead to incidents is the overtaking. In this condition, unsteady effects can become significant.

There are different methods that allow an estimation of these interaction forces, but they can be divided into two main groups: experimental techniques and numerical methods. Even though experimental techniques are generally more reliable and accurate, they are quite expensive and time consuming. Due to limited amount of time and resources during the development of a new design, numerical tools are usually preferred. Numerical methods can be divided into two groups as well: those which include viscous effects and those based on simplified hypotheses, such as potential flow methods. Potential flow methods started being used to evaluate ship-to-ship interaction in the seventies. The first methods to be developed were mainly 2D methods based on strip theory. The applicability of these methods was mainly restricted to cases where the ship could be considered slender or where longitudinal components of forces and moments were negligible. With the growing capacity of computers, 3D potential flow methods have become the standard nowadays and can provide more accurate results. As far as viscous methods are concerned for the evaluation of ship-ship interaction, the majority of these tools are Reynolds Averaged Navier-Stoke (RANS) methods. As expected, these methods provide more accurate results. Furthermore, the coupling with propeller and rudder is also taken into account. On the other hand, when simulating unsteady flows the time needed to run a simulation with these methods is much higher than its potential flow counterpart. A comprehensive review of the state-of-the-art of methods available to evaluate ship-to-ship interaction can be found in [2] and [3]. In order to replicate unsteady situations such as overtaking, a validation of the code with simpler cases has to be done. In this paper, SHIPFLOW MOTIONS, a fully nonlinear unsteady 3D potential flow method is used to replicate a benchmark case of a lightering operation in confined waters. Forces and moments are compared with captive model tests results available in [4]. The results are also compared with the ones obtained using the viscous solver of SHIPFLOW and presented in [5]. To replicate a lightering operation, model tests are run on two ships, the KVLCC2 tanker and an Aframax. The ships are tested at different speeds and different lateral positions.

NUMERICAL METHOD

The method used here is based on the potential flow theory. Under the assumptions of homogeneous, inviscid, incompressible and irrotational fluid, a scalar quantity, called velocity potential, can be introduced. The velocity potential is used to describe the motion of the fluid. It is defined setting its gradient equal to the fluid velocity:

$$\nabla\phi = \left(\frac{\partial\phi}{\partial x}, \frac{\partial\phi}{\partial y}, \frac{\partial\phi}{\partial z}\right) = (u, v, w) = \mathbf{u} \tag{1}$$

Substituting equation (1) into the continuity equation for incompressible fluids, the Laplace equation is obtained:

$$\nabla^2\phi = 0 \tag{2}$$

In order to solve this partial differential equation, boundary conditions have to be introduced. On the free surface, fully nonlinear kinematic and dynamic boundary conditions are imposed. Being $\mathbf{x} = (x, y, z)$ the position of a fluid particle on the free surface, the kinematic boundary condition is:

$$\frac{D\mathbf{x}}{Dt} = \left(\frac{dx}{dt}, \frac{dy}{dt}, \frac{dz}{dt}\right) = (u, v, z) = \nabla\phi \tag{3}$$

The dynamic boundary condition is:

$$\frac{D\phi}{Dt} = -gz + \frac{1}{2}\nabla\phi\cdot\nabla\phi - \frac{p_a}{\rho} \tag{4}$$

where g is the gravitational acceleration, ρ is the fluid density and p_a is the atmospheric pressure. $\frac{D}{Dt}$ is the material derivative and is defined as:

$$\frac{D}{Dt} = \frac{\partial}{\partial t} + \nabla\phi\cdot\nabla \tag{5}$$

Boundary conditions are applied to the floating bodies and on the bottom, ensuring that the flow does not pass through the surfaces. On the body a Neumann type condition is applied:

$$\nabla\phi\cdot n = \frac{\partial\phi}{\partial n} = n\cdot(\mathbf{u} + \omega\times\mathbf{r}) \tag{6}$$

where n is the normal to the surface pointing into the fluid domain, $\mathbf{u}$ and ω are the translational and angular velocities and $\mathbf{r}$ is the distance between the surface and the center of rotation. On the bottom an impermeability condition is applied:

$$\nabla\phi\cdot n = 0 \tag{7}$$

The *Boundary Value Problem* (BVP) described above is solved using a boundary element method. The domain is discretized with quadrilateral panels and a constant strength source distribution is placed on such panels. The fully nonlinear boundary conditions on the free surface introduce a time dependency. To solve the evolution of the free surface in time, a *Mixed Euler-Lagrangian* (MEL) method is adopted. In this method, each step is divided into two sub-steps: in the Eulerian step the BVP is solved and the velocities and the velocity potential are obtained in the domain. In the Lagrangian step, the boundary conditions are integrated in time and the position and velocity potential are updated for the next step. For a deeper insight of the MEL method see [6]. The integration in time is done with a fourth order Adam-Bashforth-Moulton method.

In order to obtain the forces acting on the ship, the pressure is integrated over the instantaneous wetted surface. The pressure is obtained using the unsteady Bernoulli equation after the fluid velocity is calculated. Once the forces are evaluated, it is possible to calculate ship motions.

To avoid unwanted wave reflection from the boundaries of the computational domain a damping zone is introduced. Since the disturbance due the presence of the hull is present only within the domain while outside the boundaries the solution is known *a priori*, the damping zone dampens out the perturbed part of the solution. In order to achieve this, damping terms are added to the free surface boundary conditions throughout the extension of the damping zone. The damping zone works also as a wave generator.

Since the computational effort required to solve large domains, fine grids and multiple bodies can be significant, a non-linear decomposition of the velocity potential is introduced to increase the efficiency of the code. Since Laplace equation is linear, each linear combination of solutions will still be a solution. With this method, the velocity potential and the free surface elevation are split into two components: an incident part, which is known *a priori* and describes the incoming waves, and a disturbance part, which represents the influence of the hulls in the domain. It must be noted though that the cases presented here do not include incoming waves but the possibility of handling such condition is included in the code. For a deeper insight of the method see [7].

As far as the discretization of the domain is concerned, an adaptive grid refinement is used in order to increase accuracy. While the grid on the hulls is kept constant during a simulation, the discretization of the free surface is adaptively refined. A refinement on the hull is initially applied, based on local curvature and a refined band is added around the waterline. The areas of the bow and stern are also refined. The grid on a KVLCC2 tanker can be seen in Figure 1. On the free surface, an adaptive refinement scheme which recursively splits a panel into four sub-panels is used to refine the grid when a criterion is met. A maximum level of refinement is specified by the user and this value corresponds to the maximum number of refinements a panel can go through. There are two criteria that trigger the refinement. One is related to the presence of the hull and one is on the curvature of the free surface. On the intersection between the free surface and the hull, the panels have the maximum level of refinement. The levels are then linearly decreased moving away from the hull. When a panel is placed completely inside the hull, it is discarded and the only panels that are further refined are those which are cut by the intersection between free surface and the hull. When this process is done, there will still be some gaps between the hull and the quadrilateral panels. In order to fill these gaps, the hanging nodes of the panels surrounding the hull are moved to the waterline, as can be seen in Figure 2. For the curvature criterion, the panels are refined if the curvature between two control points exceeds a fixed value. As can be noted, this allow a coarser discretization further away from the wave systems and a finer one in the areas of interest. In addition, a refinement region, described

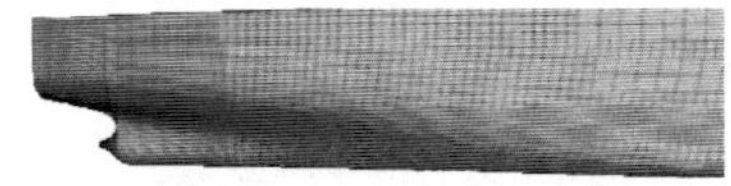

FIGURE 1: Grid resolution for KVLCC2.

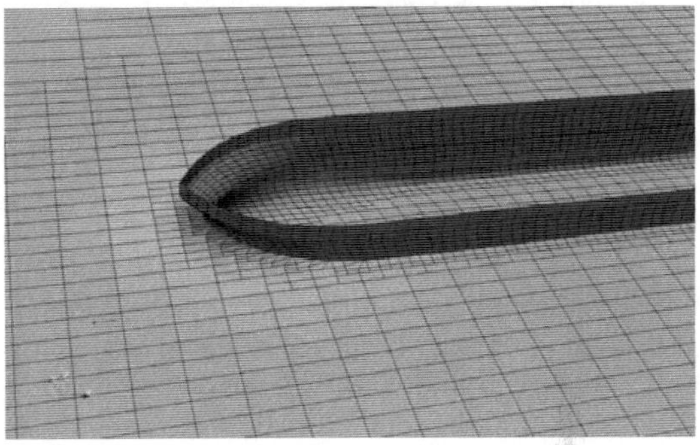

FIGURE 2: Detail of the intersection between free surface and the hull.

by a polygon, can be manually placed in area of interests, where different limits for the criteria can be applied. The adaptive algorithm is called every time the free surface changes: at each iteration and at each sub-step of the Adam-Bashforth-Moulton routine. Initially, a level one is applied on the domain and is subsequentially refined through the previous criteria. Thanks to this adaptive grid refinement it is possible to handle generic surface piercing body forms as well as multiple bodies . A comparison between the initial and final mesh for a generic case can be seen in Figure 3.

In order to decrease the computational effort, a Barnes-Hut algorithm, presented in [8], has been implemented. First, based on the distance of an area of interest, panels are grouped together to form nodes. When the BVP is solved, the panels belonging to one node are treated as a single panel. The level of approximation is specified by the user setting the ratio between the dimensions of a node and the distance at which the influence of such node fades away. The Barnes-Hut algorithm was firstly adopted for astrophysics application, where the influence of the gravity decrease with $\frac{1}{r}$. Since the the influence of the velocity potential varies with $\frac{1}{r}$, problems in accuracy can arise. To avoid this problem, a multipole expansion is adopted. Thanks to this algorithm, the computational effort needed to solve the system drops from $O(n^2)$ to $O(n\log(n))$.

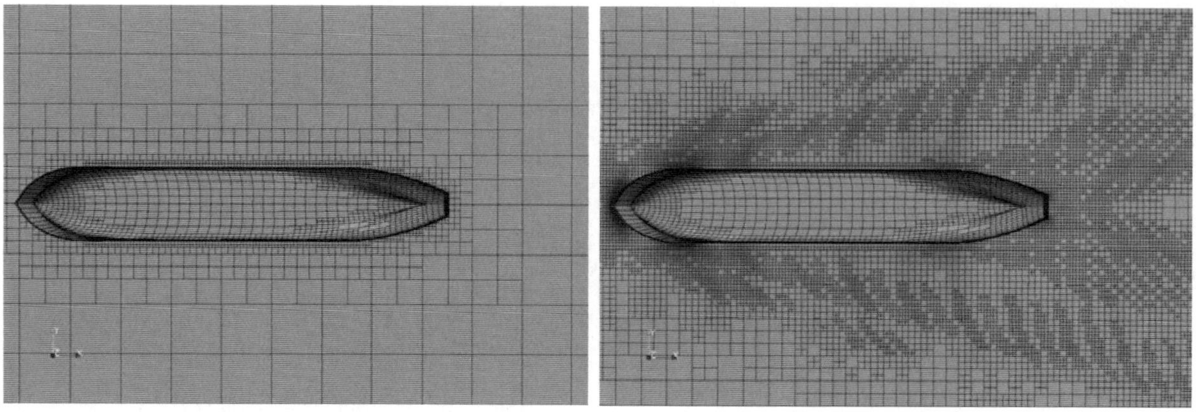

FIGURE 3: Comparison between initial and final mesh with the adaptive grid refinement.

SIMULATIONS SETUP

In the benchmark test case described in [4], a lightering operation is replicated. A *Ship To Be Lightered* (STBL) and a *Service Ship* (SS) move side by side at the same speed. During a lightering operation, the SS gradually approaches the STBL moving at the same speed. When the ships are close enough, the transfer of cargo begins. In order to replicate this scenario, captive model tests are carried out with the ships moving at the same speed and placed at different transversal positions. To properly reproduce different conditions, various combinations of drafts and water depth are tested. Furthermore, different speeds are tested as well. As far as the longitudinal relative position is concerned, the midship sections of the two ships are aligned in the cases reported here.

The STBL is a model scale of the KVLCC2 and the SS is an Aframax tanker. The scale factor is λ = 75. Both ships were equipped with rudder and propeller during the experiments. In the numerical simulations presented here, rudder and propeller are not taken into consideration and the ships are towed at the prescribed speed. In order to have a smoother transition from rest to the desired speed, an acceleration ramp is imposed to both ships.

Forces and moments are measured in each ships coordinate system. The quantities measured are: longitudinal force X, lateral force Y and yaw moment N. Since the potential flow does not take into account viscosity, frictional resistance is taken into account using the *ITTC '57* line.

A partial set of conditions is replicated here, specifically tests A, B and C. A description of the main parameters describing these conditions can be seen in Table 1. In the table, d is the water depth and δy is the distance between the ships sides.

The initial condition for case B can be seen in Figure 4. Lateral extension and water depth are chosen according to the towing tank characteristics. The longitudinal domain extends for $6L_{pp}$ upstream and $6L_{pp}$ downstream from the center line of both ships. L_{pp} is length of the KVLCC2. The damping zone is located on the last $3L_{pp}$, both upstream and downstream, to avoid reflection from the boundaries. In order to have a better resolution in the area of interest, a refinement of the grid is applied close to the hulls for a total length of $3L_{pp}$ and centered on the ship's midsection. Both hulls are discretized with approximately $20 \cdot 10^3$ panels each. On the free surface, once the full speed is reached and the solution approaches a steady state there are $10 \cdot 10^3$ panels.

TABLE 1: Main parameters of benchmark cases.

-	A	B	C
T_{STBL} [m]	12.825	12.825	12.825
T_{SS} [m]	7.5	15.0	15.0
d [m]	17.250	20.250	36.625
v [kn]	6.0	5.0	4.0
δy [m]	50.0	50.0	4.0

NUMERICAL RESULTS

In this section numerical results are presented. The time series of forces and moments have to be analyzed in order to compare numerical simulations with experimental results. The time series for the forces and moments for case B is visible in Figure 5. To obtain a stable value, a *Double Exponential Smoothing*

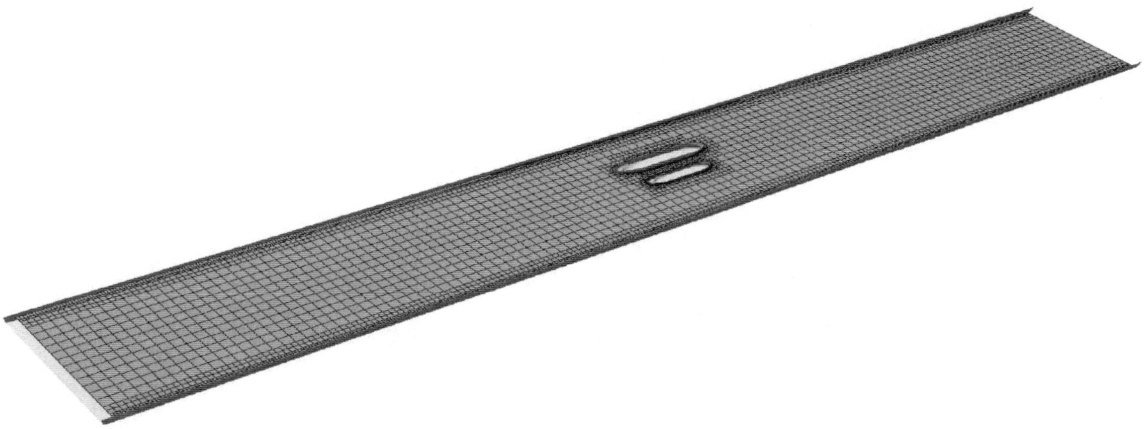

FIGURE 4: Initial configuration for case B.

(DES) is applied on the time series of the forces and moments:

$$\begin{cases} F_{DES,t} = \alpha F_t + (1-\alpha) F_{DES,t-1} \\ F^d_{DES,t} = \alpha F_{DES,t} + (1-\alpha) F^d_{DES,t-1} \end{cases} \tag{8}$$

where F_t is the value of the force at the time t, $F_{DES,t}$ the value of the the first exponential smoothing, $F^d_{DES,t}$ the double smoothing and α is the smoothing factor. In Figure 5, both the time series and the record with the DES filter are shown. As can be seen, the longitudinal forces seem to have a stable and converging behavior, with a noise due to reflected waves kept to a minimum. For the Y force and the yaw moment instead, a DES procedure is necessary in order to find a convergence.

The values obtained with the DES are compared with the ones from experimental results and from the viscous computations presented in [5]. In Table 2 these values are presented, where *SFLM* stands for SHIPFLOW MOTIONS, *SFLR* for SHIPFLOW Ranse solver and *EFD* for experimental results. A visual representation of the comparison is visible in Figure 6. As can be seen, a good agreement is reached for the Y forces, both for the Aframax and for the KVLCC2. As far as the X force is concerned, a fairly good evaluation is obtained for the Aframax, while for the KVLCC2 the prediction shows a larger discrepancy compared to the numerical results. The quantities that are compared do, however, not represent exactly the same thing: the force obtained with the potential flow includes frictional resistance through the *ITTC'57* without form factor and the viscous solver predicts the force using a double-model, *i.e.* symmetry-plane at the free surface. Both methods therefore have some limitations. In addition, it must be pointed out that since the speed is low, with a Froude number ranging between $Fn = 0.037 - 0.055$, the wave resistance will be very small compared to the frictional part. Furthermore, for Froude numbers this small, a high number of waves will be generated by the hull. The grid resolution required to properly capture these waves is too high. Even more challenging seems to be in a correct replication of the yaw moments for both hulls. A wrong sign of the moments is obtained for the Aframax while for the KVLCC2 the values obtained are aligned with the ones obtained with the viscous solver. There are two possible causes that can lead to this error. First of all, the pressure recovery at the stern can be considerably higher with a potential flow solver than with a viscous solver because of no separation. The net force and moment resulting from this will be different in magnitude. Furthermore, the center of the resultant force will be shifted from the one obtained with a viscous solver and a difference in the sign of the moment could arise. Second of all, since these cases are asymmetrical, there will be a circulation acting on the hull, resulting in a lift force. The circulation is not captured by a potential flow using only sources distribution on the panels. The same problem was faced and discussed in other papers where potential flow methods were used to evaluate the yaw moment, see for instance [9].

CONCLUSIONS

In the present paper a partial set of conditions from a benchmark of a lightering operation is replicated using a fully nonlinear unsteady potential flow code. An adaptive grid refinement is introduced in the code, which allows a more flexible discretization of the domain. In particular, regions of interests can be discretized with a finer grid, while relatively coarse grid are used in the undisturbed part. A Barnes-Hut algorithm is used to reduce

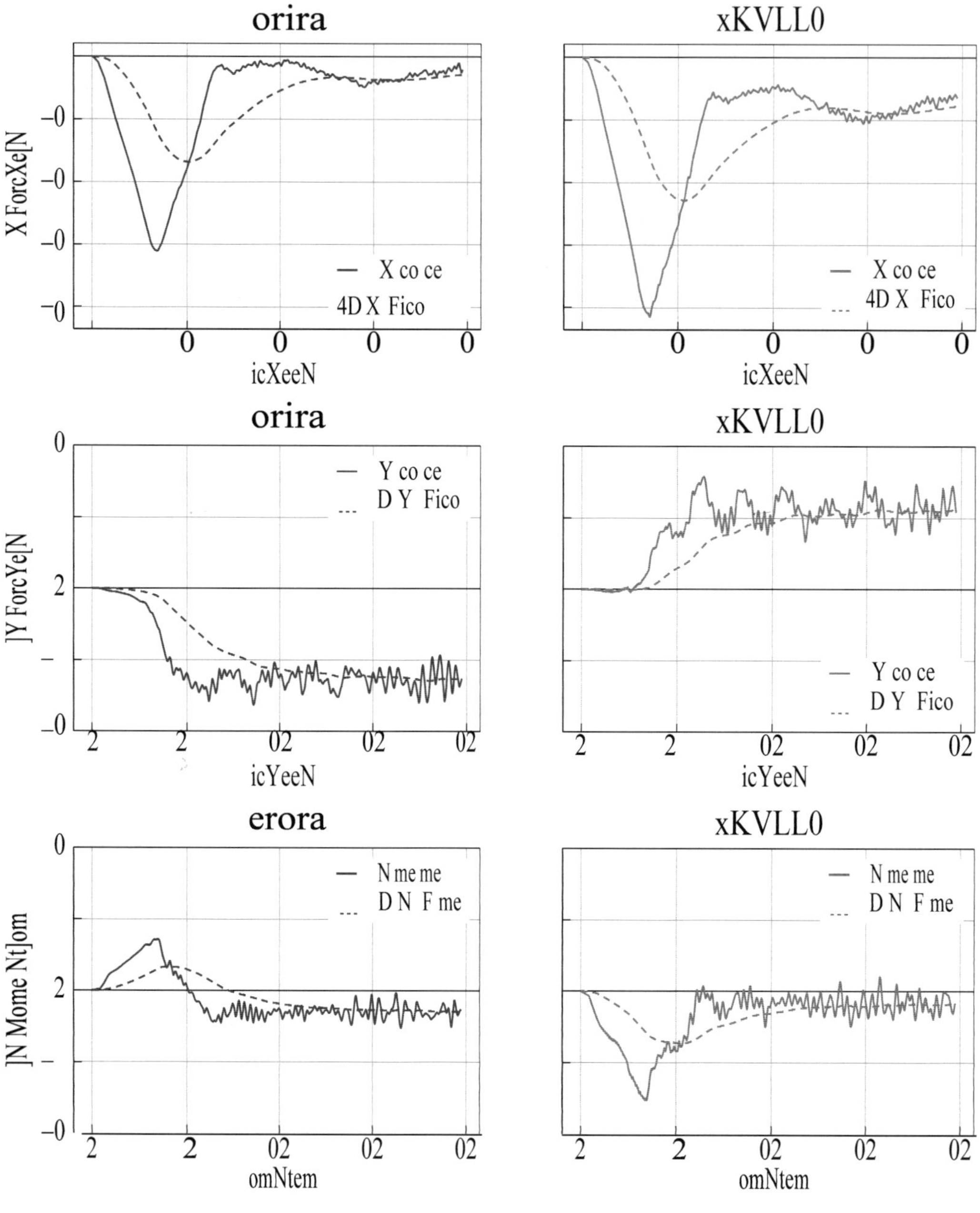

FIGURE 5: Time series of the forces and moments for case B.

the computational time.

Forces and moments are compared with both experimental results and with values obtained with viscous computations. As shown in the previous sections, forces are captured fairly well, both in magnitude and in direction. Particularly good agreement is reached for the attraction force deriving from the hydrodynamic interaction: the code can be considered a reliable tool when it comes to the prediction of such forces. There are some challenges open for the longitudinal forces though, since with such a low speed a correct prediction of the wave resistance can

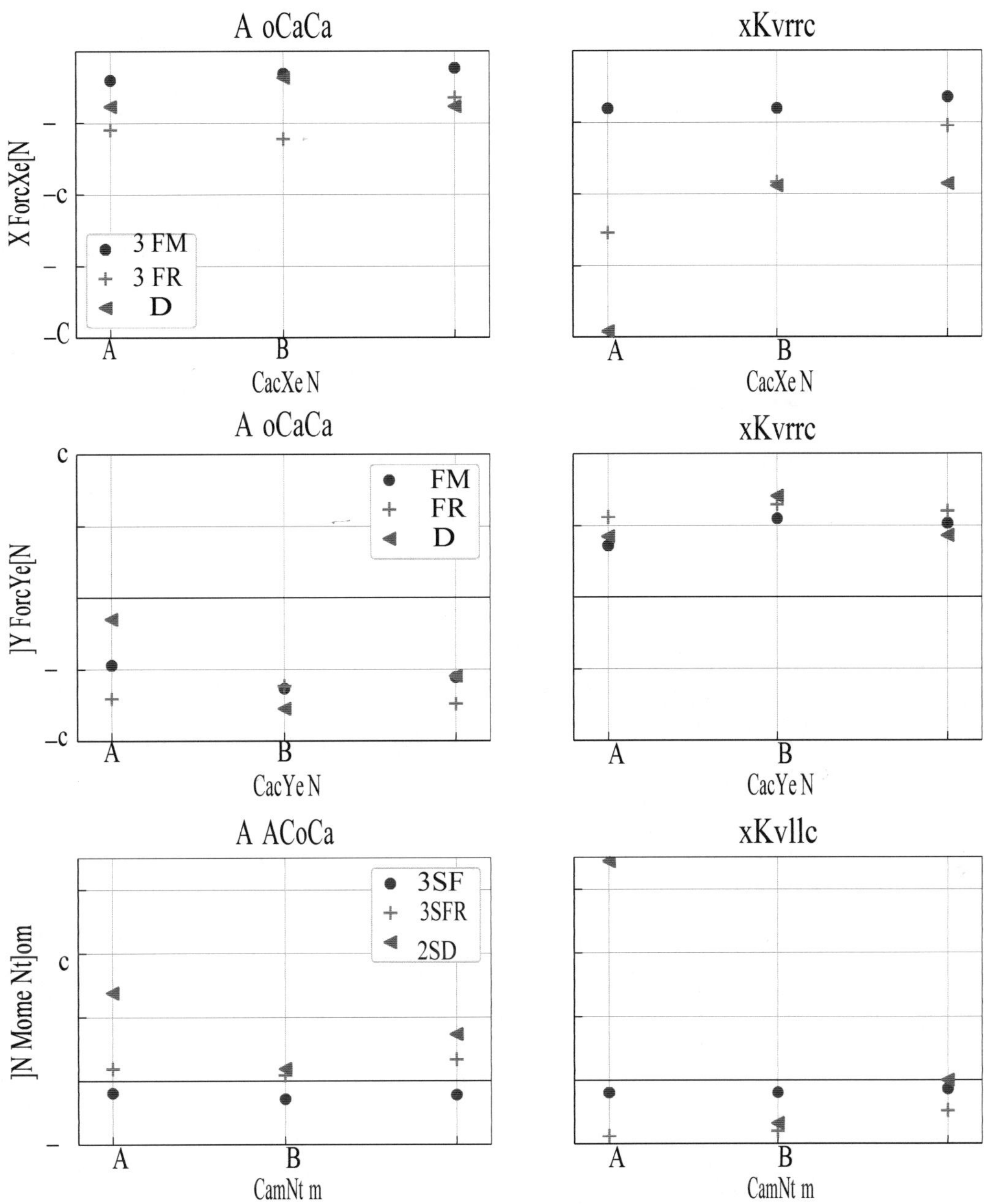

FIGURE 6: Comparison between forces and moments.

be a source of uncertainties with a potential flow code. Increasing further the grid resolution of the free surface between the hulls and downstream, could partially solve this issue. It must be stressed again that for these cases the viscous resistance is the predominant component, so it is not possible to achieve a perfect match with experimental results. As far as the evaluation of the moments is concerned, further investigation is needed. Because of the higher pressure in the stern obtained with potential flow, an error will be introduced. Modeling of the lift forces would help to account for the effects from crossflow and circulation.

TABLE 2: Forces and moments for the cases tested.

		Case A		Case B		Case C	
		SS	STBL	SS	STBL	SS	STBL
X[N]	*SFLM*	-0.41	-0.81	-0.31	-0.81	-0.24	-0.65
	SFLR	-1.10	-2.54	-1.22	-1.84	-0.65	-1.05
	EFD	-0.77	-3.92	-0.37	-1.89	-0.77	-1.86
Y[N]	*SFLM*	-0.94	0.72	-1.27	1.10	-1.11	1.03
	SFLR	-1.41	1.12	-1.23	1.29	-1.48	1.20
	EFD	-0.30	0.85	-1.55	1.41	-1.09	0.86
N[Nm]	*SFLM*	-0.19	-0.20	-0.29	-0.19	-0.22	-0.14
	SFLR	0.19	-0.88	0.09	-0.80	0.34	-0.48
	EFD	1.38	3.44	0.19	-0.68	0.73	0.00

Last but not least, the computational effort required to run these cases is relatively high: the (unsteady) transient solution and transient waves resulting from the acceleration phase take a long time before they can be considered negligible. For these specific cases, the steady viscous solver proved to be faster. Computational effort may be reduced by carefully reconsidering the grid resolution and time step size. Furthermore, some of the waves radiated from the hulls are produced by the ships heaving and pitching until a steady state is reached. Fixing or dampening these degrees of freedom could also improve the time needed to reach a steady state. The main gain deriving from using the unsteady potential flow solver though, arises when unsteady cases are simulated, such as an overtaking situation. While the increase in the computational time needed to replicate this condition will be minimal for the potential flow code, it will be much higher for a viscous solver.

REFERENCES

[1] Yuan, Z.-M., He, S., and Kellet, P. e. a., 2015. "Ship-to-ship interaction during overtaking operation in shallow water". *Journal of Ship Research,* **59**, pp. 1–16.

[2] Committee, M., 2014. Final Report and Recommendations to the 27th ITTC. Report, ITTC.

[3] Committee, M., 2017. Final Report and Recommendations to the 28th ITTC. Report, ITTC.

[4] Lataire, E., Vantorre, M., and Delefortrie, G., 2009. "Captive model testing for ship-to-ship operations". In Proceedings of MARSIM2009.

[5] Zou, L., and Larsson, L., 2013. "Numerical prediction of ship-to-ship interaction in shallow water". *Ocean Engineering,* **72**, pp. 386–402.

[6] Longuet-Higgins, M. S., and Cokelet, E. D., 1976. "The deformation of steep surface waves on water - i. a numerical method of computation". In Proceedings of the Royal Society A: Mathematical, Physical and Engineering Science, Vol. **350**, Proc. R. Soc. Lond. A.

[7] Ducrozet, G., and Engsig-Karup, H. e. a., 2015. "Ship-to-ship interaction during overtaking operation in shallow water". *Journal of Ship Research,* **59**, pp. 1–16.

[8] Barnes, J., and Hut, P., 1986. "A hierarchical o(n log n) force-calculation algorithm". *Nature,* **324**, pp. 446–449.

[9] Yuan, Z.-M., and Yeung, R. W., 2018. "Unsteady interactions among multiple ships with free-surface effects". In 32nd Symposium on Naval Hydrodynamics.

Proceedings of the 6th International Conference on Ship and Offshore Technology
ICSOT 2019
November 7-8, 2019, IIT Kharagpur, India

ICSOT INDIA 2019-10

METHODS OF SIMULATION-DRIVEN DESIGN IN THE MARITIME INDUSTRY

Heinrich von Zadow
FRIENDSHIP SYSTEMS AG
Potsdam, 14482
Germany
von.zadow@friendship-systems.com

Stefan Harries
FRIENDSHIP SYSTEMS AG
Potsdam, 14482
Germany
harries@friendship-systems.com

Paulo Macedo
KONGSBERG MARITIME
Ålesund, 6009
Norway
paulo.macedo@km.kongsberg.com

Mike Saroch
Strong Eagle Corp.
Bangkok, 10120
Thailand
saroch@caeses.com

ABSTRACT

A great effort is needed to keep up with an ever-advancing industry. This is especially true for the very complex field of maritime assets such as ships and marine structures. Whether it is a strive for higher performance or the necessity to keep up with increasing requirements, a design that is uncompromisingly safe yet the most economically attractive and environmentally friendly will always remain a key target of engineering efforts.

This paper will show different methods of process integration and optimization strategies ranging from design space exploration to meta-modeling techniques. The design and optimization platform CAESES® is used to integrate all the necessary steps into an overarching system. In this manner, a time-consuming process, traditionally involving work by many specialists using various tools, can be wrapped into an automated, closed loop. As a result, the number of variants investigated before converging towards an optimized solution is greatly increased and significant improvements can be made in shortened lead times.

An outlook towards holistic ship design from an all-encompassing viewpoint is given and different examples will be presented representative for the workflow. A comprehensive study of a RoPAX ferry, a complex geometry model of a tip rake propeller and its cavitation behavior and a SWATH hull, optimized towards the largest possible operational window will be presented as the main application cases.

Introduction

Naval Architecture today is driven by challenges and restrictions that are the highest ever experienced. Applications are complex by nature and sustainable solutions are required both, from an economic and environmental point of view. Combined with tight safety regulations and short lead times the design of ships and marine structures has grown to a multidisciplinary and elaborate field over the past decades.

Different tasks of engineering and separate stages of the design process are commonly handled by distinct departments and/or engineers. Updates on preliminary results and changes in the constraints and requirements frequently use up time as in this traditional design process, a sequential, iterative procedure is usually undergone to evolve from an initial draft towards the final design. While designers hereby used to rely mainly on experience and intuition, digitalization has taken over allowing for rational decisions with respect to a well-conceived problem. Even more so automatization takes away some of the responsibility in decision making and shifts it towards implementing constraints and rules such that they can be handled and monitored automatically. The traditional design process has changed dramatically and is in constant progression just now.

CAESES® is a general process integration and design optimization (PIDO) environment developed to meet the demands that arise from this evolution. It allows to couple external software in batch-mode to set up process chains for automated design and optimization studies. Prerequisites such as robust, variable Computer Aided Design (CAD), suitable optimization algorithms for global and local, single- and multi-objective optimization tasks, post-processing and evaluations of objective functions are taken care of while monitoring constraints throughout the entire process to ensure a feasible solution space.

This paper presents recent work from three different applications that showcase the complexity involved in state-of-the-art engineering design. Advanced parametric geometry modeling in combination with surrogate techniques allows the generation of CAD models of great value. Through this kind of high fidelity models important insight into the behavior of complex systems becomes available to the designer, yielding the necessary data for cause-and-effect analyses. Furthermore, by interlacing the geometry and its actual physical performance in the intended design condition the traditional step-by-step design procedure is replaced by more elaborate Simulation-driven Design (SDD) routines.

Simulation-driven Design Simulation-driven design is the extensive utilization of simulations to not only analyze a few manually created design candidates but, rather, to generate numerous design variants and to offer them to the design team as possible solutions from which to select the most suitable. As the term suggests the drivers of the process are the simulations which yield a quantitative assessment of the objective(s) and of the constraints.

In the maritime industry simulations may encompass Computational Fluid Dynamics (CFD), Finite Element Analysis (FEA), damage stability computations, seakeeping analyses, oil-outflow simulations, economic and life-cycle cost assessment etc., i.e., practically anything that may be of importance for a given design task.

In order to run simulation-driven design campaigns so-called PIDO environments such as CAESES® are used. PIDO environments support a design team in setting up synthesis models and in running automated optimization studies which do not need any manual interaction while variants are created and analyzed. Synthesis models can be as lean as one geometric modeling module combined with one simulation code as used for the optimization of ship hull forms with regard to resistance and propulsion and may grow as complex as integrating all relevant design aspects, say weight, payload, structural integrity, stability and safety, energy consumption, seakeeping and, very importantly, economic performance [1].

Holistic Ship Design When combining many, if not all, relevant design aspects into one binding synthesis model a so-called holistic view is taken, i.e., the maritime asset is studied comprehensively and for all chosen aspects at the same time. This is different to the view illustrated by the classical design spiral that advocates the step-by-step and iterative approach that has to be necessarily adopted – and that the overwhelming majority of design teams follow – if tools and simulation codes are not integrated and do not share data directly.

Optimization of a RoPAX ferry

The HOLISHIP project – Holistic Optimization of Ship Design and Operation for Life Cycle, undertaken by about 40 European partners from industry and academia, see www.holiship.eu – develops and runs a number of application cases representative of the many aspects that come together when looking at ship design and optimization from such a holistic perspective. One application case amongst them is the design of a RoPAX ferry which will be used as an example in this paper to highlight some of the specialties and novel approaches taken in this project. All relevant design aspects from energy efficiency, safety and environmental compatibility to production and life-cycle cost are wrapped into one integrated optimization, see Figure 1.

The combination and interaction of all different design components leads to a complex synthesis model that allows exploring a large design space and helps finding superior designs in less time. As can be seen from Figure 1 the set-up comprises the generation of geometry, here the parametric modeling of the entire hull in CAESES®, a viscous free-surface RANS simulation for the baseline complemented by potential flow analyses for variants, potential flow seakeeping analyses (added resistance in waves), weight estimates, intact and stability analyses via NAPA, cost and economic analyses.

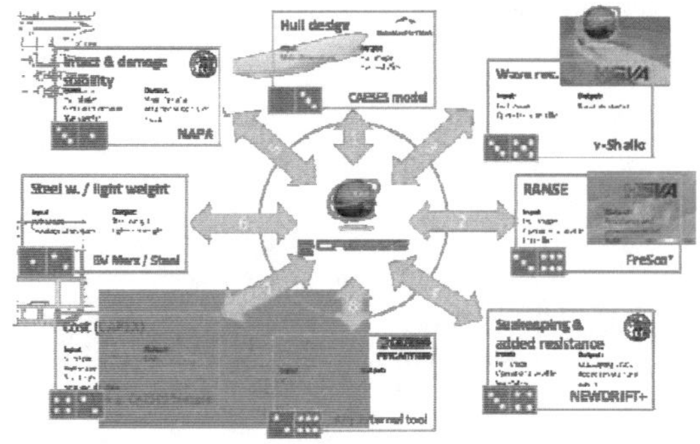

Figure 1. HOLISHIP INTEGRATION INSTANCE OF A SYNTHESIS MODEL FOR A RoPAX FERRY

Details about the application case of the RoPAX ferry are given in [2]. For additional material and further application cases from the HOLISHIP project see [3].

Response Surfaces When setting up an optimization task that involves multiple external tools, e. g. Computational Fluid Dynamics (CFD), damage stability analyses etc., the complexity of the system quickly increases. To overcome the difficulties that arise from simulations that require large amounts of computational resources, commercial licenses, dedicated hardware and/or expert knowledge, response surfaces (also known as surrogate models) can be successfully put to use in order to encapsulate the relationship between design variables and responses. This is particularly valuable for early design stages in which overall dimensions and general characteristics need to be established.

Figure 2 shows how for the RoPAX design several simulations that are both demanding to set up and time-consuming to run are replaced in the synthesis model for early design. For a comparison see Figure 1.

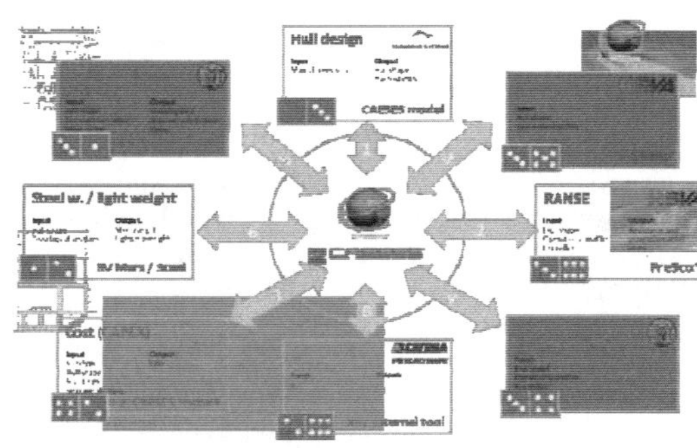

Figure 2. APPROACH OF PROVIDING INFORMATION AT AN EARLY DESIGN PHASE VIA RESPONSE SURFACES

By storing response surface models that are based on results from previously evaluated designs a prediction of high accuracy can be made for the objective function of any arbitrary geometry variant within the design space. Thus, expensive computations can be split into multiple independent tasks and difficulties arising from different system architectures do not become an issue. Combining all the separate models into one multi-objective optimization problem becomes a much simpler task and due to the significantly faster evaluation of the models a large number of designs can be evaluated, see Figure 3.

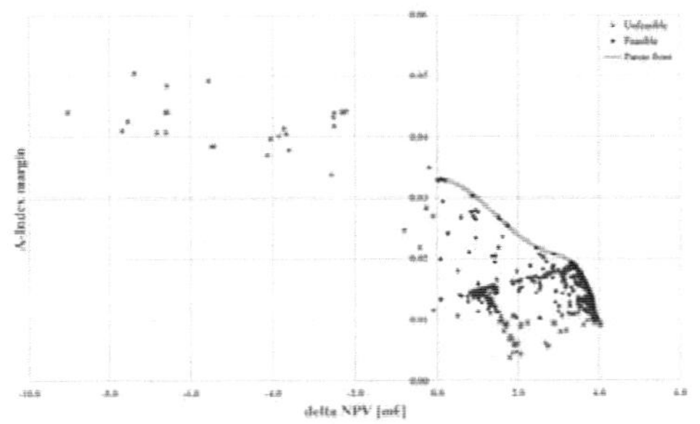

Figure 3. ATTAINED INDEX MARGIN VS. NPV VARIATION FOR THE RoPAX FERRY AS AN AGGREGATE OF COMPETING OBJECTIVES FOR SAFETY AND ECONOMICS

As a prerequisite to generation of the response surfaces a subset of design variables and their corresponding ranges needs to be selected. The selection should allow form variation in the areas that are expected to be relevant, for instance for stability and hydrodynamic performance. At the same time the resulting hull form needs to comply with owner's requirements and must allow the generation of feasible designs when considering the constraints. The CAD model of the RoPAX as set up in CAESES® allows the variation of length, beam, draft, block coefficient, midship coefficient, center of buoyancy, length of parallel mid etc. along with a set of local design variables for the bulbous bow's height, length, beam and vertical center of buoyancy. Specific export formats tailored towards the coupled software tools are available to pass on the geometry for analysis.

The generation of response surfaces in CAESES® is handled through a custom feature definition that makes use of the DAKOTA open-source optimization kit developed by Scandia National Laboratories [4] which is tightly integrated into CAESES®. Kriging is chosen as an interpolation method and all input files necessary to build and evaluate the model are created automatically. To generate a model from a previously performed Design of Experiments (DoE) an existing results table is chosen as input argument to the CAESES® feature. The measure that is to be approximated by the model needs to be specified and upon execution of the feature a file with the surrogate model is written. In the same manner any number of models can be generated, each allowing the prediction of a separate objective based on all design variables that have been modified during the DoE.

Global Optimization Figure 3 presents selected results from a global optimization for the RoPAX ferry. On the ordinate the attained index margin of damage stability (A-Index margin) and on the abscissa the Net Present Value variation (delta NPV) are given, respectively.

The A-Index margin is the difference between the required index and the achieved index computed from a probabilistic damage analysis (automatically undertaken by coupling CAESES® with NAPA). Any design that has a negative A-Index margin would be infeasible. In order to be on the safe side and to render designs that are even a bit safer than the regulative requirements a minimum A-Index margin of 0.01 was introduced. Infeasible designs are marked red.

From an economic perspective the delta NPV summarizes elegantly the superiority of new designs when compared to an acceptable baseline. Resistance in calm water and in waves, propulsive efficiency, fuel consumption and expected fuel oil prices, fixed and variable costs, revenue generated from the ferry service on a selected route and market compliant interest rates are all combined in the NPV. Designs that yield a negative delta NPV are considered infeasible since the baseline would outperform them.

A Pareto frontier of excellent design variants, i.e., designs that show superior performance for both economics and safety, is presented in Figure *3*, too. See grey line. From the many variants investigated during the RoPAX optimization not only better designs could be found but also an appreciation could be developed about cause-and-effects and about risks and potentials. Further details are given in [2] and [3].

Optimization of a Tip Rake Propeller

In search for increased efficiency of marine propellers a tip rake propeller was recently investigated [5]. The effects of different tip geometries on performance and erosion potential power were studied. While tip rake propellers and Contacted Loaded Tip (CLT) propellers show promising behavior in terms of efficiency they have proven to be challenging to design such that cavitation damage does not render the achieved improvements.

Furthermore, also the 3D modeling of such propellers imposes new challenges on the designer. Conventional modeling approaches in which blade sections are transformed onto concentric cylindrical surfaces fail for highly raked blade tips. To allow sufficient shape control in the tip area a new approach, recently developed [6], was used for the parametric modeling of such propeller blades. After an initial sampling, two separate surrogate models are incorporated to allow a detailed analysis of the design space in search for promising variants.

While initial sampling used a Reynolds Averaged Navier Stokes (RANS) approach, a set of final selected designs were further investigated with Large Eddy Simulations (LES) to allow prediction of cavitation behavior and EPP. All CFD simulations, as well as unstructured grid generation, were carried out with the commercial CFD solver Star-CCM+.

Baseline Design The KP505 propeller, initially designed for the KRISO Container Ship (KCS) is used as a baseline design in this study to allow verification of the CFD setup, see Figure 4. Discrete data tables available from SIMMAN [7] are used as a basis for the parametric radial functions of pitch, rake, skew, chord, camber and thickness. A NACA66 profile with a = 0.8 is used for the section definition.

In conventional propeller modeling rake is added by shifting of sections in axial direction. As one can easily depict, for a highly raked blade tip which is bent by either a positive or negative 90° angle, this will result in multiple sections lying flat on the same cylindrical surface. As a result control over the profile shape characteristics such as camber and thickness is lost.

To overcome these difficulties an alternative transformation of the planar sections into 3D space is used. Hereby, the 2D profiles are transformed by offsetting them in a local coordinate system that is adjusted both, to the tangent vector of the blades span direction and the local direction of inflow. The resulting sections therefore lie on arbitrary surfaces depending on the incoming flow, as well as the blades local rake.

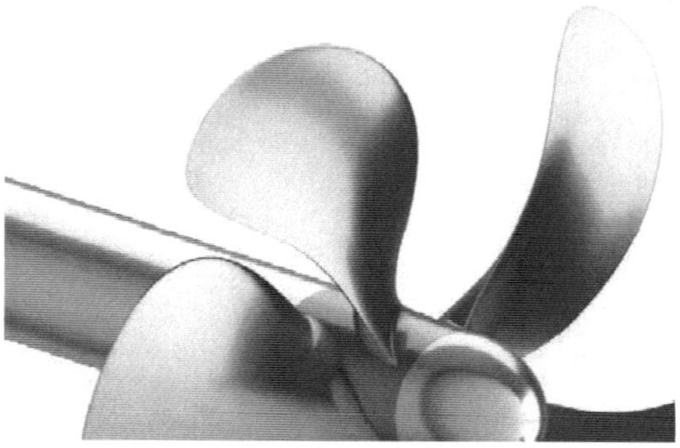

Figure 4. CAD MODEL OF THE KP505 PROPELLER

As inflow does not need to be considered homogenous but might be taken from a known wake field this approach even allows for a blade design that is dependent on the expected design condition. Not only does this make the process of designing the propeller more intuitive, as a local angle of attack with respect to the direction of the incoming flow can directly be addressed, the procedure also allows easier adjustments of an existing propeller for a different ship or design condition if an estimated wake field is at hand.

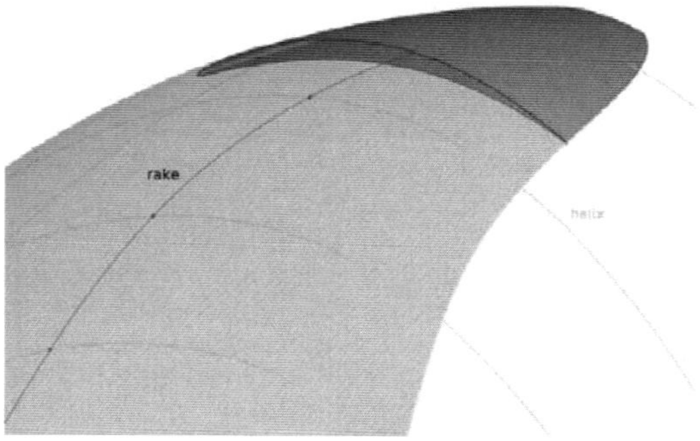

Figure 5. DEPICTION OF THE CONSTRUCTION OF A HIGHLY RAKED BLADE

In Figure 5 the construction is shown, starting from a given rake function. At any position along this function the direction of inflow is known from the rotational speed of the propeller and the given wake field. A helix can be constructed as shown exemplary for four different radii and one can imagine a surface through these curves would be perfectly aligned with the flow. Sections at any position are now offset normal to that surface and a local angle of attack can be added.

Design of Experiments To allow variant generation off of the baseline propeller geometry, 6 design variables have been introduced to the parametric model. These variables act on the radial functions of camber, angle of attack and rake. For each of these functions the value at the blade tip can be modified in a certain interval and the radial range in which this modification is applied and smoothly fades out can be adjusted. In Figure 6 three different geometry variants are shown. The KP505 baseline design along with two selected designs, one with positive and one with negative rake. The design functions that are subject to modification during the DoE are shown and the smooth introduction of tip rake starting at approximately r/R = 0.7 can be seen.

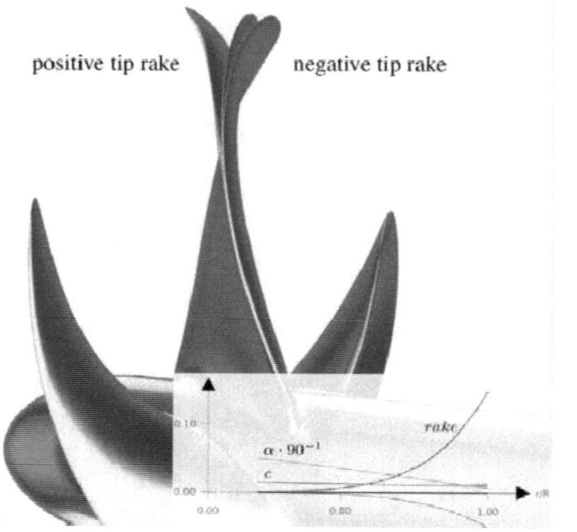

Figure 6. TWO DIFFERENT VARIANTS ALONG WITH THE BASELINE DESIGN AND PARAMETRIZED RADIAL FUNCTIONS FOR GEOMETRY VARIATION

For an initial DoE of 172 design variants RANS simulations are performed in non-cavitating open-water condition using a Latin Hypercube Sampling [8] algorithm to fill the design space. The results are used to build a surrogate model for further analysis and optimization. To ensure good prediction accuracy across the design space, a 2-fold cross-validation was carried out. The initial result pool is therefore split into two separate subsets of 86 designs each. From each subset a separate surrogate model is build and validated against the second half of the result pool. With an averaged error in prediction of open-water efficiency of 0.48 % and thrust coefficient of 0.77 % good performance of the model is shown and even higher accuracy can be expected when using the full set of results for further analysis.

Surrogate Enhanced Exploitation The differences in efficiency of the variants investigated during the initial DoE were rather low as many of the variants increased both, thrust and torque. This can be expected as no constraint was imposed as to what the delivered thrust of the propeller should be. However, by not constraining the design space at this early stage it is possible to gain valuable insight into the characteristics of the problem. In a second DoE phase the results of the first sampling are used through response surfaces and the good coverage of the design space and high model accuracy allow a much more elaborate exploitation of the solution space.

Two separate response surface are created, one for the prediction of open-water efficiency and another one to model the expected thrust of each variant. As the evaluation of these models takes place within a split-second it is faster than the CFD by multiple orders in magnitude and allows for an entirely different exploration approach. When studying the response surface model for thrust prediction one can find that by adjusting the local angle of attack in the tip region of the blade the thrust changes significantly, as depicted in Figure 7. For each combination of the remaining design variables a geometry variant can be found that features the necessary angle of attack for the required amount of thrust. A nested optimization loop based on a one dimensional optimization algorithm (Brent) [9] is set up to carry out the search.

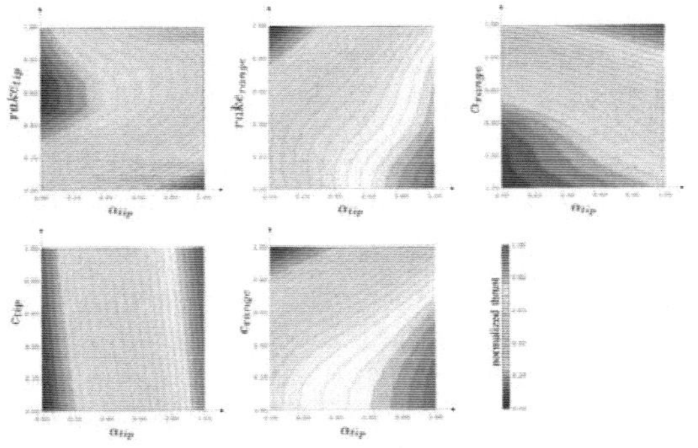

Figure 7. RESPONSE SURFACES OF THRUST COEFFICIENT; FOR ALL PERMUTATIONS THE SAME TREND OF HIGHER ANGLE OF ATTACK RESULTING IN HIGHER THRUST CAN BE IDENTIFIED

Three subsequent studies were performed based on a Sobol [10] algorithm in which designs with a maximum, minimum and variable tip rake were studied. The nested optimization loop hereby ensures that the exact same amount of thrust as it was found for the baseline design is met by each of the newly generated variants. Only if the limited range of angle of attack, as it was set during the initial DoE, does not allow matching the required thrust will a design be considered infeasible.

With each of the Sobol samples consisting of 100 designs and an approximate inner loop count of 20 to 25, this quickly adds up to thousands of design variants investigated, all within a few seconds of CPU time. While the sets of variable and maximum tip rake yield 43 and 24 valid designs respectively, no valid designs were found for the minimum amount of tip rake. Therefore another set with slightly higher tip rake was added which eventually lead to 2 feasible designs of this characteristic as well.

Cavitation assessment with LES Out of the many design variants found during the extensive DoE phase two variants of interest are selected. One of them featuring the maximum amount of tip rake and another one from the set with negative rake. While, according to the response surfaces, both variants provide the same amount of thrust at very similar open-water efficiency as the baseline design their shapes differ significantly. Hence, cavitation behavior is expected to be different and two sets of LES are carried out to investigate vapor pressure gradients in design and off-design condition for all three geometries.

Cavitation simulations use a wake field simulated with CFD and a rudder geometry as available for the KCS from SIMMAN 2008 [7]. Input parameters for design and off-design condition along with specific design variable settings are given in [5] and the following Figure 8 is referenced here to give the reader an appreciation of the flow characteristics. It can be seen that both, baseline and positive rake design display smooth sheet cavitation without gradients at low angles of 0° up to 5-10°. The negative rake design shows gradients near the leading edge which might not be a problem in design condition but indicate possible detachment of sheet cavitation in off-design condition.

Analysis of cavitation patterns at off-design condition in [5] confirmed the slightly lower and more even loading of the baseline and positive rake design and higher EPP for the negative rake design. This is in line with what can be expected as the negative rake design accelerates the flow at a higher rate and is therefore more demanding to design.

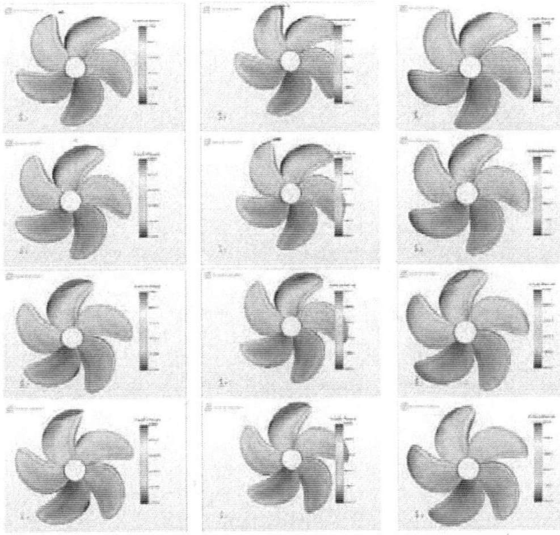

Figure 8. ISO-SURFACE OF VAPOR AND HYDRODYNAMIC PRESSURE AT DIFFERENT BLADE ANGLES IN DESIGN CONDITION. LEFT TO RIGHT: BASELINE, NEGATIVE AND POSITIVE RAKE DESIGN

Rather than aiming for an optimized performance, the goal of the study was to gather a deeper understanding of the actual correlations of shape and flow characteristics. Both, positive and negative rake designs have proven to be worthy considerations when aiming for higher propulsive efficiency with the negative rake being somewhat more sensitive to erosive detachment of sheet cavitation patterns.

Optimization of an SOV SWATH concept

In order to tackle the fiercely competitive European market of offshore wind operation, maintenance and service, ship-owners are constantly seeking lower costs, higher efficiency and profitability. The severe conditions often encountered in the North Sea have large influence on the vessel design, typically leading to larger vessels than necessary (in terms of cargo, crew and costs) in order to yield acceptable motions at sea. With an innovative SWATH (Small Waterplane Area Twin-Hull), see Figure 9Figure *8*, design as an SOV (Service Operation Vessel), which provides transit, accommodation and means of transfer to offshore wind turbines, the possibility of a wider operational window with a reduced vessel size was investigated.

The superior seakeeping capability of SWATH vessels, combined with large deck area, maneuvering capability and ample transverse stability were the main factors to compare with existing monohull SOV designs.

The primary aim of the study was to compare various designs with regard to zero-speed seakeeping performance, see [11] for further details. Four design concepts were considered, two existing monohull designs (a 62 m and an 82 m SOV by Kongsberg Maritime) and two new SWATH configurations (one symmetric and one asymmetric demi-hull of 62 m) developed and optimized in this study.

Parametric Design A fully-parametric model, developed in CAESES®, is able to capture significant variations of the SWATH sketches from members of the ship design department of Kongsberg Maritime. Among them are midship width, length and flare, waterline shape and flare, thrusters mount base inclination and diameter, bilge radii and transition smoothness between the parallel midbody and the extremities, see Figure 10.

Figure 9. VIEW OF THE SWATH SOV CONCEPT

This method also allows to easily set design dimensional constraints to the model with a Brent optimization algorithm. To create an equal-terms comparison between vessels the displacement is matched to that of the 62 m monohull SOV.

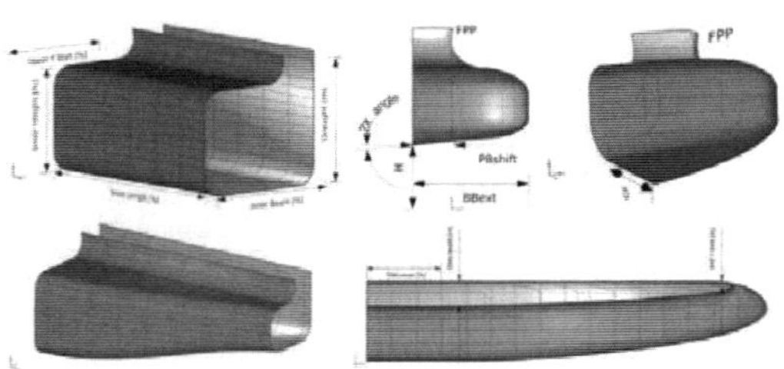

Figure 10. DESIGN VARIABLES AVAILABLE FOR THE PARAMETRIC SWATH MODEL

Seakeeping Analysis NEWDRIFT+ is coupled to CAESES® and executed in batch mode to evaluate zero-speed seakeeping performance of each design variant. It is a six degrees-of-freedom, frequency domain, 3D panel code for floating structures subject to waves developed by NTUA (National Technical University of Athens) [12].

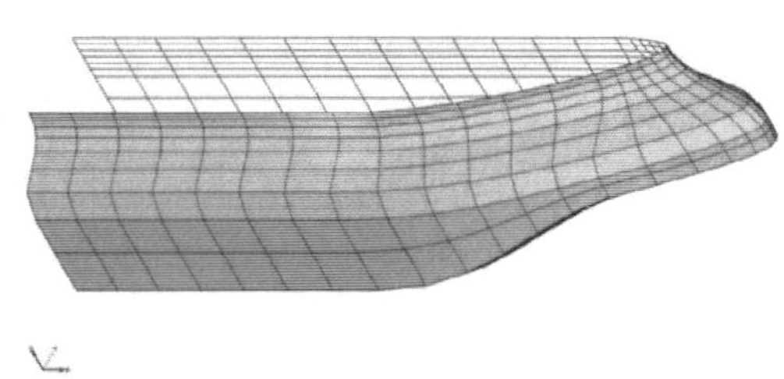

Figure 11. DEMI-HULL PANEL MESH GENERATED BY CAESES®

The panel mesh was seamlessly created from the surfaces of the parametric model with CAESES®, controlling size, density, position and export format to the integrated tool, see Figure 11.

Optimization Procedure Starting with an Exhaustive Search (subdivides the feasible domain into the given number of subintervals) of each relevant design variable, the consistency of the results could be identified and potential errors mitigated, consolidating the seakeeping simulation. In a second, multi-variable study hundreds of design variants generated with a Sobol [10] algorithm (quasi-random homogeneous filling of the design domain) are evaluated. Subsequently, using the previously mentioned tool DAKOTA, a response surface is built for finding potential best candidates. Finally, the best candidates found on the response surface are evaluated, Figure 12, and fine-tuned with a TSearch algorithm (a tangent search method that uses exploratory moves to detect the descent direction in the solution space), also built-in CAESES®, extracting the last 2.5% of potential improvement.

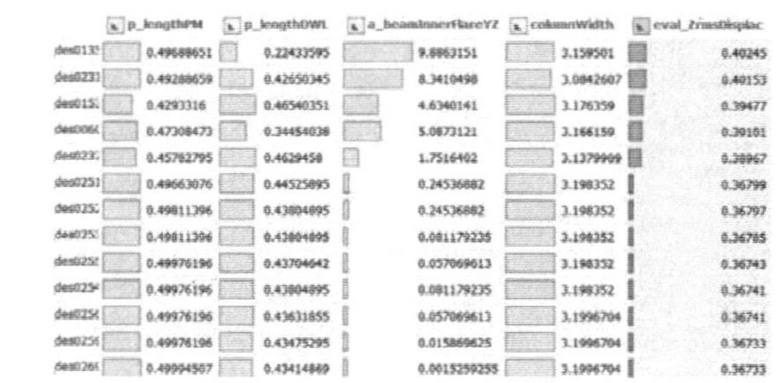

	p_lengthPM	p_lengthOWL	a_beamInnerFlareYZ	columnWidth	eval_ZrmsDisplac
des013	0.49688651	0.22433595	9.8863151	3.159501	0.40245
des0231	0.49288659	0.42650345	8.3410498	3.0842607	0.40153
des015	0.4293316	0.46540351	4.6340141	3.176359	0.39477
des006	0.47308473	0.34454038	5.0873121	3.166159	0.39101
des023	0.45782795	0.4629458	1.7516402	3.1379969	0.38967
des0251	0.49663076	0.44525895	0.24536882	3.198352	0.36799
des025	0.49811396	0.43804895	0.24536882	3.198352	0.36797
des025	0.49811396	0.43804895	0.081179235	3.198352	0.36785
des025	0.49976196	0.43704642	0.057069613	3.198352	0.36743
des025	0.49976196	0.43804895	0.081179235	3.198352	0.36741
des025	0.49976196	0.43631855	0.057069613	3.1996704	0.36741
des025	0.49976196	0.43475295	0.015869625	3.1996704	0.36733
des026	0.49994507	0.43414869	0.0015259255	3.1996704	0.36733

Figure 12. FINAL ITERATIONS OF THE DAKOTA ALGORITHM, CONVERGING TO THE FINAL SOLUTION

A custom tailored plotting feature was created to post-process results from the evaluation tool. This made it possible to capture RAOs (Response Amplitude Operators), excitation forces RMS (Root Mean Square), minima and maxima.

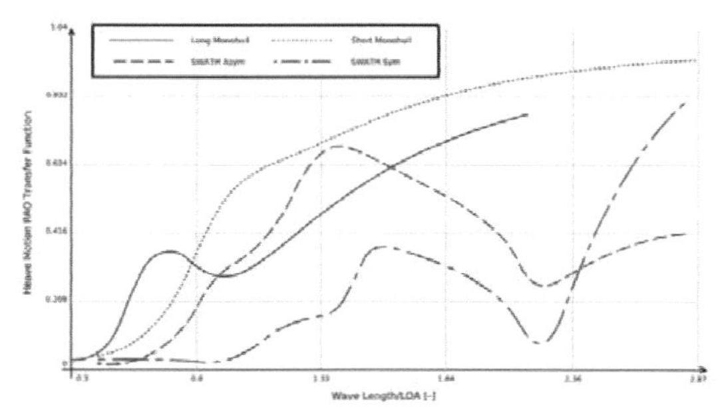

Figure 13. HEAVE RAOs COMPARISON FOR 180° HEADING ANGLE

Remarks Analyzing the results of the optimization procedure, trends were identified towards lowest possible motions. These include positioning of the displacement volume as far below the waterline as possible, combined with a long parallel midbody, allowing a lower center of buoyancy height. A flare towards the waterline is beneficial and should be combined with a matching design waterline. Bilge radii were found to have very low impact.

The freedom of the parametric design, allied with integration, optimization and post-processing capabilities of CAESES® made a very time consuming task (if manually executed) relatively simple and automatic on an early project development stage.

Conclusions

With regard to the SWATH case a final conclusion is rather difficult to state due to the quantity of uncertainties and still unknown factors to this design, such as propulsion efficiency, installed power, structural arrangement and feasibility, the center of gravity, CAPEX and OPEX costs. These elements were not considered at any depth within this preliminary study but could be incorporated by suitably (and readily) extending the HOLISHIP platform in a subsequent phase, i.e., by adding further simulation tools and increasing the relevant data sets. To summarize, in a vessel in which the mission is to provide the largest operational window for a station-keeping position, optimized SWATHs do heave, roll and pitch less in several different sea conditions, Figure 13, being an interesting alternative to monohulls, but have larger wave induced drift forces [11]. Some of the disadvantages of the design might still seem to rule-out the design on an early stage, but its advantages might pay-off along the lifetime of the vessel, thus justifying future studies on the matter.

This becomes evident when looking at the RoPAX application in which the combined effects and complex interactions of multiple design components have been covered in a comprehensive synthesis model. Many of the separate tasks involved are commonly handled by different engineers and/or departments which can be a considerable hurdle to overcome. Surrogate models have proven to be a valuable tool in splitting up the process into individual subsets that can easily be combined into a global optimization routine at a later stage again. As a result a large number of design candidates can be investigated in a short timeframe and superior designs than one would expect from an optimization that sequentially treats the involved components can be found. Additionally, the combined analysis facilitates the development of an appreciation of the interactions of the different areas and engineers gain valuable insight into cause-and-effects, risks and potentials [2], [3].

Along with given requirements, constraints and design conditions, flexible and robust parametric CAD models are the basis for the majority of tasks commonly approached with SDD. As for the SWATH and RoPAX case, this is also true for the investigated tip rake propeller [5]. The underlying CAESES® model allows detailed control over the blades section shape along the entire radial range, especially in the tip region of highly raked variants. During the study it has proven its robustness and flexibility and is therefore considered very advantageous in the demanding design of tip rake and CLT propellers. The added value from surrogate models which, upon an initial CFD based sampling, allow accurate performance prediction of any arbitrary variant has, again, significantly promoted the process. A nested optimization based on such models allows investigating a vast number of shapes that inherently fulfill critical design criteria such as the required thrust at a certain design condition. Not only does this speed up the search for promising and feasible variants or a potential optimization tremendously, but it also provides an insight and understanding into an otherwise unknown space. This insight is traditionally reserved for only the most experienced designers, of which there might be only few considering the novelty of the propeller type under consideration.

REFERENCES

[1] Harries, S. and Abt, C., 2019. "The HOLISHIP Platform for Process Integration and Design Optimization", published by A. Papanikolaou (Ed.) in A Holistic Approach to Ship Design – Vol. 1: *Optimisation of Ship Design and Operation for Life Cycle*, Springer 978-3-030-02809-1

[2] Harries, S., Cau, C., Marzi, J., Kraus, A., Papanikolaou, A. and Zaraphonitis, G., 2017. "Software Platform for the Holistic Design and Optimization of Ships". *Jahrbuch der Schiffbautechnischen Gesellschaft*

[3] Harries, S., Dafermos, G., Kanellopoulou, A., Florean, M., Gatchell, S., Kahva, E. and Macedo, P. 2019. "Approach to Holistic Ship Design – Methods and Examples". *18th International Conference on Computer and IT Applications in the Maritime Industries (COMPIT 2019)*, March, pp. 224-245

[4] DAKOTA, Sandia National Laboratories. URL: `https://dakota.sandia.gov`

[5] Sánches Castro, L. F., von Zadow, H. and Vesting, F., 2019. "Tip Geometry Effects on Performance and Erosion for Tip Rake Propellers". *Sixth International Symposium on Marine Propulsors (SMP 2019),* May, pp. 35-43.

[6] Praefke, E., Stoye, T. and Abt, C., 2017. "A generalized description of hydrodynamic parts based on aerodynamic profile sections". *Fifth International Symposium on Marine Propulsors (SMP 2017),* June, pp. 596-604.

[7] SIMMAN, 2008. "Workshop on Verification and Validation of Ship Maneuvering Simulation Methods". `http://www.simman2008.dk`

[8] Helton, J. C. and Davis, F. J., 2003. "Latin hypercube sampling and the propagation of uncertainty in analyses of complex systems". *Reliability Engineering System Safety,* pp. 23-69

[9] Brent, R. P., 1973. "Algorithms for Minimization without Derivatives", Prentice-Hall, Englewood Cliffs, ISBN-13: 978-0-486-41998-5

[10] Sobol, I. M., 1967, "Distribution of points in a cube and approximate evaluation of integrals". *U.S.S.R Comput. Maths. Math. Phys. 7,* pp. 86–112

[11] Macedo, P., 2018. "SWATH SOV Hull Concept and Optimisation for Seakeeping". MS Thesis, Chalmers University of Technology, Gothenburg, Sweden.

[12] Papanikolaou, A., 1985. "On Integral-Equation-Methods for the Evaluation of Motions and Loads of Arbitrary Bodies in Waves". Omgemoeir-Archiv. 55, pp. 17-29

Proceedings of the 6th International Conference on Ship and Offshore Technology
ICSOT 2019
November 7-8, 2019, IIT Kharagpur, India

ICSOT INDIA 2019-11

PROCESS-STRUCTURE & PROPERTY CORRELATION OF FRICTION STIR WELDING OF MARINE GRADE AA5083 THROUGH FINITE ELEMENT METHOD

Mrinal Sahu
Department of Metallurgical and Materials Engineering, National Institute of Technology, Raipur, Chhattisgarh, PIN-492010, India,
msahu.phd2017.met@nitrr.ac.in

Atanu Paul
Department of Ocean and Naval Architecture, Indian Institute of Technology, Kharagpur, West Midnapore, West Bengal, PIN-721302, India,
apaul20072007@gmail.com

Subhas Ganguly
Department of Metallurgical and Materials Engineering, National Institute of Technology, Raipur, Chhattisgarh, PIN-492010, India,
sganguly.met@nitrr.ac.in

1. ABSTRACT

The marine grade AA5083 have an attractive property of light weight, which significantly reduces the overall ship weight, enabling to operate at elevated speed with lower fuel consumption and fabulous maneuverability. Thus, it is considered as the materials for future of ship building industries. However, the ability to join AA5083 made structures remain the primary challenge and has brought widespread attention of research. Friction stir welding (FSW) is a ''green technology'' of weldingthat has already proved its immense potential in joining of low & high melting point materials. Although the technique has gained sufficient advancement, either experimentally or computationally, but the fundamental knowledge about thermomechanical process involved during FSW and the process property correlation from a simulation is still uncertain. Thus, in the present investigation, we have proposed a 3D finite element based thermo-mechanical model for evaluation of temperature, residual stress and strain during FSW of marine grade AA5083. An attempt to correlate the effect of welding speed and tool rotational speed with the simulated thermal history and residual stresses has been made. In addition to this, a mutual relationship between the longitudinal strain, evolved from proposed model, and the experimental microhardness data is established, which further strengthen the validation of the proposed FE model. The influence of welding parameters on the tensile strength is analyzed and metallurgical characteristics are evaluated from the EBSD study of the different zones of the samples is discussed in light of process-property correlation.

Keywords: Marine grade AA5083; FEM; Thermo-mechanical model; Friction Stir Welding; Process-property correlation.

NOMENCLATURE

UTS	Ultimate Tensile Strength
D	Tool Shoulder Diameter
P	Tool Pin Pitch
d_t	Tool Pin Tip Diameter
d_r	Tool Pin Root Diameter
RPM	Tool Rotational Speed
v	Tool Welding Speed
E	Elastic Youngs Modulus
μ	Poisson's Ratio
K	Thermal heat Transfer Coefficient
C_P	Specific Heat
ρ	Mass Density
α	Coefficient of Thermal Expansion
T_E^{max}	Peak Temperature from Experiment
T_S^{max}	Peak Temperature from Simulation
Δ	Deviation in Experimental and Simulated Peak Temperature

2. INTRODUCTION

Friction Stir Welding (FSW) technique is devised by Wayne Thomas at The Welding Institute (TWI) Ltd, Cambridge, in the year 1991[1]. From the date of its invention, the process has established a world-wide attention in industries which includes automotive, shipbuilding, railways, defense, aerospace etc. and also in research & development cells. The technique is effectively utilized in the joining of materials with low melting temperature like aluminium, brass, copper, and also in high melting temperature materials like steels, nickel and titanium based alloys[2][3][4]. The process has numerous advantages over other conventional techniques like it does not involve any utilization of consumables, filler wire, shielding gases[5]. Moreover to this, since the deformation of material from welding takes place at temperature lower than the melting temperature, it yields less distortion than the fusion welding processes and thus the problems associated with the solidification of a fused material are avoided[6]. The usage of aluminium alloys is increased exponentially, especially in naval, aerospace, automotive industries because of its magnificent property of low weight and excellent corrosion resistance. Shipbuilding is one of the most prominent sectors which has elevated the use of aluminium, especially of series 5XXX and 6XXX. The increasing demand of light weight and less expensive ships has attracted the researchers to explore the knowledge about this "green" technology, FSW.

FSW process employs a non-consumable revolving tool. The tool comprises of two parts as the shoulder and the pin with variety of geometries. This rotating tool plunges on to the workpiece, till the shoulder establishes connection with the available surface of the workpiece. When the temperature raised (because of friction) is high enough, so as to plasticize the material, a horizontal welding speed velocity is provided to the tool along the joint line, producing defect free &high quality of welds. The material gets soften because of the localized heating from the pin and the rotating tool in combination with transverse tool travel, transfers the soften material from front to the rear of pin. The selection of correct tool geometry also plays a vibrant role in governing the quality of the weld joints[7]. The material suffers severe plastic deformation at elevated temperature, which produces ultrafine and equiaxed grains which further produces welds of excellent mechanical properties[8].

FSW is an extremely complex process which involves parallel physical phenomena like heavy plastic deformation, turbulent material mixing, mechanical plunging & stirring, surface interaction between the tool and the work piece, dynamic heat generation and microstructural alteration. In order to enhance the precision in understandingthe mechanics and thermodynamics of the process and to minimize the experimental work, various computational frameworks are adopted. The perfect utilization of these modeling techniques not only saves the time of the research, but also provides a precise & deep knowledge about the investigation to the researchers. Finite Element Method (FEM) is the most prominent & promising computational simulation techniques utilized by the researchers for the modeling of FSW. Various other modeling techniques have been found in the past literature like Bjorneklett et al. proposed a heat flow model which utilizes both the dynamic thermal chemistry and diffusion theory for the alloy design and optimization of welding conditions for Al-Zn-Mg extrusions[9]. Frigaard et al. has proposed a three-dimensional (3-D) mathematical heat movement model of FSW for calculation of the microstructure evolution and hardness distribution[10]. Dong et al. has developed distinguished numerical models for coupled frictional heat generation; development of plastic flow slip zone; and three-dimensional heat and material movement in order to understand different characteristics of the complex thermomechanical phenomena'sinvolved in friction stir welding[11]. Chao et al. has developed three-dimensional nonlinear finite element welding simulation code—WELDSIM, for the investigation of the influence of temperature-dependent material properties on the time dependent temperature, residual stress and distortion associated with the welding process[12]. Chen et al. has studied the temperature-time history and thermomechanics involved in the similar welding of aluminum alloy 6061-T6 in butt configuration by developing a three-dimensional finite element model [13]. Schmidt et al. has developed an analytical model for heat generation of FSW. He has assumed different contact status between the rotating tool surface and the workpiece and studied the mechanism behind the contact conditions as sliding, sticking or combination of both [14]. Dickerson et al. has reported about the 10% loss of the total heat generated into the tool at steady state, by developing a transient thermal model. He has suggested that the presence of grooves on the tool shank increases the thermal efficiency and stability of the welding process[15]. Buffa et al. has developed a numerical model to study the effect of tool geometry in deciding the microstructure of heat affected zone and nugget zone. He has also proposed a thermomechanical viscoplastic model to examine the flow pattern of material and the grain size of the welded zone[16]. In addition to this, Buffa has also proposed a 3D continuum rigid-viscoplastic FE model which determines the strain and temperature distribution in heat affected and weld zone. The model clearly calculates the non-symmetricity of the FSW technology and correlates the process parameters with the plunge force[17]. Soundararajan et al. has loaded a mechanical load on tool and implemented a uniform contact conductance in the proposed finite element thermo-mechanical model for evaluating the stress at the interface of workpiece and backplate[18]. Uyyuru et al. has proposed a nonlinear finite element code DEFORM, which investigates the temperature division, strain rate and, residual stress across the weld zone in terms of material flow[19].

In last 20 years, a bulk of investigation has been carried out in the field of simulation of welding process among which, most of them are predominantly analytical or numerical simulations. Only a few of the literatures are available [20][13][18][17] which has reported the practical and much more realistic simulation of the process. This is because of the difficulty in complete mapping of actual and accurate real-time welding onto the simulation platform. Thus, in the present investigation, an attempt has been made to model the FSW process in much more realistic manner and overcome the formerdifficulties of simulation. In this research work, a 3-D nonlinear thermo-mechanical model is developed, based on finite element method, which not only provides a realistic physical environment of the actual process but also adds quality in process property correlation. The simulation is carried out in ANSYS APDL using Coupled Field element, both for tool and workpiece. The reason behind the usage of ANSYS simulation package is that the temperature outputs can be determined at each desired time stage. The model is utilized in evaluating the temperature contours, residual stresses and strains generated during FSW at different input parameters. Because this transient temperature field affects the stress distribution, thus an attempt to correlate the influence of varying process parameters with the simulated temperature-time history and the residual stresses has been made. The inverse relation between hardness and strain is utilized and longitudinal strains evaluated from FEM simulation is correlated with the microhardness data from experiment, which further helps in validating the proposed FE model. The successful validation of the proposed 3D thermo-mechanical model establishes a processing door, within which welds of superior quality can be produced, not only for similar friction stir welding but also for dissimilar joining.

3. MATERIALS AND METHODS

The present work utilizes both experimental and computational approach. Detailed discussion of each approach is provided below.

3.1 Experimental Details

Two AA5083 plates, each having dimensions of 150X60X5 mm^3 are welded in butt configuration. A dedicated upright milling machine is used for FSW of the plates. The chemical composition of the as-received cold rolled plates is shown in Table 1, while, the mechanical properties are presented in Table 2.

Table 1. CHEMICAL COMPOSITION OF AA 5083 IN WEIGHT %

Mg	Mn	Cr	Si	Fe	Cu	Al	Others
4.67	0.70	0.12	0.04	0.02	0.08	95.99	Remaining

The plates are arranged in butt configuration and were strongly clamped upon the steel backing plate that was itself sited upon the traverse table of the milling machine.

Table 2. MECHANICAL PROPERTIES OF AA 5083

Temper Designation	0.2% Proof Stress (MPa)	UTS (MPa)	Elongation (%)	Hardness (Hv)
O (Annealed)	125-200	275-350	14	130

The tool used is made of AISI 310 Austenitic Grade Stainless Steel with two different specifications which is mentioned in the Table 3.

Table 3. TOOLS DESIGNATION AND SPECIFICATIONS

Tool Number	D	Tool Pin			
		Geometry	p	d_t	d_r
T1	27	Conical Threaded	3	5	7
T2	27.19	Conical Threaded	3	5.8	7.2

The inner face of the shoulder is not horizontal but is inclined by 3 deg to create a conical surface. The direction of welding is similar to the rolling direction of the plates and in every case no tilt degree is provided to the tool.A K-Type thermocouple is mounted on the top surface of the plates using punch and hammer. Thermocouple is placed on retreating side at 75 mm from one end and at 20 mm from the weld centerline (as shown in Figure 1). The thermocouples are so mounted that it will not be affected by the tool travel. The thermocouple senses the temperature and the data are fed into a data logger. In this work the tool rotation and welding speeds were varied along with the tool geometry.

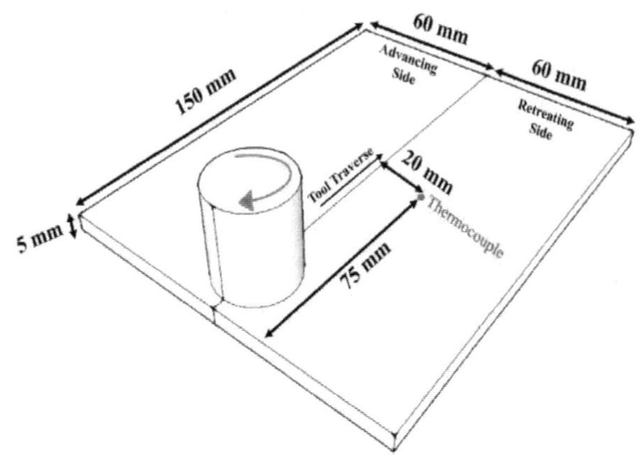

Figure 1. SCHEMATIC REPRESENTATION OF WELDING SETUP OF FSW

Welding trials were performed at two level of tool rotation rates (710 & 1000 rpm) and two level of traverse speeds (28 & 56 mm/min), thereby creating 8 welds matrix for all possible parameter combination which is shown in Table 4.

Table 4. SAMPLE DESIGNATION AND EXPERIMENT MATRIX

Sample Designation	Tool Designation	Rotational Speed (RPM)	v (mm/min)
S1	T1	1000	28
S2	T1	710	28
S3	T2	1000	28
S4	T2	710	28
S5	T1	1000	56
S6	T1	710	56
S7	T2	1000	56
S8	T2	710	56

A constant plunge force of 5 KN is applied for all the FSW trials. The benefit of this matrix is that it allows the study to be more specific and precise by investigating independently the effect of each parameter upon the weld quality and establish the process parameter-property correlation. The tensile tests trials are done at room temperature on an INSTRON 1342 Servo Hydraulic Material Testing Machine having static and dynamic loading capacity of 0 to 250 KN with a frequency upto 15 Hz & a crosshead velocity of 2.5 mm/min is maintained. All the testing samples were cut with wire cut EDM with required dimensions as per ASTM 18 standards. The schematic picture of the tensile test sample is represented in figure 2(a). The Vickers microhardness tests were performed with Vickers's Hardness Testing Machine having model number OLYMPUS U-PMTVC 4F06790 Japan. The micro hardness is calculated at every 1mm on either side of weld centerline from the center of welded zone to the base metal zone using a 10-g load and dwell time is maintained for 10 seconds. A schematic illustration of the microhardness sample is represented in figure 2(c). Also, EBSD metallographic samples are prepared after electropolishing with A2 electrolyte. The stoichiometric composition for a 1000 ml of the above-mentioned etching solution is as follows: 80 ml Perchloric acid, 90 ml of D.I. Water, 730 ml ethanol and 100 ml of Butoxy Ethanol ($C_6H_{14}O_2$). The temperature of the solution is maintained at -5 to -10℃ and is pumped on to the samples with a voltage of 39 volts for 60 seconds. The schematic diagram of the sample for metallurgical analysis is represented in figure 2(b).

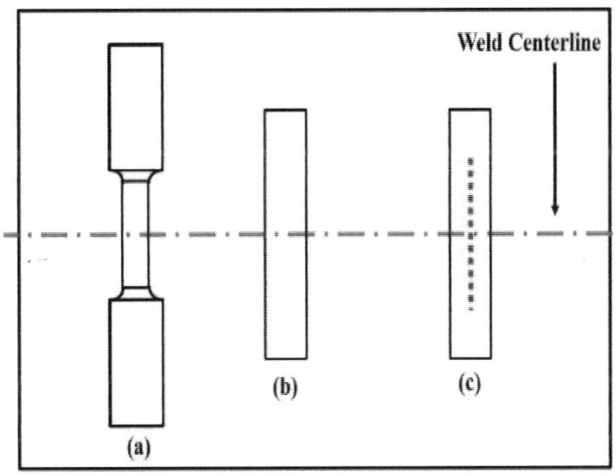

Figure 2. DUMMY ILLUSTRATION OF THE SAMPLE (A) TENSILE TEST SPECIMEN, (B) TEST SAMPLE FOR METALLURGICAL ANALYSIS, (C) SAMPLE FOR MICROHARDNESS.

3.2 Computational Details

A direct thermo-coupled analysis is performed and instead of using a moving heat source, a rotating tool with a constant traverse velocity is used in order to make the simulation more genuine and accurate.

Table 5. TEMPERATURE DEPENDENT MATERIAL PROPERTIES FOR WORKPIECE MATERIAL AA5083 USED IN THERMO-MECHANICAL MODEL [20]

Temperature (°C)	K (W/m °C)	Cp (J/Kg m³)	ρ (Kg/ m³)
20	112.5	942.1	2673.9
80	122.7	984.2	2642.7
180	131.6	1039.6	2629.4
280	142.3	1081.2	2611.5
380	152.5	1136.6	2589.3
480	159.5	1178.2	2567
580	177.2	1261.4	2549.2

The assumptions made while defining the input conditions and creating geometry during simulation are (i) the frictional coefficient is assumed to be constant with a value of 0.5 at the interface of tool and workpiece[10]; (ii) the geometry of tool pin is assumed to be straight cylindrical; (iii) the heat loss due to radiation is neglected as it is comparatively less than the conductive and convective losses. The modeling is done in two parts (A) Workpiece and tool modeling, and the other is (B) Contact Modeling. Both the parts are described separately below. The temperature reliant material properties (specific heat, thermal conductivity and density) used in the model, both for workpiece and the tool is shown on the table 5 & 6 respectively. The workpiece and tool are considered to have isotropic material properties.

Table 6. MATERIAL PROPERTIES FOR TOOL MATERIAL AISI 310 AUSTENITIC GRADE STAINLESS STEEL USED IN THERMO-MECHANICAL MODEL

E (GPa)	μ	K (W/m °C)	C_p (J/Kgm³)	ρ (g/cm³)	α (10⁻⁶/°C)
520	0.22	20	600	7800	17

(A) Workpiece & tool modeling: The two plates of same dimensions as that used in experimental trials are modeled by creating two rectangle volumes. The tool dimensions are also like the one used during experimental trials. The tool height measures the same dimension as that of the shoulder diameter. A coupled-field element SOLID226, is used for modeling of both the workpiece plates and the tool. Since the model primarily focuses on the generation of heat and temperature contours at the vicinity of weld centerline, therefore instead of using tetrahedral mesh, a fine hexahedral mesh is used which eliminates the problem of nonphysical thermal oscillations because of dependent mesh orientation [21].

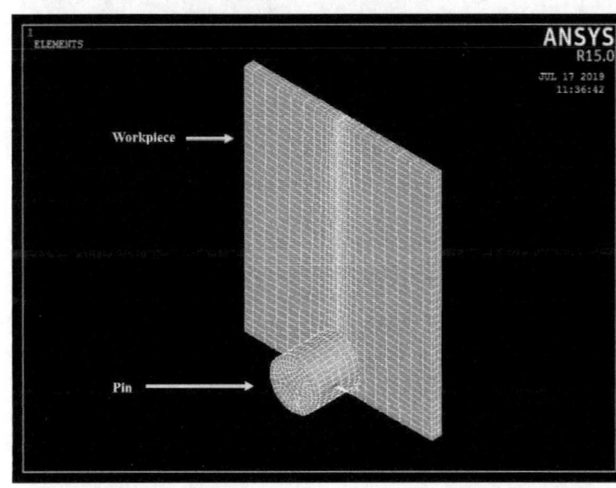

Figure 3. WORKPIECE AND TOOL MODELING

(B) Contact modeling: Two contact pairs are provided in the model. First is the contact between the plates and the next is the contact established at the interface of tool and workpiece. Moreover, to this, a rigid surface constraint is provided for establishing the contact between the top surface of the tool and the pilot node created at the top surface of the tool.

(i) Contact pair for plates: A usual surface-to-surface contact pair is setup between the plates using the elements TARGE170 and CONTA174 (shown in Figure 4), as the problem is highly associated with geometrical, material, and contact nonlinearities. A large thermal contact conductance (TCC) with a value of 2e6 W/m² °C is defined[22] to establish a strong thermal contact between the plates and simulate continuous bonding between them.

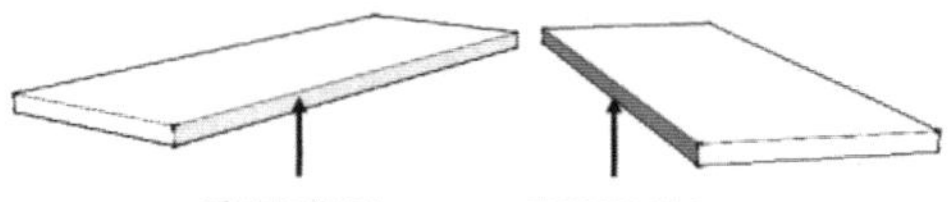

Figure 4. SCHEMATIC ILLUSTRATION OF CONTACT PAIR MAINTAINED BETWEEN THE PLATES

A bonding temperature of 300 °C is provided, since the peak temperature lies in the range of 60 to 90 percent of the melting temperature of the workpiece material[23].

(ii) Contact pair for tool and workpiece: A normal surface-to-surface contact pair is setup to establish connection between the tool and workpiece. CONTA174 element is used for modeling of top surface of the workpiece and the portion of the tool which will establish contact with the top surface of the workpiece is modeled using TARGE170 element (as shown in this Figure 5). This contact pair is specified with two real constants. The first real constant is the FHTG, which is assigned a value of 1, so that there is 100% or complete conversion of frictional energy into heat.

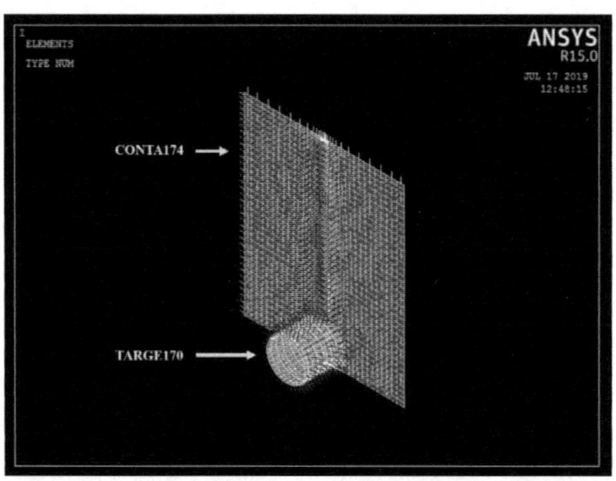

Figure 5. SCHEMATIC ILLUSTRATION OF CONTACT PAIR ESTABLISHED BETWEEN TOOL AND WORKPIECE

The next real constant is FWGT which is assigned with fractional value of 0.95. This implies that 95% of the frictional heat is transferred into the workpiece and only 5% of the heat flows into the tool[21]. This contact pair is assigned with a low thermal contact conductance (TCC) value of 10 W/m^2 °C because most of the frictional heat is transferred on to the workpiece. In addition to this, coefficient of friction 0.5 is defined for this contact pair[10].

(iii) Rigid constraint for pilot node: A pilot node of TARGE170 element is created and is linked at the center of the top surface of the tool. This pilot node controls the entire motion of the tool. This pilot node is rigidly bonded with the nodes of top surface of the tool, which is made of CONTA174 element.

(C) Boundary Conditions:

(i) Thermal Boundary Conditions: It is assumed that the heat loss not only takes place from convection, but also from radiation. It is assumed that the heat loss due to radiation is neglected as it is noticeably less as compared to the conductive and convective losses. Other than this convective heat transfer coefficient values are assigned, in order to account the heat losses due to convection. For all external surfaces other than, the bottom surface of workpiece, the value of the convective heat transfer coefficient is 30 W/m^{2}°C and for the bottom surface of workpiece high heat-transfer coefficient (about 10 times the convective coefficient) of 350 W/m^2 °C is assumed [24]. This is so because, the workpiece is always placed upon the backing plate and backing plates have substantial effect on the thermal distribution of workpiece. Thus, to account the loss of heat from the backing plate, the lowermost surface of workpiece is assigned with high thermal conductivity value. Apart from this, a uniform temperature of 25 °C is assigned as initial temperature boundary condition to the entire model.

(ii) Mechanical Boundary Conditions: The entire simulation is split into three load steps, each of which is a replica of the three steps during the experimental FSW process as plunge, dwell, and traverse. Workpiece is kept fixed in all the load steps of simulation for which the bottom of the workpiece is restricted with motion along z direction. With the aim to introduce the effect of rigid clamping of the workpiece, the sides of the workpiece are restricted with motion along x, y & z direction. The transient analysis is done with large deformation effect in addition to ramped loading. The maximum time step is restricted to 0.03, in order to save the simulation time.

4. RESUTS AND DISCUSSION:

4.1 Temperature-Time History: The figure 6 represents the temperature vs. time curve obtained from the experimental welding trials.

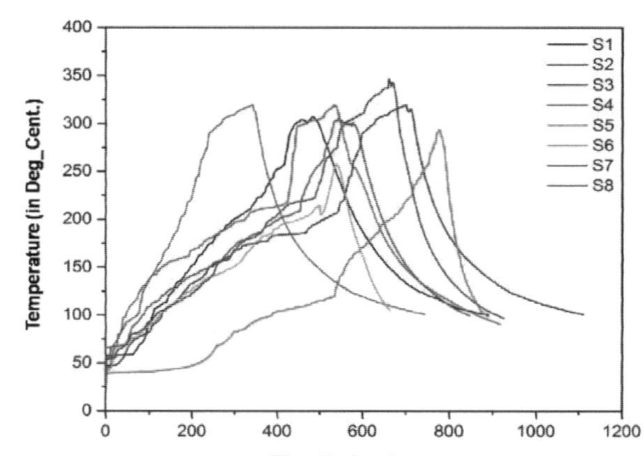

Figure 6. TEMPERATURE-TIME HISTORY OF WELDED SAMPLE FROM EXPERIMENTAL TRIALS

The peak temperature raised during welding is one of the most important factors which governs the evolution of microstructure & mechanical properties & thus, eventually the quality of the welds. It is undoubtedly visible from the temperature contours on figure 6 that the peak temperature in every experimental trial is lower than the melting temperature of the alloy i.e., 630 °C. This result gives the strong evidence of the FSW as a solid-state joining process. In contrast to the experimental results, the figure 7 shows the simulated temperature time history. The experimental results hold decent corroboration with the results originated from the simulation (in figure 7). It is clearly evident from the figure 7 that the maximum temperature of the sample S1 with 1000 rpm is much higher than the peak temperature of sample S4 with 710 rpm. Similar observation was recorded in the case of experimental results (figure 6) of the same specimen i.e., in S1 and S4.

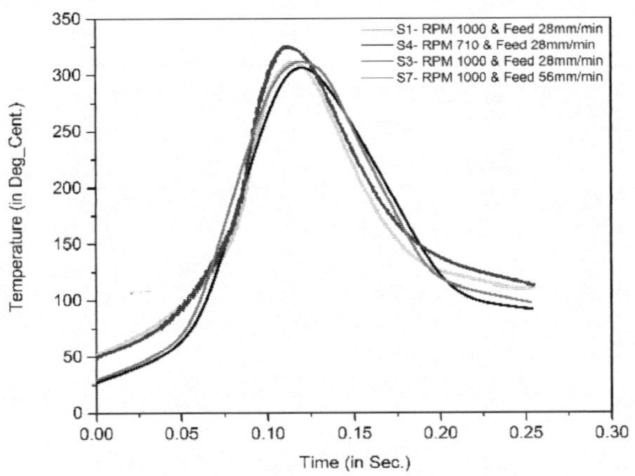

Figure 7. SIMULATED TEMPERATURE-TIME HISTORY OF WELDED SAMPLES OBTAINED FROM THERMO-MECHANICAL MODEL

4.1.1 Influence of rotational speed (RPM) on the Temperature-Time History: On varying only the rotational speed of the tool and keeping the welding speed rate (28 mm/min) constant, it is obviously evident from the temperature time profile on figure 6 that the highest temperature of sample S1 is 307.25 °C and the peak temperature of sample S4 is 293.691 °C. The increased temperature of sample S1 as compared to sample S4 is mainly because of high frictional heat generation. Large power input causes the faster rotation of tool which causes high frictional heat generation which is the main cause of high temperature of sample S1 as compared to S4. The table 7 presents the detail comparison of experimental and simulated temperature profiles showing the influence of rotational speed in the evolution of temperature profile. The deviation in the profiles is accounted in an appreciable range of less than 4% (in Table 7). Such a small deviation significantly indicates that the simulation model closely represents the real experimental situation of FSW. Thus, the model has been accepted for subsequent analysis of process-structure & property correlation.

Table 7. EFFECT OF TOOL ROTATIONAL SPEED ON PEAK TEMPERATURE, ON KEEPING THE WELDING SPEED RATE CONSTANT

Sample	RPM	v (mm/min)	T_E^{max} (°C)	T_S^{max} (°C)	Δ (%)
S1	1000	28	307.205	317.67	3.40
S4	710	28	293.721	304.508	3.67

4.1.2 Influence of welding velocity (Welding speed) on the Temperature-Time History: Table 8 shows the comparison of experimental and simulation results in terms of welding speed on the thermal behavior of the joints. On varying the welding speed only and keeping the tool rotational speed constant (1000 RPM), it is found that the peak temperature of sample S7 with 56 mm/min of welding speed is 307.205 °C and the peak temperature of sample S3 with 28 mm/min of welding speed is 319.271 °C. It is noticeable that the welding speed mainly affects the movement of the plasticized material from the front to rear side of the rotating pin. The fast movement of the tool may cause improper disposal of molten material leading to the welding defect and inferiors the joint performance. As the tool moves faster, the time available to transfer the heat is less and hence lower the temperature is evolved. In this event also, the simulation results (shown in figure 7) holds a decent correlation with the temperature outcomes attained from experimental trials. It can be clearly evident from the figure 7 that the peak temperature of the sample S3 with 28 mm/min of welding speed is much higher than the peak temperature of sample S7 with 56 mm/min of welding speed. The error percentages are accounted less than 5% (shown in Table 8) which further supports the acceptance of the proposed model.

Table 8. EFFECT OF WELDING SPEED RATE ON PEAK TEMPERATURE, ON KEEPING THE TOOL ROTATIONAL SPEED (RPM) CONSTANT

Sample	RPM	v (mm/min)	T_E^{max} (°C)	T_S^{max} (°C)	Δ(%)
S3	1000	28	319.271	328.38	2.85
S7	1000	56	307.205	311.67	4.45

4.2 Tensile Properties: The engineering stress- strain curve of all 8 welds are plotted in Figure 8. On comparing with the tensile curve of the base metal, it is found that, in all cases, the

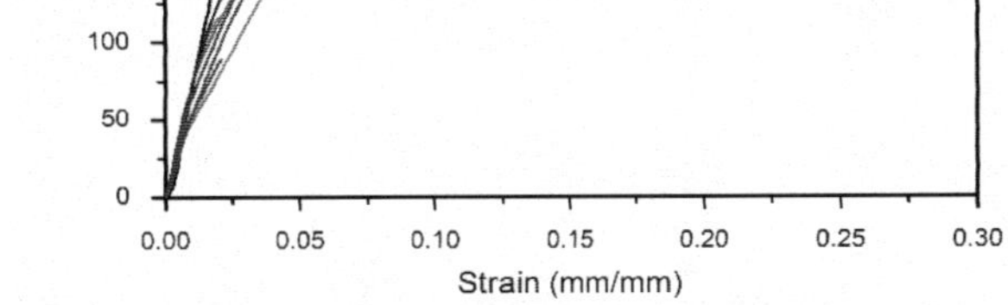

Figure 8. STRESS-STRAIN CURVE OF WELDED SAMPLES

tensile curves of samples lie below the tensile curve of the parent metal which strongly concludes that joint strength of all the fabricated weld butt joints is significantly lower than the strength of the base metal. The joint efficiency and the tensile strength of all the welded samples are summarized in the Table 9 and it is observed that the joint efficiency is calculated with a maximum value of 60.28% in case of sample S8 and a minimum value of 35.22% for sample S3, with no defect was recorded in the joint. However, in some cases very poor joint efficiency has been recorded (Sample S6 & S7) which might be because of the development of worm hole kind of defect, which deteriorates the mechanical properties.

Table 9. JOINT EFFICIENCY AND TENSILE STRENGTH OF THE WELDED SAMPLES

SNo.	Sample	UTS (MPa)	Welding Efficiency (%)
01	S1	173.38	48.95
02	S2	153.38	43.30
03	S3	124.74	35.22
04	S4	209.98	59.28
05	S5	146.75	41.43
06	S6	52.22	14.75
07	S7	93.76	26.47
08	S8	213.45	60.26
09	Base Metal	354.18	100.00

4.2.1 Influence of tool rotational speed on tensile properties: Figure 9 represents the effect of rotation speed on the weld property of the FSW joint. Rotational speed is responsible for the heat generation as well as stirring and mixing of the material at the welds zone. The figure clearly demonstrates that the tensile strength of the weld joints is not only affected by the tool rotational speed but the tool shoulder diameter also plays a vital role in deciding the mechanical properties.

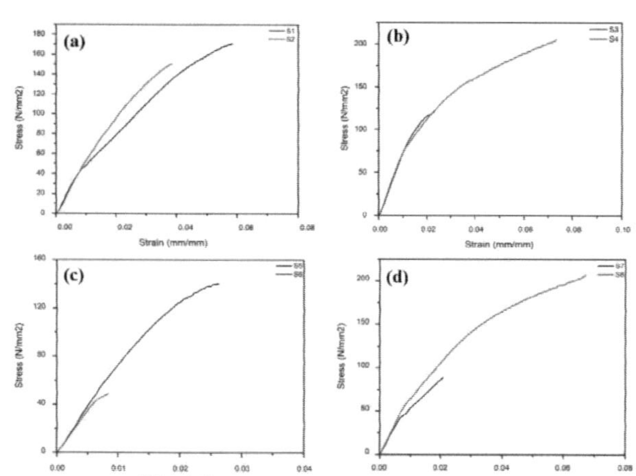

Figure 9. EFFECT OF TOOL ROTATIONAL SPEED IN THE ASSESSMENT OF ENGINEERING STRESS- STRAIN CURVE (A) SAMPLE S1 & S2 WELDED WITH TOOL T1 AT 710 RPM AND 1000 RPM RESPECTIVELY WITH 28 MM/MIN OF FEED RATE (B) SAMPLE S3 & S4 WELDED WITH TOOL T2 AT 710 RPM AND 1000 RPM RESPECTIVELY WITH 28 MM/MIN OF FEED RATE, (C) SAMPLE S5 & S6 WELDED WITH TOOL T1 AT 710 RPM AND 1000 RPM RESPECTIVELY WITH 56 MM/MIN OF FEED RATE, (D) SAMPLE S7 & S8 WELDED WITH TOOL T2 AT 710 RPM AND 1000 RPM RESPECTIVELY WITH 56 MM/MIN OF FEED RATE.

With the increase in rotational speed (RPM), the tensile strength increases for the samples S1 (shown in figure 9(a)) & S5 (shown in figure 9(c)) which are welded by tool T1. But the reverse relation is noticed in case of samples S3 (shown in figure 9 (b)) & S7 (shown in figure 9(d)) for which tool T2 is used for their welding. This is because, large shoulder diameter in combination with high rotational speed produces large amount of heat which results into recrystallization and grain growth which ultimately hinders the refinement of the grains and thereby lowers down the strength of the weld zone.

4.2.2 Influence of welding speed on tensile properties: Figure 10 shows the effect of welding speed on the tensile behavior of the joints. It is clearly evident from the figure that samples S1, S2, S3 & S4 with lower welding speed produces joints of higher tensile strengths as compared to the samples S5, S6, S7, S8 with higher welding speed and lower value of tensile strength.

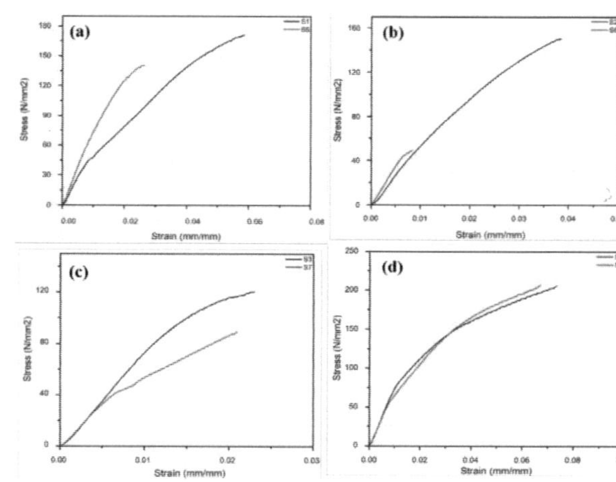

Figure 10. EFFECT OF WELDING SPEED ON THE ENGINEERING STRESS- STRAIN PLOTS (A) SAMPLE S1 & S5 WELDED WITH TOOL T1 AT 710 RPM AND FEED RATE OF 28 MM/MIN AND 56 MM/MIN RESPECTIVELY, (B) SAMPLE S2 & S6 WELDED WITH T1 AT 1000 RPM AND FEED RATE OF 28 MM/MIN AND 56 MM/MIN RESPECTIVELY , (C) SAMPLE S3 & S7 WELDED WITH T2 AT 710 RPM AND FEED RATE OF 28 MM/MIN AND 56 MM/MIN RESPECTIVELY, (D) SAMPLE S4 & S8 WELDED WITH T2 AT 1000 RPM AND FEED RATE OF 28 MM/MIN AND 56 MM/MIN RESPECTIVELY.

This is so because welding speed is responsible for the movement of the molten material from the front to back of the rotating pin. The faster the tool moves forward the dumping of molten material becomes improper. The welding speed also influences the heat generated from the tool movement. Thus, we can say that the faster the tool moves the lower the temperature is evolved. If the forward motion is too slow then there will be over heating leading to worm holes and grain coarsening which results in lowering of strength.

4.3 Microstructure and microhardness profile: The figure 11(a) shows the variation of microhardness with transverse distance from weld centerline recorded experimentally while figure 11(b) shows the variation of strain obtained from simulation model as against the transverse distance. Figure 11(a) clearly indicate the work hardened state of the parent metal with a hardness value of around 125-135 VHN as against the usual range of hardness of 95-100 VHN under annealed condition in AA5083 alloys. It can be noticed that lower hardness value lies in the region near to the weld centerline.

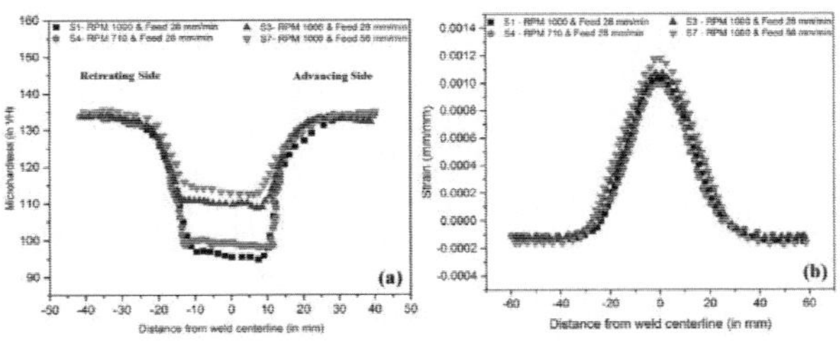

Figure 11. (A) MICROHARDNESS VARIATION AS A FUNCTION OF TRANSVERSE DISTANCE FROM WELD CENTERLINE OBTAINED FROM EXPERIMENT; (B) STRAIN VARIATION AS A FUNCTION OF TRANSVERSE DISTANCE FROM WELD CENTERLINE OBTAINED FROM THERMO-MECHANICAL MODEL.

Not only this, the region of lower hardness value is almost limited within 10-15mm on either side of weld centerline. In all the specimens, the hardness values of the nugget zone (NZ) has been recorded in the range of 90-100 VH which is approximately 20-25% lower than the hardness value range of base metal (BM) which is of about 130-135 VH. This results holds strong agreement with the previously reported work in this alloy[25] in terms of the variation, breadth of profile and especially the region of lower hardness value. It is also evident from figure 11(a) that both the parameters, i.e. tool rotational speed and welding speed have significant impact upon the plateau of the hardness profile. It is clearly observed that with higher welding speed, the hardness value increases. This is because, higher welding speed produces much more finer grains which results in the increase in hardness values [26][25]. Also, the tool rotational speed is in inverse relation with the hardness values. This is so because, higher rotational speed not only causes the refinement of the grains, but the excess energy left after the refinement process promotes the grain growth. Thus, the hardness value falls with increase in the tool rotational speed. It can also be noted that the profile is somewhat asymmetric. The hardness values measured on the retreating side is around 5-6% higher than that of the measured hardness values on advancing side (from figure 11(a)). This is so because, retreating side experiences higher thermal gradient which causes a considerable increase in the transverse residual stresses on that side as compared to the residual stresses on the other side. Figure 11(b) shows the simulated strains generated along the transverse distance from the weld centerline during the FSW process. Since, the highest temperature and the strain occur at the nugget regions and in turn resulted into attainment of maximum grain refinements in this zone. Therefore, the expected microstructure of this region is refined equiaxed grains with low dislocation density which shows the lower hardness than that of the elongated restrained base metal region. Nevertheless, thus the simulated strain profile well correlates with the hardness profile.

The variation of microhardness discussed above is well corroboratingwith the microstructure evolved from the EBSD study (shown in figure 12). The softened NZ and the unaffected BM zone is the main region of interest, with the aim to correlate the microhardness values and the microstructure. The microstructure is originated from the EBSD mapping of the BM zone, the HAZ and the NZ represented in the figure12(a), (b), and (c) respectively. The image clearly indicates that the NZ of the weld consists of small and highly dense equiaxed grains of order 8-10 μm. The region HAZ, which is only influenced by the thermal flux, has grains of order 18-28 μm while the unaffected BM zone has more coarse grains of order 35-50 μm. These values are in strong agreement with the results obtained by H Larsson et.al [27]. Since, it's an alloy of 5XXX series, so main contribution on its hardness is from rolling or work hardening. This work hardening on the region of parent metal dominates over the hardness due to FSW phenomenon in the region of nugget zone.Figure 12 (a) & (c) resembles the band contrast image of the unaffected BM zone and the NZ respectively. The high deformation of the weld zone can be clearly explained from the grain boundary miss-orientation map shown on figure 12 (e).

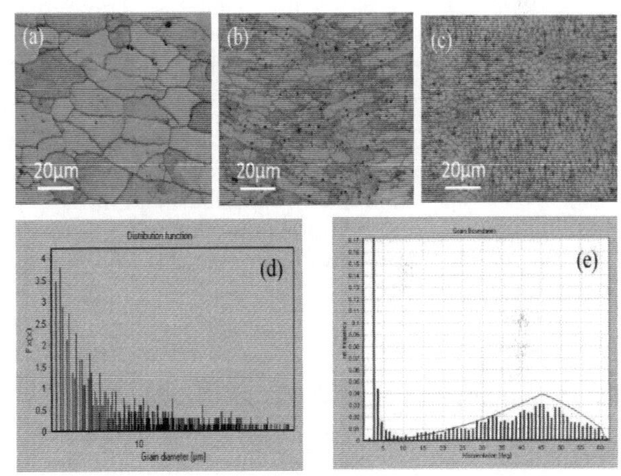

Figure 12. EBSD MAPPING OF SAMPLE S7; BAND CONTRAST IMAGE OF (A) BASE METAL ZONE (BMZ); (B) HEAT AFFECTED ZONE (HAZ); (C) NUGGET ZONE (NZ); (D) GRAIN SIZE DISTRIBUTION MAP MEASURED FROM WELD CENTERLINE; (E) GRAIN BOUNDARY MISS-ORIENTATION DISTRIBUTION MAP MEASURED FROM WELD CENTER

It is worthy to note, this analysis of miss-orientation map corroborates the simulated strain profile (figure 11(b)) which indicate that the nugget zone is the most potential region for dynamic recrystallization to occur and explain the miss-orientation of grains at this region. A statistical distribution of the grain boundary miss-orientation histogram provides a strong evidence of the high deformation of the NZ. It is clearly visible that the concentration of the lower angled grain boundaries (LAGBs) is higher near the NZ and as we go away from the weld centerline, this concentration diminishes. Correspondingly, the concentration of the higher angled grain boundaries (HAGBs) starts increasing and the concentration of HAGBs reach to a peak value near the base metal region. The

LAGBs resembles the high deformed status of the material whereas the HAGBs shows low deformed status of the material. Not only this, the grain size distribution histogram (shown in figure 12 (d)) is also in good agreement with the miss-orientation distribution plots (shown in figure 12 (e)), meanwhile the grains in smaller in size are comparatively in high deformed state which is because of excessive refinement of the grain[28]. The large concentration of the fine equiaxed grains inhibits the alignment of the grains. This results in increase in grains with lower value of orientation or low angle grain boundaries (LAGB). It can be noted from the images, that FSW process has altered the microstructure drastically. The heavily worked grains in the province of parent metal are completely replaced by the ultrafine and equiaxed grains in the region of NZ. The microstructure resembles a good agreement with the previous research reported on the FSW of aluminium alloy of 5XXX, 6XXX and 7XXX series[31][32].

Previous research reports that the deformation induced in the material during FSW causes dynamic recrystallization and that deformation is confined to a narrow region of not more than 5 mm[33][34]. In the present discussion of microhardness, it is clearly evident from the figure11(a) that the recrystallized region or the region of lower value of microhardness is extended to a distance of 10mm. The microstructural changes on the region other than the NZ is only from the heat generated because of the friction between tool and workpiece. The extremely work hardened grains of parent material alters its stability and when these unstable grains are coupled with the thermal fluxes, the formation of new equiaxed grains occurs.

4.4 Residual Stresses: The variation of residual stress on either side of weld centerline is plotted in figure 13 along the X-X line on the simulation results, shown in figure 14, 15, 16 & 17, obtained from the proposed thermo-mechanical FE model. The generated profile are in-general in decent agreement with the those found in the literature[35] in terms of the position and breadth of NZ and the region of tensile stresses. Figure 13(a) illustrates that the magnitude of maximum longitudinal residual stress is around 60-80% higher than the magnitude of transverse residual stresses (shown in figure 13(b)). The positive tensile stresses are within the region of 25mm, which is in accordance with the shoulder diameter. This is mainly because of the action of high shear forces on the peripheral region of tool shoulder.

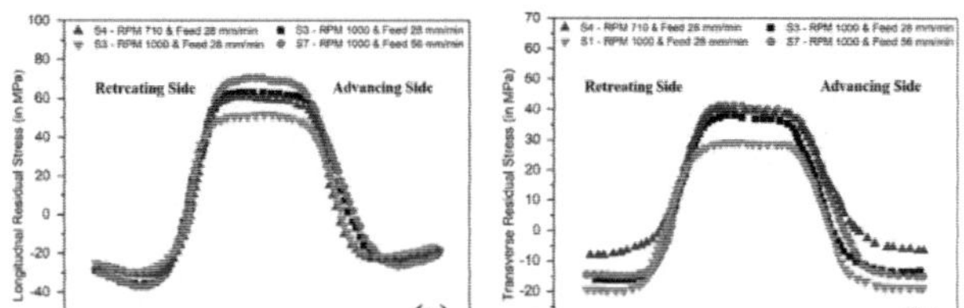

Figure 13. VARIATION OF (A) LONGITUDINAL AND, (B) TRANSVERSE RESIDUAL STRESS WITH TRANSVERSE DISTANCE ON EITHER SIDE OF WELD CENTERLINE OBTAINED FROM THERMO-MECHANICAL MODEL

Not only this, but this tensile stress regions are also because of the action of the plunge force is in the axial path of the tool. The peak tensile stresses are acting at the vicinity of the tool shoulder are also reported in the previous journal[36]. The tensile stress within the NZ is not constant and varies within the region of 7-8mm on either side of weld centerline. This variation might be because of the penetration of the tool pin which induces plastic flow and this is how the thermo-mechanical phenomena linked with the FSW process is correlated with the residual stress data so generated from the proposed thermo-mechanical FE model. As we travel apart from the weld centerline, these tensile stresses at the vicinity of NZ, are normalized by the compressive stresses generated near the base metal. The positive stresses cause the expansion of the material and at the same time this expansion is restricted by the fixtures provided at the end of the weld plates. It is clearly visible from the simulation results (represented in Figure 14(a), 15(a), 16(a) & 17(a)) that most part of the parent metal is under the region of compression or negative stresses and this negativity increases near the end of plate (transverse distance away from weld centerline). These high compressive stresses are because of the presence of fixtures. These fixtures restrict the propagation of stresses which results in accumulation of large negative stresses in the region apart from weld centerline. It can be mentioned from the simulation results that the transverse residual stresses (shown in figure 14(b), 15(b), 16(b), 17(b)) spreads in larger area as compared to the longitudinal residual stresses (shown in figure 14(a), 15(a), 16(a), 17(a)). This is because the longitudinal residual stress depends mainly upon the tool movement or position of the tool. The tool is not static. It is moving along the weld centerline and continuously changing its coordinates. The tool movement is responsible for the creation of the stress along the weld line (i.e. longitudinal direction). The positive stress causes the expansion of the material. When this expansion takes place, the time available for the propagation of the stress wave-front is very less. As soon as the stress wave-front begins to propagate, the tool proceeds and takes the new position where the former propagated wave-front has reached. Thus, the propagated wave-front disappears and a new stress wave-front of new tool position begins. The region behind the tool is not constrained and thus the stresses so generated are free for propagation. Hence the stress propagates more and this is the foremost reason behind the large spreading of the transverse residual stress than the longitudinal ones.

4.4.1 Influence of welding speed on residual stresses: Figure 14, 15 and the graph plotted on figure 13, shows the effect of welding speed on the longitudinal and transverse residual stresses. It is clearly evident that the welding speed is in direct relation with the residual stress i.e. higher welding speed induces higher residual stresses. Sample S7 with welding speed 56 mm/min is having higher longitudinal stress of 68.9 MPa (shown in Figure 14(a)) as compared to the sample S3 which has lower welding speed of 28 mm/min, resembling a peak longitudinal stress value of 61.3 MPa (shown in Figure 15(a)). Not only longitudinal residual stress, but transverse residual stress also shows the similar behavior. Sample S7 with welding speed 56 mm/min is having higher transverse residual stress of 35.3 MPa (shown in Figure 14(b)) as compared to the sample S3 which has lower value of welding speed of 28 mm/min, and resembles a peak transverse residual stress value of 38.2 MPa (expressed in Figure 15(b)). This is because of the lower availability of the heat input from faster tool travel which results in inappropriate deposition of the plasticized material.

Time available to transfer the available heat is low and this restricts the thermal expansion upon cooling which is the main cause of lowering of residual stress with higher welding speed.

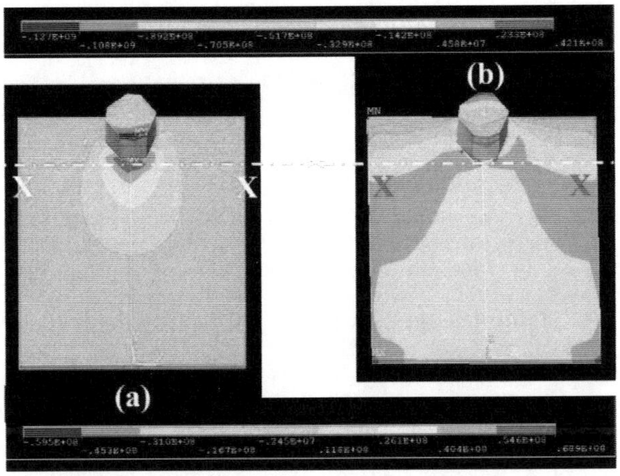

Figure 14. SIMULATED (A) LONGITUDINAL AND, (B) TRANSVERSE RESIDUAL STRESS OF SAMPLE S7 WITH 1000 RPM & 56 MM/MIN OF WELDING SPEED.

Such support in correlation with the heat input is also observed in the previous literature[25][35]. But this trend is not always true, because a higher energy input can also lead to lowering of friction coefficient which causes the improper dumping of the molten material[37].

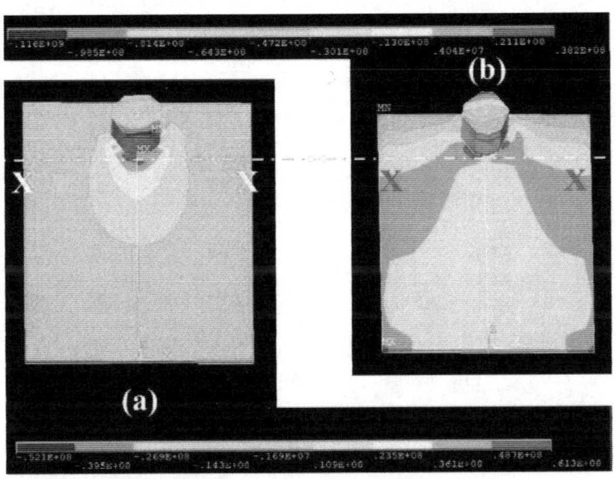

Figure 15. SIMULATED (A) LONGITUDINAL AND, (B) TRANSVERSE RESIDUAL STRESS OF SAMPLE S3 WITH 1000 RPM &28 MM/MIN OF WELDING SPEED

4.4.2 Influence of tool rotational speed (RPM) on residual stresses: From figure 13, 16 & 17 illustrates the marginal effect of tool rotational speed on peak residual stresses. In connection to this, tool rotational speed is somewhat in inverse relation with the tool rotational speed, i.e. increased rotational speed results in degradation in the magnitude of maximum residual stress. Sample S4 with a rotational speed of 710 RPM is having greater longitudinal stress of 60.1 MPa (shown in Figure 16(a)) as compared to the sample S1 which has higher rotational speed of 1000 RPM, resembling a peak longitudinal stress value of 53.8 MPa (shown in Figure 17(a)).

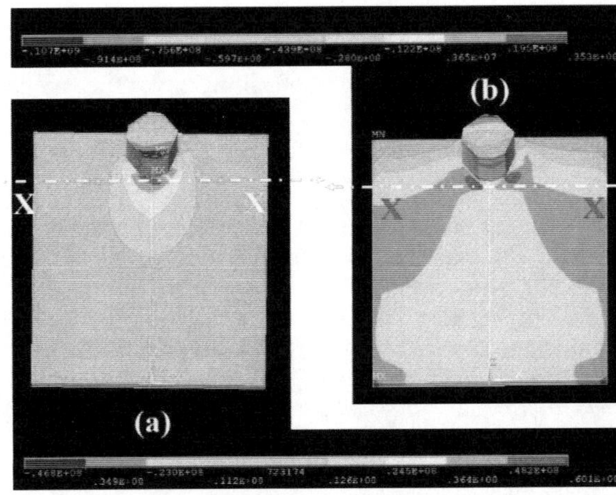

Figure 16. SIMULATED LONGITUDINAL AND, (B) TRANSVERSE RESIDUAL STRESS OF SAMPLE S4 WITH 710 RPM & 28 MM/MIN OF WELDING SPEED

The effect of tool rotational speed is not only limited to longitudinal residual stress, but transverse residual stress also follows the same behavior. Sample S4 with lower rotational speed 710 RPM has higher transverse residual stress of 35.3 MPa (shown in Figure 16(b)) as compared to the sample S1 which has elevated rotational speed of 1000 RPM resembles a peak transverse residual stress value of 30.0 MPa (shown in Figure 17(b)).

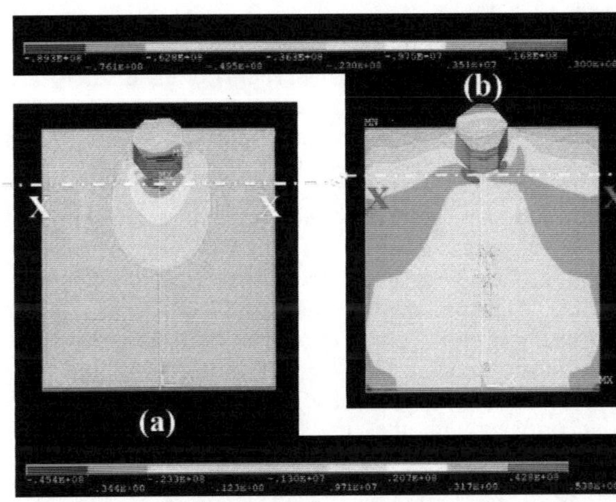

Figure 17. SIMULATED LONGITUDINAL AND, (B) TRANSVERSE RESIDUAL STRESS OF SAMPLE S1 WITH 1000 RPM &28 MM/MIN OF WELDING SPEED

This observation is supported from the statement that, higher rotational speed not only causes the refinement of the grains, but the excess energy left after the refinement process is utilized in the grain growth phenomenon. Thus, there is a slight decrease in the residual stress value with the rise in tool rotational speed.

5. CONCUSIONS:

1. The peak temperature generated in temperature-time history is in direct relation with the tool rotational speed and in indirect relation with the welding speed. The simulation out comes attained from the proposed thermo-mechanical model holds strong correlation with the peak temperature acquired from the

experimental trials. The deviation percentages are accounted below 5% and this permits the acceptance of the proposed FE model in relationto temperature-time history. The model provides a crystal-clear evidence of the influence of tool rotational speed and the welding speed velocity on the temperature-time history.

2. The tensile results show that joint strength of all the fabricated weld butt joints is significantly lower than the strength of the base metal. The joint efficiency is calculated with a maximum value of 60.28% in case of sample S8 and a minimum value of 35.22% for sample S3, with no defect was recorded in the joint. However, very poor joint efficiency is also recorded for sample S6 & S7 which might be because of defects like worm hole. Tensile strength of the weld joints is affected by both the tool rotational speed and the tool shoulder diameter. The tensile strength is in inverse relation with the welding speed.

3. The micro-hardness values of the NZ lies in the range of 90-100 VH while the hardness value varies in range of 130-135 VH for BMZ. The hardness values on the retreating side are around 5-6% higher that of the advancing side because of higher temperature gradient on the retreating side. The softened NZ is almost confined within 10-15 mm on both sides of weld centerline. The NZ consists of small and highly dense equiaxed grains of order 8-10 μm whereas the HAZ has grains of order 18-28 μm and the unaffected BM zone has more coarse grains of order 35-50 μm.The microstructure in the weld zone is dominated by the fine equiaxed grains which are because of recrystallization. This recrystallization causes the lowering of the hardness value of the NZ than the hardness of the parent metal zone. The grain boundary miss-orientation map and the grain size variation histogram further support the recrystallization of NZ.

4. The maximum residual stress values is in direct relation with the welding speed whereas tool rotation speed has very marginal inverse relation on the peak residual stress. The magnitude of longitudinal residual stress is around 60-80% greater than the magnitude of transverse residual stresses. The positive tensile stresses are at the vicinity of tool shoulder and are confined within the region of 25mm. The parent metal is mostly confined with the compressive residual stresses. In addition to this, the simulation results show that transvers residual stresses spreads in larger area as compared to the longitudinal residual stresses. This is because of free propagation of stress wave front in case of transverse residual stress.

5. The simulation results from the proposed thermo-mechanical model confirms that with the correct implementation of the thermal, mechanical loading and boundary conditions, the proposed model can be successfully used in correlating the simulation results in terms of temperature histories, residual stress and, strain can be useful in optimizing the FSW process.

6. REFERENCES

[1] W.M. Thomas, E.D. Nicholas, J.C. Needham, M.G. Murch, P.Temple-Smith, and C.J. Dawes: Friction Stir Butt Welding. 1991,International Patent No. PCT/GB92/02203.

[2] A. Kouadri-Henni and L. Barrallier, "Mechanical properties, microstructure and crystallographic texture of magnesium AZ91-D alloy welded by friction stir welding (FSW)," *Metall. Mater. Trans. A Phys. Metall. Mater. Sci.*, vol. 45, no. 11, pp. 4983–4996, 2014.

[3] R. Ayer, H. W. Jin, R. R. Mueller, S. Ling, and S. Ford, "Interface structure in a Fe-Ni friction stir welded joint," *Scr. Mater.*, vol. 53, no. 12, pp. 1383–1387, 2005.

[4] A. P. Reynolds, E. Hood, and W. Tang, "Texture in friction stir welds of Timetal 21S," *Scr. Mater.*, vol. 52, no. 6, pp. 491–494, 2005.

[5] H. Jin, S. Saimoto, M. Ball, and P.L. Threadgill: Mater. Sci. Technol.,2001, vol. 17 (12), pp. 1605-14.

[6] S. Mistikoglu, "Recent Developments in Friction Stir Welding of Al-alloys," no. July, 2014.

[7] R. S. Mishra and Z. Y. Ma, "Friction stir welding and processing," *Mater. Sci. Eng. R Reports*, vol. 50, no. 1–2, pp. 1–78, 2005.

[8] K. V. Jata, "Friction Stir Welding of High Strength Aluminum Alloys," *Materials Science Forum*, vol. 331–337. pp. 1701–1712, 2009.

[9] B. I. Bjørneklett, Grong, O. R. Myhr, and A. O. Kluken, "A process model for the heat-affected zone microstructure evolution in Al-Zn-Mg weldments," *Metall. Mater. Trans. A Phys. Metall. Mater. Sci.*, vol. 30, no. 10, pp. 2667–2677, 1999.

[10] Ø. Frigaard, Ø. Grong, and O. T. Midling, "A process model for friction stir welding of age hardening aluminum alloys," *Metall. Mater. Trans. A*, vol. 32, no. 5, pp. 1189–1200, 2001.

[11] P. Dong, F. Lu, J. K. Hong, and Z. Cao, "Coupled thermomechanical analysis of friction stir welding process using simplified models," *Sci. Technol. Weld. Join.*, vol. 6, no. 5, pp. 281–287, 2013.

[12] X. K. Zhu and Y. J. Chao, "Effects of temperature-dependent material properties on welding simulation," *Comput. Struct.*, vol. 80, no. 11, pp. 967–976, 2002.

[13] C. M. Chen and R. Kovacevic, "Finite element modeling of friction stir welding - Thermal and thermomechanical analysis," *Int. J. Mach. Tools Manuf.*, vol. 43, no. 13, pp. 1319–1326, 2003.

[14] H. Schmidt, J. Hattel, and J. Wert, "An analytical model for the heat generation in friction stir welding," *Model. Simul. Mater. Sci. Eng.*, vol. 12, no. 1, pp. 143–157, 2004.

[15] T. Dickerson, Q. Shi, and H. R. Shercliff, "Heat flow into friction stir welding tools," *4th Int. Symp. Frict. stir welding, USA*, no. May, pp. 1–11, 2003.

[16] G. Buffa, J. Hua, R. Shivpuri, and L. Fratini, "Design of the friction stir welding tool using the continuum based FEM model," *Mater. Sci. Eng. A*, vol. 419, no. 1–

2, pp. 381–388, 2006.

[17] G. Buffa, J. Hua, R. Shivpuri, and L. Fratini, "A continuum based fem model for friction stir welding - Model development," *Mater. Sci. Eng. A*, vol. 419, no. 1–2, pp. 389–396, 2006.

[18] V. Soundararajan, S. Zekovic, and R. Kovacevic, "Thermo-mechanical model with adaptive boundary conditions for friction stir welding of Al 6061," *Int. J. Mach. Tools Manuf.*, vol. 45, no. 14, pp. 1577–1587, 2005.

[19] R. K. Uyyuru and S. V. Kailas, "Numerical Analysis of Friction Stir Welding Process," *J. Mater. Eng. Perform.*, vol. 15, no. 5, pp. 505–518, 2006.

[20] X. K. Zhu and Y. J. Chao, "Numerical simulation of transient temperature and residual stresses in friction stir welding of 304L stainless steel," *J. Mater. Process. Technol.*, vol. 146, no. 2, pp. 263–272, 2004.

[21] Y. J. Chao, X. Qi, and W. Tang, "Heat Transfer in Friction Stir Welding—Experimental and Numerical Studies," *J. Manuf. Sci. Eng.*, vol. 125, no. 1, p. 138, 2003.

[22] P. Kumar Thimmaraju, K. Arkanti, and G. Chandra Mohan Reddy, "Parametric Analysis of Friction Stir Welding Process," *Mater. Today Proc.*, vol. 5, no. 13, pp. 26981–26989, 2018.

[23] R. Mohan, N. R. Rajesh, S. K. S, M. Student, M. Engineering, and G. E. C. B. Hill, "Finite element modeling for maximum temperature in friction stir welding of AA1100 and optimization of process parameter by taguchi method," *Int. J. Res. Eng. Technol.*, vol. 3, no. 5, pp. 728–733, 2014.

[24] P. Prasanna, B. S. Rao, and G. K. M. Rao, "Finite element modeling for maximum temperature in friction stir welding and its validation," *Int. J. Adv. Manuf. Technol.*, vol. 51, no. 9–12, pp. 925–933, 2010.

[25] M. J. Peel, A. Steuwer, P. J. Withers, T. Dickerson, Q. Shi, and H. Shercliff, "Dissimilar friction stir welds in AA5083-AA6082. Part I: Process parameter effects on thermal history and weld properties," *Metall. Mater. Trans. A*, vol. 37, no. 7, pp. 2183–2193, 2006.

[26] M. Peel, A. Steuwer, M. Preuss, and P. J. Withers, "Microstructure, mechanical properties and residual stresses as a function of welding speed in aluminium AA5083 friction stir welds," *Acta Mater.*, vol. 51, no. 16, pp. 4791–4801, 2003.

[27] E.D. Nicholas: in ICAA-6: 6th International Conference on AluminiumAlloys, Toyohashi, Japan, 1998, Japan Institute of Light Metals.

[28] M. M. Z. Ahmed, S. Ataya, M. M. El-Sayed Seleman, H. R. Ammar, and E. Ahmed, "Friction stir welding of similar and dissimilar AA7075 and AA5083," *J. Mater. Process. Technol.*, vol. 242, pp. 77–91, 2017.

[29] B. Hammert, K. W. Jacobsent, and V. Milmanf, "Stacking fault energies in aluminium," vol. 10453.

[30] S. Gourdet and F. Montheillet, "An experimental study of the recrystallization mechanism during hot deformation of aluminium," *Mater. Sci. Eng. A*, vol. 283, no. 1–2, pp. 274–288, 2000.

[31] Y. S. Sato and H. Kokawa, "Distribution of tensile property and microstructure in friction stir weld of 6063 aluminum," *Metall. Mater. Trans. A Phys. Metall. Mater. Sci.*, vol. 32, no. 12, pp. 3023–3031, 2001.

[32] H. Jin, S. Saimoto, M. Ball, and P. L. Threadgill, "Characterisation of microstructure and texture in friction stir welded joints of 5754 and 5182 aluminium alloy sheets," *Mater. Sci. Technol.*, vol. 17, no. 12, pp. 1605–1614, 2013.

[33] K.V. Jata and S.L. Semiatin: Scripta Mater., 2000, vol. 43 (8), pp. 743-49.

[34] S. Tanaka and M. Kumagai: in 3rdInternational Symposium on FSW(CD-ROM), Kobe, Japan, 2001, TWI.

[35] H. Lombard, D. G. Hattingh, A. Steuwer, and M. N. James, "Effect of process parameters on the residual stresses in AA5083-H321 friction stir welds," *Mater. Sci. Eng. A*, vol. 501, no. 1–2, pp. 119–124, 2009.

[36] K.K. Masubuchi, Analysis ofWelded Structures, Pergamon Press, Oxford, New York, 1980.

[37] P. A. Colegrove, H. R. Shercliff, R. Zettler, P. A. Colegrove, H. R. Shercliff, and R. Zettler, "Model for predicting heat generation and temperature in friction stir welding from the material properties Model for predicting heat generation and temperature in friction stir welding from the material properties," vol. 1718, 2013.

Proceedings of the 6th International Conference on Ship and Offshore Technology
ICSOT 2019
November 7-8, 2019, IIT Kharagpur, India

ICSOT INDIA 2019-12

EFFECT OF CONDUCTOR FIXITY AT SPIDER DECK & CELLAR DECK LEVEL ON ULTIMATE STRENGTH OF JACKET STRUCTURE

Nitin Nehra
Institute of Engineering and Ocean technology (IEOT)-ONGC
Panvel, Maharashtra, 410221,India
nehra_nitin@ongc.co.in

Praveen Bhat
Institute of Engineering and Ocean technology (IEOT)-ONGC
Panvel, Maharashtra, 410221,India
bhat_praveen@ongc.co.in

ABSTRACT
ONGC operates more than 300 fixed offshore steel platforms in western offshore of India. These Platforms are installed in a water depth ranging from 20m to 90m in western offshore. More than 45% of these platforms have already outlived their design lives but are still required for extended period of production and thus require life extension studies for ensuring safe and continued usage. Many of these platforms have to be analyzed using non-linear ultimate strength analysis while carrying out the life extension studies. IEOT- ONGC has carried out a study on the effect of fixity of conductors at spider deck & cellar deck level on ultimate strength of Jacket structure in terms of Reserved Strength Ratio (RSR) and on Pile Unity Check and axial load. This process will be helpful for documenting the structural integrity of the existing platforms. The study has been carried out on selected three representative unmanned well platforms; A, B & C, installed in different water depths with different structural configurations.
The welding (fixing) of the conductors at spider deck and cellar deck level may be quite useful in some of the platforms in critical condition of overstressing of the piles. This is the first of its kind study for documenting the structural adequacy of the aged platforms for continued usage.

NOMENCLATURE
RSR Reserve Strength Ratio
UC Unity Check

INTRODUCTION
ONGC operates more than 300 fixed offshore steel platforms in western offshore of India, installed in a water depth of 20m to 90m in western offshore. These platforms are designed as per provisions of API RP 2A (WSD)[1]. These platforms support conductor/wells for facilitating the drilling and production of fossil fuels to the surface of Earth. A typical offshore jacket platform and enclosed conductors are presented in Figure 1.

Figure 1. CONDUCTORS IN AN OFFSHORE PLATFORM

These conductors are self-supported bottom founded (boundary condition at mud line is Free Free Fix Free Free Fix) casing structures and are, in general, enclosed within a conductor bay area of offshore platforms, Figure 1. The conductors are placed within conductor guides as shown in Figure 2. Many of the jacket platforms operating in western offshore of India have undergone modifications for additional facilities like clamp-on wells, risers, riser protectors, deck extensions, etc. All these modifications along with change in provisions of design code have resulted in increased loading on the structures. These platform are required to operate for an extended period of life, however, it is important that their structural integrity to be established for continued operations.

(a)

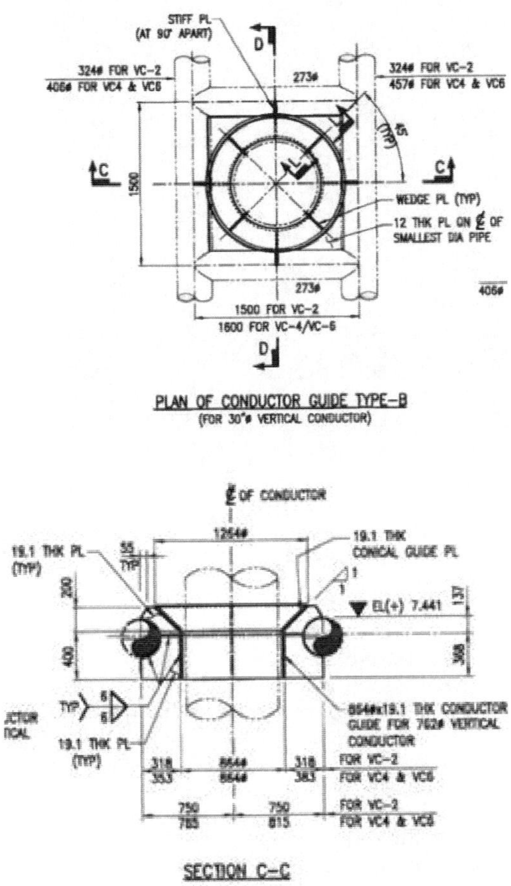

(b)

Figure 2. (a) CONDUCTOR GUIDE AROUND AN OFFSHORE CONDUCTOR (b) CONDUCTOR AND GUIDE SCHEMATIC

The genesis of this paper is in context of exploring the feasibility to use the extra stiffness of conductors for withstanding environmental loads. The conductors are not designed to provide any stiffness to the global structural integrity of the jacket platform however; the lateral loads (due to marine environment) coming on to the conductors are transferred to the main jacket structure (boundary Condition Fix Fix Free Free Free Free) at all horizontal levels. In this study, an attempt has been made to assess the increase in global structural load carrying capacity after considering welding of conductors at Spider deck and Cellar deck levels (Boundary conditions Fix Fix Fix Fix Fix Fix) i.e. rigidly fixed with structure.

IEOT-ONGC has carried out a study on the effect of welding of conductors at spider deck & cellar deck level on ultimate strength of Jacket structure [3] i.e. in terms of Reserved Strength Ratio (RSR) and on Pile Unity Check. This method was found to be helpful for improving the structural integrity of the existing platforms as a result of conductor support providing additional structural stiffness.

ANALYSIS AND METHODOLOGY

The study has been carried out on three representative unmanned well platforms at Mumbai High North field: A, B & C, installed in different water depths and of different structural configurations (Table 1 and Fig 3) for south direction wave. The in-place static global analysis has been carried out for 100-year extreme storm condition, for 8 directions of wave, current and wind with other design loads and design marine growth thickness. In-place and Ultimate strength analyses have been carried out on the selected three platforms.

Table 1. PLATFORM SALIENT DETAILS

Platform	A	B	C
Year of Installation	1984	1984	1976
Water depth	73 m	70 m	65 m
No. of conductors	6	9	6
No. of main piles	4	4	4
No. of skirt piles	2	2	4
Diameter of main pile	1.219 m	1.422 m	1.219 m
Diameter of skirt pile	1.219 m	1.219 m	1.219 m
Topside Load (MT)	1011	1090	907
Working Point Size	15.3mx6.1m	15.3mx6.1m	15.3mx6.1m

(a)

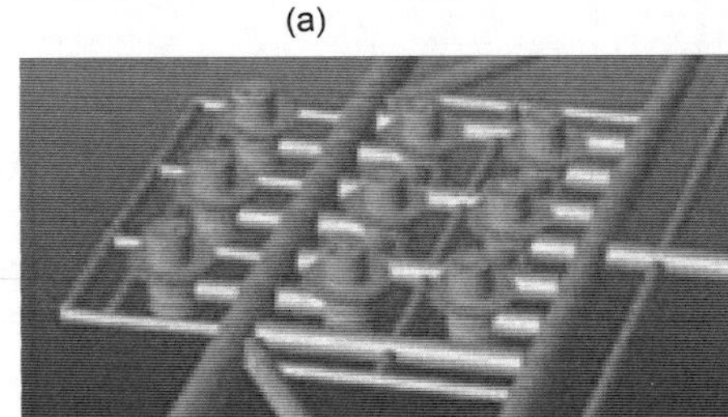

(b)

Figure 3. (a) SACS MODELS FOR PLATFORMS A, B & C (b) TYPICAL CONDUCTOR GUIDES

RESULTS

In place static analysis carried out using SACS software show that pile UC ratio improves by 2 to 8% after welding the conductor at cellar deck & spider deck levels. The result of soil pile interactive analysis show that the UC for the piles of Row-1 are improved after welding of conductors at cellar & spider deck, whereas there is very less effect on Row-2 Piles. Following table shows the % improvement in UC for the critical direction i.e. south for the selected platforms (Table 2).

Table 2. Percentage Improvement in Pile Unity Check for Platforms A, B & C

PILE	% improvement in UC after welding of conductors to the jacket frame		
	Platform A	Platform B	Platform C
A1	8.49	8.0	3.10
A2	0.37	0.12	0.11
B2	0.47	0.12	0.00
B1	7.51	8.1	2.77
SP1	8.09	5.95	2.34
SP2	7.44	5.88	0.00
SP2	-	-	0.00
SP1	-	-	1.79

In ultimate strength analysis carried out, it was observed that RSR improves after welding of conductors at cellar deck & spider deck. For platform A situated in water depth of 73m and with a pile configuration of 4+2 (4 Main piles + 2 Skirt Piles), the failure mode is governed by combination of primary X-brace joint and pile failure and the increase in RSR of platform A is 0.46% (Figure 4) .

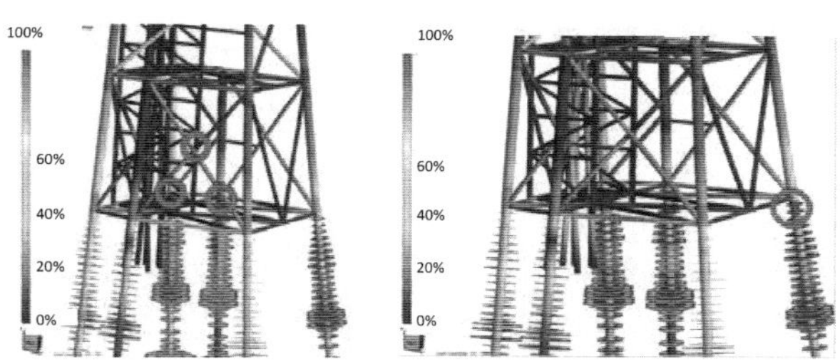

(a) (b)

Figure 4. Ultimate strength Results (Plastic capacity utilization) for Platform A: (a) before welding, (b) after welding

However, the similar configured platform B (with 4+2 pile configuration) situated in 70m water column was showing an increase in RSR by 9.17% (Figure 5) and the failure mode was governed by primary X-brace joint failure.

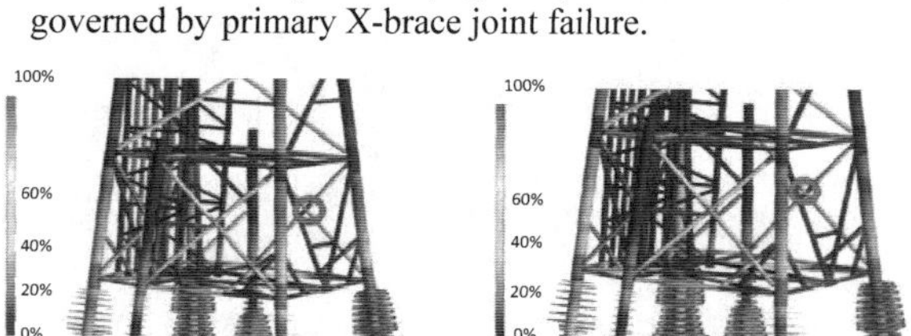

(a) (b)

Figure 5. Ultimate strength Results (Plastic capacity utilization) for Platform B: (a) before welding, (b) after welding

For platform C which is situated in 65m water depth and with pile configuration of 4+4, an increase in RSR was found to be 6.8% (Fig 6).

(a) (b)

Figure 6. Ultimate strength Results (Plastic capacity utilization) for Platform C: (a) before welding, (b) after welding

CONCLUSIONS

The analytical study for welding of the conductors at spider deck and cellar deck level is observed to be quite useful for some of the platforms in critical condition of overstressing of the piles. This will be helpful in documenting the structural integrity (SIA) of aged platforms. This is the first of its kind study for documenting the structural adequacy of the aged platforms for continued usage.

ACKNOWLEDGEMENT

The views expressed in the paper are those of authors only. We acknowledge the support and resources provided by ONGC required for carrying out this study. The study has been immensely beneficial in understanding the pertinent issues relevant to the structural behavior of jacket structures and is of vital importance in purview of ONGC's operational requirements to carry out life extension studies of existing plat-forms. We extend our sincere thanks to Shri Dinesh Kumar, ED-HoI, IEOT-ONGC for his generous support and encouragement. We are also immensely grateful for the continuous motivation received from Shri S. Karunanithi, GM-Head of Structures Section, IEOT-ONGC.

REFERENCES

[1] API RP 2A-WSD, 2014: American Petroleum Institute Recommended Practice 2A (Working Stress Design), 2014

[2] Technical report,2012. "Re-qualification methodology for fixed offshore platforms in west coast of India." DOC IEOT-LIFE- 001, REV-4, IEOT, ONGC, 2012

[3] Technical report, N Nehra, 2017."Study on effect of welding of conductors at Spider deck and Cellar deck levels on ultimate strength of Offshore structures", EOT/STR/21/2016-17

Proceedings of the 6th International Conference on Ship and Offshore Technology
ICSOT 2019
November 7-8, 2019, IIT Kharagpur, India

ICSOT INDIA 2019-13

NUMERICAL INVESTIGATION OF ROLE OF ANGLE OF ATTACK OF DUCT IN DUCTED PROPELLER WITH PBCF

Pritam Majumder
Research Scholar
Shillong, Meghalaya, 793003
India
majumderpritam08@gmail.com

Dileep Kumar Avanapu
M. Tech scholar
Shillong, Meghalaya, 793003
India
1dileepkumaravanapu@gmail.com

Subhendu Maity
Assistant professor
Shillong, Meghalaya, 793003
India
subhendu.maity@nitm.ac.in

ABSTRACT

Propeller boss cap fin (PBCF) is a well known energy saving devices (ESD), appreciated by various researchers for their capability of destroying hub vortex. Propeller duct also used for highly loaded propeller vessel due its ability to destroy blade tip vortex. But combined study of PBCF and propeller duct to improve propeller efficiency has not been investigated as per concern. In this paper numerical investigations have been performed to observe significance of duct angle of attack in the combine effect of PBCF and propeller duct. Here 19A duct is combined with aero foil shape PBCF with NACA4412 profile as ESD in VP1304 controllable pitch propeller. For numerical study, angle of attacks are varied for 5° interval and three duct models are generated with 0°, 5° and 10° angle of attack. Results shows improvement of propeller efficiency is reducing gradually with the increase of angle of attack and maximum 7.05% efficiency improvement is obtained at 0° angle of attack compare to parent propeller. This may be due to restricted flow at the tail section of the duct with the increase of angle of attack. Besides that fluctuation of pressure, various velocity components, formation of vortexes will also be investigated.

NOMENCLATURE

ρ	Fluid density
$\bar{u}_i$	Average velocity in each principal direction
x_i	Cartesian coordinate
F_i	Body force
G_i	Gravitational force
$\bar{p}$	Mean pressure
$\overline{\tau_{ij}}$	Mean viscous stress tensor
T	Thrust
Q	Torque

INTRODUCTION

Propeller is the most principle device in marine vehicle to generate propulsive force as well as ship motion. A major part of the produced energy from the propeller losses due to formation of cavitation, vortex from blade tip and boss, hull interaction and hence cause for extra power consumption. ESDs keeps prominent role to overcome the losses energy from propeller and hence helps to improve propeller efficiency. In this paper significance of angle of attack of duct in the combined effect of PBCF and propeller duct to enhance propeller efficiency has been investigated numerically.

Borros et al.[1] combining CFD and analytical semi-empirical (ASE) approach analyzed the flow around the duct hull combination body to realize more physical concepts of experimental work. The effect of duct in sea keeping has been observed numerically as well as model test by Bhattacharyya et al.[2] representing duct as a flat foil with same lifting properties. Result showed there were negligible changes in pitch and heave motion, acceleration and added resistance, but with a continuous reduction nature. It helps to conclude about the probability of the impact of duct in sea keeping. Duct size also keep prominent role in sea keeping. Gaggero *et al.* [3] found the initiation of the cavitation and tip vortex for conventional propeller as well as duct propeller for different loading condition. Due to the limitation of axial momentum theory to capture variation influence of thrust coefficient on propeller efficiency and cavitation phenomenon, Bontempo *et al.*[4] analyzed propeller performance with decelerating duct combing axial momentum theory with nonlinear semi-analytical actuator disk model. It has been observed propeller cavitation can be improve with the variation of NACA duct profile with chamber length, chord length, thickness of annular section.

Model scale as well as full scale condition, influence of Reynolds number and hull generated wake on PBCF studied by Kawamura et al. [5] numerically. Significant effects on propeller efficiency with PBCF in full scale condition had been noticed in combined effect of Reynolds number and hull wake from numerically obtained results and 2.32% efficiency enhance according to their observation. Based on RANS, hydrodynamic efficiency improvement of controllable pitch propeller with PBCF was evaluated by Xiong et al. [6]. Influence of aerofoil shape PBCF and axial as well as angular location for installation were also studied to detect the optimum shape of the PBCF and optimum position respectively. Propeller efficiency improved dominantly with the help of aerofoil shape PBCF. Through factor design method parametric study performed by Lim et al. [7] to achieve a correlation between different design parameters of propeller boss cap fin in full scale approach. For improvement of propeller efficiency, installation angle and position of fin kept prominent impact among the six selected design parameters. Besides the numerical observation experimental study also performed to check the effect of divergence cap along with PBCF to improve propeller performance. Results showed divergence cap with PBCF offer inconsequential effect by reducing propeller efficiency in major cases. Along with CFD simulation the capability of PBCF to improve propeller efficiency has been investigated by Mizzi *et al.* [8] in full scale with some optimization in shape. Based coupling effect of numerical model with optimization design parametric module 120 various design of PBCF has been studied. There is significant improvement in propeller efficiency by 2.5% gain in thrust coefficient, 1.4% rises in torque coefficient and hence 1% escalation in efficiency. PBCF prominently reduces hub vortex by creating restriction in the root section as well as decelerating the velocity and enhance efficiency.

GEOMETRIC MODELLING

For numerical study, an open source controllable pitch propeller VP1304 [9], developed by SVA Potsdam Model Basin in Germany in 1998 is used and shown in fig 1. The details of geometry are given in table 1

Table 1. GEOMETRIC PARAMETER DETAILS OF VP1304 PROPELLER [9]

Type	Symbol	Unit	Value
Diameter	D	m	0.250
Pitch ratio (r/R=0.7)	$P_{0.7}/D$	-----	1.635
Area ratio	A_E/A_O	-----	0.779
Chord length	$C_{0.7}$	m	0.104
Skew	θ	°(deg)	18.837
Hub ratio	D_h/D	-----	0.300
No. of blade	Z	-----	5
Rotation	Direction	-----	Right
Revolution per second	n	/sec	15

With a concept that aerofoil shape geometry can generate smoother flow in case of a turbine, aerofoil shape PBCF with NACA4412 geometric profile is introduced in propeller boss as shown in the fig 5. The number of fin is kept as number of propeller blades and PBCF radius is kept 28% of propeller radius. Each PBCF are placed exactly in between two adjacent propeller blades

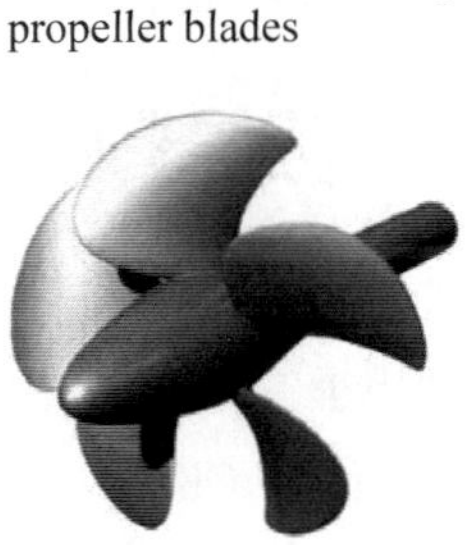

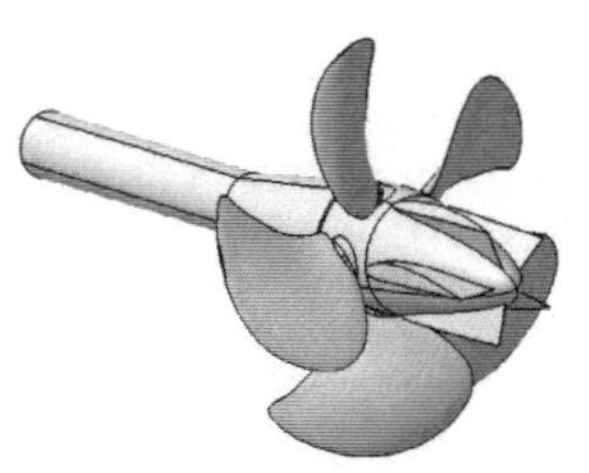

Figure 1.VP1304 PROPELLER [8]. Figure 2. PROPELLER WITH AEROFOIL FIN

In this study numerical investigations have been performed with 19A duct profile along with NACA4412 profile PBCF to investigate its effect in improvement of propeller efficiency. Here geometric modification is done in 19A duct profile with an interval of 5° difference and a comparative study is carried out with various hydrodynamic parameters such as KT, KQ and η for the duct geometry with 0°, 5°, and 10° angle of attack as shown in fig 2.

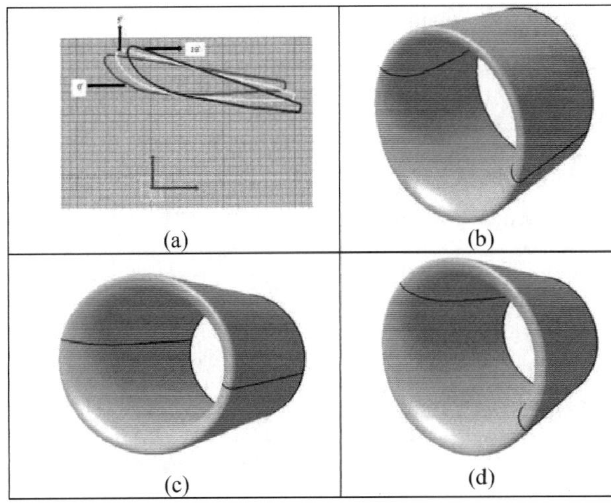

Figure 2 (a) 2D sketch of different angle of attack of 19A duct; 3D sketch of duct for different angle of attack (b) 0°, (c) 5°, and (d) 10°.

COMPUTATIONAL DETAILS

Computational domain

For numerical simulation computational domain are chosen in such a way that it does not affect the numerical calculation and details are shown in fig 3.

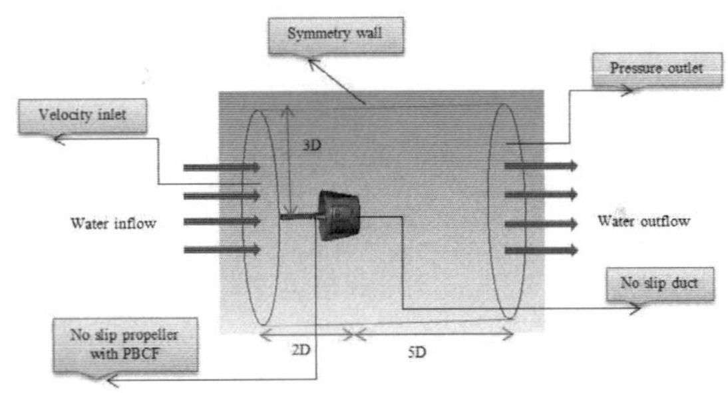

Figure 3 COMPUTATIONAL DOMAIN

Velocity at inlet and pressure at outlet are two major boundary conditions. Along that propeller with PBCF and duct are kept as no slip boundary condition. Since the surrounding walls are far enough from the solid wall, they are considered as symmetric wall.

Mesh

For any numerical simulation mesh generation is the most important part. Here due to complex geometry and less computational time requirement, cut cell mesh has been preferred which is a hybrid type mesh as shown in fig 4. Along that some local refinement also performed at sharp edge and corner for capturing more accurate result.

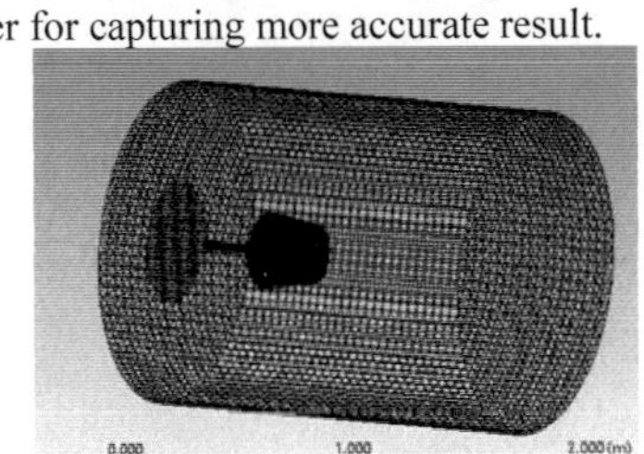

Figure 4. MESH GENERATION

Governing equations

Based on Reynolds Average Navier Stoke (RANS) equation numerical simulation performed for unsteady flow condition for incompressible flow. The principle governing equations are as follows.

$$\frac{\partial}{\partial x_i}(\rho \bar{u}_i) = 0 \quad (1)$$

$$\frac{\partial}{\partial t}(\rho \bar{u}_i) + \frac{\partial}{\partial x_j}(\rho \bar{u}_i \bar{u}_j + \rho \overline{u_i' u_j'}) = -\frac{\partial \bar{p}}{\partial x_i} + \frac{\partial \bar{\tau}_{ij}}{\partial x_j} + G_i + F_i \quad (2)$$

SST k-ω turbulence model are used to solve Reynolds stresses. Moving reference frame (MRF) approach are utilize for numerical simulations with 0.075 sec time step size.

RESULTS AND DISCUSSION

In this investigation three principle hydrodynamic parameters are investigate to observe effect of angle of attack in combined effect of PBCF and propeller duct to improve its efficiency. These are thrust coefficient (K_T), torque coefficient (K_Q) and efficiency (η) and express as follows

$$K_T = \frac{T}{\rho n^2 D^4} \quad (3)$$

$$K_Q = \frac{Q}{\rho n^2 D^5} \quad (4)$$

$$\eta = \frac{K_T}{K_Q} \times \frac{J}{2\pi} \quad (5)$$

where,

$$J = \frac{V_A}{nD} \quad (6)$$

Validation study

Here obtained results validated with the experimental results of SVA [9], which has been interpolated by Owen *et al* [10]. It is noticed that there is a good agreement between numerical and experimental results as shown in table 2

Table 2 MRF method

Adv. Ratio	K_T	Error (%)	K_Q	Error (%)	η	Error (%)
1	0.4245	5.56	0.1012	3.33	0.6679	2.73
0.8	0.4853	5.01	0.1178	0.34	0.5247	4.56
0.6	0.6339	0.77	0.1451	4.01	0.4171	2.88

Grid dependency test

Three different types (coarse, medium and fine) of mesh has been used to study the mesh sensitivity and found with the increasing number of mesh more stable and more accurate result has been obtained as shown in fig 5.

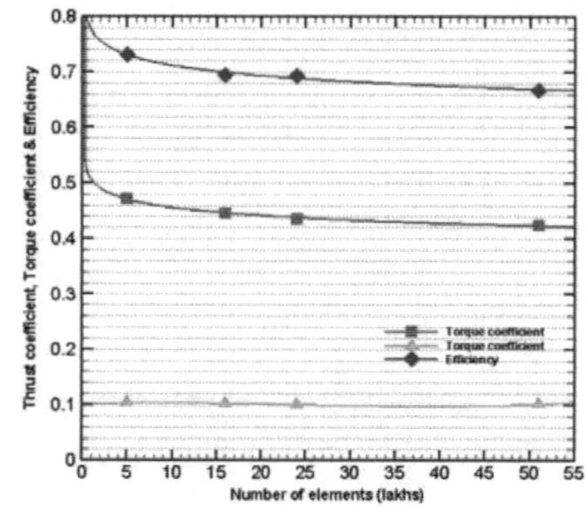

Figure 5. GRID DEPENDENCY TEST

Case study for combined effect of duct and PBCF

Table 3: CASE STUDY FOR COMBINED DUCT AND PBCF AT VARIOUS J

Case	J	K_T	K_Q	η
Propeller with 19A duct	0.2	3.46	4.86	0.022
	0.4	1.156	0.6837	0.1076
Propeller with 19A duct along with PBCF	0.2	1.1733	2.51616	0.01488
	0.4	0.9571	0.629354	0.09687

To study the effect of PBCF in ducted propeller in efficiency improvement some case study has been studied at lower advance ratio at particular angle of attack (0°) of duct as shown in table 3. It has been noticed that at lower advance ratio (J=0.2) PBCF does not keep prominent rule to improve efficiency compare to without PBCF in ducted propeller. At lower advance ratio fluid velocity are not significant to overcome the effect of boundary layer and skin friction from solid wall and hence increase produced drag. But with increase of advance ratio these trends reduce and efficiency increase gradually.

Role of angle of attack

Here numerical simulation performed in open water environment without considering ship hull, which may effects the observations. In these investigation three different angles of attacks has been investigated to obtain maximum efficiency improvement. The obtain result of table 4 clearly shows with the increase of angle of attack propeller efficiency reduces gradually and maximum efficiency improvement obtained at 0° angle of attack around 7.05% compare to parent propeller. Though PBCF and propeller duct reduces hub vortex as well as tip vortex respectively, but with the increase of angle of attack the blockage at the tail section of the duct is also increases, which drastically reduces mass flow rate as well as propeller efficiency.

Table 4. ROLE OF ANGLE OF ATTACK OF DUCT ON COMBINED EFFECT OF PBCF AND 19A DUCT

Angle of attack	Advan. ratio(J)	K_T	K_Q	η	% increase
0°	1	0.4196	0.096	0.6959	7.05
	0.8	0.7394	0.167966	0.5609	2.01
5°	1	0.441	0.1031	0.6811	4.76
	0.8	0.7526	-0.17433	0.5499	0.02
10°	1	0.5128	0.1247	0.6544	0.66

	0.8	0.8028	0.193505	0.5285	3.86% reduce

Graphical representations of thrust torque and efficiency improvement are shown in fig. 6

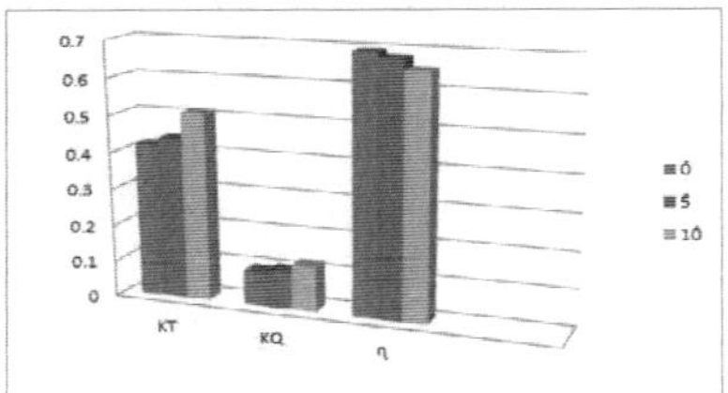

Figure 6. GRAPHICAL REPRESENTATION OF EFFECT OF ANGLE ATTACK ON COMBINED EFFECT OF PBCF AND DUCT

Pressure

To investigate details of flow feature with different angle of attack in combined effect of PBCF and propeller duct, fluctuation of pressure at different location has been studied. Along with propeller with PBCF AND duct, pressure distribution three different plane as shown in fig 7.

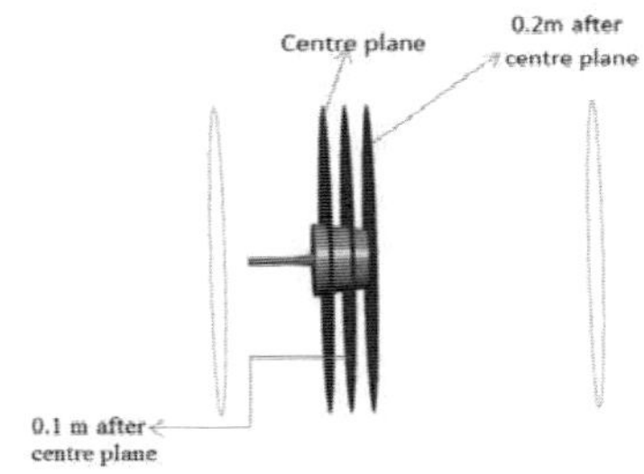

Figure 7. VARIOUS PLANE LOCATIONS

From the fig 8 it has been noticed that with the increase of angle of attack a wider high pressure zone appear on the pressure side of the propeller, but simultaneously a high pressure also develop on the inner wall of the propeller duct. This increase pressure on the inner wall of duct generates obstruction in the direction of the flow. As a result the blockage effect increases with the increase of the angle of attack in the combine effect of PBCF and propeller duct. Besides that fluctuation of pressure also observe in different plane as mention before and noticed 0.2 m after the propeller center plane generates maximum pressure compare to others two plane and this trend increase with the increase of angle of attack.

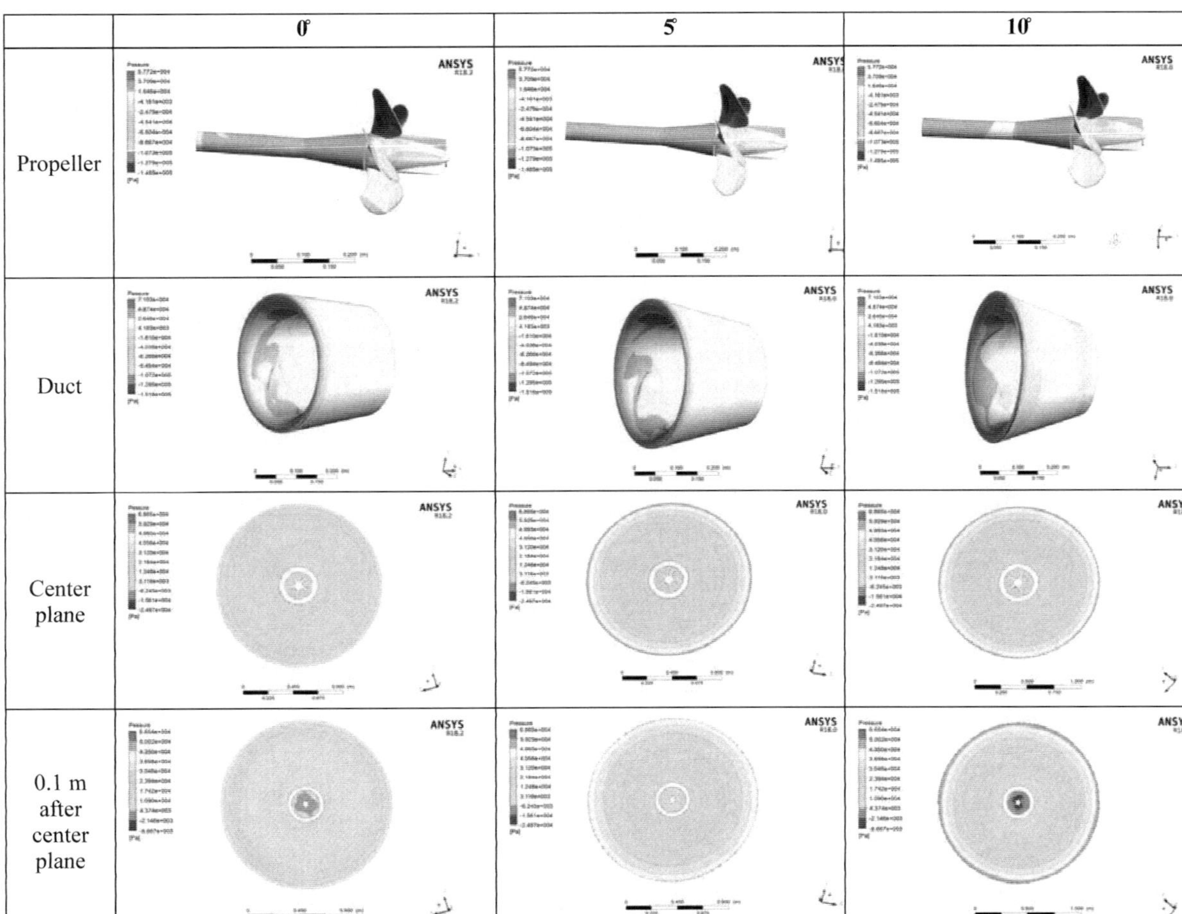

0.2 m after center plane	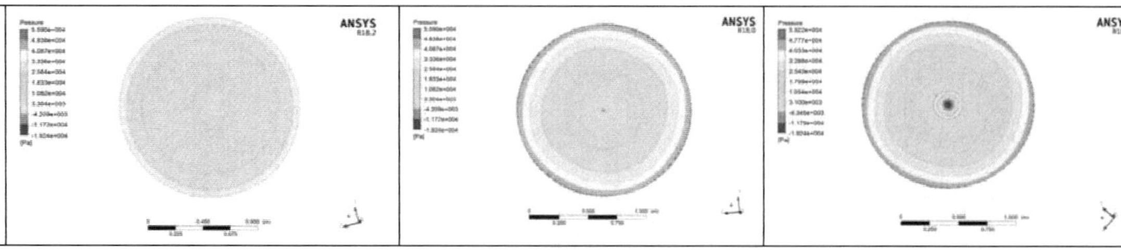

Figure 8. PRESSURE DISTRIBUTION IN DIFFERENT POSITION.

Velocity distribution

Along with pressure distribution velocity fluctuation also studied at various locations. It is noticed that velocity are gradually increasing with the increase of distance from center plane. Maximum velocity obtained at center plane around 59.70m/sec and 0.1m after it becomes 60.90 m/sec and 0.2 m after obtained maximum with 62.26 m/sec. From the fig 9 it's seen that increasing tendency of velocity are proportional to the angle of attack and here at 10° angle of attack maximum obtained velocity is 62.26m/sec. besides that a low velocity zone also appear surrounding the propeller with the increase of angle of attack at the center plane which gives evidence of presence of vortexes at that zone.

Conclusions

This paper investigates the role angle of attack to improve propeller efficiency in combined effect of PBCF and propeller duct. Three different angles of attack have been studied with 5° interval with 19A duct and NACA4412 shape PBCF. Obtained results shows with the increase of angle of attack the propeller efficiency reduces gradually and at 0° angle of attack we obtained the maximum efficiency improvement with 7.05% enhancement. This is because with the raising angle of attack the blockage effect increases significantly which reduce mass flow rate at the exit part of the duct. Besides that fluctuation of pressure at various positions also studied and found, increasing angle of attack helps to rises pressure on the back side but simultaneous increase pressure in the duct inner wall causes for overall efficiency reduction. Along that velocity distribution also studied at different location to obtain more details clarification.

	0°	5°	10°
Center plane	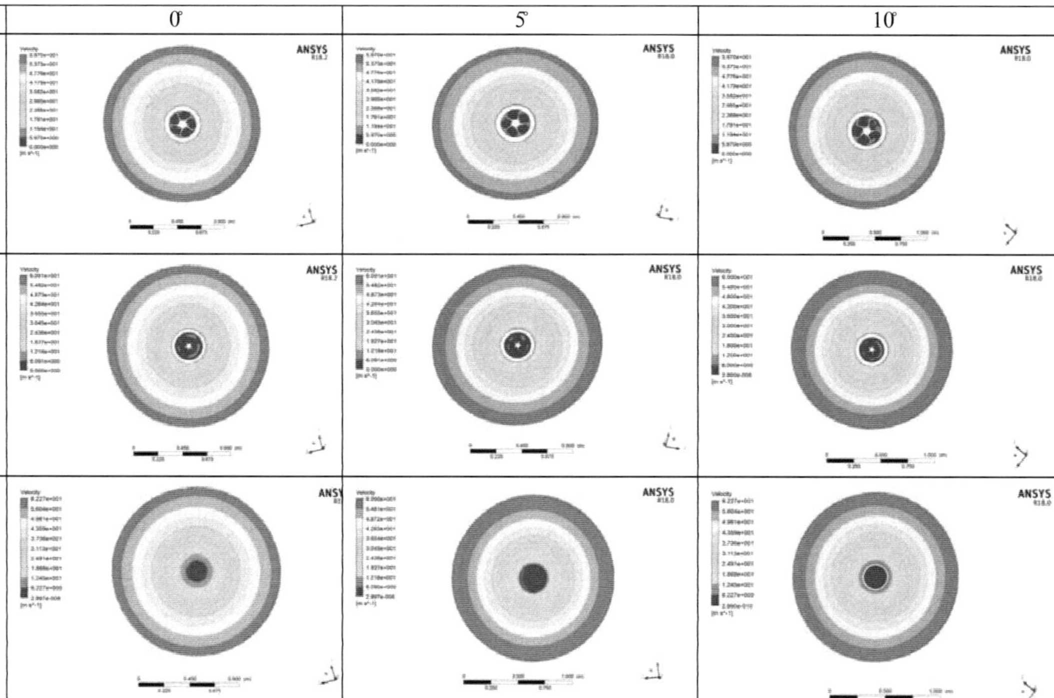		
0.1 m after center plane			
0.2 m after center plane			

Figure 9. VELOCITY AT VARIOUS POSITION

REFERENCES

[1] Barros E A., Dantas J.L.D., 2012. "Effect of a propeller duct on AUV maneuverability". Ocean Engineering, 42, pp. 61–70.

[2] Bhattacharyya A., Steen S., 2014. "Influence of ducted propeller on sea keeping in waves". Ocean Engineering, 91, pp. 243–251.

[3] Gaggero S., Tani G., Viviani M., Conti F., 2014. "A study on the numerical prediction of propellers cavitating tip vortex". Ocean Engineering, 92, pp.137–161.

[4] Bontempo R., Cardone M., Manna M., 2016 Performance analysis of ducted marine propellers. Part I – Decelerating duct. Applied Ocean Research 58, pp. 322–330

[5] Kawamura T., Ouchi K., and Nojiri T., 2012. “Model and full scale CFD analysis of propeller boss cap fins (PBCF)”. J Mar Sci Technol, 17, pp. 469–480.

[6] Xiong Y., Wang Z., and Qi W., 2013. “Numerical Study on the Influence of Boss Cap Fins on Efficiency of Controllable-pitch Propeller”. Journal of Marine Science and Application, 12, pp. 13-20.

[7] Lim S., Kim T., Lee D., Kang C., and Kim S., 2014. “Parametric study of propeller boss cap fins for container ships”. Int. J. Nav. Archit. Ocean Eng, 6, pp. 187-205.

[8] Mizzi K., Demirel Y. K., Banks C., Turan O., Kaklis P., and Atlar M., 2017. “Design optimization of Propeller Boss Cap Fins for enhanced propeller performance”. Applied Ocean Research, 62, pp. 210–222.

[9] Schiffbau-Versuchsanstalt Potsdam GmbH, Potsdam propeller test case (PPTC). 2011. “Open Water test with the model propeller VP1304”. Report no. 3752

[10] D Owen., Y K Demirel., E Oguz., T Tezdogan., and A Incecik., 2018. “Investigating the effect of biofouling on propeller characteristics using CFD”. Ocean Engineering, 159, pp. 505–516.

ICSOT INDIA 2019-14

DEVELOPMENT OF CONDITIONAL ASSESSMENT SYSTEM FOR BERTHING STRUCTURES – A CASE STUDY

Raghava Kumar Vanama
Research Scholar, IITB
Mumbai, Maharashtra-400076
India
174040002@iitb.ac.in

Balaji Ramakrishnan
Associate professor, IITB
Mumbai, Maharashtra-400076
India
rbalaji@iitb.ac.in

Balasubramannian K
Managing director, Hitech Concrete
Solutions Chennai Pvt. Ltd.
Chennai, Tamil Nadu 600095
kbs@hitechconcrete.net

Suresh Patil
Chief Engineer, DPT
Kandla, Gujarat, 370210
India
jekptdesign@gmail.com

ABSTRACT

Present Study mainly focusses on the development of a conditional assessment system for concrete structures in marine environment with a case study of cargo berths (CB) at Deendayal port trust (DPT), Kandla. DPT is highest revenue generating and busiest port in India located on the Gulf of Kutch where the maximum tidal range is nearly 8m, which make the field non-destructive testing (NDT) challenging. The assessment system proposed in this study is based on the damage level classification (DLC) of structure, evaluated by a set of widely used non-destructive tests. Study further investigates the usage of DLC system in comparison with the Condition Rating (CR) method developed by Verma et al. Several filed and laboratory NDT were carried out at 180 locations of CB 7-10 and observed that the condition of the berthing structures, indicated by a method based on CR system and DLC system are giving the similar results irrespective of their different test methodologies. The condition assessment system based on the damage level classification requires less efforts compared to CR system and found to be reliable, quick, and efficient, hence can be used to monitor the performance of marine structures in Kandla, India.

INTRODUCTION

Chloride induced corrosion is the prime reason behind the degradation of embedded rebar in reinforced concrete (RC) marine structures. Along with the chloride attach, RC members present in the intertidal zone and splash zone are imperil to alternate wetting and drying cycles because of continuous waves and tides. This process escalates the corrosion of embedded rebar causing corrosion cracks, spalling of concrete, degradation of mechanical properties of rebar[1,2] and bond with concrete [3–5], thereby reducing the overall strength and service life of the structure. These aggressive marine environment and heavy operating load conditions, most of marine structures deteriorating faster resulting in lesser service life than they have been designed for.

Estimating the service life of RC structure demands ample of field and laboratory testing. Many researchers in past have described their models and methods for estimating the service life. However, there are very few models that are applicable for the structures in marine environment, in particularly the berthing structures, where heavy loads are also expected. Verma et al.[6] have proposed a model based on Tuutti[7], for estimating the service life of RC structures based on corrosion rating (CR) indices. However, estimating these indices requires lot of field and laboratory tests, viz., estimation of chloride content at rebar depth, core testing, rapid chloride migration test (RCMT), depth of cover concrete, carbonation test etc. that requires sophisticated laboratory equipment. Present study proposes a damage index model that gives the idea of condition of structure based on the widely used NDT methods, viz., half-cell potential test, UPV test, Rebound hammer test etc. Proposed model also considers the mass loss of rebar due to corrosion and extent of physical damage caused to the structural members in terms of cracking and spalling of concrete etc.

Present study demonstrates the effectiveness of CR system and damage index model for assessing the condition of marine structure with a case study of cargo berthing structures at Deendayal port trust, Kandla.

STUDY AREA

Deendayal Port Trust (DPT) is one of the major and busiest ports in India, located on the Gulf of Kutch in Gujarat on the west coast. DPT facilitates different type of cargo handling with the help sixteen berthing structures. Present study focusses on the conditional assessment of cargo berth (CB)-7 to CB-10, located at 22°59'32.8"N 70°13'27.2"E (Figure 1). In the Gulf of Kutch region, the maximum tidal range is about 8m and the tidal currents reaches up to 3 m/s [8,9]. The tidal range in this region is the second largest among the other part of the Indian coastline because of its complex shape and sudden seabed variations on the head[8]. The depth varies from 60 m near the mouth to 20 m at the eastern end of the Gulf.[10]

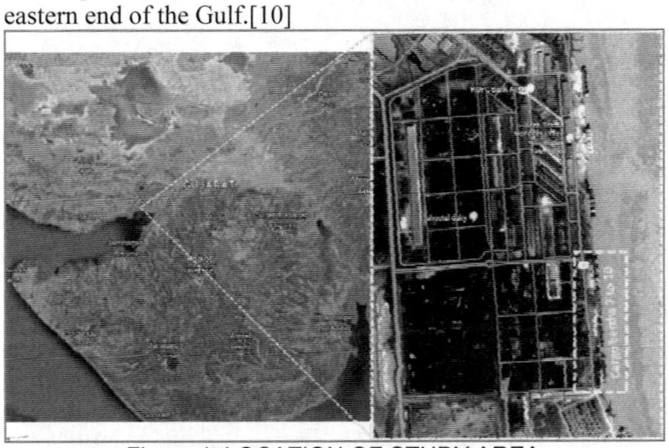

Figure 1. LOCATION OF STUDY AREA

Challenges encountered during field tests

Present study involves few field tests to assess the condition of berthing structures. Field measurements and testing become challenging in this region in terms of reaching the location and accessing the structural members due to the busy cargo berths and high tidal range. Some of the field tests requires electricity and potable water that are difficult to possess under the berthing structure, where small boats are often hired for accessing the required test location. Moreover, the movement of high speed navy boats, tug boats and passing vessels makes the field tests more challenging.

Overcoming all these challenges required rigorous planning before proceeding for the testing activities on field. Firstly, high and low tides on that particular day was understood from the available tide tables. Accordingly, the sequence of testing was decided (from deck beams to piles or vice versa). Secondly, a boat that is just sufficient to accommodate test and boat operating team, the electric generator and potable water required for operating the testing instruments was hired. Sufficient number of lifejackets, torch lights, helmets and other safety measures were taken during testing. Ropes were used to tie the boat to nearby structural piles to minimizing the rolling caused due to the movement of high speed navy/tug boats and passing vessels.

ASSESSMENT METHODOLOGY

Present study investigates the practical usage of CR system and damage index model for the structures present in the marine environment through a case study at DPT, Kandla. Conditional assessment methodology adopted and the required tests need to be carried out to utilize the models are presented schematically in Figure 2.

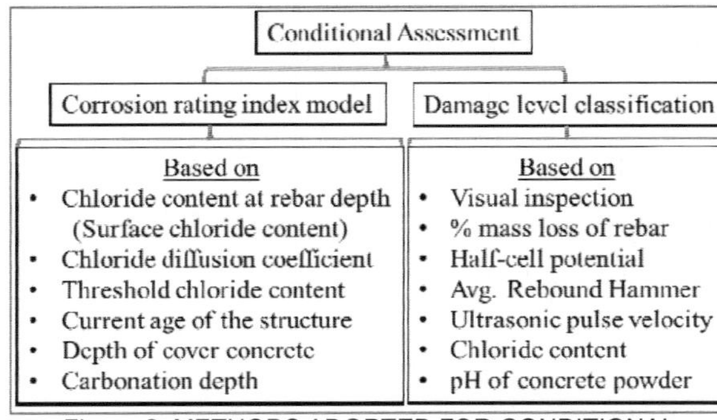

Figure 2. METHODS ADOPTED FOR CONDITIONAL ASSESSMENT

Corrosion rating index model

This model was proposed by Verma et al.[6] based on the corrosion model demonstrated by Tuutti [7]. This method involves the estimation of corrosion initiation period and the present age condition of the structure. Arriving at the corrosion initiation period involves the estimation of chloride diffusion coefficient, chloride content at the rebar location and requires idea of chloride threshold value. It is to be noted that chloride threshold value is influenced by several parameters such as surface characteristics and metallurgical composition of steel, binder materials, type of reinforcement etc.[11]

Damage level classification model

This model is based on the test results obtained from the most commonly used NDT testing, viz., half-cell potential test, UPV test, rebound hammer test, chloride and pH levels of concrete, cracks and spalling observed during visual inspection and extent of mass loss due to corrosion. Unlike CR model, this method is flexible with no compulsion on carrying out all the above mentioned tests.

In this model, score on the scale of 10 is assigned based on the performance of material/structure for the each of the given NDT tests, as given in Table 1. Level of structural damage is assessed based on the average score obtained from the NDT tests carried out and then the structural condition is evaluated, as mentioned in Table 2. Thereby, this method is simple and requires minimal time for evaluating the safety of structure and gives an overall summary of several diversified NDT tests in terms of damage level.

Table 1. ADOPTED SCORING SCHEME BASED ON TEST RESULTS

Criteria		Score					
		10	8	6	4	2	0
Visual inspection	Cracks	Nil	Minor	Minor	Medium	Large	Large
	Spalling	Nil	Nil	Minor	Minor	Medium	Heavy
Reinforcement corrosion		Nil	< 5%	5-10%	10-20%	20-30%	> 30%
No. of points having half-cell potential values <-300mV		Nil	< 5%	5-15%	15-30%	30-50%	> 50%
No. of points having UPV values < 3 km/sec		Nil	< 5%	5-10%	10-25%	25-40%	> 40%

Criteria	Score					
	10	8	6	4	2	0
Avg. rebound hammer value	> 45	40-45	30-40	20-30	10-20.0	< 10
Chloride content (kg/m3)	< 0.6	< 0.6	< 0.6	> 0.6	> 0.6	> 0.6
pH of concrete powder	> 12	11.5 -12	10-11.5	10-11.5	9-10.0	< 9

Table 2. DAMAGE LEVEL CLASSIFICATION SCHEME OF STRUCTURE

Damage		Structure condition	Remarks
Index	**Level**		
10.0 - 9.0	0	Good	No visible damage but greater part in the initiation period
9.0 - 7.5	1	Minor Damage	Minor visible damage, if left unattended may lead to loss in structural integrity
7.5 - 6.0	2	Moderate Damage	Moderate visible damage, loss in structural integrity initiated, need to be attended as soon as possible
6.0 - 4.0	3	Severe Damage	Severe damage visible, falling pats may be dangerous but the loss of serviceability and safety of the structure are minimal, need to attended at the earliest
4.0 - 2.0	4	Heavy damage (Reduced safety)	Loss of serviceability and reduced safety of the structure, immediate action has to be taken
2.0 - 0.0	5	Heavy damage/ Collapse	Repair or rehabilitation may not be possible, visible. Demolition may be thought upon for safety of nearby structures and public

NON-DESTRUCTIVE TESTING

For arriving the parameters required for assessing the condition of berthing structures by CR method and damage index model, a set of field and laboratory test haven conducted as mentioned in Figure 3.

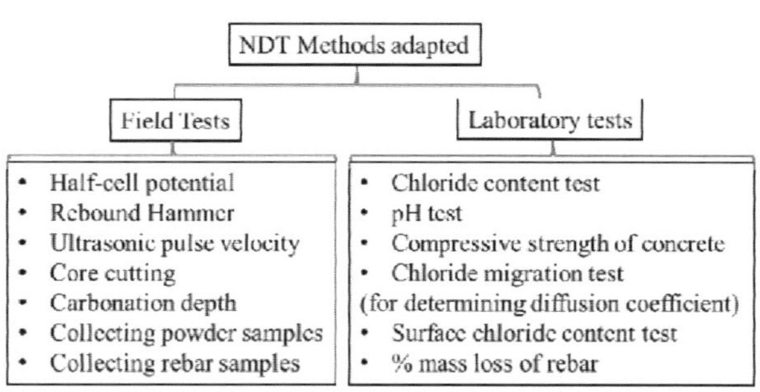

Figure 3. TESTS PERFORMED FOR CONDITIONAL ASSESSMENT OF BERTHING STRUCTURES

Field tests

Six field tests were carried out as per their respective standards, viz., Half-cell potential[12], Rebound Hammer[13], Ultrasonic pulse velocity[14], Core cutting (Figure 4), Carbonation depth (by phenolphthalein indicator), Chloride content and pH test (Figure 5), at 36 locations (Figure 6 to 9) across the structural members of cargo berths 7 to 10.

Figure 4 CORE CUTTING FROM VARIOUS STRUCTURAL MEMBERS OF CB

Figure 5 DIFFERENT NDT TESTS CARRIED OUT ON CB STRUCTURES

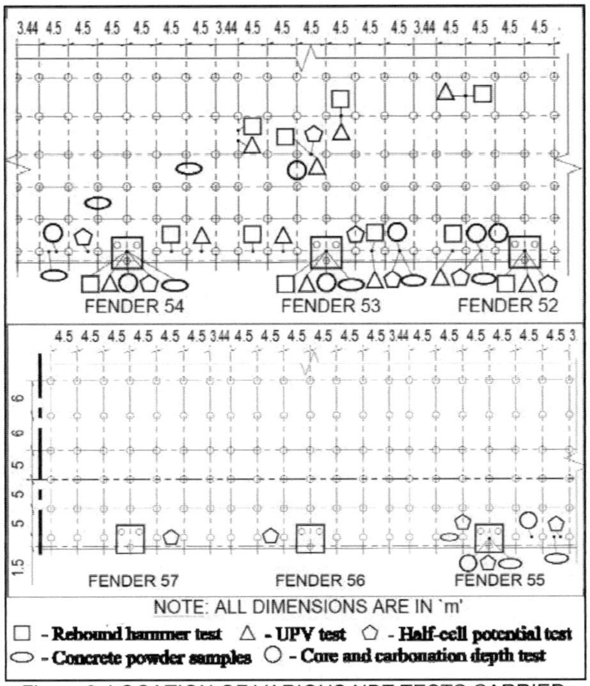

Figure 6. LOCATION OF VARIOUS NDT TESTS CARRIED OUT IN CB-7

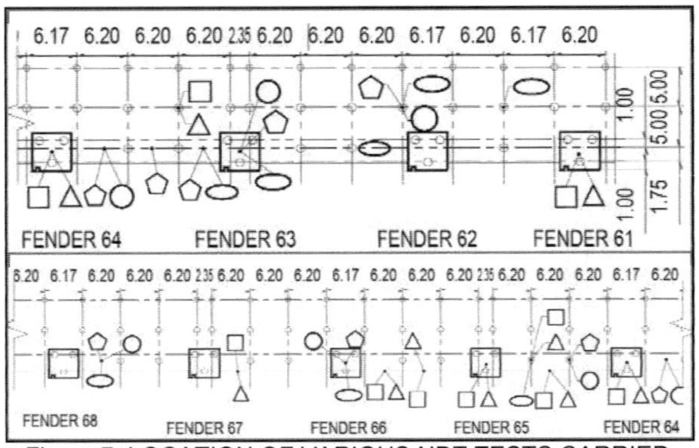

Figure 7. LOCATION OF VARIOUS NDT TESTS CARRIED OUT IN CB-8

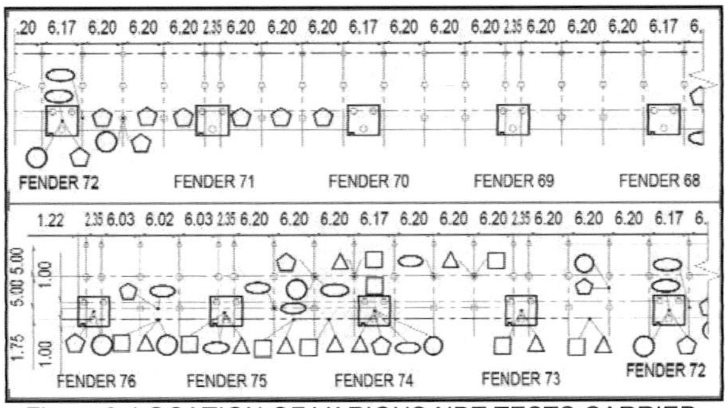

Figure 8. LOCATION OF VARIOUS NDT TESTS CARRIED OUT IN CB-9

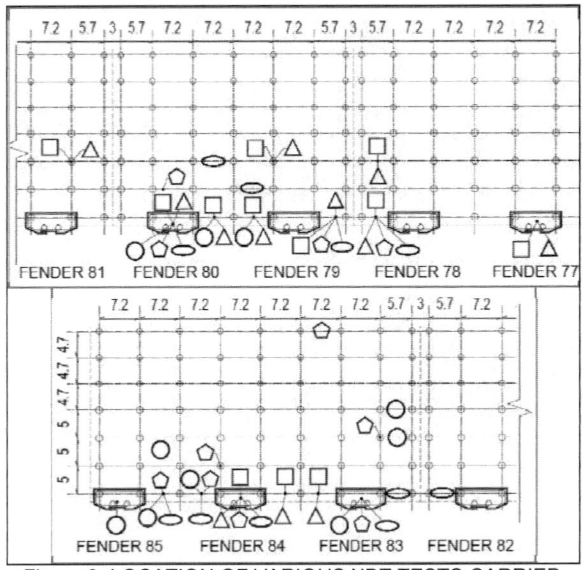

Figure 9. LOCATION OF VARIOUS NDT TESTS CARRIED OUT IN CB-10

Laboratory tests

Chloride content at rebar level and the chloride content of the concrete powder samples collected from field are tested in accordance with Indian standard[15], as shown in Figure 10.

Figure 10. ESTIMATION OF CHLORIDE CONTENT IN CONCRETE

Compressive strength of concrete was estimated in accordance with IS 516 [16] based on the results obtained from compression test carried out on core samples collected from field. L/d correction and cylinder to cube strength conversions were appropriately considered for obtaining the equivalent cube compressive strength of concrete. pH of the concrete powder samples was obtained from the pH meter.

Rapid chloride migration tests has been performed for various core specimens obtained from CB 7 to 10 in accordance with NT BUILD 492[17], as shown in Figure 11. As the sample diameter (69 mm) is lesser than the prescribed dimension (100 mm)[17], appropriate silicon rubber sleeves of 69 mm inner diameter were used. Specimens were preconditioned for longer durations till the chlorides present in the specimens were extracted. Specimens were supplied with appropriate voltage[17] and monitored the resistivity offered by concrete specimen against the chloride ingression using a patent device[18].

Figure 11. RCMT TEST SET-UP

RESULTS AND DISCUSSION

Field and laboratory test results required for estimating the corrosion initiation time are presented in Table 3, for CB 7 to 10. Condition rating of the cargo berth at present age is estimated based on the chloride content and depth of carbonation as given in Table 4. Further, residual life of the structure is estimated by the model based on Tuutti [7] and presented in Figure 12 for CB 7-10. However, arriving at the diffusion coefficient of concrete which was already contaminated with chlorides was challenging and required longer preconditioning and extraction of chlorides from specimen.

NDT results were incorporated in the proposed DLC model and presented in Table 5. From the Tables 4 and 5, it can be observed that the condition of the berthing structures, indicated by a method based on CR system and the method based on damage index system are giving the similar results irrespective of their different test methodologies.

This exercise shows that, with limited field NDT results, the condition of the marine structure can be effectively assessed to represent their actual structural condition using DLC method. Thereby, proposed model is helpful in reducing the time and efforts required for effective conditional assessment of a port structure like cargo berths.

Table 3. ESTIMATION OF CORROSION INITIATION PERIOD

Parameter	CB 7	CB 8	CB 9	CB 10
Concrete cover (c) m	0.05	0.05	0.05	0.05
Diffusion coefficient (D) m^2/s (x 10^{-12})	3.04	2.23	2.23	2.12
Threshold chloride content (C_{th}) kg/m^3	0.60	0.60	0.60	0.60
Surface chloride content (C_s) kg/m^3	1.92	2.55	2.49	4.31
Corrosion Initiation time (T_{ini})Years	12.76	12.56	12.95	8.55

Table 4. CONDITION OF BERTHING STRUCTURES BASED ON 'CR' METHOD

CR index	Years from construction				Action required
	CB 7	CB 8	CB 9	CB 10	
0	0.00	0.00	0.00	0.00	Excellent condition
2	12.76	12.56	12.95	8.55	Corrosion initiated, required regular inspection
5	26.50	**28.00**	**28.00**	**20.00**	Maintenance is required to increase the service life
6	**34.00**	33.15	33.02	23.82	Maintenance is must for continuous use, likely to repair
7	39.31	38.29	38.04	27.63	Structure must be closed for maintenance
8	44.62	43.44	43.05	31.45	Poor condition not likely to be repaired
9	49.93	48.59	48.07	35.27	Replacement of structures
Note: Numbers in bold represents the present age/condition of CB					

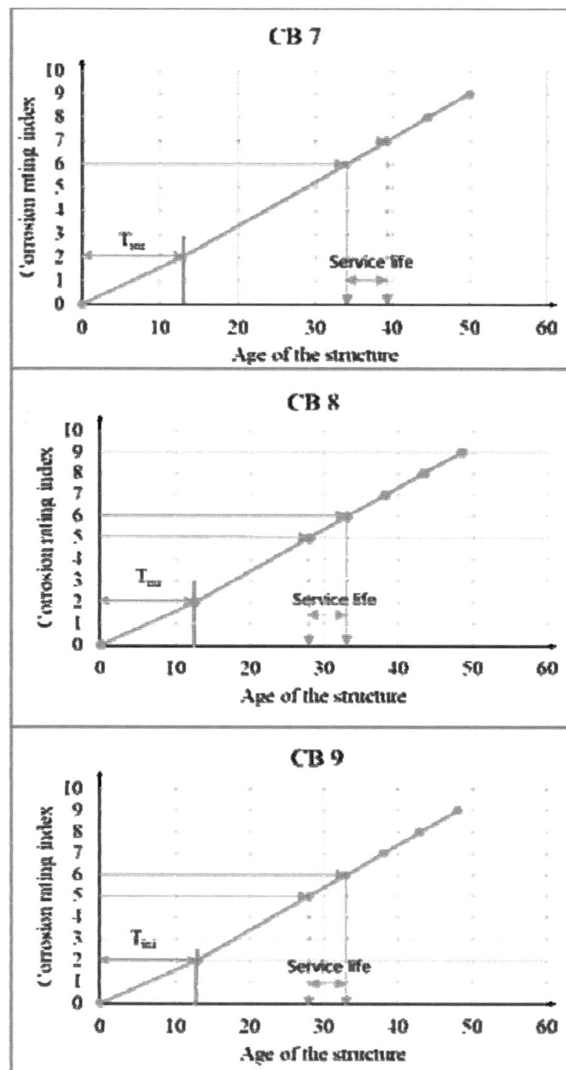

Figure 12. RESIDUAL LIFE ESTIMATION OF BERTHING STRUCTURES

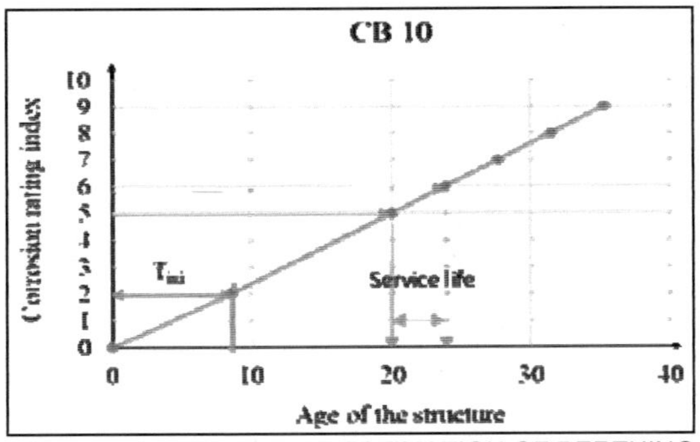

Figure 12. RESIDUAL LIFE ESTIMATION OF BERTHING STRUCTURES (Contd.)

Table 5. DAMAGE LEVEL CLASSIFICATION OF CARGO BERTHS

Criteria		Score			
		CB 7	CB 8	CB 9	CB 10
Visual insp.	Cracks	2	2	2	0
	Spalling	2	2	2	0
No. of points having half-cell potential values < -300mV		0	4	4	0
No. of points having UPV values < 3 km/sec		8	8	8	8
Avg. rebound hammer value		6	6	6	6
Chloride content (kg/m^3) (Cl)		4	4	4	4
pH of concrete powder		4	4	4	4
Damage Index (average score)		3.71	4.29	4.29	3.14
Damage Level		4	3	3	4
Structure Condition		Heavy Damage	Severe Damage	Severe Damage	Heavy Damage

CONCLUSION

Conducting field non-destructive tests are always challenging for busy port structures. Procedure followed in the present study gives a possible solution to overcome these challenges and successfully conduct the field tests in the marine environment safely. Present study demonstrates the effectiveness of CR system and proposed DLC model for assessing the condition of marine structure with a case study of cargo berthing structures at Deendayal port trust, Kandla. The major conclusion of the study are summarized as follows.

- Conditional assessment models for estimating the service life of structure (like CR system) needs the reliable data related to chloride diffusion coefficient of concrete, which is difficult to obtain for a chloride contaminated concrete.
- Estimating the corrosion initiation period (in CR system) required lot of data related to the type of materials used for construction of structure, which is difficult obtain for old ports due to lack of as-built drawings and information on mixture designs adopted at the time of construction. Thereby, numerous assumptions need to be made for obtaining the condition and residual life of the structure using CR system
- The damage level classification model proposed in the present study is only based on the widely used NDT methods that are relatively easily to perform and avoids tedious and time consuming laboratory testing like RCMT of chloride contaminated concrete.

It can be concluded from the present study that the condition of the berthing structures 7 to 10, DPT, Kandla, indicated by a method based on CR system and DLC methods are giving the similar results irrespective of their different test methodologies. Thereby, present comparative study demonstrates the effectiveness of developed DLC model in the marine environment that is helpful in reducing the time and efforts required for effective conditional assessment of a port structure like cargo berths.

REFERENCES

[1] Imperatore, S., Rinaldi, Z., Drago, D., 2017. "Degradation relationships for the mechanical properties of corroded steel rebars", *Constr. Build. Mater., 148*, April, pp. 219–230.

[2] Almusallam, A.A., 2001. "Effect of degree of corrosion on the properties of reinforcing steel bars", *Constr. Build. Mater.*, 15, Jan, pp. 361–368.

[3] Lin, H., Zhao, Y., Ožbolt, J., Hans-Wolf, R., 2017. "The bond behavior between concrete and corroded steel bar under repeated loading", *Eng. Struct.*, 140, Feb, pp. 390–405.

[4] Coccia, S., Imperatore, S., Rinaldi, Z., 2016. "Influence of corrosion on the bond strength of steel rebars in concrete", *Mater. Struct. Constr.*, 49, Dec, pp. 537–551.

[5] Almusallam, A.A., Al-Gahtani, A., Aziz, A., "Effect of reinforcement corrosion on bond behavior", *Constr. Build. Mater.* 10, Oct, pp. 123–129.

[6] Verma, S.K., Bhadauria, S.S., Akhtar, S., 2013 "Estimating Residual Service Life of Deteriorated Reinforced Concrete Structures", *Ame. Journ. Civil Engg. Arch.*, 1, Jun, pp. 92–96.

[7] Tuutti, K., 1982. *Corrosion of steel in concrete.* Swedish Cement and Concrete Research Institute, Stockholm.

[8] Kumar, S.S., Balaji, R., 2015 "Coastal and Shelf Science Effect of bottom friction on tidal hydrodynamics along Gulf of Khambhat, India", *Estuar, Coast. Shelf Sci.*, 154, Jan, pp. 129–136.

[9] Satheeshkumar, J., Balaji, J., 2015. "Inter-comparison of hydrodynamic models of Gulf of Khambhat", *Intern. Journ. Ocean Climate Sys.*, 6, Feb, pp. 35–46.

[10] Satheeshkumar, J., Balaji, J., 2016. "Tidal power potential assessment along the Gulf of Kutch, Gujarat, India", *Asian Wave Tidal Energy Conf., Aug.*

[11] Mahima, S., Moorthi, P.V.P., Bahurudeen, A., Gopinath, A., 2018. "Influence of chloride threshold value in service life prediction of reinforced concrete structures", *Sādhanā*. 43, Feb, pp. 1–19.

[12] ASTM C876: Standard Test Method for Corrosion Potentials of Uncoated Reinforcing Steel in concrete.

[13] IS 13311 (Part 2)-2018: Non-destructive testing of concrete – Method of test- Rebound hammer

[14] IS 13311 (Part 1)-2018: Non-destructive testing of concrete – Method of test- Ultrasonic pulse velocity

[15] IS 6925-2013: Methods of test for determination of water soluble chlorides in concrete admixtures

[16] IS 516- 2016: Photography- Camera shutters- Timing

[17] NT Build 492: Chloride migration coefficient from non-steady-state migration experiments. Nordtest method, 1999:1-8

[18] Vanama, R.K. and Balaji, R. (2019). "An interface circuit for an accelerated corrosion test apparatus". Indian patent application No. 201921020474, filed May 23, 2019.

Proceedings of the 6th International Conference on Ship and Offshore Technology
ICSOT 2019
November 7-8, 2019, IIT Kharagpur, India

ICSOT INDIA 2019-15

SHIP RESISTANCE PREDICTION USING ARTIFICIAL NEURAL NETWORKS

Rishav Raj
Naval Construction Wing,
Applied Mechanics Department,
IIT Delhi
New Delhi, 110016
India
rajrishav743@gmail.com

ABSTRACT
Ship design is a complex process undertaken in the form of a design spiral. Calculation of resistance of a vessel is one of the key steps of this design spiral. The traditional methods of calculation of resistance involve model tests or regressions equations. The major disadvantages of these methods are that they are time and cost intensive and the generalisation ability is low.

In this paper, the feasibility of using artificial neural networks for determining the residuary resistance of a ship has been studied. The data for prediction of resistance is obtained from experiments carried out on catamaran hull forms by Insel and Molland. The input features are slenderness ratio, B/T ratio, S/L ratio and Froude number. It was found that a three hidden layered network predicted the resistance of a ship with good accuracy. The mean relative error for different optimisers was also obtained and Adam resulted in the least error. Hence, we can conclusively say that artificial neural networks are a useful tool for predicting the residuary resistance. The main advantage it offers is that neural networks can adapt to new data easily.
Keywords: resistance, artificial neural networks

NOMENCLATURE

L	Length of the vessel
B	Breadth of the vessel
T	Draught of the vessel
S	Separation between the two hulls
C_R	Coefficient of residuary resistance
C_T	Coefficient of total resistance
C_F	Coefficient of frictional resistance
P_E	Effective Power
$F_{n.}$	Froude number
$L/\nabla^{1/3}$	Slenderness ratio

INTRODUCTION
Ship design is a complex process. Many engineering disciplines are involved in the design process. A ship is required to float and propel in the harshest of seas. Along with this basic requirement, a warship is required to fight and survive in case of any damage to its structure. The Design Spiral is a systematic approach to repeatedly update the design to arrive at the best compromise possible.

The very first step in the design spiral of a ship is to determine the proportions and powering requirements. This is vital for the selection of the main engine of the ship. The power required is obtained by calculating the resistance of the ship.

The figure below represents how power is transmitted to the propeller via the engine. EHP is the effective power(P_E) obtained from the total coefficient of resistance(R_T) at the given velocity.

$$P_E = R_T V \quad (1)$$

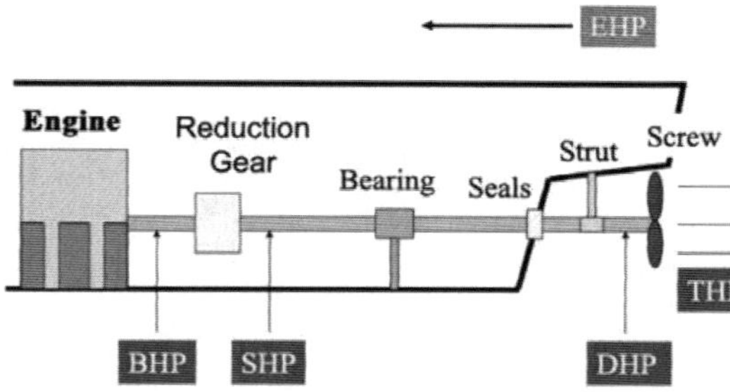

Figure 1. POWER TRAIN

An advantage of artificial neural networks over statistical methods is their ability to adapt to new data and it can quickly be retrained as new data becomes available. In addition, the same artificial neural network topology can be trained with other data for different vessel types. Retraining the artificial neural network with additional data is generally simpler and quicker than recomputing a statistical model.[1]

Resistance Data
The resistance data used for this study was obtained from a series of experiments carried out on a systematic series of hull forms tested in both monohull and catamaran configurations.

The experiment and the resistance data are described in full in Insel and Molland (1992) [2] and Molland et al. (1994,1995) [3]. The hull forms were tested in various configurations by changing the parameters such as slenderness ratio($L/\nabla^{1/3}$), B/T ratio, S/L ratio. The table below gives the model details.

Table 1. HULL MODEL NOTATIONS AND PARAMETERS

Slenderness Ratio	B/T=1.5	B/T=2.0	B/T=2.5
6.3	-	3b	-
7.4	4a	4b	4c
8.5	5a	5b	5c
9.5	6a	6b	6c

In this study, only the residuary resistance has been for prediction using neural networks. Total resistance (C_T) is composed of frictional resistance (C_F) and the residuary resistance (C_R). The frictional resistance can be obtained using the ITTC -57 equation.

$$C_R = C_T - C_F \tag{2}$$

The coefficient of residuary resistance varies nonlinearly against Froude number. Figure 2 shows the curve of residuary resistance. The humps and hollows are typical of such a curve. A hump is a region of high resistance and a hollow is a region of low resistance. Ships would like to operate at low resistance so that the power requirements are low. The hull of a ship forms waves when moving through water. The most significant of these are the bow waves and the stern waves. When the bow and stern waves interact such that a destructive interference results we get a hollow and when constructive interference results we get a hump. ScanIt was used to obtain data points from the graph to get the joint distribution of input output pairs. A sample data set is shown in the table below.

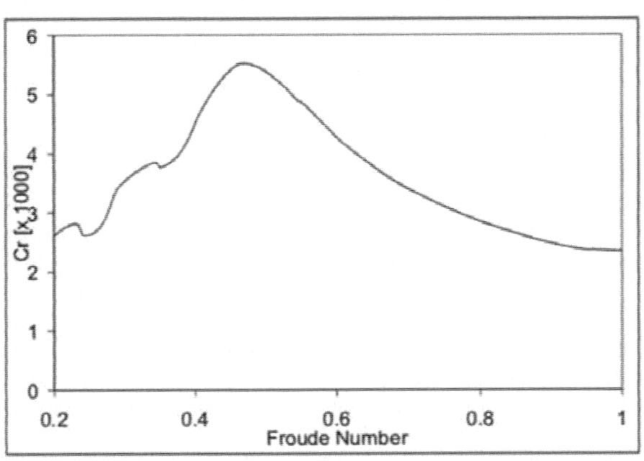

Figure 2: RESIDUARY RESISTANCE CURVE

Table 2: COEFFICIENT OF RESISTANCE(3b)

Froude Number	Monohull	S/L=0.2	S/L=0.3	S/L=0.4	S/L=0.5
0.2	2.971	3.192	3.214	2.642	2.555
0.25	3.510	4.540	3.726	4.019	3.299
0.30	3.808	5.303	4.750	4.464	3.938
0.35	4.800	6.771	5.943	5.472	4.803
040	5.621	8.972	7.648	7.085	6.589
0.45	8.036	12.393	12.569	10.934	9.064
0.50	9.038	14.874	14.237	12.027	10.112
0.55	8.534	15.417	12.275	10.535	9.394
0.60	7.626	12.818	10.089	8.962	8.361

NEURAL NETWORK ARCHITECTURE

An artificial neural network is an information processing system. It is a network of neurons. Neurons are mathematical functions which are built along the lines of the biological neurons. It takes an input, performs mathematical operations and returns the output. A neural network is characterised by the pattern of connection between neurons, the training algorithm, activation function, optimiser.

Selection of Input variables

The predictive ability of a network depends on the selection of key input variables. While there can be many a parameter in any experiment, to model a neural network it should be kept in mind to only consider parameters whose impact on the value of the predicted result are profound. In the original experiment length to displacement ($L/\nabla^{1/3}$) was found to be the predominant hull parameter, resistance decreasing with increasing $L/\nabla^{1/3}$. The other hull factors were B/T and L/B. The results show that the effects of L/B and $L/\nabla^{1/3}$ are similar. The effect of S/L was also studied in the experiment. As separation increased resistance was seen to decrease. So, the primary input features used in the modelling of the network are $L/\nabla^{1/3}$, B/T, S/L and Froude number. The output variable is the coefficient of residuary resistance.

Developing a network architecture

The total number of data points obtained using ScanIt software is 7513. This is split into training and test sets with 80% of data in favour of the training set. The training set is further split into a validation set.

Training set: The training set is used to adjust the weights of the network by backpropagation.

Validation set: It is a further subset of the test set and is used to tune parameters other than the weights such as the learning rate.

Test set: This contains new set of data on which the network has not been trained. It tests the predictive accuracy of the network.

The number of layers and the number of neurons in each layer can be varied to obtain the best solution. Since a network with three hidden layers can approximate any function the number of hidden layers taken is three. The number of neurons has been varied in the first hidden layer. Research on neural network suggests that for every parameter there should be at least 10 labelled data points. Based on this consideration the maximum number of neurons are fixed. The neurons in the second layer was varied and the error in prediction was seen to vary as shown in the figure below.

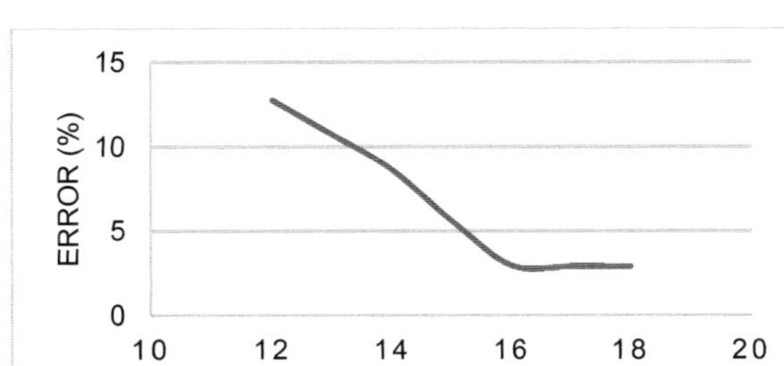

Figure 3: NO. OF NEURONS IN FIRST HIDDEN LAYER AND ERROR

Based on this the network architecture is fixed. Increasing the number of neurons in the first hidden layer beyond 16 does not decrease the error by a significant amount but would increase the computation cost. Hence the network architecture used for prediction is 16, 12 and 10 neurons in the hidden layers.

The activation function is Rectified Linear Unit (ReLu). The loss function is Mean Squared Error function because it is a regression problem. The optimizers are varied and the mean error obtained for different optimizers are plotted.

RESULTS

The network was developed in Python using Keras. Keras is a high-level neural networks API, written in Python and capable of running on top of TensorFlow. The following figures show the results obtained. Figure 4 shows that the mean squared error decreases with increase in epochs. The predicted values when plotted against the actual values should ideally be a straight line with slope equal to unity. Figure 5 shows that the results are very close to the ideal.

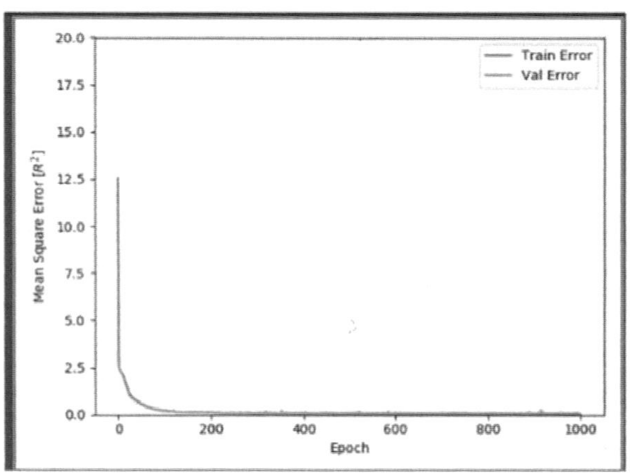

Figure 4: MEAN SQUARED ERROR

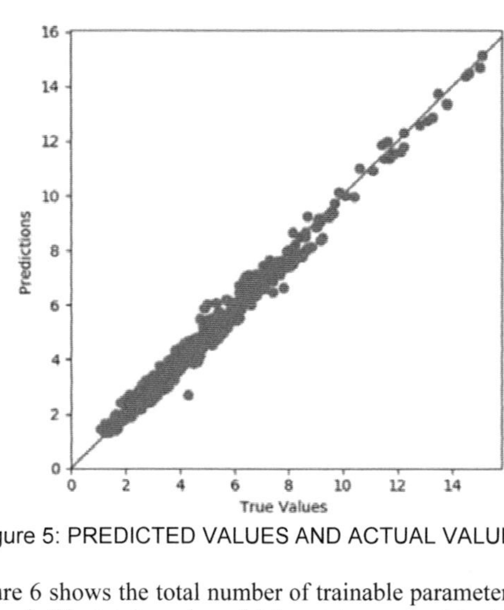

Figure 5: PREDICTED VALUES AND ACTUAL VALUES

Figure 6 shows the total number of trainable parameters in the network. The total number of data points are consistent with the number of parameters in that for every parameter there should be at least ten training data set. Figure 7 shows the mean error for different optimizers.

Layer (type)	Output Shape	Param #
dense (Dense)	(None, 16)	80
dense_1 (Dense)	(None, 12)	204
dense_2 (Dense)	(None, 10)	130
dense_3 (Dense)	(None, 1)	11

Total params: 425
Trainable params: 425
Non-trainable params: 0

Figure 6: TOTAL PARAMETERS

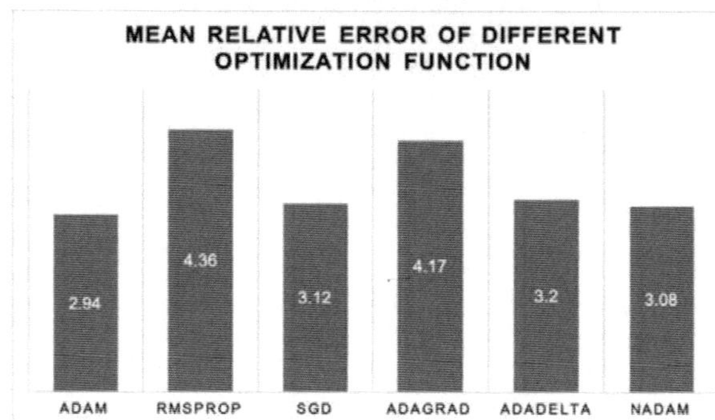

Figure 7: ERROR FOR DIFFERENT OPTIMISERS

CONCLUSIONS

It can be concluded from the results that the artificial neural networks provide a new approach to the prediction of ship resistance.

With sufficient data, the network can be used to predict almost anything and that is coherent with the Universal approximation theorem. The predictive ability of the network increases if the number of data points are increased as there is more data to train the network and hence adjust the weights and other parameters such as the learning rate. The results show that a three hidden layered network with 16,12 and 10 neurons in each can predict the resistance with an accuracy of over 97 percent.

The results are compared for different optimizers. The error is least with Adam. Adam is an adaptive learning rate method. It uses squared gradients to update the learning rate like RMSprop. Its similarity with SGD with momentum stems from using changing average of gradient descent and not gradient descent. This helps in the reduction of vertical oscillation which shifts the prediction away from the mean error and increases the time for convergence.

Further scope of work can include increasing the accuracy of the network using techniques such as Dropout Regularization.The data that has been trained in this paper using ANN is for NPL round bilge high speed displacement hull forms having a fixed block coefficient of 0.4. To improve the accuracy of this network it should be trained with data of different types of hull forms with varying hull form coefficients. The model proposed here will be limited in its ability to predict coefficient of residuary resistance of hull forms different from NPL round bilge high speed displacement hull form.

REFERENCES

[1] P. Couser, A. Mason, G. Mason, C. SMith, Br. Konsky, Artificial Neural network for hull resistance prediction. Computer applications and information technology in the maritime industries (COMPIT), 391–402 (2004).

[2] M. Insel, A. F. Molland, An Investigation Into Resistance Components of High Speed Displacement Catamarans. Trans RINA. 134, 1–20 (1992).

[3] A.F. Molland, P.Courser, Resistance experiments on a systematic series of high speed displacement catamaran forms: Variation of length-displacement ratio and breadth- draught ratio ,1994.

[4] Mason, A., Couser, P., Systems, F. D., Mason, G., Smith, C. R., and Konsky, B. R. V. (2005). Optimisation of Vessel Resistance using Genetic Algorithms and Artificial Neural Networks. In 4th International Conference on Computer Applications and Information Technology in the Maritime Industries (COMPIT 2005), (pp. 440–454). Hamburg, Germany.

[5] Deep Learning: Overview of Neurons and Activation Functions,2018.URL https://medium.com/@srnghn/deep-learning-overview-of-neurons-and-activation-functions-1d98286cf1e4.

[6] Neural networks and back-propagation explained in a simple way,2018.URL https://towardsdatascience.com/neural-networks-and-back-propagation-explained-in-a-simple-way-c6ee791499fa.

[7] Keras: The Python Deep Learning library, URL https://keras.io.

Proceedings of the 6th International Conference on Ship and Offshore Technology ICSOT 2019
November 7-8, 2019, IIT Kharagpur, India

ICSOT INDIA 2019-16

PREDICTION OF SPIRAL PATH CHARACTERISTICS OF AUTONOMOUS UNDERWATER GLIDERS

R V Shashank Shankar
DepartmentofApplied Mechanics,
Indian Institute of Technology Delhi
New Delhi, Delhi, 110016
India
Email

Sai Sabarish M
Department of Applied Mechanics,
Indian Institute of Technology Delhi
New Delhi, Delhi, 110016
India
saisabarishm7@gmail.com

R Vijayakumar
Department of Ocean Engineering
Indian Institute of Technology Madras
Chennai 600036
Vijay2028@iitm.ac.in

ABSTRACT

Underwater Gliders are very efficient mobile sensor platforms that can operate for months at a go and can travel thousands of kilometres. They embody simple, yet very effective mechanisms that make them perfect for oceanographic studies as well as oil spill detection operations. Gliders are underwater vehicles designed to locomote in water without using any external propulsion mechanism, rather they use buoyancy variation to propel themselves forward. The speed of the vehicle is restricted by the operation efficiency of the buoyancy tank. They manoeuvre using two mechanisms: deflection of an external rudder, and/or rotation of internal weights. Gliders follow a saw tooth profile and a spiral profile for executing motions in 2D and 3D planes. Pitch and roll corrections are used to place the glider into a 3D motion.

This paper aims to characterize a Robotic Fish glider studied in published works previously and predict the path the glider transfers into after the pitch and roll mechanism corrections using the deflection of the tail. The effect of rudder mechanism is studied by solving the highly non-linear equations of motion using Newton-Raphson method and fsolve recursive algorithm that were modelled in MATLAB and is validated using published results. The paper also aims to identify the most efficient algorithm that can be used to solve the established non-linear equations of spiralling motion.

NOMENCLATURE

B_0	Body Frame
I_0	Inertial Frame
π_0	Flow frame
$V=[V_1,V_2,V_3]^T$	Translational velocity in body frame ($m\ s^{-1}$)
$\Omega=[p,q,r]^T$	Angular velocity in body frame ($rad\ s^{-1}$)
$b=[x,y,z]^T$	Fish position in inertial frame
$\theta=[p,q,r]^T$	Fish attitude in inertial frame
m_s	Static mass of fish (kg)
m_h	Hull mass of fish (kg)
m_r	Roll control mass (kg)
m_p	Pitch control mass (kg)
r_p	Position of pitch control mass (kg)
r_r	Position of roll control mass (kg)
F_{ext}	Total external force created by the wings in the flow frame
T_{ext}	Hydrodynamic moments in the flow frame
α	Angle of attack
β	Angle of sideslip
M^{β}_{My}	Yaw moment with respect to βv^2
δ	Angle executed by Rudder
φ	Vehicle roll angle
θ	Vehicle pitch angle
ψ	Vehicle yaw angle
m_b	Ballast mass of fish
M	Mass of the displaced fluid by the fish
m_0	Excess mass
M	Generalised inertia matrix
m_1,m_2,m_3	Diagonal elements of mass Matrix
I_1, I_2, I_3	Diagonal elements of inertial matrix ($kg\ m^2$)
C_{q1}	Roll moment coefficient with respect to v^2
C_{q2}	Pitch moment coefficient with respect to v^2
K_{q3}	Yaw moment coefficient with respect to v^2 (m s/rad)

R_{bf}	Rotation matrix between body and flow frames
R_{eb}	Rotation matrix between body and inertial frames
C_{Do}	Drag coefficients with respect to v^2
C^{β}_{SF}	Side force coefficient with respect to βv^2 (rad $^{-1}$)
C^{δ}_{SF}	Side force coefficient with respect to δv^2
C_{L0}	Lift coefficient with respect to v^2
C_{M0}	Pitch moment with respect to v^2 (m)
M^{β}_{MR}	Roll moment with respect to βv^2
M^{α}_{MP}	Pitch moment with respect to αv^2
C^{α}_{D}	Drag coefficients with respect to αv^2 (rad $^{-2}$)
$C_{\alpha L}$	Lift coefficient with respect to αv^2 (rad $^{-1}$)

INTRODUCTION

Underwater gliders designed to locomote in water without using any external propulsion mechanism. Instead, gliders use buoyancy variation to descend, and surface and hydrodynamic lift forces experienced by the wings attached to the glider is used to propel itself forward. The lack of power-hungry active external propulsion is one of the reasons for the efficiency of the glider. The speed of the vehicle is restricted by the operation efficiency of the buoyancy tank. This operating mechanism results in a sawtooth pattern in the vertical plane as it moves along the horizontal plane.

Several studies have been conducted on the saw-tooth profile of the glider. P. Bhatta and N. E. Leonard[1] and N. E. Leonard and J. G. Graver [2] established the dynamic models of gliders that used a linearly traversing internal mass for steering itself. N. E. Leonard and J. G. Graver [2] also examined the saw tooth motion's stability parameters and established a feedback control procedure to stabilise it.

In addition to the sawtooth pattern, there is one more characteristic motion profile that is exhibited by the glider that can be described as a spiral. This spiral path is also an effect of the hydrodynamic forces experienced by the glider and is exceptionally efficient. This spiral motion is used to change the heading course of the glider. Though it is commonly observed, it has always been hard to mathematically quantify this motion as it is a very complicated dynamic model and the intricate multifaceted relationship that the hydrodynamic forces make it all more complicated to formulate a relationship between the control input and the resulting spiralling motion. P. Bhatta and N. E. Leonard [1] tried to bring out the numerical solution to the spiral motion and L. Kan et al. [3] presented simulations to the above-mentioned equations, and Mahmoudian et al. [4] used perturbation theory to offer an approximate solution. F. Zhang et al. [5] offer a simplified recursive algorithm to solve the equations analytically where the dynamic equations of a robotic fish were derived, and its spiral path was studied using fixed control inputs by solving the resulting equations using Newton's method. Shashank et al [6] developed solution for Zhang et al work based on fsolve algorithm and further predicted the effect of combined use of rudder and a roll control mechanism on the turning parameters.

This paper uses the equations of motion derived by F. Zhang et al. [5] using two different iterative solution methodology, namely the Newton-Raphson method and the MATLAB built-in non-linear equation solver fsolve. Newton-Raphson method was further enhanced using Jacobian Free Newton-Krylov method. The paper uses the hydrodynamic coefficients and the control inputs used by F. Zhang et al. [5] for verification of the results obtained. The results obtained were then compared and tabulated. Further, the run-time of the methods was also considered to find the most efficient method to solve these equations.

DYNAMIC MODELING OF UNDERWATER GLIDER

F. Zhang et al. [5] dynamically modelled a robotic fish that has an internal moving mass and a tail which also acts as a rudder. A simple point mass dynamic model with internal mass used for pitch control is used to describe the six degrees of motion of the glider. In this paper, the glider including all the internal moving masses is deemed as a rigid body system. The masses' position will thereby determine the speed and angle of pitch and glide. Figure 1 describes the mass distribution in the underwater glider.

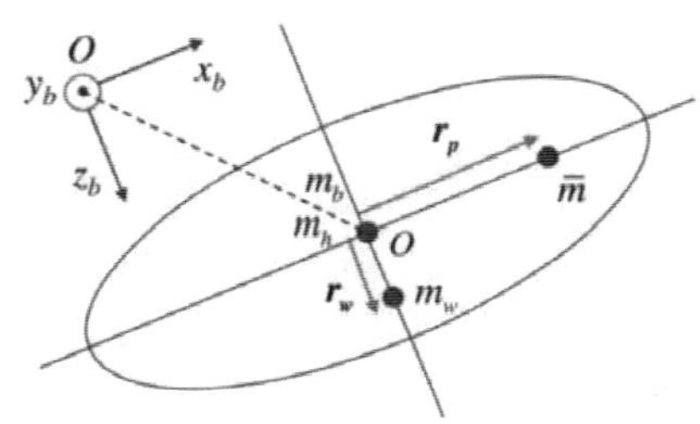

Figure 1. MASS DISTRIBUTION OF THE GLIDING ROBOTIC FISH[5]

The stationary body mass m_s involves three components: the hull mass m_h, which is assumed to be uniformly distributed, point mass m_w which is used to account for the non-uniform mass distribution located at a distance r_w and ballast mass m_b at the centre of gravity. The movable mass **I** located at r_p generates a pitching moment.

Excess mass m_0 is defined as $m_{I} = m_{n} + \mathbf{I} - \mathbf{I}$. The glider will be positively buoyant if $m_0 > 0$ and negatively buoyant if $m_0 < 0$.

As commonly used in the field of engineering, R represents the rotation matrix that translates the body-fixed reference frame to the inertial frame by following the 3-2-1 intrinsic rotation sequence [7].

Let ';::i=(';::1 ';::2 ';::3)1 and 1i=(11 12 13)1 represent the translational velocity and angular velocity, respectively. R is defined by the Euler angles with roll angle Φ, pitch angle θ and yaw angle ψ. R is thereby defined as,

cos n cos n   sin n sin n cos n - cosn sinn   cos n sin n cos n + sinn sinn
1 = n n 1 n n nn 1 ψ   cos n cos n + sinn sin n sinw   −sin ncos n + cos n sin n sinn 1
−sin n   sin n cos n   cos n cos n

(1)

It is assumed that the tail fin is rigid and pivots at a joint between the body and the tail axis Oz_b.

The dynamic model of tail fin enabled robotic fish was then calculated as

$$';:n_1 = n^{r-} (n';:_1 \times 1_1 + I_I 1 I^1 1 + I_{n11}) \times \quad (2)$$

$$\ln_1 = \overset{r-}{} (-11_1 + 11_1 \times 1_1 + n';:_1 \times ';:_1 + 1_{n11} + I_1 1 1_l \; (I^1 1) + I_1 1 1_l \times (I^1 1)) \quad (3)$$

where $n = (I_n + I_I) I_3 + m_l$, where I_3 is a 3 x 3 identity matrix and m_f is the added mass matrix calculated using strip theory.

HYDRODYNAMIC MODEL

To understand the hydrodynamic behaviour of the glider, all fixed velocity coordinates are translated to the body-fixed frame of reference using rotation matrix R_{bf}

$$1_{n1} = 1 \begin{matrix} \cos I \cos I & -\cos I \sin I & -\sin I \\ \sin I & \cos I & 0 \\ \sin I \cos I & -\sin I \sin I & \cos I \end{matrix} I \quad (4)$$

Where the angle of attack $I = \arctan(\frac{1_I}{1_n})$ and sideslip angle $I = \arcsin(\frac{1_I}{\|1_1\|})$.

The hydrodynamic forces (Lift L, Drag D and Side force SF) and moment (roll M_1, pitch M_2 and yaw M_3 moments) are translated from velocity frame to body-fixed frame, which results in

$$1_{n11} = I_{11}(-1\ 11 - I)^1 \quad (5)$$

$$1_{n11} = I_{11}(n\ n\ n)^1 \quad (6)$$

The hydrodynamic forces and moments are defined as,

$$1 = -11\ 1(I_{11} + I_1 I + I_1^I I) \quad (7)$$

$$I_n = \frac{1}{2} 11\ 1(I_{1n} I + I_{11} I)^I \quad (8)$$

$$1 = \frac{1}{2} 11\ 1(I_{11} + I_l I) \quad (9)$$

$$M_l = 11\ 1(\ _I 1 + I_{II}) \quad (10)$$

$$n = \frac{1}{2} 1 I_n\ II_{II} + I_{II} I + \ _I 1\ I \quad (11)$$

$$n = \frac{1}{2} 1 I_n (I_I\ I + \ _{II}\ 1 + I^I_I\ I) \quad (12)$$

where Kq_1, Kq_2, Kq_3 are rotation damping coefficients, and all constants with C in their notations are hydrodynamic coefficients which were evaluated by F. Zhang et al. [8] and used in the calculation.

STEADY-STATE SPIRALLING EQUATION

The paper investigates the steady state spiralling motion numerically using the Newton-Raphson method and the fsolve recursive algorithm which is an inbuilt function in MATLAB. The spiralling motion of glider is controlled by rudder deflection angle, internal mass position and excess mass I I 1_l I_I I. The equations (3-4) are now set to zero to derive the nonlinear steady-state spiralling equations given below

$$0 = m_2\sin\beta\ V\cos\varphi\cos\theta\ \omega_{3i} - m_3\sin\alpha\cos\beta\ V\sin\varphi\cos\theta\ \omega_{3i} - m_0 g\sin\theta + L\sin\alpha - SF\cos\alpha\sin\beta - D\cos\alpha\cos\beta$$

$$0 = m_1\cos\alpha\cos\beta\ V\sin\varphi\cos\theta\ \omega_{3i} + m_2\sin\beta\ V\sin\theta\ \omega_{3i} - D\sin\alpha\cos\beta - SF\sin\alpha\cos\beta - L\cos\alpha + m_0 g\cos\varphi\cos\theta \quad (14)$$

$$0 = (I_2 - I_3)\sin\varphi\cos\theta\cos\varphi\cos\theta\ \omega^2_{3i} + (m_2 - m_3)\sin\beta\sin\alpha\cos\beta\ V^2 - m_w g r_w \sin\varphi\cos\theta \pm 11\ IIn1\ _I\ nn1nn1nn1\ n) \pm 11\ 1(I_I\ \beta - \ _I\ nn1n1\ n) n1n1\ nn1\beta - 11\ 1(I_I\ \beta + \ _I\ n1nnn1nn\ n + I^I_{I_I} I) nn1I \quad (15)$$

$$0 = (I_1 - I_3)\sin\theta\cos\varphi\ \omega^2_{3i} + (m_3 - m_1)\cos\alpha\cos\beta\sin\alpha\cos\beta\ V^2 - m_w g r_w \sin\theta - I11In1nn\ n1nn + -11\ 1(I_I\ \beta + -11\ 1(I_I\ \beta - \ _I\ nn1n1\ n) nn1\beta + -11\ 1(In_I + + I_I\ I + \ _I\ nn1\ nn1nn1\ n) n1nI \quad (16)$$

$$0 = (I_2 - I_1)\sin nn\ \cos\varphi\cos\theta\ \omega^2_{3i} + (m_1 - m_2)\cos\alpha\cos\beta\sin\beta\ V^2 + mgr_p\sin\varphi\cos\theta - 11\ 1(In1 + I_I\ I + \ _I\ nn1\ nn1nn\ n) + -11\ 1(I_I\ \beta - \ _I\ nn1\ n) nn1I\cos\beta - 11\ 1(I_I\ \beta + \ _I\ n1nnn1nn\ n + I^I_{I_I} I) n1nI$$

It is assumed that the mass matrix and inertia matrix have the following form:

$$I = 1 \begin{matrix} I & 0 & 0 \\ 0 & I & 0 \\ 0 & 0 & I \end{matrix} I \quad (18)$$

$$I = 1 \begin{matrix} 1 & 0 & 0 \\ 0 & 1 & 0 \\ 0 & 0 & 1 \end{matrix} I \quad (19)$$

There are two important parameters in the spiralling motion: the turning radius R and the vertical speed $V_{vertical}$. By projecting the robot trajectory onto the horizontal plane and the velocity into the vertical direction, we have the following approximations for the turning radius and the vertical speed

$$I = \ln\ln(n - I)/1_1 \quad (20)$$

$$1_{1n111111} = 1\ nn1(n - I) \quad (21)$$

ITERATIVE METHODS

Newton-Raphson Method

Newton Raphson method is one of the most widely used root-locating formulae. If the initial guess is at the root x_i, a tangent is extended from the point [xi, f(xi)]. The point where this tangent crosses the x-axis represents the improved estimate of the root.[9]

Similarly, the Newton-Raphson method can be extended to multi equation form. It uses multivariable Taylor series to account for the fact that more than one independent variable contributes to the determination of the root. Rearrangment and algebraic manipulation of the Taylor series can be employed to solve for (22)

(13)

1,l 1,l/ − 1,l 1,l/

,11 = ,1 − I .l .l l,l 1,l (22)

1,l 1,l/ − 1,l 1,l/

,11 = ,1 − I .l .l l,l 1,l (23)

The denominator of each of these equations are referred to as the determinant of the Jacobian of the system.

fsolve ALGORITHM

fsolve algorithm is a combination of Levenberg-Marquardt and trust-region methods. These methods are based on nonlinear-least square algorithms. [10]

The function solves the nonlinear equations with input of the function matrix and the initial guesses. If needed, the optimisation options can be used to switch between the methods mentioned above.

RESULTS AND DISCUSSIONS

Newton-Raphson Method

Newton Raphson's method was used to solve the highly non-linear spiral equations with the initial guesses where In n 1 1 1 I II were set as zero. The iterations were run for different control inputs of II I l1 I I and conditions are given below in Table 1

Table 1. INPUT CONDITIONS

Condition Number	Inputs		
	m0	r_p	delta
1	25	0.3	45
2	25	0.4	45
3	25	0.5	45
4	25	0.6	45
5	25	0.7	45
6	10	0.5	45
7	15	0.5	45
8	20	0.5	45
9	30	0.5	45
10	35	0.5	45
11	40	0.5	45
12	25	0.5	30
13	25	0.5	35
14	25	0.5	40
15	25	0.5	50
16	25	0.5	55
17	25	0.5	60

The input conditions mentioned above were used to calculate the roots of the equations which is represented in Figure 2.

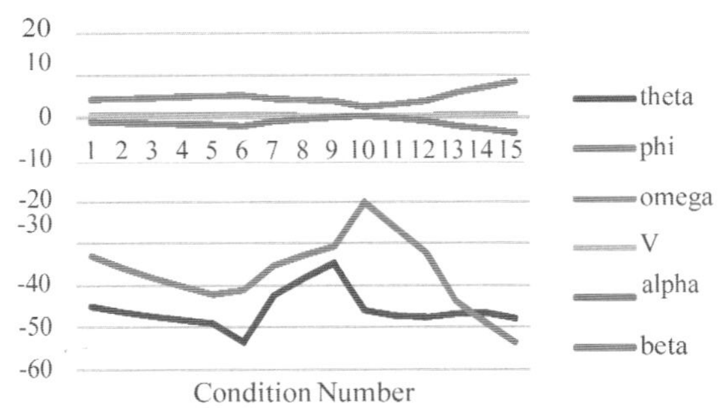

Figure 2. ROOTS OBTAINED USING NEWTON-RAPHSON METHOD

It should be noted that two iterations did not converge with the initial guesses given, and this instead exhibits the fundamental flaw of Newton Raphson method where there is an overwhelming dependence on the initial guesses to calculate the root.

Turning radius and Vertical Speed was then calculated by using Eqn Nos. (20), (21) and is shown in Figure 3.

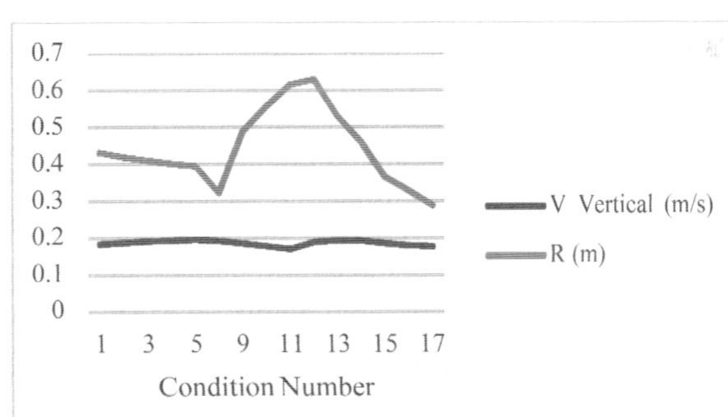

Figure 3. TURNING RADIUS AND VERTICAL SPEED USING NEWTON-RAPHSON METHOD

Roots of the equations calculated using fsolve algorithm is shown in Figure 4.

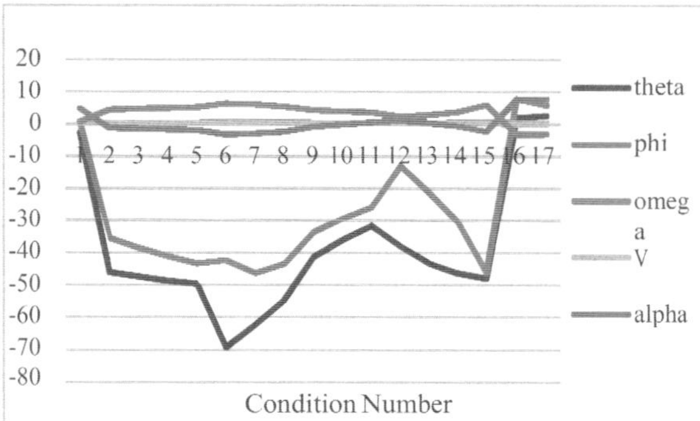

Figure 4. ROOTS OBTAINED USING fsolve ALGORITHM

Turning radius and Vertical Speed calculated using these roots is shown in Figure 5.

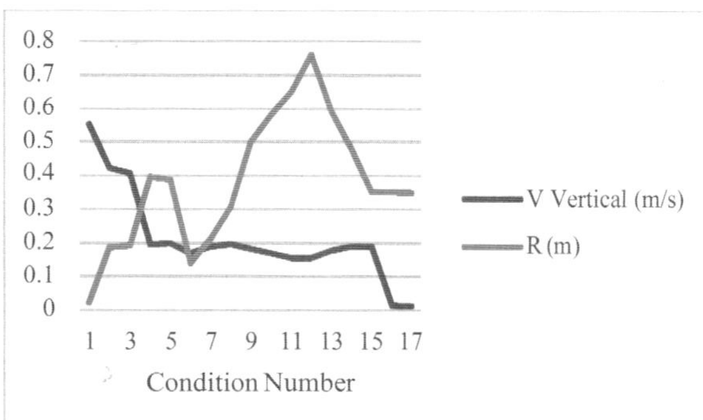

Figure 5. TURNING RADIUS AND VERTICAL SPEED USING fsolve ALGORITHM

Discussions on the results obtained

The percentage of error calculated by comparing the predicted values using Newton-Raphson method with that of Zhang, F et al. [5] is given in Figure 6.

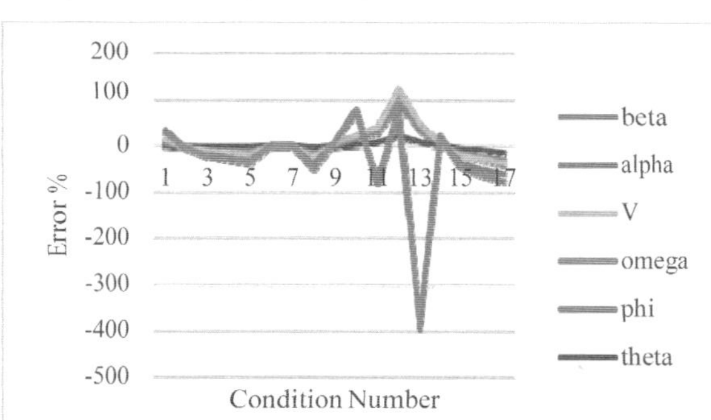

Figure 6. PERCENTAGE OF ERROR OF VALUES OBTAINED FROM NEWTON-RAPHSON METHOD IN COMPARISON WITH VALUES OF ZHANG, F et al. [5]

Two iterations fail in the Newton-Raphson method and five iterations have high value of error. This can be due to the initial guesses provided maybe inadequate.

After this, the roots were calculated using fsolve algorithm and is represented in Figure 7.

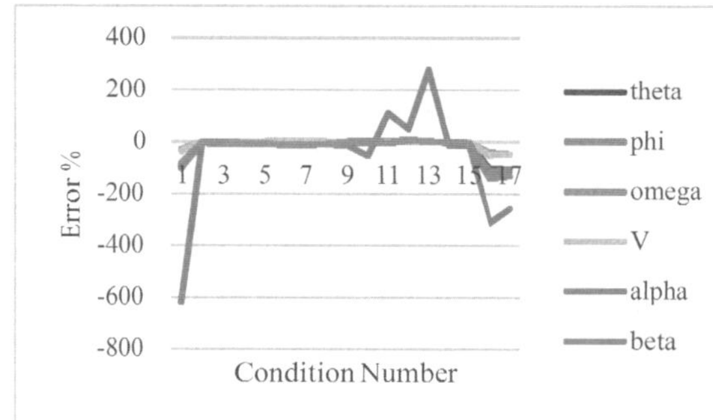

Figure 7. PERCENTAGE OF ERROR OF VALUES OBTAINED FROM fsolve RECURSIVE ALGORITHM COMPARISON WITH VALUES OF ZHANG, F et al. [5]

It is also seen that fsolve algorithm converges on the iterations that Newton-Raphson method failed to achieve, but the high value of error percentage persists especially in **I** in few input conditions.

It is also to be noted that Zhang, F et al. [5] uses Newton's iterative method to solve the equations, but the initial guesses provided in the paper does not work with the conventional Newton-Raphson method used in this project. It can hence be concluded that the method used by Zhang, F et al. [5] is not conventional Newton's method, rather is a Quasi-Newton method which uses a combination of multiple methods to overcome the issues plaguing the Newton-Raphson method.

It was seen that the Newton-Raphson method is twice as faster in computational speed when compared to fsolve algorithm. D. A. Knoll, D. E. Keyes [11] provide a faster and efficient alternative to use Newton-Raphson method. This method is called the Jacobian Free Newton-Krylov (JFNK) method, which as the name suggests bypasses the calculation of Jacobians and hence reduces the time involved. Using JFNK, the computational time for Newton-Raphson method reduced by half making it more than four times faster than fsolve algorithm.

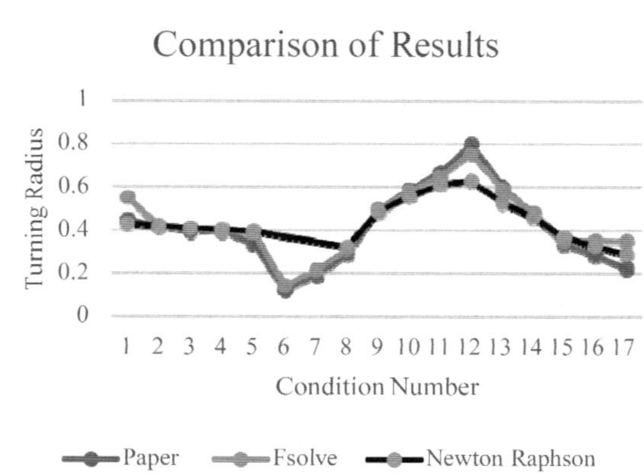

Figure 8. COMPARISON OF RESULTS OF DIFFERENT ALGORITHMS WITH THAT OF F. ZHANG ET AL. [5]

Figure 8 depicts the turning radius calculated using the roots obtained and substituting in equation (20) , it is seen the fsolve algorithm results closely match that of the published results for most condition numbers whereas Newton-Raphson method has significant errors in specific condition numbers and does not converge in condition numbers 6 and 7.

CONCLUSION

In this paper, the dynamic model of a robotic gliding fish used by F. Zhang et al. [5] was studied. The numerical model derived consists of a tail rudder effect on the roll of the body and the radius of the spiral path and the vertical speed executed. The equation of motions for the glider was derived and two numerical methods were used to solve it has been discussed and MATLAB programs were written for the same. The results obtained from both the methods have been compared with that of F. Zhang et al. [5].

It can be concluded that –

i. Newton-Raphson inherently is more efficient in computational time when compared to fsolve algorithm and can be further enhanced using the JFNK method but fails to converge in certain conditions.

ii. fsolve algorithm is relatively less error-prone and converges in all input conditions that were studied.

REFERENCES

[1] P. Bhatta and N. E. Leonard, "Nonlinear gliding stability and control for vehicles with hydrodynamic forcing," Automatica, 2008.

[2] N. E. Leonard and J. G. Graver, "Model-based feedback control of autonomous underwater gliders," IEEE J. Ocean. Eng., 2001.

[3] L. Kan, Y. Zhang, H. Fan, W. Yang, and Z. Chen, "MATLAB-based simulation of buoyancy-driven underwater glider motion," J. Ocean Univ. China, 2007.

[4] N. Mahmoudian, J. Geisbert, and C. Woolsey, "21 - Dynamics and control of underwater gliders in steady motions. Technical report.," 2009.

[5] F. Zhang, F. Zhang, and X. Tan, "Tail-Enabled Spiraling Maneuver for Gliding Robotic Fish," J. Dyn. Syst. Meas. Control, vol. 136, no. 4, p. 041028, 2014.

[6] Rayaprolu SS, Rajagopalan V. "Effect of Rudder and Roll Control Mechanism on Path Prediction of Autonomous Underwater Gliders". In: International Conference in Ocean Engineering. Springer; 2018.

[7] M. S. Triantafyllou and F. S. Hover, MANEUVERING AND CONTROL OF MARINE VEHICLES. 2003.

[8] F. Zhang, J. Thon, C. Thon, and X. Tan, "Miniature underwater glider: Design, modeling, and experimental results," in Proceedings - IEEE 0International Conference on Robotics and Automation, 2012.

[9] S. Chapra, Applied Numerical Methods with MATLAB® for Engineers and Scientists. 2015.

[10] "fsolve," Mathworks. [Online]. Available: https://fr.mathworks.com/help/optim/ug/fsolve.html#bqbnh0l-1.

[11] D. A. Knoll and D. E. Keyes, "Jacobian-free Newton-Krylov methods: A survey of approaches and applications," Journal of Computational Physics. 2004.

Proceedings of the 6th International Conference on Ship and Offshore Technology
ICSOT 2019
November 7-8, 2019, IIT Kharagpur, India

ICSOT INDIA 2019-17

IMO SECOND GENERATION INTACT STABILITY CRITERIA FOR PARAMETRIC ROLL RESONANCE (LEVEL-1 AND LEVEL-2 CRITERIA)

Dr. Shivaji Ganesan T
Indian Register of Shipping
Mumbai, Maharashtra, 400072
India
shivaji@irclass.org

Amresh Negi
Indian Register of Shipping
Mumbai, Maharashtra, 400072
India
amresh.negi@irclass.org

ABSTRACT
Immediately, after the release of international intact stability code (IS 2008 code [1]) IMO started to work on second generation intact stability criteria which would incorporated the ship dynamics in realistic sea waves [2]. Following detail discussions [3-5] at IMO subcommittee level, it was concluded that 2nd generation intact stability assessment will be based on multi-tier approach for five different failure modes (i.e. Pure loss of stability, parametric roll resonance, surf-riding and broaching to, dead ship, excessive accelerations) with each failure mode containing two prescriptive vulnerability checks (i.e. Level-1, Level-2) and third level (Level-3 direct stability assessment). This paper focus on parametric roll resonance phenomena which is a common phenomenon of slender large fast ship. Vulnerability formulas and criteria checks prescribed by IMO[5] are programmed using FORTRAN programming language. Level-1 vulnerability check is based on magnitude of stability change and level-2 vulnerability assessment consist of two checks: 1) *first check requires calculation of metacentric height variations along with critical ship speed* 2) *second check requires calculation of maximum roll amplitude in various head and following seas acted upon by a regular incident wave of length equal to ship length with different wave amplitude. Finally, all numerical results for a given loading conditions pertaining to both checks are presented and discussed in detail.*

NOMENCLATURE

GM_C	metacentric height of the loading condition in calm water including free surface correction (m);
ΔGM_1	the amplitude of the variation of the metacentric height (m):
C_m	midship section coefficient of the fully loaded condition in calm water;
L	Length of the ship (m);
B	moulded breadth of the ship (m).
A_k	total overall projected area of the bilge keels (no other appendages) (m^2);
V	volume of displacement corresponding to the loading condition (m^3);
D	moulded depth at side to the weather deck (m);
V_D	volume of displacement at a waterline equal to D and at zero trim (m^3);
d	draft amidships corresponding to the loading condition under consideration (m);
A_W	water plane area at the draft equal to d (m^2);
S_W	0.0167 wave steepness;
d_H	$d+\delta d_H$ (m);
d_L	$d-\delta d_L$(m);
δd_H	$Min(D-d, 0.5\times L.S_W)$ (m);
δd_L	$Min(d-0.25d_{full}, 0.5\times L.S_W)$ (m);
d_{full}	draft corresponding to the fully loaded departure condition (m).
I_H	moment of inertia of the waterplane at the draft d_H and at zero trim (m^4);
I_L	moment of inertia of the waterplane at the draft d_L and at zero trim (m^4);
V_s	the service speed (m/s);
T_ϕ	the roll natural period in calm water;
H_i	wave height (m);
λ_i	wave length (m);
g	gravity;

INTRODUCTION
Ship should possess adequate intact stability to carry out its mission safely and comfortably. Adequate intact stability is one of the most basic requirements which assures the safety of the ship and its cargo. To assess whether the ship is inherited with sufficient intact stability, IMO released the international stability regulation 2008 intact stability (IS) code through MSC circular 267(85) [7] which came into force in July 2010. After the release of so called 'First generation Intact stability (2008 IS code), some delegations raised concern over insufficient addressing of dynamic stability failure in waves which are dangerous from stability viewpoint. To address this critical issue, the preamble and Part-B chapter:5 of 2008 IS code provides some insight for special consideration to be taken into account of dynamic stability phenomena in waves which are i) righting lever variation ii) resonant roll in dead ship condition iii) broaching and other maneuvering related phenomena. In this paper we discuss about parametric roll resonance stability failure mode which is related to righting lever variations in waves. During 1950s, several casualties regarding severe rolling were reported on fishing vessels and on small coastal cargo vessels when sailing in following seas. Experimental and

numerical investigations by Paulling and Rosenberg [8] and Paulling [9] clearly illustrated the effects of waves on transverse stability in following and stern quartering seas. Even before these two works, the effect of transverse stability variations for a ship travelling in longitudinal waves is studied by Kempf [10], Graff and Heckscher [11]. Based on the above mention works, parametric roll is defined as an unstable roll motion which can grow to large amplitude under certain circumstances if proper measures are not taken to mitigate its growth, it can lead to ship capsizing or loss of its cargo.

Parametric roll or traditionally 'auto parametrically excited roll' is a phenomena due to periodic variations of transverse stability in waves, characterized by increase of stability for wave trough at amidships and decrease of stability for wave crest located at amidships. Till late 1990s, it was presumed that parametric roll is a major concern for small vessels sailing in following seas, but after the incident of APL China (Post-Panamax container carrier) on its trans-pacific voyage, the fundamental understanding of parametric rolling has come under rigorous scrutiny. A detail investigation by France et al., [12] on APL China incident reveals that roll amplitude upto 35deg along with significant pitch and yaw motion is observed in head seas. These severe motions is due to considerable change in transverse stability in head or near head seas which is a direct result of finer hull form accompanied with large bow flare and overhanging stern. Numerical and experimental studies of Shin et al. [13] found out that large container ships in head seas are vulnerable to parametric rolling when the frequency of transverse stability variations occurs at approximately twice the natural frequency of roll motion with encounter wave's length ranging between 0.8-1.2 times of ship length. Outcome of the investigation raised concerns over long standing heavy weather maritime practice of sailing in head sea at reduced forward speed. To accommodate dynamic stability failure, IMO initiated work on new generation of intact stability which is based on physics of the phenomena resulting in stability failure mode called as 'second generation intact stability criteria' following a multi-tier approach which was proposed by ABS [14].

In this paper, all mathematical formula and its numerical implementation to perform Level-1 and Level-2 vulnerability check as outlined in IMO document [6] will be presented. Vulnerability formulas and criteria's are programmed in-house at IRS research centre using INTEL FORTRAN. Verification and validation of developed numerical code is carried out at regular intervals with available industry standard stability software's MAXSURF and also with IMO SDC documents results[16]. Level-1 vulnerability check is based on magnitude of stability change for a single longitudinal wave passing the ship with wave length equaling the ship length. Level-2 vulnerability assessment consist of two checks: i) Check-1 requires calculation of metacentric height (GM) variations in waves along with critical ship speed as a weighted average in a set of 16 waves derived from IACS recommended wave scatter table or any other site specific scatter table ii) Check-2 is based on probabilistic approach requiring calculation of maximum roll amplitude in head and following seas of wave length equal to ship length with different wave amplitude for various operational speeds. This vulnerability assessment requires complex numerical simulations of ship motions in both following and head seas. Finally, all numerical results for a given loading conditions pertaining to both checks are presented and compared with published results. As of now level-3 direct stability assessment will be taken up as a future research scope.

PHYSICAL BACKGROUND

If a ship sails through waves of length comparable to ship length, the submerged part of the ship hull changes drastically. Suppose, if the ship is rolled when wave trough is at amidship, increased transverse stability provides stronger pushback or restoring to its equilibrium upright position. As the ship returns to the upright position, its roll rate is greater, since there was an additional pushback from the increased stability. If at that time, the ship has the wave crest at amidship, the stability is decreased and the ship will roll further to the opposite side because of the greater speed of rolling and less resistance to heeling. Then, if the wave trough reaches the amidship section when the ship reaches its maximum roll, transverse stability increases again and the cycle starts again. Note that there was one half of the roll cycle associated with the passing of an entire wave. So, there are two waves passing the ship during each roll period which states the roll period is about twice of the encountered wave period. However, roll damping plays a vital role in parametric roll development. If the 'loss' of energy per cycle caused by damping is more than the energy 'gain' caused by the changing stability in longitudinal seas, the roll amplitude will not increase and the parametric resonance will not be developed. Once the energy "gain" per cycle is more than the energy "loss" due to damping, the amplitude of the parametric roll starts to grow. In other words, if the roll damping moment is higher than the certain roll damping threshold value, then parametric roll resonance is not possible. If the roll damping moment is below the threshold, then parametric roll resonance can take place. Schematic of parametric roll development is shown in Belenky et al. [15]

PARAMETRIC ROLLING: LEVEL 1

A ship is considered not to be vulnerable to the parametric rolling failure mode if:

$$\frac{\Delta GM_1}{GM_C} \leq R_{PR} \qquad (1)$$

where,

$$\begin{aligned} R_{PR} &= 1.87, if\ ship\ has\ a\ sharp\ bilge\ keel^{*} \\ &= 0.17 + 0.425\left(\frac{100A_k}{LB}\right), if\ C_m \geq 0.96; \\ &= 0.17 + (10.625 \times C_m - 9.775)\left(\frac{100A_k}{LB}\right), if\ 0.94 < C_m < 0.96; \\ &= 0.17 + 0.2125\left(\frac{100A_k}{LB}\right), if\ C_m \leq 0.94; and\ \left(\frac{100A_k}{LB}\right) \leq 4.0, \end{aligned}$$

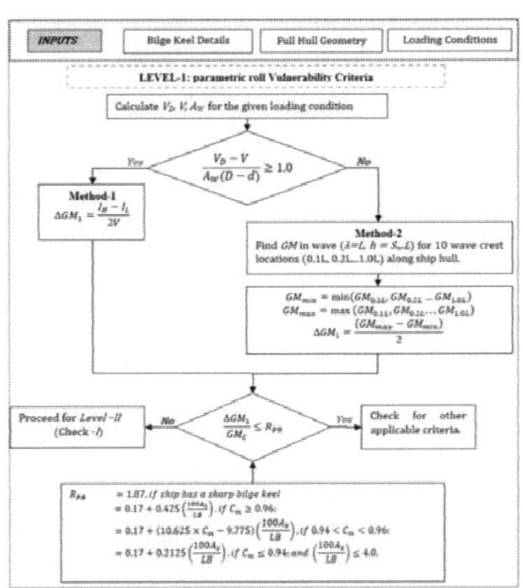

Figure 1 FLOWCHART: LEVEL-1 VULNERABILITY CRITERIA

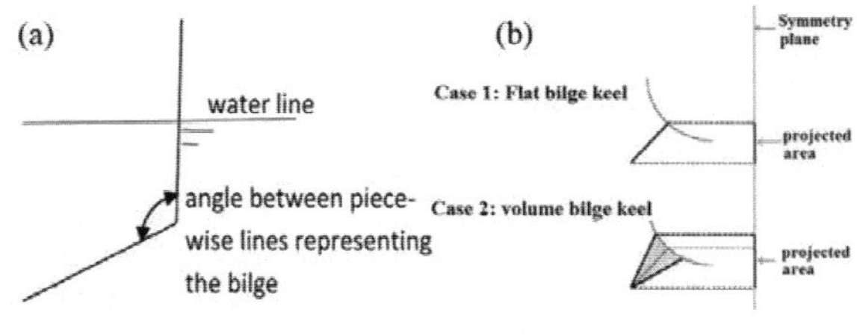

Figure 2 A) SHARP KEEL DEFINITION B) PROJECTED AREA OF FLAT AND VOLUME BILGE KEELS (COURTESY. [16])

For all standard flared hull in Level-1 criterion [6], the change in amplitude of metacentre ΔGM is calculated by a simplified method (method-1) in which the moment of inertia of waterplane in waves is approximated with that in calm water but with lowest draught d_L and highest draught d_H. This is because the restoring variation in longitudinal waves depends on geometry of bow and stern sections. In case of tumblehome topside ships method-2 is adopted to determine ΔGM. Fig. 1 outlines the overall procedure to perform level-1 parametric roll vulnerability check.

*sharp bilge keel see fig.2 used here means that bilge radius is smaller than 1% of the ship's breadth and the angle between piece-wise lines representing the bilge is smaller than 120°.

Numerical example

Level-1(Method-1): Ship consider for this example is a C11-Post Panamax container ship. Fig.3 displays the 3D perspective view and body plan view of container hull considered. Loading condition of this hull is given in Table.1 calculated vertical prismatic coefficient is greater than 1.0 $\left(\frac{V_D-V}{A_W(D-d)} \geq 1.0\right)$ so a simple calculation of moment of inertia I_H & I_L for two different drafts d_H and d_L are sufficient to determine ΔGM_1. Calculated values of GM and other parameters are presented in Table 2. For this loading condition, from Table .2 $\frac{\Delta GM_1}{GM_C}$ is greater than the required index R_{PR}, so the ship is judged as vulnerable to parametric roll resonance.

Table.1 PRINCIPAL DIMENSIONS, HYDROSTATIC DATA

Parameter	Input value
L_{pp}(m)	262.0
L_f(m)	262.0
B(m)	40.0
D(m)	24.45
d(m)	11.5
LCG(m)	125.52
C_B	0.559
C_M	0.950
GM(m)	1.965
T_ϕ(s)	25.1
VCG(m)	18.4
l_{BK}/L_{pp}	0.292
b_{BK}/B	0.010
V_s(m/s)	12.165

Table.2 LEVEL-1 CALCULATED PARAMETERS

Parameters	Present value	IMO SDC 4/5/1/Add.1
I_H (m^4)	1073029.8	1,076,500
I_L (m^4)	794571.93	799,900
R_{PR}	0.3561	0.3561
$\frac{\Delta GM_1}{GM_C}$	1.05	1.045

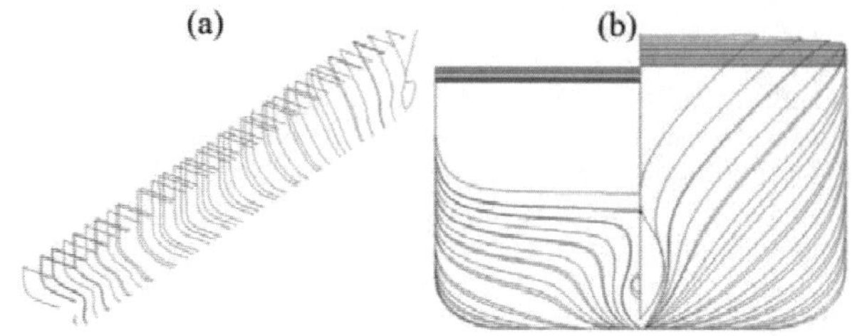

Figure 3 A) 3D-PERSPECTIVE VIEW B) BODY PLAN VIEW OF C11-POST PANAMAX CONTAINER SHIP

Level-1(Method-2): Suppose, consider a case for which the vertical prismatic coefficient$\left(\frac{V_D-V}{A_W(D-d)}\right)$ is less than 1.0, then the calculation of change in magnitude of metacentre height variation become little complicated. For a given loading condition, ΔGM_1 is calculated as one-half the difference between the maximum and minimum values of the metacentric height considering the ship to be balanced in sinkage and trim on a series of waves with following characteristics: wavelength $\lambda = L_{pp}$; wave height $h = S_W \times L_{pp}$ (S_W = 0.0167); and the wave crest is to be centered amidships, and at 0.1L, 0.2L, 0.3L, 0.4L and 0.5L forward and 0.1L, 0.2L, 0.3L and 0.4L aft thereof including free surface correction. Fig.4 displays the corresponding GM values for different wave crest location along the ship length. Calculated values of relevant parameters are presented in Table 3.

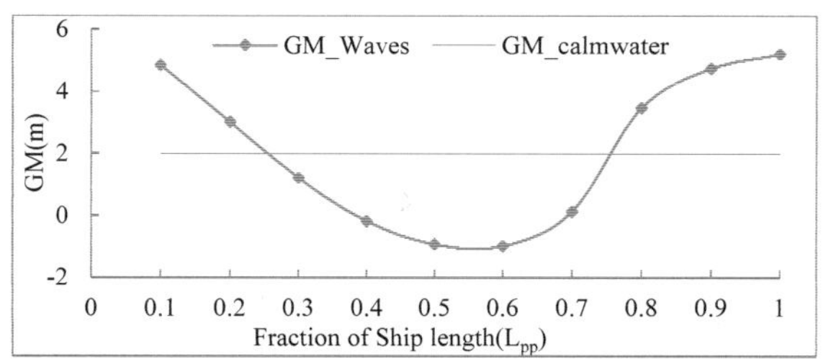

Figure 4 METACENTRIC HEIGHT VARIATIONS FOR DIFFERENT WAVE CREST LOCATION ALONG SHIP LENGTH

Table.3 LEVEL-1 PARAMETERS VALUE FOR DIFFERENT WAVE LOCATION

Parameters and factors	Present value
GM_{min}	-0.9861
GM_{max}	5.202
$\Delta GM_1 = \frac{(GM_{max} - GM_{min})}{2}$	3.094
R_{PR}	0.3561
$\frac{\Delta GM_1}{GM_C}$	1.575

From Table 3, it is observed that $\frac{\Delta GM_1}{GM_C}$ is greater than the required index R_{PR}. So the ship is judged as vulnerable to parametric rolling for this loading condition. By doing the above exercise, we can conclude that two methods to calculate the magnitude change in metacentric height variations are consistent and reliable.

PARAMETRIC ROLLING: LEVEL 2

Level-2 vulnerability assessment consist of two checks: i) Check-1 requires calculation of metacentric height (GM) variations in waves along with critical ship speed as a weighted average in a set of 16 waves derived from IACS recommended scatter table or any other site specific scatter table ii) Check-2 is based on probabilistic approach requiring calculation of maximum roll amplitude in head and following seas of wave length equal to ship length with different wave amplitude for various operational speeds.

Level-2(check-1)

A ship is considered not to be vulnerable to the parametric rolling failure mode if:

$$C1 \leq R_{PRO} \tag{2}$$

where $C1 = \sum_{i=1}^{N} W_i C_i$, and $R_{PRO} = 0.06$. Here C_1 is calculated as a weighted average from a set of 16 waves where, W_i = weighting factor;

$$C_i = \begin{cases} 0 \;\; for\; GM(H_i, \lambda_i) > 0 \; if \; \dfrac{\Delta GM(H_i,\lambda_i)}{GM(H_i,\lambda_i)} < R_{PR} \; or \; V_{PRi} > V_s \\ 1 \qquad\qquad\qquad\qquad\qquad\qquad\qquad\qquad else \end{cases}$$

$\Delta GM(H_i, \lambda_i)$ = one-half the difference between the maximum and minimum values of the metacentric height calculated for the ship (m), corresponding to the loading condition, considering the ship to be balanced in sinkage and trim on a series of waves characterized by H_i& λ_i

$GM(H_i, \lambda_i)$= the average value of the metacentric height calculated for the ship (m), corresponding to the loading condition, considering the ship to be balanced in sinkage and trim on a series of waves characterized by a H_i& λ_i

V_{PRi} = the reference ship speed (m/s) corresponding to parametric resonance conditions, when $GM(H_i, \lambda_i) > 0$

$$= \left| \frac{2\lambda_i}{T_\phi} \cdot \sqrt{\frac{GM(H_i,\lambda_i)}{GM_C}} - \sqrt{g \frac{\lambda_i}{2\pi}} \right|$$

Numerical procedure: Ship and the loading conditions considered for this check is same as the previous case. Methodology adopted to evaluate the level 2 check-1criteria is shown in Fig. 5. Entire process is subdivided into three sections.

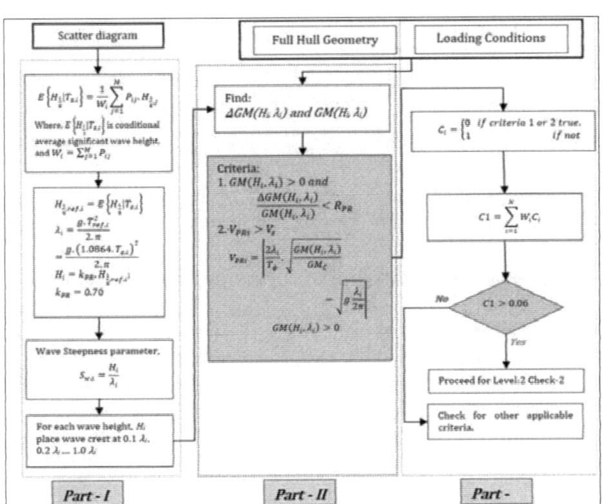

Figure 5 FLOWCHART: LEVEL-2 CHECK 1

Part-1: Here the inputs are only the scatter table from which the different wave cases are determined along with wave length (λ_i), wave height (H_i), weightage (W_i), and wave steepness($S_{w,i}$) using the below numerical formula as given in [16],

$$H_{\frac{1}{3},ref,i} = E\left\{H_{\frac{1}{3}} | T_{z,i}\right\} = \frac{1}{W_i}\sum_{j=1}^{M} P_{ij} . H_{\frac{1}{3},j}$$

weighting factor $W_i = \sum_{j=1}^{M} P_{ij}$;

wave height, $H_i = k_{PR} . H_{\frac{1}{3},ref,i}$;

wave steepness, $S_{w,i} = \frac{H_i}{\lambda_i}$;

wave length, $\lambda_i = \frac{g.T_{ref,i}^2}{2.\pi} = \frac{g.(1.0864.T_{z,i})^2}{2.\pi}$

Part-2: Here the input geometry and loading conditions are required for calculating the change in magnitude of metacentric height ($\Delta GM(H_i, \lambda_i)$), average value of metacentric height ($GM(H_i, \lambda_i)$) and reference ship speed (V_{PRi}) for different wave cases (i.e. series of 16 wavesIACS scatter table) with wave length λ_i and the associated zero up-crossing period T_z, is used for calculations of GM variations which are presented in Table 4. *Part-3:* Using the R_{PR}(=0.3561)from previous level-1 calculations, and V_S=12.165 m/s the dangerous wave cases among 16 are identified is shown in Table 4. Then, the weighted average is calculated as summation over 16 wave cases. Since the calculated C_1=0.436417 is more than the threshold value R_{PRO}, ship is judged as vulnerable to parametric roll which direct us to level 2 check-2.

Table 4: DANGEROUS WAVES CASES IDENTIFIED

Wave steepness $S_{w,i}$	$\frac{\Delta GM(H_i,\lambda_i)}{GM(H_i,\lambda_i)}$	$GM(H_i,\lambda_i)$	V_{PRi}	Weight W_i	C_i
0.0155	0.023763	1.93959	4.142191	0.000013	0
0.0133	0.040557	1.938756	4.667983	0.001654	0
0.0154	0.093711	1.933916	4.906389	0.020912	0
0.0166	0.131882	1.952909	4.818701	0.092799	0
0.0167	0.198589	2.048191	4.260029	0.199218	0
0.0166	0.220631	1.947332	3.820994	0.248788	0
0.0162	0.523859	1.974888	2.784678	0.208699	1
0.0156	0.643739	2.096851	1.032768	0.128984	1
0.0149	0.659472	2.178445	1.005801	0.062446	1
0.0140	0.634935	2.221136	3.267499	0.02479	1
0.0132	0.5901	2.234013	5.724957	0.008367	1
0.0123	0.519459	2.148571	7.788205	0.002473	1
0.0115	0.421646	1.962384	9.072806	0.000658	1
0.0107	0.315155	1.788852	10.27383	0.000158	0
0.0100	0.229531	1.693559	12.19345	0.000034	0
0.0094	0.168978	1.660005	14.9534	0.000007	0

Level-2(check-2):

For a ship to be considered not to be vulnerable to the parametric rolling failure mode if:

$$C2 \leq R_{PRO}(0.06) \tag{3}$$

where

$$C2 = \frac{\left[\sum_1^3 C2_h(Fn_i) + 0.5C2_h(0) + \sum_1^3 C2_f(Fn_i) + 0.5C2_f(0)\right]}{2N+1}$$

$C2_h(Fn_i)$ = $\sum_{i=1}^{N} W_i C_i$ for the ship in head waves with a speed equal to V_i;

$C2_f(Fn_i)$ = $\sum_{i=1}^{N} W_i C_i$ for the ship in following waves with a speed equal to V_i;

$$C_i = \begin{cases} 1 & \text{if } \textbf{max. roll amplitude} \text{ exceeds 25 deg.} \\ 0 & \text{else} \end{cases}$$

Fn_i = $V_i / \sqrt{(Lg)}$;

V_i = $V_s K_i$, means the ship speed (m/s)

$$k_i = \begin{cases} 1.0 & i = 1 \\ 0.866 & i = 2 \\ 0.5 & i = 3 \end{cases}$$

Here 'N' is the total number of wave cases for which the maximum roll angle is evaluated for a combination of ship speed and heading. The value of C2 is calculated as an average of values of C2(F_{ni}), each of which is a weighted average from the set of waves given by IACS wave scatter table, for each set of Froude numbers and wave directions. Fig.6 displays the procedure to estimate the index C2. For all calculation, the ship is balanced in sinkage and trim on a series of waves with the following characteristics: wavelength, λ=L; wave height, h_i=0.01iL where, i=0,1,2,...,10. For each wave height h_i, the maximum roll angle for different ship speed V_i in head and following waves can be evaluated using three methods. To determine the maximum roll amplitude Newton's second law of motion is used to define the roll motion of the vessel with simplest one degree of freedom motion which includes four moments acting on the ship:

1.inertia, including added inertia (or added mass, A_{44}) as a part of hydrodynamic forces; $M_{IN} = (I_{xx} + A_{44})\ddot{\phi}$
2.roll damping, which expresses energy loss from roll motions in creating waves, vortexes and skin friction; $M_D = (I_{xx} + A_{44}) \times (2\alpha\dot{\phi} + \gamma\dot{\phi}^3)$
3.roll restoring (stiffness) is modelled with GZ curve; $M_R = \rho\nabla gGZ(t, \phi)$
4.transverse wave forces are absent for a ship in exact following or head long-crested seas; $M_{EX} = 0.0$
Combining all of the above terms (1)-(4) we get the uncoupled single degree of freedom roll motion equation:

$$\sum M = 0.0$$

$$(I_{xx} + A_{44})\ddot{\phi} + (I_{xx} + A_{44}) \times (2\alpha\dot{\phi} + \gamma\dot{\phi}^3) + M_R = \rho\nabla gGZ(t, \phi) = 0.0 \quad (4)$$

I_{xx}, A_{44} can be calculated by empirical formulae given in [4]. This equation. (4) is a differential equation relating the roll motion with the roll rate ($\dot{\phi}$) and the roll acceleration ($\ddot{\phi}$). These quantities are related through differentiation: the angular velocity is a derivative of roll and the angular acceleration is a derivative of angular velocity. The solution of a differential equation yields roll motion amplitude. Table 5 list the methods to determine the moment coefficients in eqn.4 which are required to determine the maximum roll amplitude. For the prediction of restoring moment at smaller heel angle GM may be sufficient but for the calculation of restoring moment at higher heel angles, GZ curve need to be included in the calculations.

Table 5: METHOD FOR DETERMINING MOMENT COMPONENT

Coefficients	Method	
Inertial term , $I_{xx} + A_{44}$	Trivial. based on empirical formulae	
Nonlinear roll damping coefficients α & γ	Trivial	**Ikeda's simplified method** Roll decay model tests RANSE based CFD solvers
Nonlinear restoring $GZ(t, \phi)$	Non-Trivial	**Method-1**(GZ curves in waves fitted with bi-cubic spline) **Method-2**(GZ curve in waves appx. using only GM values in waves) **Method-3**(Calm GZ fitted with a quintic curve)

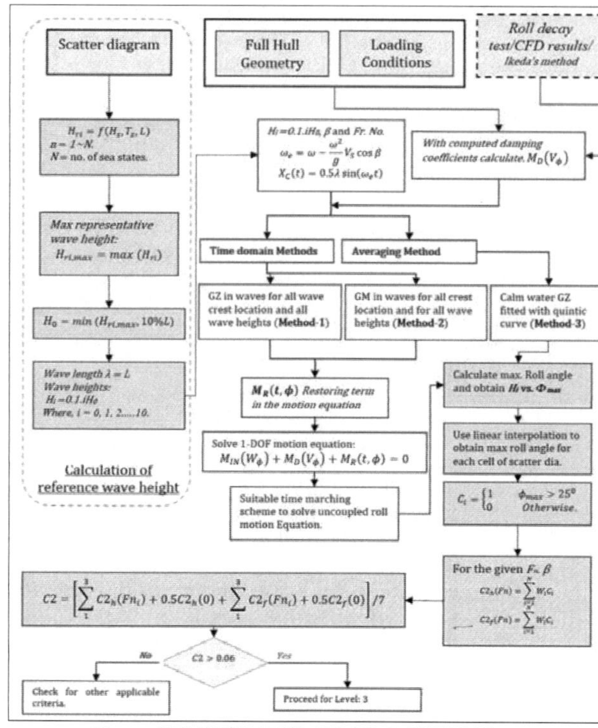

Figure 6 FLOWCHART: LEVEL-2 CHECK 2

Systematic numerical procedure adopted in evaluation of the stability failure index is as follows:

Step-1(Determination of reference wave height H_{ri}): For all possible significant wave heights H_s and zero-crossing wave period T_Z, in the wave scatter diagram(Standard IACS Scatter table) using the below formula

$$H_{\%eff} = 4.0043\sqrt{m_0}$$

$$m_0 = \left[\begin{array}{l} \int_{0.01\omega_L}^{\omega_L} \left\{\frac{\frac{\omega^2 L}{g} sin\left(\frac{\omega^2 L}{2g}\right)}{\pi^2 - \left(\frac{\omega^2 L}{2g}\right)^2}\right\}^2 A\omega^{-5} e^{-B\omega^{-4}} d\omega + \\ \int_{\omega_L}^{3\omega_L} \left\{\frac{\frac{\omega^2 L}{g} sin\left(\frac{\omega^2 L}{2g}\right)}{\pi^2 - \left(\frac{\omega^2 L}{2g}\right)^2}\right\}^2 A\omega^{-5} e^{-B\omega^{-4}} d\omega \end{array}\right] \quad (5)$$

Effective significant wave height as given by [16], $H_{ri} = min[H_{\%eff}, 0.1 \times L_{pp}] = 11.9365$ (m).

Step-2(Determination of roll damping coefficients): Using simplified Ikeda's method as outlined in [16], equivalent linear roll damping coefficient $B_{44}(\phi_a)$ as a function of the roll amplitude ϕ_a can be obtained. Fig.7 displays the computed damping coefficient along with IMO [16] published results. Here the roll damping coefficient (B_{44}) and circular roll frequency (ω=2π/T_ϕ) are normalized as follows,

$$\hat{B}_{44} = \frac{B_{44}(\phi_a)}{\rho\nabla B^2}\sqrt{\frac{B}{2g}}; \hat{\omega} = \omega\sqrt{\frac{B}{2g}} \quad (6)$$

where ρ denotes water density, ∇ displacement volume, B beam, T_ϕ roll period and g is gravity acceleration. The prediction method used here divides the roll damping into five different components. The roll damping coefficient B_{44} is expressed as summation of frictional (B_F), the wave (B_W), the eddy (B_E) and the bilge keel (B_{BK}) components at zero forward speed, and at forward speed, the lift (B_L) component as given in below equation 7:

$$B_{44} = B_F + B_W + B_E + B_{BK} + B_L \qquad (7)$$

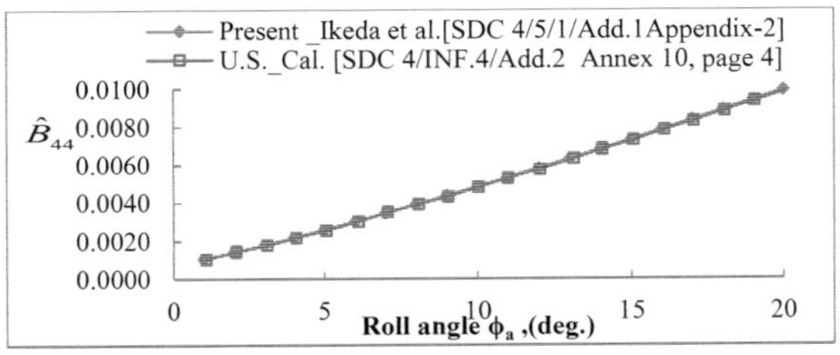

Figure 7 IKEDA SIMPLIFIED NON-DIMENSIONAL ROLL DAMPING

For the given loading condition of ship, the equivalent linear and cubic damping coefficients are,

linear roll damping coefficient, $\alpha = \frac{\omega_\phi}{\pi} a = 1.58 \times 10^{-03}$

cubic roll damping coefficient, $\gamma = \frac{4c}{3\pi^2}\left(\frac{2\pi}{\omega_\phi}\right)$=3.022

Step-3(Hydrostatic restoring computation): For a given loading condition, restoring coefficients are obtained for waves of height equal to 0.1H_{ri}, 0.2H_{ri}, 0.3H_{ri}, 0.4H_{ri}, 0.5H_{ri}, 0.6H_{ri}, 0.7H_{ri}, 0.8H_{ri}, 0.9H_{ri}, 1.0H_{ri} with its wave crest centered at amidships, and at 0.1L, 0.2L, 0.3L, 0.4L and 0.5L forward and 0.1L, 0.2L, 0.3L and 0.4L aft thereof. Three different methods are used to calculate the restoring coefficients from the computed GZ curve database which are to be used in motion equation given by Eqn.4 for the prediction of maximum roll amplitude. Fig.8 displays the comparison of GZ curve between present method and MAXSURF for wave height 1.0H_{ri} with crest located along the length of ship. It can be observed that the present method produced almost identical results with industry state-of-the-art stability software MAXSURF results.

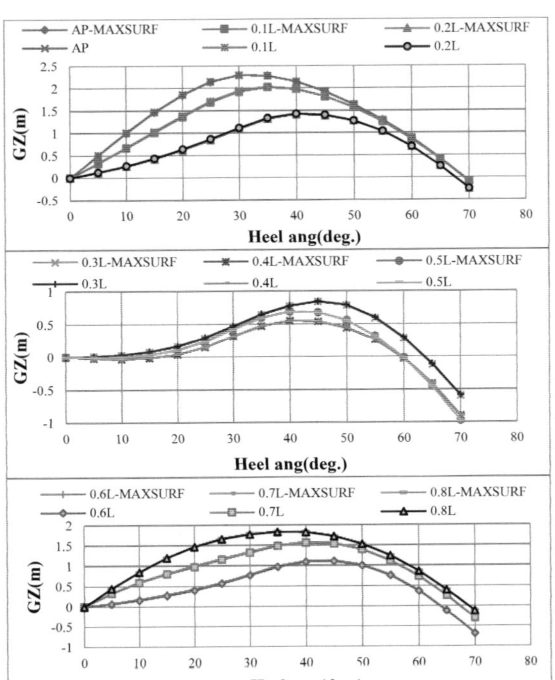

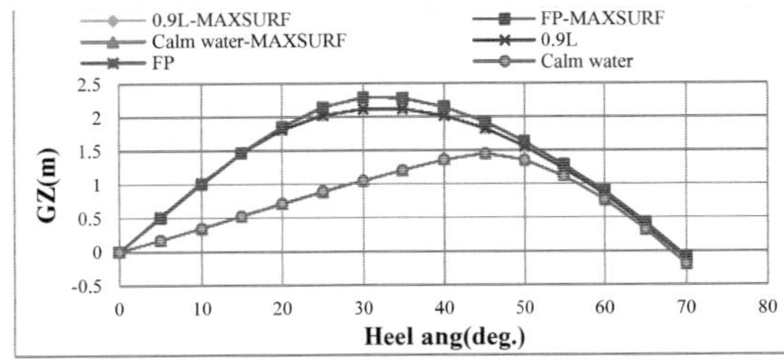

Figure 8 GZ COMPARISON FOR WAVE HEIGHT H_W= 1.0H_{RI} FOR DIFFERENT WAVE CREST LOCATIONS

Method-1(Calculation of restoring moment based on bi-cubic spline fitting): The GZ curves in waves is calculated using the "conventional" static algorithm (in which forces and moments are balanced in trim and sinkage), but the waterplane is not flat it is determined from the intersection of wave and the hull surface which is commonly known also as "wave-pass" calculations. Here an in-house FORTRAN code is developed to compute the restoring arm curve and it is validated with commercially available state-of-the-art hydrostatic software packages MAXSURF. Fig.9 displays the sample 3-D surface plot of restoring arm variation for two different wave height (0.5H_{ri}, and 0.9H_{ri}) for different wave crest locations along ship length.

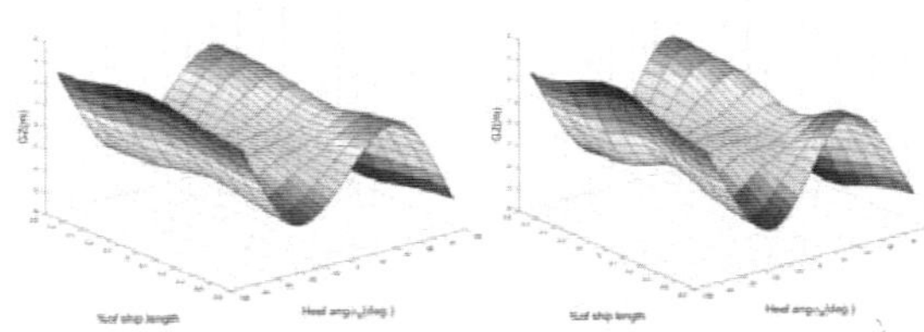

Figure 9 3D-PLOT OF RESTORING ARM VS HEEL ANGLE FOR DIFFERENT WAVE HEIGHT WITH WAVE CREST LOCATED ALONG THE SHIP LENGTH.

Method-2(Calculation of restoring moment based on calm water GZ): GZ in waves can be approximated with calm water GZ_0 and modulated with GM in waves $GZ(t,\phi) = \frac{GM(t)}{GM_0} GZ_0(\phi)$. Fig. 10 displays the plot of GM for various wave crest location for all 10 wave heights.

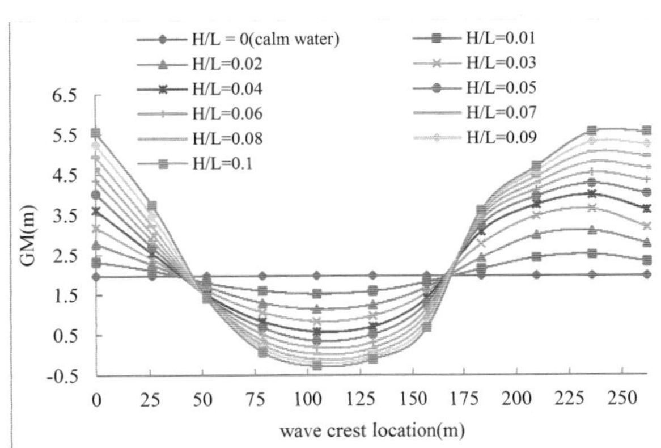

Figure 10 METACENTRIC HEIGHT FOR DIFFERENT WAVE HEIGHT WITH WAVE CREST LOCATED ALONG THE SHIP LENGTH.

Method-3(Calculation of restoring moment based on quintic curve fitting): The calm water GZ curve is fitted with quintic curve, to determine l_3 and l_5, $GZ = GM(\phi + l_3\phi^3 + l_5\phi^5)$. Fig.11 displays the calm water GZ curve and the fitted quintic curve with a least square method. The coefficients thus calculated are used in the averaging method based on the analytical formula to predict the maximum roll amplitude.

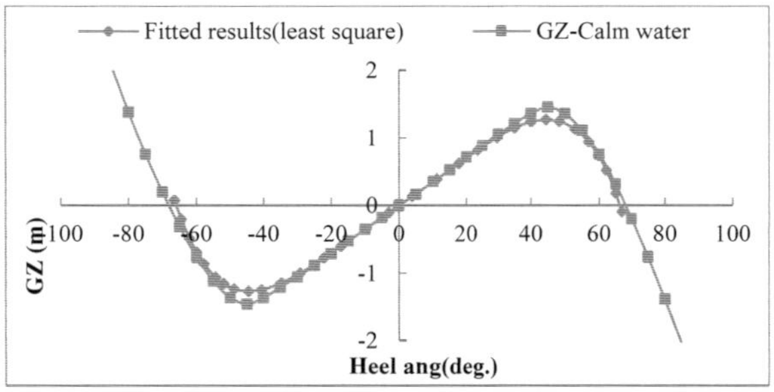

Figure 11 QUINTIC CURVE FITTED TO CALM WATER GZ CURVE

Step-4(roll amplitude and long term stability index calculations): Solving the parametric roll differential equation for 10 different wave height (using time domain simulations for both method-1 and method-2). Following Newton's second law, the 1-DOF equation [16] of roll motion is expressed as.

$$M_{IN}(\ddot{\phi}) + M_D(\dot{\phi}) + M_R(t, \phi) = 0.0 \quad (8)$$

$$\left.\begin{aligned} \ddot{\phi} &= \frac{\left(-M_D(\dot{\phi}) - M_R(t,\phi)\right)}{M_{IN}} \\ \dot{\phi} &= \int \ddot{\phi}\, dt \\ \phi &= \int \dot{\phi}\, dt \end{aligned}\right\} \quad (9)$$

The above equation is a differential equation which can be considered as an initial boundary value problem. The solution to this differential equation is sought with any time marching numerical scheme. Here we use a higher order time marching scheme such as 4th order Runge–Kutta (RK4) method and Adams–Bashforth predictor corrector (ABM) method.

Method-1(time domain simulations): Inertia and damping moments are calculated using formula proposed in Annex 19 of [17], whereas the calculation for restoring moment is based on fitting of bi-cubic spline to the GZ curves in waves. Since $M_R(t, \phi)$is a function of both time't' and heel angle 'ϕ' for any given instance of time, the heel angle and wave crest location ($X_c(t)=0.5\lambda \sin\omega_e t$, λ is wave length, ω_e is encounter wave frequency) is known a-prior, with a help of any standard interpolation method GZ for the given input t & ϕ is determined. Once all roll moments are known for that instant of time, roll acceleration is determined from Newton's law of motion equation given by equation (8), by using a suitable time integration scheme the velocity and displacement are determined from equation (9). In the present method 4th ABM scheme is used to time integrate the motion equation with 4th Runge Kutta scheme used as starter for initial time steps. Sample time histories of roll motions are shown in fig.12. Similarly for all wave height range 0~1.0H_{ri} numerical simulations are performed and the maximum roll amplitude is determined from the converged solution. Details of numerical simulation such as time step size, maximum number of time steps are followed as given in [17] document. Fig. 13 displays the maximum roll amplitude for all wave height range.

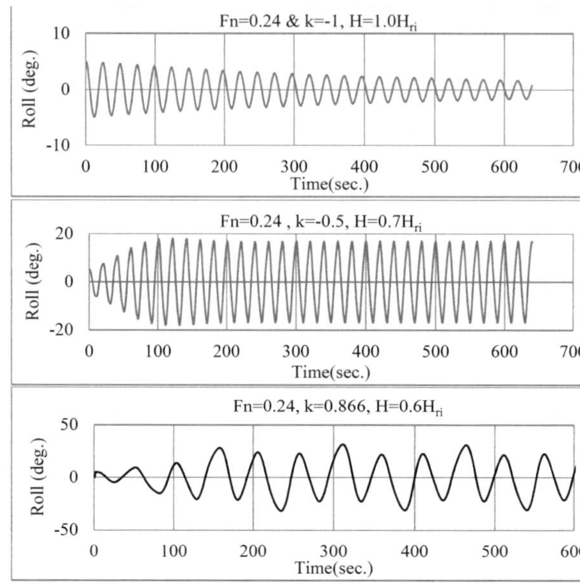

Figure 12 ROLL MOTION TIME HISTORIES

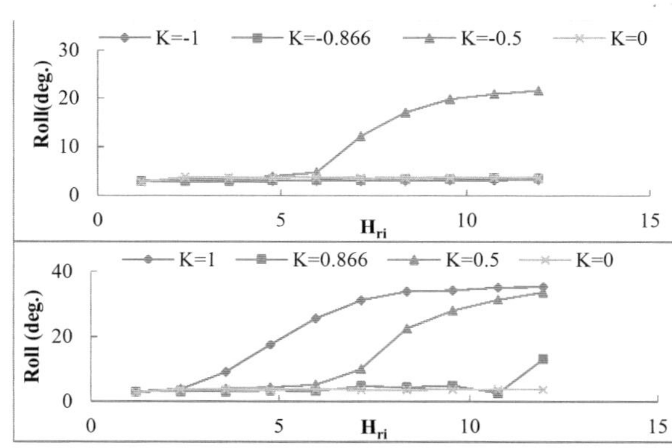

Figure 13 METHOD-1 RESULTS OF ROLL MOTION AMPLITUDE FOR A RANGE OF EFFECTIVE WAVE HEIGHTS

Once the maximum roll amplitude for 10 effective wave height (0.1~1.0H_{ri}) is obtained, roll amplitude corresponding to all possible short-term sea states given in the scatter diagram can be calculated with suitable interpolation. Then the probability to meet dangerous sea states where the roll amplitude is greater than the critical angle can be calculated for various heading and speed. Table 6 show the sample list of the short term value $C2_h(Fn)$ for each heading and speeds.

Table 6: SHORT TERM FAILURE INDEX FOR SEA STATE H_S=14.5M AND T_Z=8.5 H_{EFF}=9.7622M

Heading		Roll (deg)	Weighting factor	$C2_i(Fn_i)= W_iC_i$
Following sea	K=0.5	36.37		0.000001
	K=0.866	26.12	0.000001	0.000001
	K=1	4.02		0
Beam sea		3.68	0.000001	0
Head sea	K=0.5	3.86		0
	K=0.866	5.00	0.000001	0
	K=1	3.76		0

Average of individual components (C2) is checked against the required value R_{PRO}. If the ship fails to comply with this check, these data could be used for the operational limitations.

$$C2 = \frac{\left[\sum_1^3 C2_h(Fn_i) + 0.5C2_h(0) + \sum_1^3 C2_f(Fn_i) + 0.5C2_f(0)\right]}{7} = 0.0032$$

Since, C2= 0.0032 which is less than the required index 0.06, the ship is judged as non-vulnerable to parametric roll resonance.

Method-2(time domain simulations): Inertia and damping moment calculations are similar to method-1, but the restoring moment is calculated by an approximation where the GZ in waves is derived from calm water GZ_0 augmented with GM in waves. For a given instant of time and wave crest position, the GZ can be determined by $GZ(t,\phi) = \frac{GM(t)}{GM_0} GZ_0(\phi)$. Numerical integration scheme adopted in this case is same as method-1. Fig. 14 displays the time histories of roll motion using approximated GZ in waves for restoring moment. Maximum roll amplitude for all effective wave height is displayed in Fig. 15. Interpolation procedure identical to earlier method is used to determine all short term indexes for all non-zero weighted sea states.

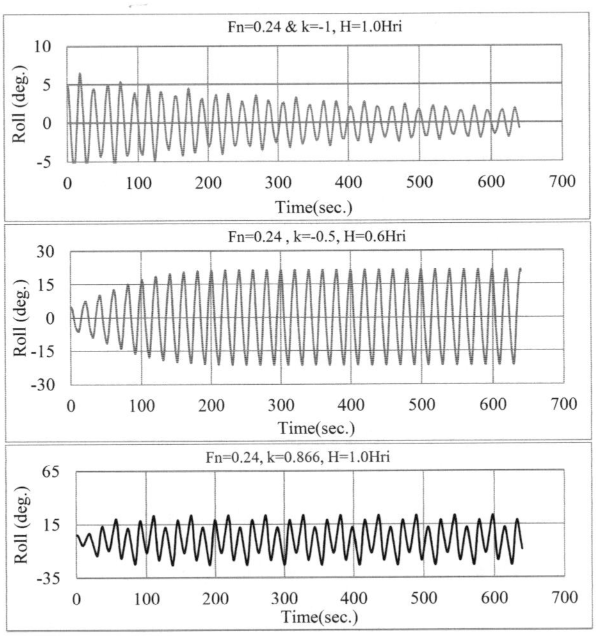

Figure 14 ROLL MOTION TIME HISTORIES

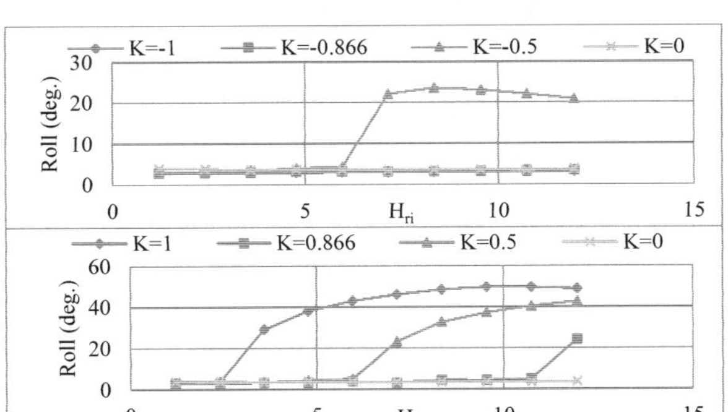

Figure 15 METHOD-2 RESULTS OF ROLL MOTION AMPLITUDE FOR A RANGE OF EFFECTIVE WAVE HEIGHTS

$$C2 = \frac{\left[\sum_1^3 C2_h(Fn_i) + 0.5C2_h(0) + \sum_1^3 C2_f(Fn_i) + 0.5C2_f(0)\right]}{7} = 0.0252$$

Since, C2= 0.0252 which is less than the required index 0.06 for this loading condition the ship is judged as non-vulnerable to parametric roll resonance.

Method-3(averaging method): Averaging method is one of the classical method for analyzing problems in nonlinear oscillations and also has the provision to manipulate the nonlinear system analytically. This method has been applied to mathematics and engineering problems for about two hundred years. In this method, the inertia and linear restoring terms are assumed to be regarded as leading ones and the others such as damping and restoring variation terms are assumed to have smaller magnitude. As a result, the first order solution becomes a harmonic solution with the amplitude and phase. Then the other terms are averaged within one cycle so that the amplitude and phase of the first order solution may slowly change. If we set this change rates to be zero, the amplitude and phase of the steady parametric roll can be obtained as given in[16].

$$\left\{\frac{\pi^2\hat{\omega}\left(3A^2\hat{\omega}^2\gamma+8\alpha\right)}{(2\pi^2-A^2)\omega_\phi^2}\right\}^2 + \left\{\frac{(6A^2\omega_\phi^2-8\pi^2\omega_\phi^2)}{4(\pi^2-A^2)\omega_\phi^2}\frac{GM_{mean}}{GM} + \frac{(-5\pi^2A^4l_5\omega_\phi^2-6\pi^2A^2l_3\omega_\phi^2+8\pi^2\hat{\omega}^2-8\pi^2\omega_\phi^2)}{4(\pi^2-A^2)\omega_\phi^2}\right\}^2 = \left(\frac{GM_{amp}}{GM}\right)^2 \quad (10)$$

Above equation 10 is a 12th order polynomial with respect to roll amplitude 'A' which can be regarded as a sixth order polynomial, with respect to 'A^2', then the latter (the sixth order) shall be used when considering the numerical accuracy and efficiency. Bairstow method is used for the solution of simplified six order polynomial equation (equation.10) to determine the roll amplitude. An overview of entire numerical process is shown in Fig.16. By solving the 12th order algebraic equation, maximum roll amplitude is obtained for 10 wave heights. GM_{amp} and GM_{mean} are the amplitude of metacentric height for different height variation and the mean metacentric height for different wave crest location along ship length which is similar to Level -2 check-1 calculations. Fig. 17 displays the roll amplitude variation for different wave height. Adopting a similar numerical procedure as previous methods for different headings and speed, short term indexes for all non-zero weighted sea states can be determined.

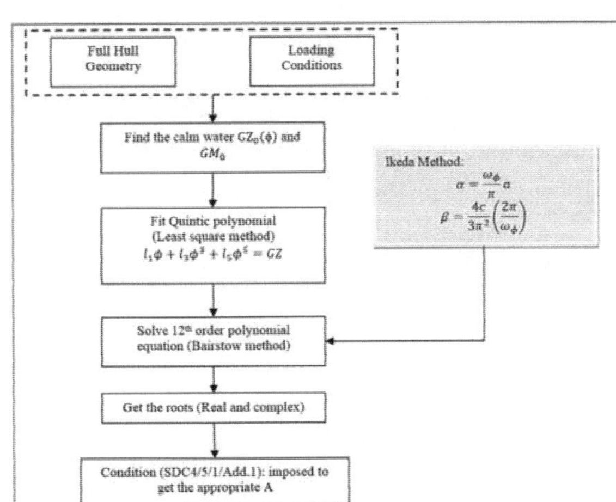

Figure 16 NUMERICAL PROCESS FOR AVERAGING METHOD

$$C2 = \frac{\left[\sum_1^3 C2_h(Fn_i) + 0.5C2_h(0) + \sum_1^3 C2_f(Fn_i) + 0.5C2_f(0)\right]}{7} = 0.8226$$

Since,C2= 0.8226 which is greater than the required index 0.06 for this loading condition the ship is judged as vulnerable to parametric roll resonance.

Table 7 summarize the numerical results (i.e. index C2) calculated using three different methods. Both time domain results(Method-1 and Method-2 of Level-2 check-2) are consistent with the output i.e. the attained index is less than the required index which imply that the ship is not vulnerable to parametric rolling whereas the averaging method's outcome index is higher than the require index R_{PRO} thereby suggesting further vulnerability check(i.e. level-3 assessment).

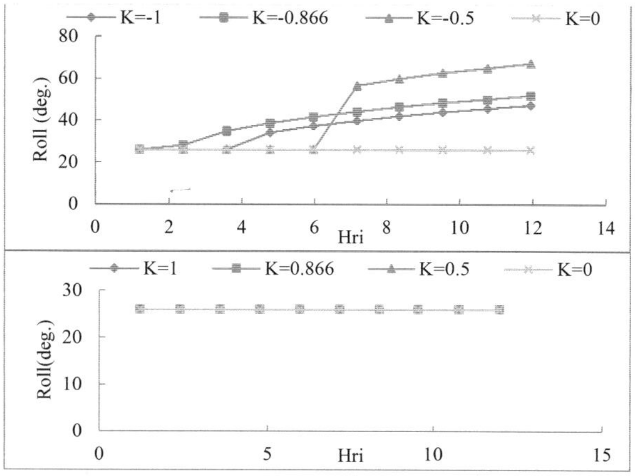

Figure 17 METHOD-3 RESULTS OF ROLL MOTION AMPLITUDE FOR A RANGE OF EFFECTIVE WAVE HEIGHTS

Table 7: SUMMARY OF RESULTS

Methods	$C2$
Time domain –GZ (Method -1)	C2= 0.0028 (C2<0.06, ship is not vulnerable to parametric roll) ([17] Annex, page 19)
Time domain –GM (Method -2)	C2= 0.0252(C2<0.06, ship is not vulnerable to parametric roll)
Averaging method (Method -3)	C2= 0.8226(C2>0.06, ship is vulnerable to parametric roll)

CONCLUSION

The main remarks from this work are summarized below:

- Numerical code for parametric roll criterion check (level 1 and level 2) developed by IMO Sub-Committee until SDC-5 is presented in detail.
- Numerical results for both levels are presented for a C11-postpanamx container ship. Level-1 and Level-2 (check -1) confirms that for the particular loading condition the considered ship is vulnerable to parametric roll resonance.
- Both time domain and averaging methods are used to determine maximum roll amplitude.
- In both time domain simulations, numerical procedure to determine roll amplitude is similar but the modeling of GZ (i.e. restoring term) in waves differ. Both formulations yield consistent result stating that the ship is not vulnerable to parametric roll in a long term analysis. In case of averaging method which is a conservative approach, the result indicates that the ship is vulnerable to parametric roll. Although the AVM is computationally efficient, quintic fitting of GZ curve is debatable.
- Time domain simulations results are consistent with one other but the averaging method results demand further check i.e., level-3 direct numerical simulations.
- Overall in terms of numerical efficiency and accuracy, time domain method-2 is preferred over other two methods.

ACKNOWLEDGEMENTS

Authors would like to thank Mr. Girish N (HOD) and Dr. Suhas in-charge of Strategic research group Indian register of Shipping for their continuous support in this research. Also authors would like to thank Prof. Naoya Umeda of Dept. of Naval Architecture and Ocean Engineering, Osaka University, for allowing access to the necessary document from IMO working group website for this work. We would also thank Prof. Debabrata Sen of Dept. of Naval Architecture and Ocean Engineering, IIT Kharagpur for providing us the higher order time integration numerical scheme in FORTRAN.

REFERENCES

[1] INTERNATIONAL CODE ON INTACT STABILITY, 2008 (2008 IS CODE)

[2] IMO Document SLF 51/17, "Report to the Maritime Safety Committee", 28 August 2008, London, UK

[3] Belenky, V., J. O. de Kat, and N. Umeda (2008), "Toward Performance-Based Criteria for Intact Stability," Marine Technology, 45(2)

[4] SLF 51/4/4 2008, REVISION OF THE INTACT STABILITY CODE Further proposal for so-called new generation intact stability criteria Submitted by Germany

[5] IMO Document SLF 52/3/6, DEVELOPMENT OF NEW GENERATION INTACT STABILITY CRITERIA, 20 November 2009, Japan.

[6] IMO Document SDC 2/WP.4. FINALIZATION OF SECOND GENERATION INTACT STABILITY CRITERIA -Report of the working group (Part 1) 19 February 2015.

[7] MSC.267(85) RESOLUTION ADOPTION OF THE INTERNATIONAL CODE ON INTACT STABILITY, 2008 (2008 IS CODE) 4 December 2008.

[8] Paulling J. R. and Rosenberg R.M. (1959). "On unstable ship motions resulting from nonlinear coupling", Journal of Ship Research, Vol. 3, No 1, pp. 36-46.

[9] Paulling, J. R. (1961). "The transverse stability of a ship in a longitudinal seaway". Journal of Ship Research, vol. 4, no. 4, pp. 37-49.

[10] Kempf, G, "Die Stabilitätsbeanspruchung der Schiffedurch Wellen und Schwingungen", Werft-Reederei-Hafen, vol. 19, 1938

[11] Graff, W. and E. Heckscher, "Widerstands und Stabilitäts versuche mit drei Fischdampfermodellen", Werft-Reederei-Hafen, vol. 22, 1941 (also DTMB Translation No.75, June 1942)

[12] France, W. N., Levandou, U. M., Treakle, T. W., Paulling, J. R., Michel, R. K.., Moore, E. C., 2003. An Investigation of Head-Sea Parametric Rolling and its

Influence on Container Lashing Systems, Marine Technology 40.
[13] Shin, Y.S., Belenky, V.L., Paulling, J.R., Weems, K.M. and Lin, W.M., 2004. Criteria for parametric roll of large containerships in longitudinal seas. *Trans. Soc. Naval Arch. Marine Eng.(SNAME), Washington, DC*, pp.14-47.
[14] ABS, 2004, "Guide for the assessment of parametric roll resonance in the design of container carriers.
[15] Vadim Belenky, Christopher C. Bassler, Konstantinos J. Spyrou. Development of Second Generation Intact Stability Criteria. Hydromechanics Department Report, NSWCCD-50-TR-2011/065. December 2011
[16] IMO Document SDC 4/5/1/Add.5. FINALIZATION OF SECOND GENERATION INTACT STABILITY CRITERIA - Report of the correspondence group (part 6), 11 November 2016.
[17] IMO Document SDC 5/INF.4/Add.1. FINALIZATION OF SECOND GENERATION INTACT STABILITY CRITERIA - Information collected by the Correspondence Group on Intact Stability , 15 November 2017.
[18] IMO Document SDC 6/5. FINALIZATION OF SECOND GENERATION INTACT STABILITY CRITERIA – report of Correspondence Group (part-1), 1 November 2018.

Proceedings of the 6th International Conference on Ship and Offshore Technology
ICSOT 2019
November 7-8, 2019, IIT Kharagpur, India

ICSOT INDIA 2019-18

SCALE EFFECTS OF SLOSHING ON A SWAY EXCITED RECTANGULAR SHIP TANK

Shwetha Shri
Research student
National Institute of Technology, Karnataka,
Surathkal
India
shwethashri23@gmail.com

T. Nasar
Assistant professor
National Institute of Technology, Karnataka,
Surathkal
India
t.nasar@gmail.com

ABSTRACT

Sloshing behavior of liquid in containers is one of the fundamental fluid-structure interactions. Liquid sloshing is associated with various dynamical systems and engineering problem, such as the liquid cargo slosh in ships and trucks of transporting liquid, oil oscillation in large tanks, and water oscillation in a reservoir due to ground and wind acceleration, etc. Oscillation of water in a partially filled container due to the external forces or external disturbances is called as sloshing. An experimental study has been carried out to study the scale effects on sloshing dynamics of liquid in a partially filled sway excited rectangular tank. The model tanks considered in this study are 1:43, 1:57 and 1:86. The aspect ratio (h_s/L) of 0.1625, 0.325, 0.4875 are adopted which represents 25%, 50% and 75% of liquid fill levels of tank depth respectively. A sloshing tank is positioned on the shake table such that during the sway excitations, the sloshing oscillations occurs along the longitudinal axis. Time histories of sloshing oscillations and sloshing force are measured using load cells and wave probes, respectively. The effect of excitation frequencies and excitation amplitudes on sloshing force and sloshing oscillations is discussed here.

Key words: sloshing, scale effect, oscillation

NOMENCLATURE

f_n Sloshing frequency
g Acceleration due to gravity
h_s Static water depth
L Length of the tank
n Surface mode number

1. INTRODUCTION

Oscillation of the water in the partially filled vessel or container due to the external forces or external disturbances is called sloshing. Sloshing is observed in the many of the engineering applications such as oil carrying ships, liquid transporting trucks and trains, space crafts and rockets, water oscillation in reservoirs and liquid storage tanks when subjected to earthquakes and others. Exporting and importing oils through the countries is the one of the main trading. Transporting oil and water are mainly through the water waters ways (through ships) and land base (through trucks and trains). Sloshing occurs in the above mentioned transporting ways. Sloshing is severe problem in the oil carrying ship because of oscillation of the oil inside the tank. And it is one of the main reason for instability and damage of the oil carrying ships. Knowing the sloshing effect will helpful for engineers to design the oil carrying ships and transit truck tanks. There are so many experiments have been going on to get solve the sloshing problems. Considerable studies have been carried out on the phenomena of sloshing in the late Eighties. Akyildiz and Unal [1] compared the observed experimental pressure distribution in a roll excited tank with the predicted pressure time traces. Armenio and La Rocca [2] investigated sloshing oscillation behaviour in a roll excited tank by both experimental and numerical approaches. Frandsen [3] reported the multiple resonance conditions in a combined sway and heave excitations. Kim et.al [5] considered scale effects on three-dimensional sloshing flows where a series of model tests were conducted for three differently scaled tanks. The key sloshing load parameters were analyzed by a statistical approach, and the results from the tests at different scales (15%, 70% and 95%) were compared. Mei-rong et.al [6] studied the characteristics and the variation of the sloshing elevation and the pressure in an elastic tank. Nakayama and Washizu [7] used nonlinear approach to investigate the liquid motion and the resulting sloshing pressure in a container subjected to forced oscillations using finite element analysis. Nasar et.al [8, 9] carried out an experimental work and studied the wave induced sloshing pressure / impact pressure on regular beam sea condition. Xue et.al [10] reported the impact of the vertical baffles on the sloshing frequency with shake table tests.

The present study focuses on finding out the sloshing oscillation and sloshing force in three different scaled sway excited tanks of rectangular ship tanks.

2. EXPERIMENTAL INVESTIGATION

The set up for the experimental work comprises of a servo-hydraulic shake table along with a digital amplifier. The shake table is driven by a signal conditioner which is connected to a hydraulic power pack as shown in Figure 1. The frequency and amplitude of excitation are controlled with the help of a Scientech (wave transformer). The shake table is designed in such a way that the load cells are used to measure the inertial forces experienced by the masses kept on the platforms. The subtraction of the two force signals gives the hydrodynamic forces by the fluid. LVDT was used to measure the horizontal displacement of the shake table. The instrumentation is used to capture the free surface response, base shear forces, and table motion. The partially liquid filled tank is mounted on the shake table and the table is excited harmonically at the required excitation frequency. The sloshing motion of liquid is captured by capacitance type wave probes. The excitation frequencies varied between 0.4566 Hz to 2.794 Hz which encompasses upto fifth mode sloshing frequencies. The excitation frequencies ranging between 0.4566 Hz to 1.9757 Hz for 1:43 scale tank, 0.4566 Hz to 2.2813 Hz for 1:57 scale tank and 0.4566 Hz to 2.794 Hz for 1:86 scale tank. There are 17 discrete frequencies and two excitation amplitudes of 4mm and 8mm are considered for each of the scaled tank. The experimental parameters such as sloshing oscillations, sloshing force, table displacements and table accelerations are measured by various transducers.

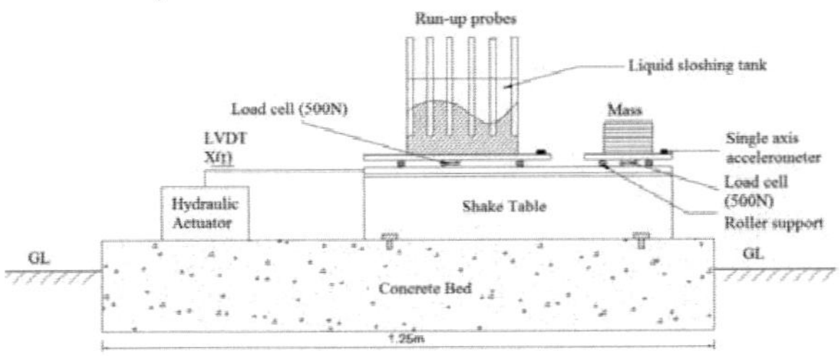

Figure 1. CROSS-SECTION OF LIQUID TANK ON SHAKE TABLE

The liquid oscillation behaviour in the partially filled tank depends on the excitation frequency and amplitude, size and geometry of the tank of the tank, fill depth and internal baffle arrangements if any. Ibrahim [4] gave the linear approximation of resonant liquid sloshing frequencies (in Hz) for the nth mode in a rectangular tank as follows:

$$f_n = \frac{1}{2\pi}\sqrt{\frac{n\pi g}{L}\tanh\left(\frac{n\pi h_s}{L}\right)} \quad (1)$$

where L is the length of the tank, h_s is the static liquid depth, and n is the surface mode number.

If the excitation frequency is very close to the sloshing frequency of liquid, then the liquid sloshes with greater oscillations for the same amplitude of excitation.

2.1. Model details:

Three rectangular tanks of scale 1:43, 1:57 and 1:86 are fabricated using acrylic sheets of 12mm thickness (Nasar [8]). The specifications of the tanks are given in Table 1. Six capacitance probes are fixed inside the tank at regular spacing and among that two are fixed at end walls. Single axis accelerometer is used to measure the table acceleration.

Table 1. SPECIFICATIONS OF THE LIQUID TANKS

		Dimensions in mm		
Sl. No.	Designation of tank	Tank 1	Tank 2	Tank 3
1.	Length (l)	1000	750	500
2.	Width (b)	400	300	200
3.	Height (h)	650	487.5	325

Figure 2 shows the photographic view of the rectangular tanks.

Figure 2. PHOTOGRAPHIC VIEW OF RECTANGULAR TANKS

3. RESULTS AND DISCUSSIONS

3.1 Sloshing oscillation:

The steady state sloshing oscillation for an initial period of 15 seconds is reported herein for three scaled tanks. Figure 3 presents the sloshing oscillation time histories for 25%, 50% and 75% fill level depths subjected to excitation amplitude of 4mm. Figures 3(a) to 3(c) represent the sloshing oscillation time histories of 1:86 scale tank, Figures 3(d) to 3(f) represent the sloshing oscillations time histories of 1:57 scale tank, and 3(g) to 3(i) represent the sloshing oscillations time histories of 1:43 scale tank for the excitation amplitude of 4mm. In each of the figure the sloshing oscillation responds for the wave excitation $f = f_1, f_3$ and f_5 are superposed. In the similar manner of representation figure 4 illustrates the sloshing response for the excitation amplitude of 8mm. The non-linear behavior of free surface of the liquid is evident.

Higher sloshing oscillations are observed in the order of $f = f_1$, $f = f_3$ and $f = f_5$ i.e at odd mode sloshing frequencies irrespective of the fill levels and excitation amplitudes. As seen from the graphs for 8mm excitation amplitude sloshing oscillation is higher compared with 4mm excitation amplitude irrespective of the fill levels. As dimensions of the tanks increases the number of peak responses are less. Sloshing oscillations decreases in an order of 1:86, 1:57 & 1:43 for 25% and 50% fill levels. Whereas for 75% fill level sloshing oscillations increases in the order of 1:86, 1:57 & 1:43 irrespective of excitation amplitudes.

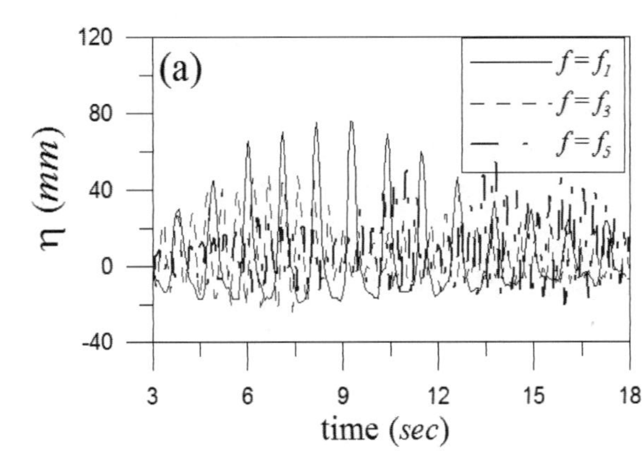

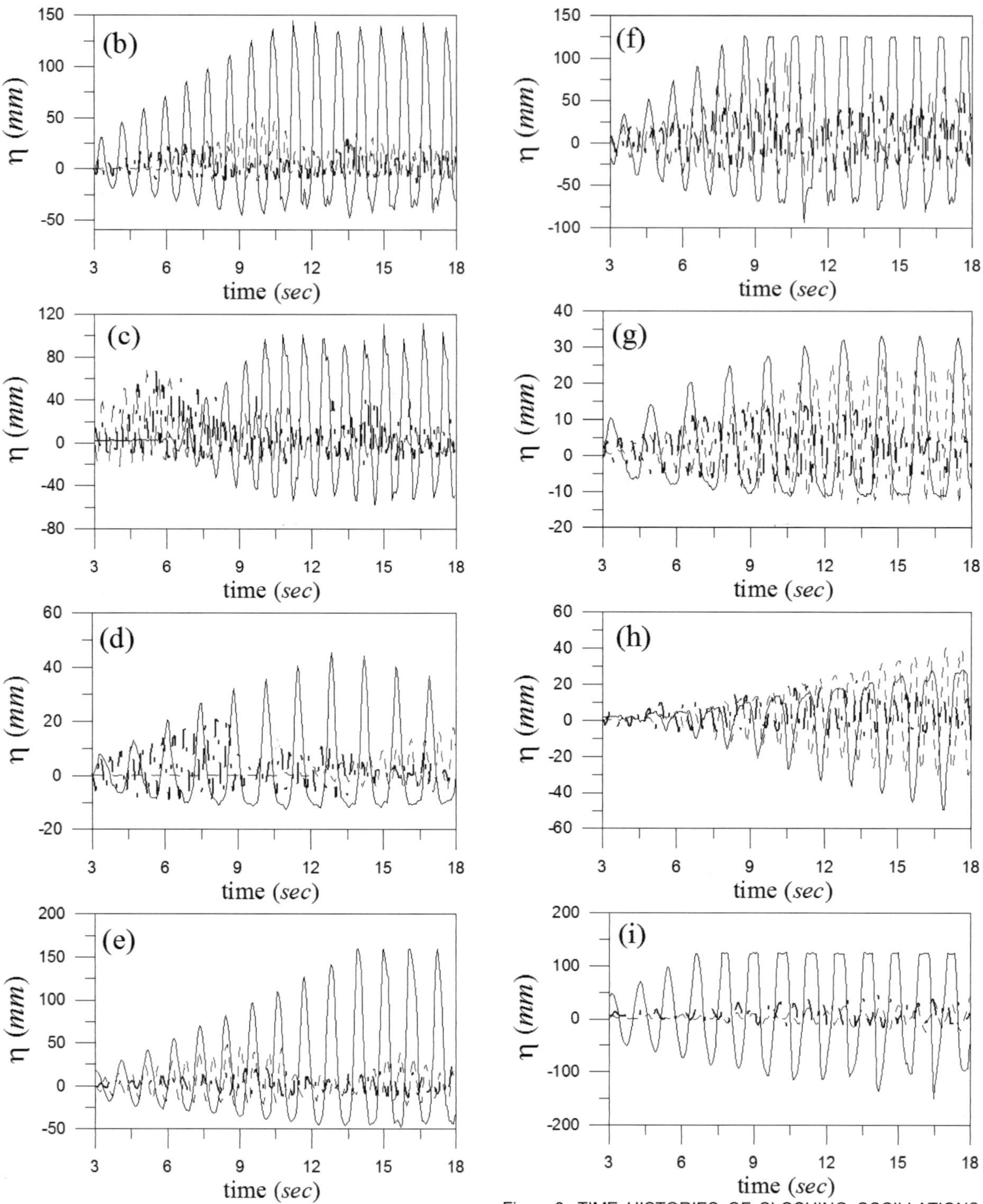

Figure.3: TIME HISTORIES OF SLOSHING OSCILLATIONS CORRESPONDING TO EXCITATION AMPLITUDE OF 4*mm*: (a) h_s/L = 0.1625 FOR 1:86 SCALE TANK, (b) h_s/L = 0.325 FOR 1:86 SCALE TANK, (c) h_s/L = 0.4875 FOR 1:86 SCALE TANK, (d) h_s/L = 0.1625 FOR 1:57 SCALE TANK, (e) h_s/L = 0.325 FOR 1:57 SCALE TANK, (f) h_s/L = 0.4875 FOR 1:57 SCALE TANK, (g) h_s/L = 0.1625 FOR 1:43 SCALE TANK, (h) h_s/L = 0.325 for 1:43 SCALE TANK AND, (i) h_s/L = 0.4875 FOR 1:43 SCALE TANK

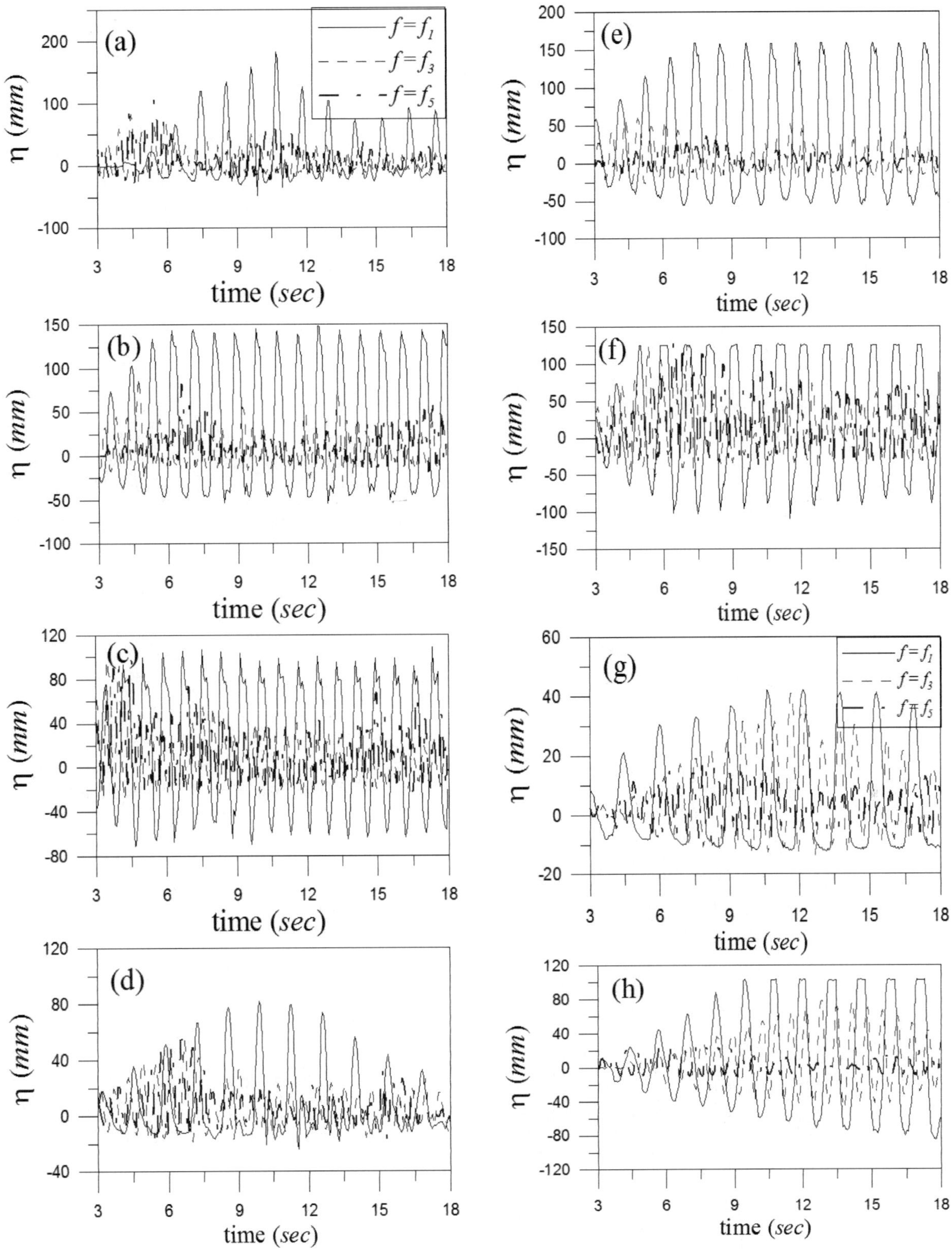
(a)
$f=f_1$
$f=f_3$
$f=f_5$
η (mm)
time (sec)
(b)
(c)
(d)
(e)
(f)
(g)
(h)

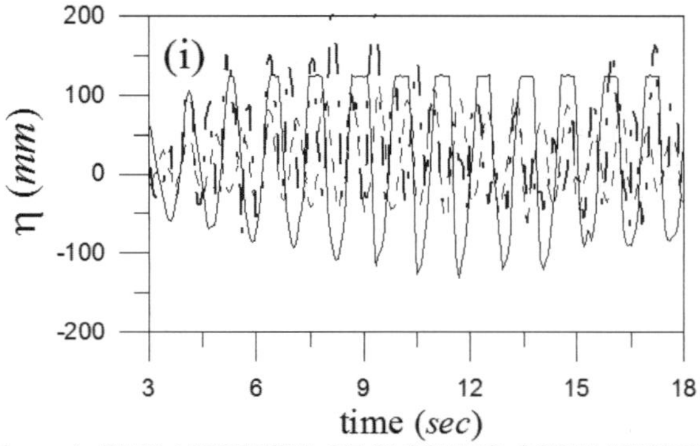

Figure.4: TIME HISTORIES OF SLOSHING OSCILLATIONS CORRESPONDING TO EXCITATION AMPLITUDE OF 8*mm*: (a) h_s/L = 0.1625 for 1:86 SCALE TANK, (b) h_s/L = 0.325 FOR 1:86 SCALE TANK, (c) h_s/L = 0.4875 FOR 1:86 SCALE TANK, (d) h_s/L = 0.1625 FOR 1:57 SCALE TANK, (e) h_s/L = 0.325 FOR 1:57 SCALE TANK, (f) h_s/L = 0.4875 FOR 1:57 SCALE TANK, (g) h_s/L = 0.1625 FOR 1:43 SCALE TANK, (h) h_s/L = 0.325 for 1:43 SCALE TANK AND, (i) h_s/L = 0.4875 FOR 1:43 SCALE TANK

3.2 Sloshing force:

Force due to the participating fluid is called as sloshing force. The portion of the fluid that generates the dynamic sloshing force is called as participating fluid. Sloshing forces are significant in design of partially filled excited tanks. Figure 5 and Figure 6 represents the temporal evolution of sloshing force for the excitation amplitude of 4mm and 8mm respectively. Sloshing force time histories for 25%, 50% and 75% fill levels to three scaled rectangular tanks (i.e. 1:86, 1:57 & 1:43) with an excitation at odd mode sloshing (i.e. f_1, f_3,and f_5).

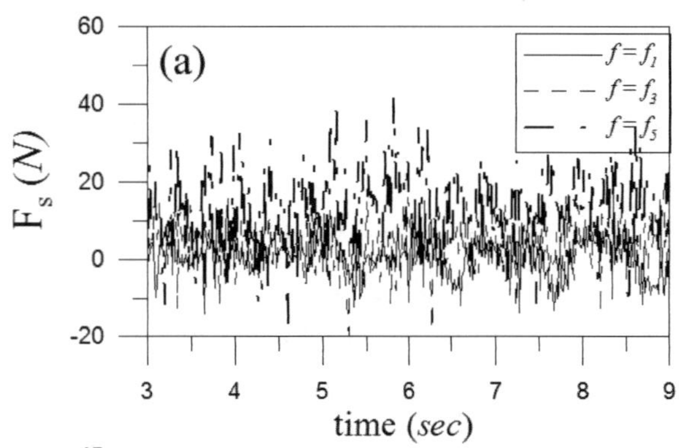

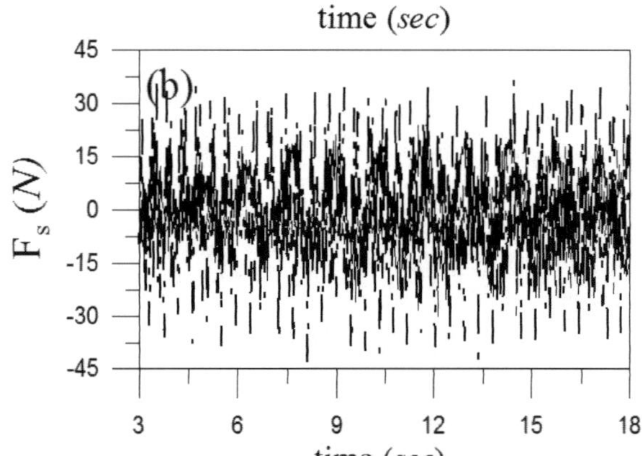

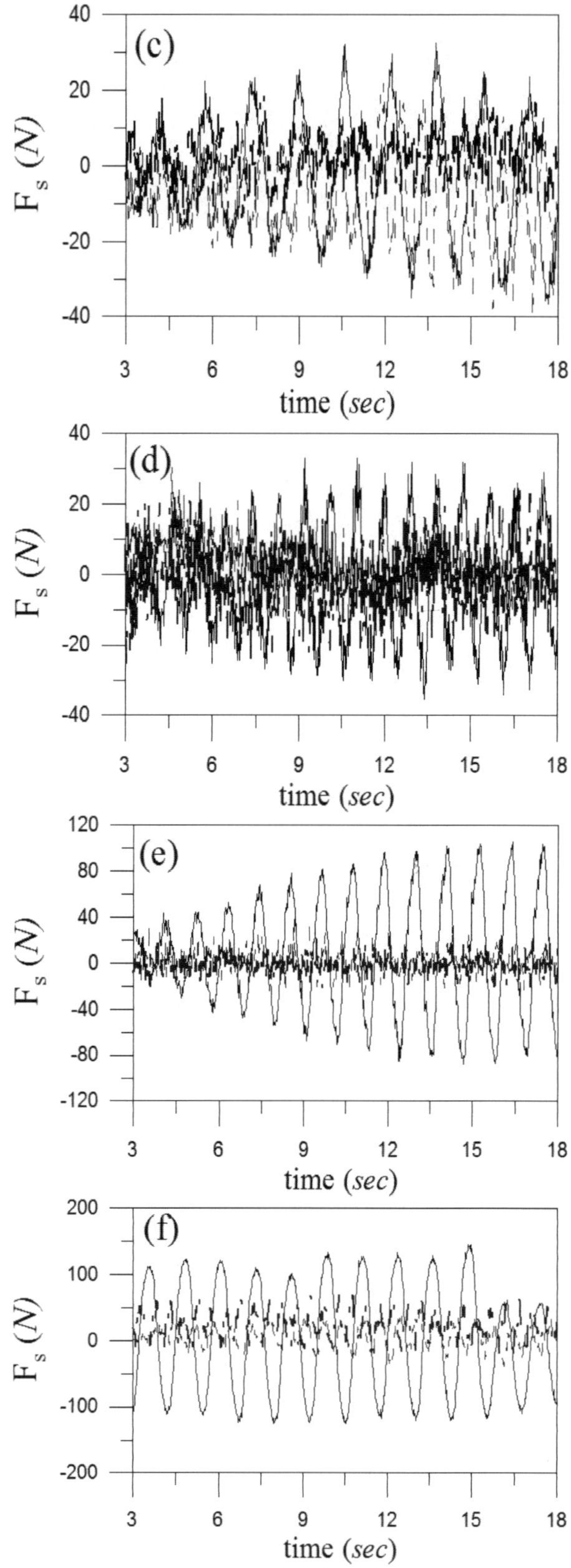

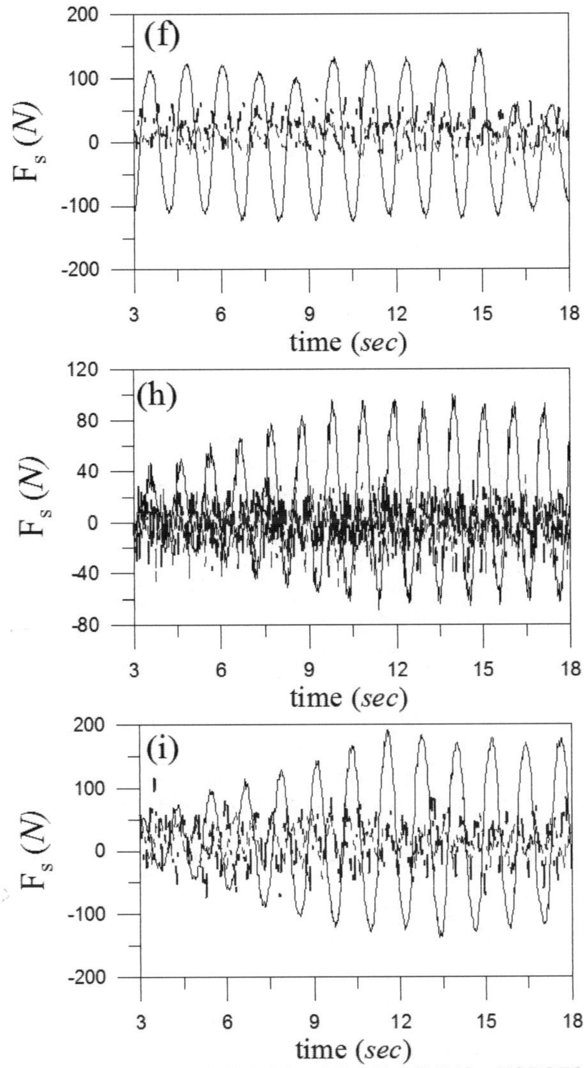

Figure.5: TIME HISTORIES OF SLOSHING FORCES CORRESPONDING TO EXCITATION AMPLITUDE OF 4*mm*; (a) h_s/L = 0.1625 FOR 1:86 SCALE TANK, (b) h_s/L = 0.1625 FOR 1:57 SCALE TANK, (c) h_s/L = 0.1625 FOR 1:43 SCALE TANK, (d) h_s/L = 0.325 FOR 1:86 SCALE TANK, (e) h_s/L = 0.325 FOR 1:57 SCALE TANK, (f) h_s/L = 0.325 FOR 1:43 SCALE TANK, (g) h_s/L = 0.4875 FOR 1:86 SCALE TANK, (h) h_s/L = 0.4875 FOR 1:57 SCALE TANK, (i) h_s/L = 0.4875 FOR 1:43 SCALE TANK

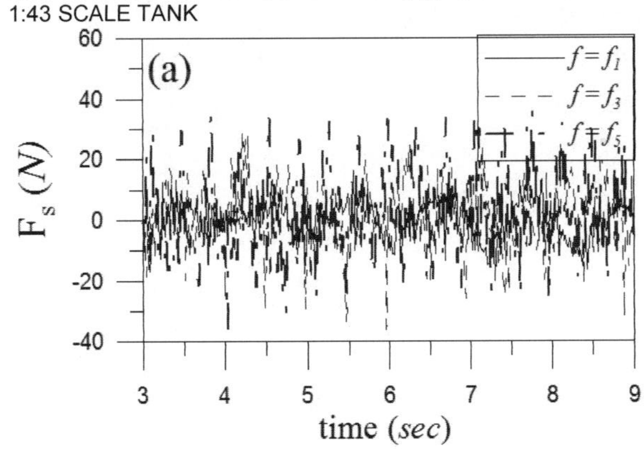

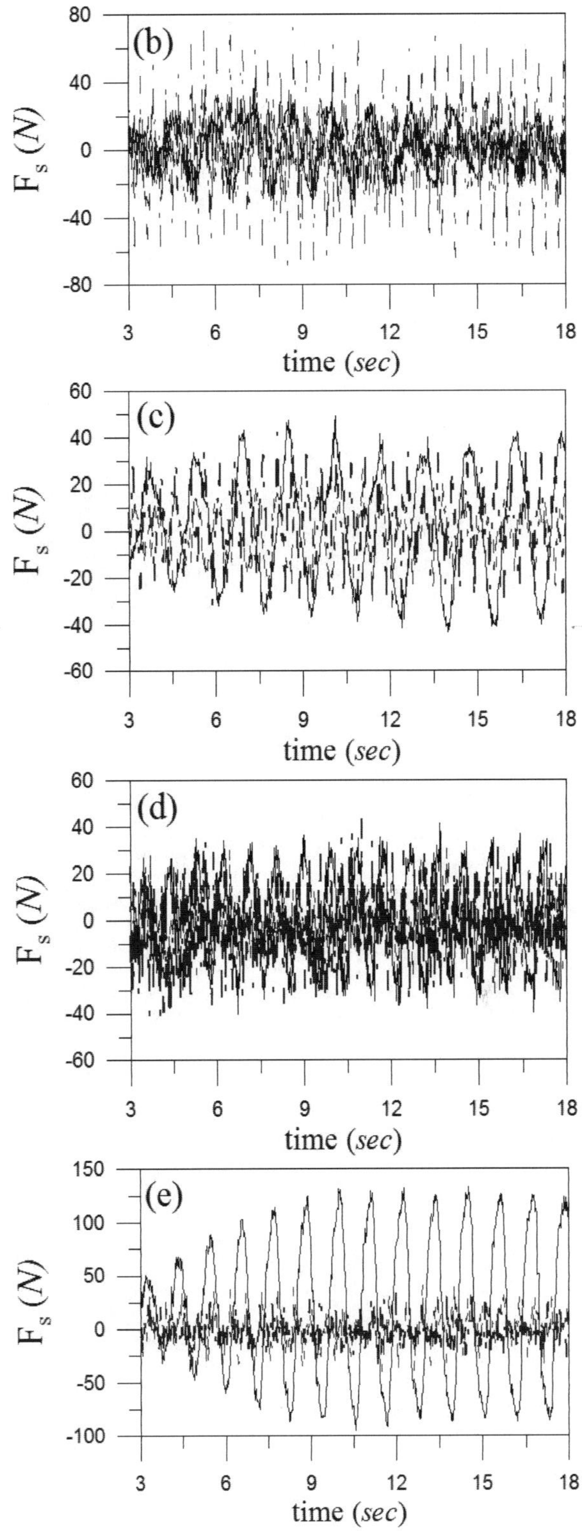

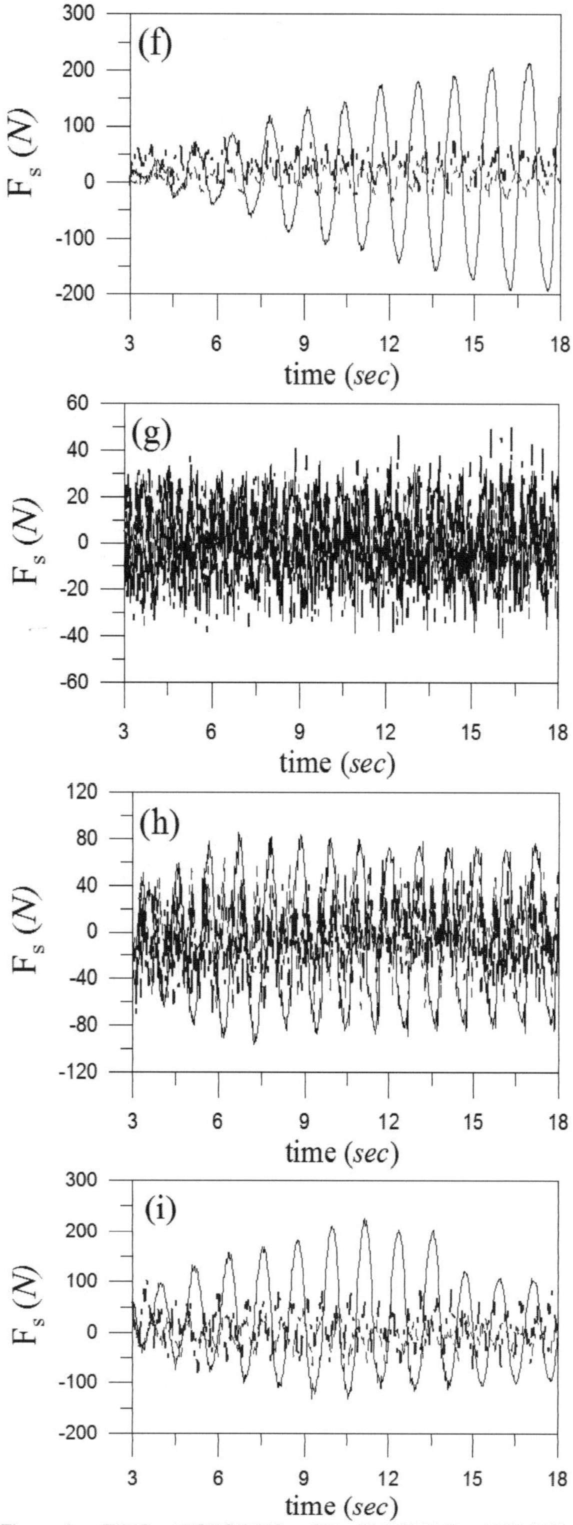

Figure.6: TIME HISTORIES OF SLOSHING FORCES CORRESPONDING TO EXCITATION AMPLITUDE OF 8*mm*; (a) h_s/L = 0.1625 FOR 1:86 SCALE TANK, (b) h_s/L = 0.1625 FOR 1:57 SCALE TANK, (c) h_s/L = 0.1625 FOR 1:43 SCALE TANK, (d) h_s/L = 0.325 FOR 1:86 SCALE TANK, (e) h_s/L = 0.325 FOR 1:57 SCALE TANK, (f) h_s/L = 0.325 FOR 1:43 SCALE TANK, (g) h_s/L = 0.4875 FOR 1:86 SCALE TANK, (h) h_s/L = 0.4875 FOR 1:57 SCALE TANK, (i) h_s/L = 0.4875 FOR 1:43 SCALE TANK

It is evident that excitation at first mode sloshing ($f = f_1$) is critical, because maximum sloshing force can be seen at $f = f_1$ irrespective of the fill levels and excitation amplitudes considered. The decreasing order of sloshing force are $f = f_1$, $f = f_3$ and $f = f_5$ mode sloshing frequencies irrespective of the fill levels and excitation amplitudes. As the excitation amplitude increases, the sloshing force also increases irrespective of the fill levels. As seen from graphs, maximum sloshing force is observed for 50% fill level compared with 25% and 75% fill levels. Sloshing force increases in an order of 1:86, 1:57 & 1:43 for 25%, 50% and 75% fill levels irrespective of excitation amplitudes.

3.3 Maximum free surface response (η_{max}):

The maximum free surface response within the steady state oscillation gives the maximum sloshing run-up in the sway excited rectangular sloshing tank for given excitation amplitude and excitation frequency. The Normalised maximum free surface response (η_{max}/A) for the different frequency ratio (f/f_1) are projected in Figures 7, 8 and 9 for 1:86 scale, 1:57 scale and 1:43scale tanks respectively.
Experimental results are compared with numerical works of Nasar [11]. A good correlation and maximum responses at odd mode sloshing frequencies are predicted exactly.
For all the scaled tanks, maximum free surface response is higher for 50% fill level (h_s/L = 0.325) than the response obtained for 75% fill level (h_s/L = 0.4875) and 25% fill level (h_s/L = 0.1625). It can be seen from graphs the first mode ($f = f_1$)is the critical mode of sloshing for sway excited rectangular tanks with partially filled water depths. Maximum free surface response is observed in the order $f = f_1$, $f = f_3$ and $f = f_5$ i.e at odd mode sloshing frequencies. A 1:43 scaled tank showed more response when compared to 1:57 scaled tank and 1:86 scaled tank. On comparison of three different fill levels, the higher sloshing oscillations are observed in the order of 50%, 75% and 25% fill levels.

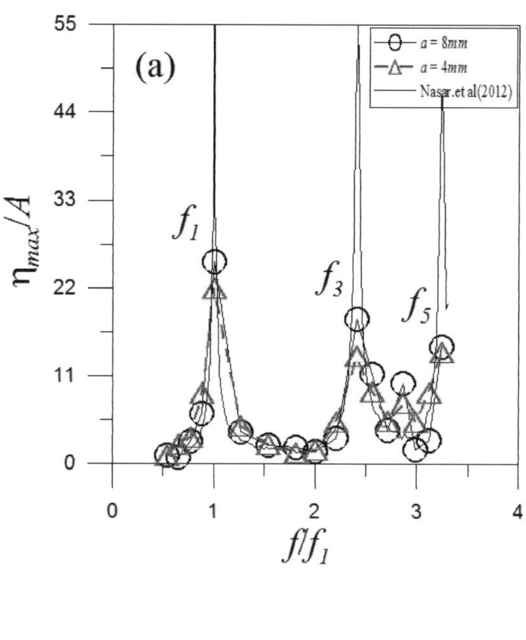

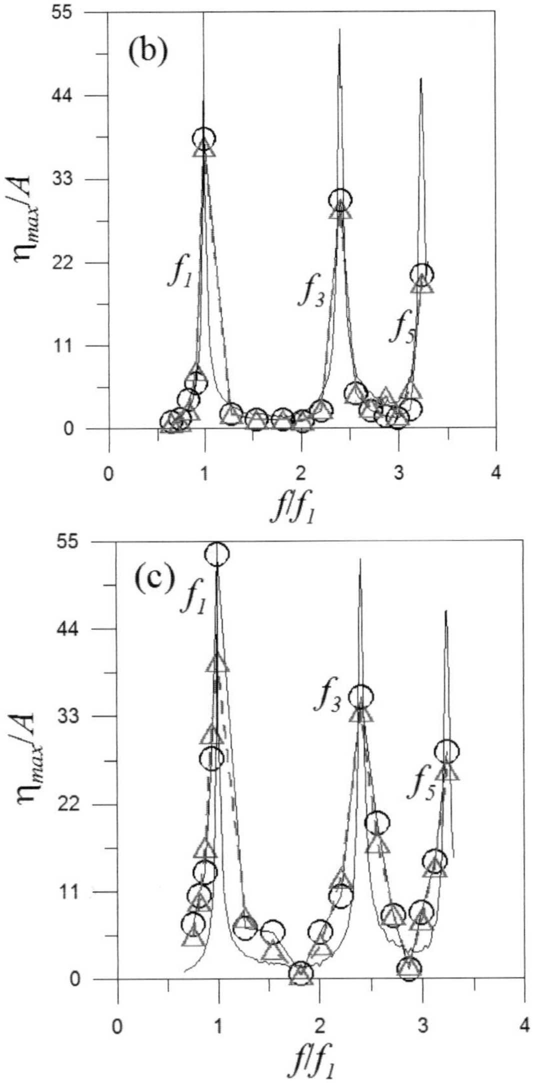

Figure.7: VARIATION OF η_{max}/A WITH DIFFERENT FREQUENCIES RATIO (f/f_1) FOR h_s/L=0.1625; (A) 1:86 SCALE TANK, (B) 1:57 SCALE TANK, AND (C) 1:43 SCALE TANK

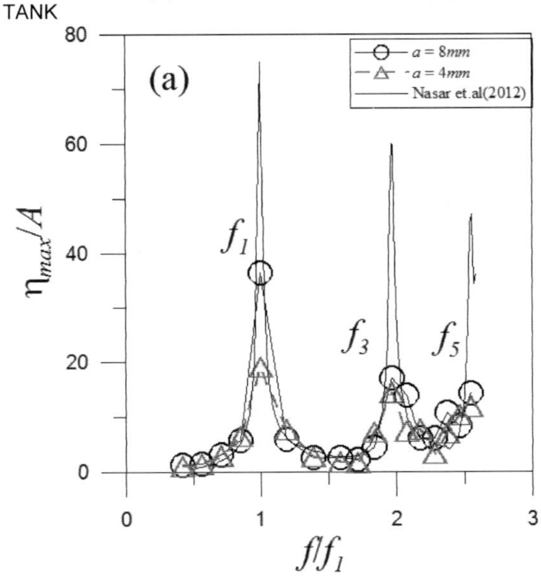

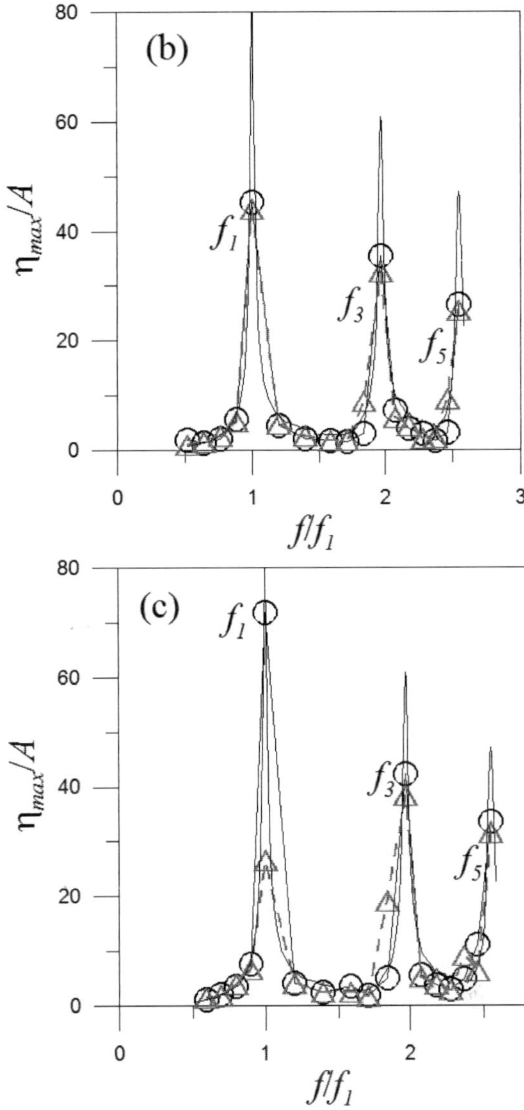

Figure.8: VARIATION OF η_{max}/A WITH DIFFERENT FREQUENCIES RATIO (f/f_1) FOR h_s/L=0.325; (A) 1:86 SCALE TANK, (B) 1:57 SCALE TANK, AND (C) 1:43 SCALE TANK

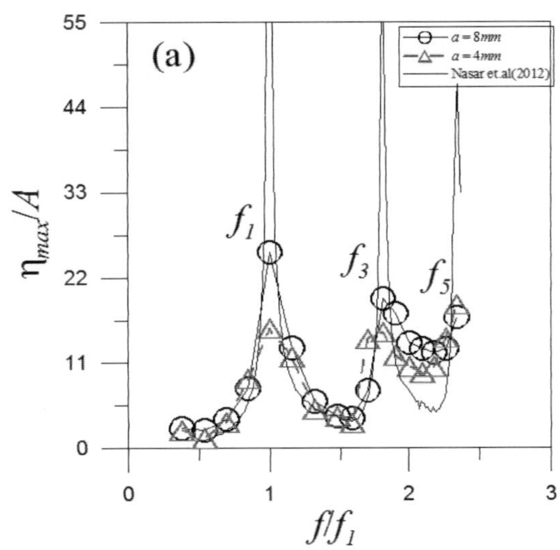

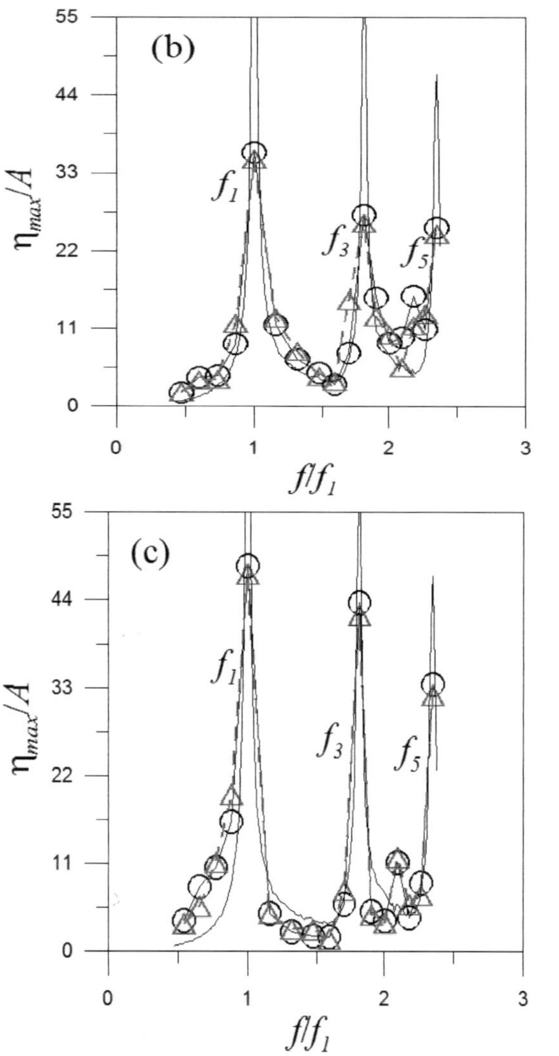

Figure.9: VARIATION OF η_{max}/A WITH DIFFERENT FREQUENCIES RATIO (f/f_1) FOR h_s/L=0.4875; (A) 1:86 SCALE TANK, (B) 1:57 SCALE TANK, AND (C) 1:43 SCALE TANK

3.4 Root mean square surface elevation (η_{rms}):

The normalised root mean square of sloshing (η_{rms}/A) for the various excitation frequency ratio (f/f_1) are presented in Figures 10, 11 and 12 for the scaled tanks of 1:86 scale, 1:57 scale and 1:43scale tanks, respectively.

By considering all the scaled tanks, normalised root mean square surface elevation is higher for 50% fill level (h_s/L = 0.325) than the response obtained for 75% (h_s/L = 0.4875) and 25% fill level (h_s/L = 0.1625). Normalised root mean square surface elevation is observed in the decreasing order of $f = f_1$, $f = f_3$ and $f = f_5$ i.e at odd mode sloshing frequencies. A 1:43 scaled tank showed more response when compared to 1:57 and 1:86 scaled tank. The trend was same as the normalised maximum free surface response.

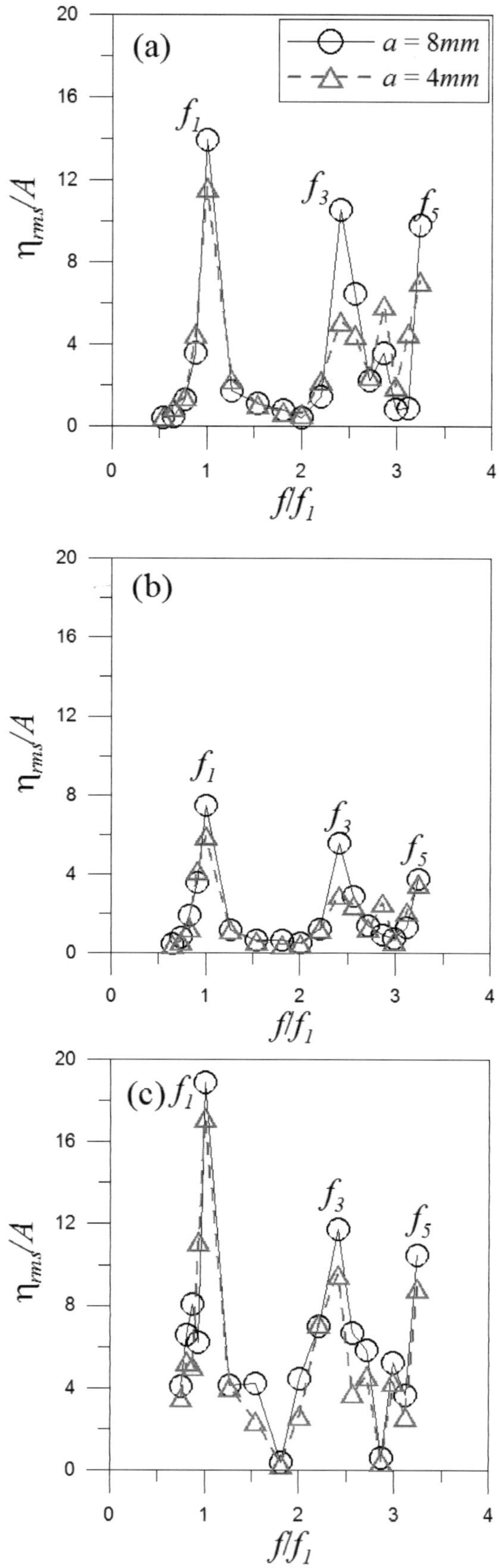

Figure.10: VARIATION OF η_{rms}/A WITH DIFFERENT FREQUENCIES RATIO (f/f_1) FOR h_s/L=0.1625; (A) 1:86

SCALE TANK, (B) 1:57 SCALE TANK, AND (C) 1:43 SCALE TANK

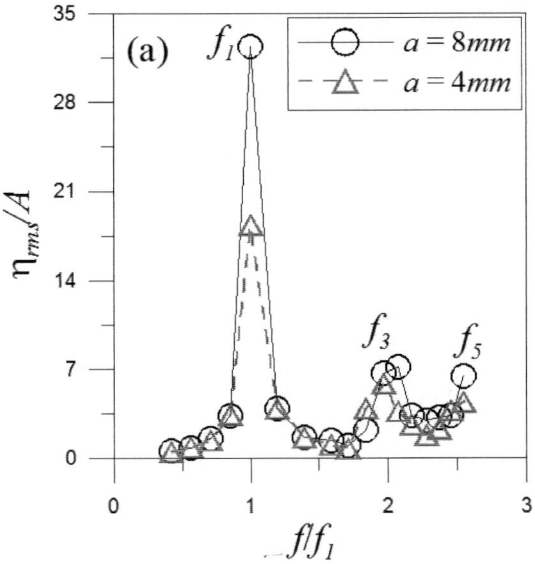

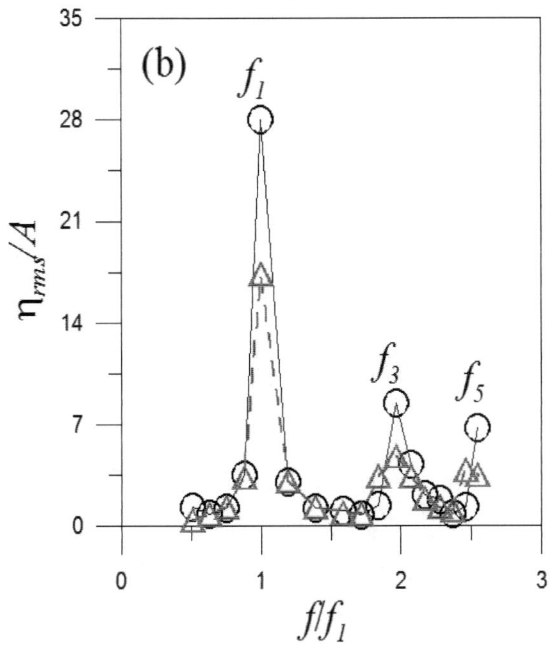

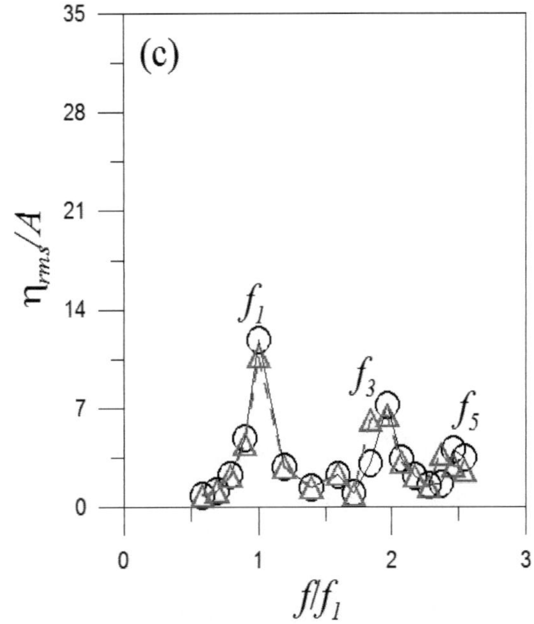

Figure.11: VARIATION OF η_{rms}/A WITH DIFFERENT FREQUENCIES RATIO (f/f_1) FOR h_s/L=0.325; (A) 1:86 SCALE TANK, (B) 1:57 SCALE TANK, AND (C) 1:43 SCALE TANK

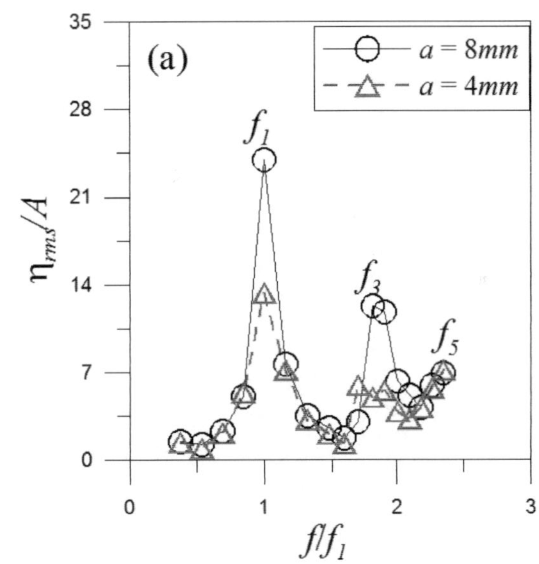

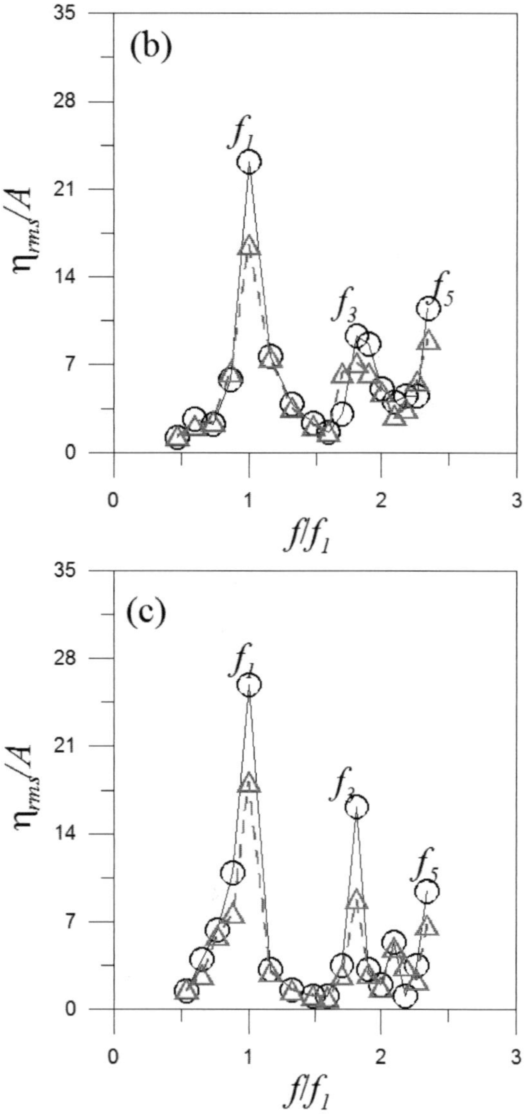

Figure.12: VARIATION OF η_{rms}/A WITH DIFFERENT FREQUENCIES RATIO (f/f_1) FOR h_s/L=0.4875; (A) 1:86 SCALE TANK, (B) 1:57 SCALE TANK, AND (C) 1:43 SCALE TANK

4. CONCLUSIONS

In this paper, the sloshing behavior in the sway excited rectangular tank has been presented by using shake table experiments. The partially filled r tank has been subjected to excitation frequencies ranging between 0.4566 *Hz* to 2.794 *Hz,* which overs upto fifth mode sloshing frequency. Two excitation amplitudes of 4mm and 8mm are considered. Sloshing behaviors have been reported for 1:86, 1:57 and 1:43 scaled tanks with three different fill levels. The experimental parameters such as sloshing oscillations, sloshing force are discussed for partially filled scaled sloshing tanks. The key conclusions drawn from the experimental study is given below:

- As the dimensions of the sloshing tank increased the number of the peak responses decreased.
- Sloshing oscillations decreases in an order of 1:86, 1:57 & 1:43 for 25% and 50% fill levels and for 75% fill level sloshing oscillations increases in the order of 1:86, 1:57 & 1:43 irrespective of excitation amplitudes.
- The experimental work iterates that first sloshing mode is the critical mode irrespective of scaled ratios of sloshing tank, aspect ratios.
- Normalised Maximum free surface response and Normalised root mean square of sloshing elevation is observed in the order $f = f_1$, $f = f_3$ and $f = f_5$ i.e at odd mode sloshing frequencies.
- It can be concluded that the 50% filled condition experiences violent sloshing irrespective of the scaling tanks, excitation amplitudes and excitation frequencies.

REFERENCES

[1] Akyildiz, H., and Ünal, E., 2005. "Experimental investigation of pressure distribution on a rectangular tank due to the liquid sloshing". Ocean Engineering, 32(11–12),1503–1516.

[2] Armenio, V., and La Rocca, M. 1996. "On the analysis of sloshing of water in rectangular containers: Numerical study and experimental validation". Ocean Engineering, 23(8), 705–739.

[3] Frandsen, J.B., 2004. Sloshing in excited tanks. J. Computational Physics. 196, 53–87.

[4] Ibrahim, R.A., 2005. Liquid Sloshing Dynamics—Theory and Applications. Cambridge University Press, New York.

[5] Kim, S., Kim, Y., and Lee, J. 2017. "Comparison of sloshing-induced pressure in different scale tanks." Ships and offshore structures. 12(2), 244-261.

[6] Mei-rong, J., Bing, R., Guo-yu, W and Guo-yu, W. 2014. "Laboratory investigation of the hydroelastic effect on liquid sloshing in rectangular tanks." J. Hydrodynamics. 26(5), 751-761.

[7] Nakayama, T., Washizu, K. 1980. "Nonlinear analysis of liquid motion in a container subjected to forced pitching oscillation".International. J. Numerical. Methods Eng. 15, 1207–1220.

[8] Nasar, T., Sannasiraj, S. A. and Sundar, V. 2008. "Experimental study of liquid sloshing dynamics in a barge carrying tank." Fluid Dynamics Research, 40(6), 427-458.

[9] Nasar, T., Sannasiraj, S. A., and V., Sundar. 2012. "Liquid sloshing dynamics in a barge carrying container subjected to random wave excitation". Journal of Naval Architecture and Marine Engineering, 9(1), 43-65.

[10] Xue, M. A., Zheng, J., Lin, P. and Yuan, X. 2017. "Experimental study on vertical baffles of different configurations in suppressing sloshing pressure." Ocean Eng., 136, 178-189.

[11] Nasar, T., Sannasiraj, S. A., and V., Sundar. 2012. "A Numerical study: Liquid sloshing dynamics in a tank due to uncoupled sway, heave and roll ship motions". Journal of Naval Architecture and Marine Engineering, 10(2), 119-138.

Proceedings of the 6th International Conference on Ship and Offshore Technology
ICSOT 2019
November 7-8, 2019, IIT Kharagpur, India

ICSOT INDIA 2019-19

"IoS OPEN PLATFORM" ADAPTIVE SCHEME TO IMPLEMENT MARITIME BIG DATA

Sumithran Sampath
General Manager
ClassNK Mumbai
Maharashtra,India

Dr M Abdul Rahim
Regional Manager
South Asia
ClassNK Mumbai

Hirofuni Takano
Director of Innovation
Development Division
ClassNK, Tokyo

ABSTRACT

As shipping evolves into a "Big Data" industry, ClassNK and its subsidiary Ship Data Center Co., Ltd. (ShipDC) are supporting the maritime community by providing the platforms to reap the benefits in partnership with the consortium partners in Japan.

KEY WORDS
Big data; Internet of Shipping; Ship Data Centre

INTRODUCTION

ClassNK and its subsidiary Ship Data Center Co. Ltd. (ShipDC) have investigated how the industry can utilize the so-called big data, teaming up with shipping companies, shipbuilders, suppliers, ICT companies, and other related parties. For sharing the data in the industry, shipping companies, as the data source, have pointed out the necessity of a common rule for data property and distribution as well as a secure data center. Responding to these needs, ClassNK and ShipDC have designed a scheme in which shipping companies are able to provide the data with less concern, and data users in the industry can utilize data for development and improvement of their products and services.

The paper describes the above-mentioned scheme of "IoS Open Platform (IoS-OP)" consisting of the data center with various functions to encourage the data use and "the common rules" to store, exchange, and utilize the data, as the model of data sharing and distribution in the industry.

THE ERA OF DATA

Big Data has been called "the oil of the 21st century" to emphasize the wide spread of data flow brought about by the new era of digitization. 'Internet of Things' is expected to comprise between 20[1)]-40[2)] billion devices by 2020 depending on who is doing the forecasting, but keeping pace with data flows is certainly likely to be challenging.

Once considered a very "low data" industry, the maritime industry is now vigorously waking up to the new digital age. Thanks to rapid advances in the development of information and communication technologies, it is now possible to collect large volumes of data on a diverse range of items related to ship operations. The information acquired from sensors of equipment, machinery and any other onboard devices can be recorded as digitalized data. The generated data connected to the internet as IoT can be accumulated as Big Data, which can be the basis of digitalization in the maritime industry. Remote access monitoring, condition-based maintenance, data analytics and forecasting are significantly improving and optimizing numerous functions in operations and ship management. As a result, the international shipping industry is beginning to embrace the tangible opportunities that the growth of big data presents.

In addition, regulatory requirements also pose the necessity of data collection. The implementation of fuel consumption data reporting regulations has been ongoing in an effort to reduce GHG emissions and the EU-MRV (European Union - Monitoring, Reporting, Verification) regulations for ships operating in the EU area began in 2018. The upcoming IMO DCS regulations require all globally operating ships over 5,000GT to collect fuel consumption data and create an annual fuel consumption data report to submit to their flag administration or recognized organization for verification.

While more shipping companies need and are willing to share information with a view to reaping the benefits of big data or complying with the international or local regulations, the approach to data capture remains very fragmented as seen in Fig. 1. Similar data is routinely sent to several vendors and analysis is still being carried out almost entirely on a ship-by-ship basis, in processes that are both time-consuming and inefficient.

To make larger gains, an effective platform capable of centralizing and managing such diverse data was considered essential. However, creating and maintaining this kind of platform is costly, time-consuming and unrealistic for some organizations. Furthermore, special care needs to be given to the handling of data to ensure confidentiality of information; hence it is also necessary to establish a secure yet effective platform from an impartial perspective.

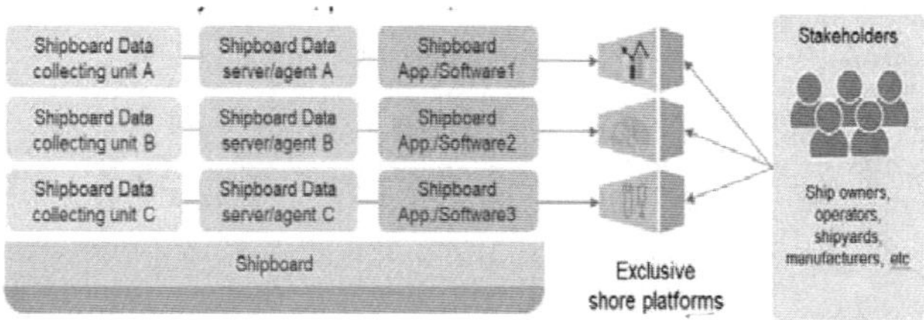

Fig.1 Fragmented data capture

TIMELINE AND THOUGHTS

ClassNK has long served the maritime industry through its technical and third-party service responding to industry needs. Its longstanding commitment to helping shipping realize the potential of big data is summarized as follows. In Dec. 2015 Ship Data Center Co., Ltd. (ShipDC) was established as a separate entity from ClassNK. It started to store shipping data in May 2016 as a trial, and simultaneously, ShipDC started to receive marine weather information from Japan Weather Association. Through Japan Weather Association's free provision of real-time marine weather information such as offshore wind (direction, speed), waves (height, frequency, direction) and ocean currents (direction, speed), the comprehensive analysis of voyage data from vessels at sea and marine weather information was made possible.

As seen in Fig. 2, in Aug. 2016, the trial for calling the data started. While ShipDC was preparing for the data center operation and technical trial to transmit the ship data to shore, it discussed with industry players how to best utilize the data related to ships. In 2017 the related conferences were held with more than 700 attendees in total, and the "Internet of Ships Open Platform Promotion Council" was established to deepen the discussion.

Without data collection beyond the border of companies, the data cannot become real Big Data. However, a giant platformer's monopoly on data and data use for their own business, which is often seen today in other industries, is not appropriate for industry platforms containing highly confidential information. There must be clear rules for fair data use between data owners and data users, and confidentiality of the data has to be strictly guarded. With this kind of framework, players especially who own the data can be willing to pass and share it. The conclusion from the discussion confirmed the necessity of common rules for data property and distribution, and a fair, reliable, and independent scheme

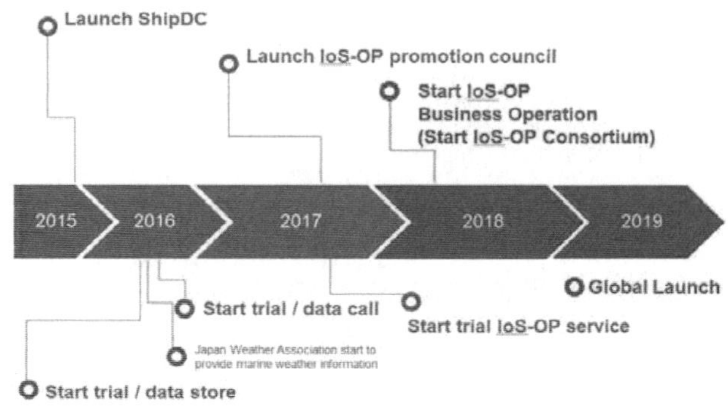

Fig.2 Summary of ShipDC activities

IOS-OP AND IOS-OP CONSORTIUM

In May 2018, for satisfying the above-mentioned requirements, the "Internet of Ships Open Platform (IoS-OP)" was launched[3)] as the common platform to share and distribute the operation data of ships to enable shipbuilders, manufactures and other stakeholders to access the data without infringing on data providers' interest. IoS-OP consists of the data center service and the common rules for data distribution agreed among the industry. The initiative is aiming to co-create data-driven new values, new solutions, and foster innovation.

In order to operate IoS-OP as the neutral platform, and avoid any monopoly of the data, an association was formed by the member of ShipDC users which is called as "IoS-OP Consortium". ClassNK is also a member of IoS-OP Consortium and any important decisions to operate IoS-OP shall be made by the prescribed procedures of IoS-OP Consortium. This ensures sound and permanent management of the IoS-OP. Initial members of the IoS-OP Consortium consist of 46 shipping companies, shipbuilders, marine manufacturers, ICT and other organizations based in Japan[3)].

On IoS-OP, data will be collected from multiple vessels, regardless of class or company, through data collection devices onboard. Companies will be free to choose what they want to share and to specify whether they want it shared with engine makers, equipment manufacturers, shipyards or other stakeholders who might benefit.

During 2018 the IoS-OP was available to the IoS-OP Consortium members located mainly in Japan. Having confirmed the validity in the period, it was globally launched in Apr. 2019[4)]. At that time the number of members had grown to

55 organizations.

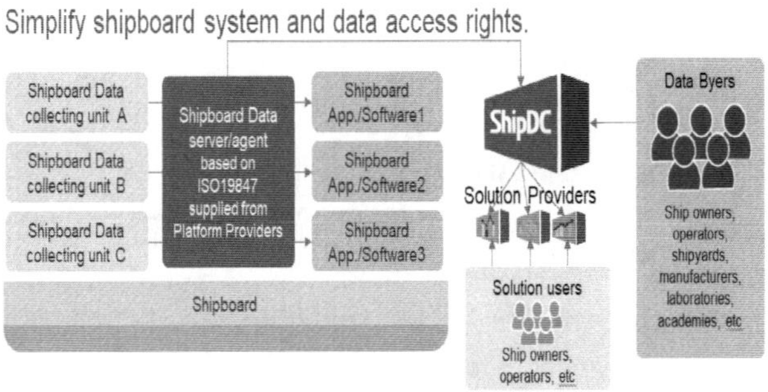

Fig.3 Simplified and integrated data capture

OUTLINE AND ROLES ON IOS-OP

The players on IoS-OP are described as seen in Fig. 4.

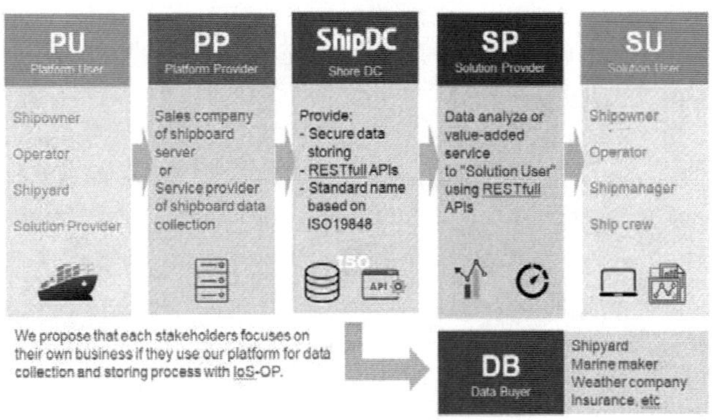

Fig.4 Outline of IoS-OP

Platform User (PU):
The PU should mainly be a ship owner or operator. A shipyard and Solution Provider (explained later) may also take this role. The PU bears the cost of data collection (data ownership). Costs here means shipboard server cost, data communication cost and data storage cost, etc. The PU manages data access rights and data collection of the data they provide and authorize solution user and data set range.

Platform Provider (PP):
The PP is a service provider of data collection to the PU, which should be a sales company of shipboard servers or service provider of shipboard data collection.

ShipDC (DC):
Through devices and services of the PP, the data shall be transmitted to the data center operated by DC. DC is ShipDC itself storing the collected data securely and providing RESTfull API for the data distribution. Furthermore, it harmonizes the data captured from ships, as seen in Fig. 3, by converting the specific data format of a software service provider or system into a standardized ISO format: "ISO19848:2018 Ships and marine technology — Standard data for shipboard machinery and equipment", which was originally developed by Smart Ship Application Platform (SSAP). SSAP is the project of Japan Ship Machinery and Equipment Association[5].

Solution Provider (SP):
The SP provides the data analysis or any value added service utilizing the transmitted data such as remote maintenance, performance report, condition monitoring, and so on. SP will use ShipDC's RESTfull API as the data access interface.

Solution User (SU):
The beneficiary from SP and their service is categorized as SU, who should be a ship owner, operator, ship manger, or crew.

Data Buyer (DB)
The DB uses the data for improvement of its own product/building ship, big data analysis, performance analysis, and so on under the authorization of PU. DB should be a shipyard, marine manufacturer, weather company, or insurance company.

For ensuring fair and transparent data use, membership in IoS-OP Consortium and company registration are required for the PP, SP and DB. On IoS-OP, each stakeholder can focus on their own business by using the common platform for data collection and storing process.

RULES FOR IOS-OP

IoS-OP provides common and individual rules corresponding to the role and relationship of each stakeholder in order to distribute IoS-OP data among stakeholders fairly. For the time being, nine sets of rules are stipulated (see Fig. 5).

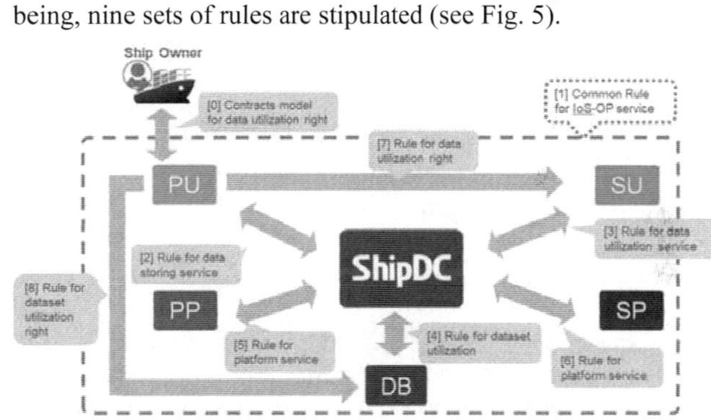

Fig.5 Rules for IoS-OP

[0] Contracts model for data utilization right
The contract model is for ship owner and PU to reduce the related negotiation of data ownership and data utilization rights.

[1] Common Rule for IoS-OP service
The rule is a basic one to be applied to all IoS-OP users. It will be applied between ShipDC and each stakeholder.

[2] Rule for data storing service (for PU)
[3] Rule for data utilization service (for SU)
[4] Rule for dataset utilization service (for DB)
[5] Rule for platform service (for PP)
[6] Rule for platform service (for SP)

These rules contain requirements for participation qualification, registration condition, use condition, observance matters and prohibited items by each role.

[7] Rule for data utilization right between PU and SU

[8] Rule for dataset utilization right between PU and DB

These rules are for relative transaction between PU and SU/DB for data/dataset utilization right.

SHIPDC'S SERVICE ON IOS-OP

The primary role of ShipDC is data center operation of IoS-OP, centralizing the data from ships to shore. However it provides additional service for smooth distribution including data standardization.

On IoS-OP ShipDC provides the following functions to receive data from various equipment, facilities, and systems onboard:

- Automatic data registration from attached data files based on the file naming rule
- Support multiple onboard devices
- Support data other than text-format data files
- Support confidential data for maker

In order to accept and process various data, ShipDC has introduced a data structure as seen in Fig. 6.

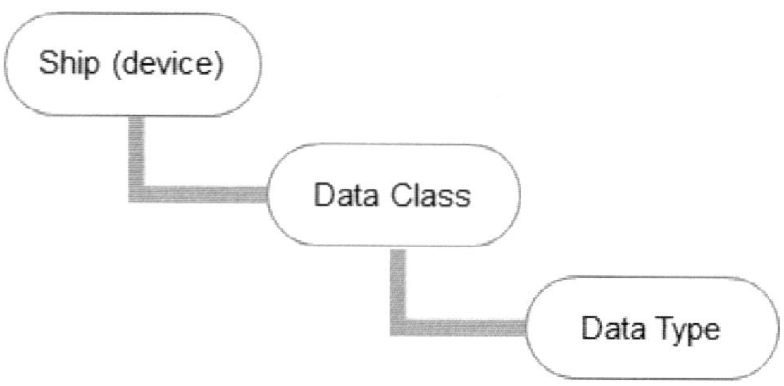

Fig.6 Data structure concept

"Data Class" defines the attribute of the data as seen in Fig. 7.

As the information hub for the maritime industry, security measures are taken for safe and secure distribution of data such as protection from unauthorized writing by specifying senders of data transmission mail, data key distribution by encrypted file, communication protection in data call, enhanced authentication for data access interface.

For data access control ShipDC provides sub-user setting function to grant access authority based on data ownership and schemes to detect/correct error data. ShipDC issues and distributes data keys with an access range set by ShipDC based on the request of data ownership holder. Users access data using the data key via software/application authorized by ShipDC. For data quality improvement, it also provides an error data detection/notification scheme to mechanically detect/notify errors including email non-delivery, data file damage, numerical data blanks and notification.

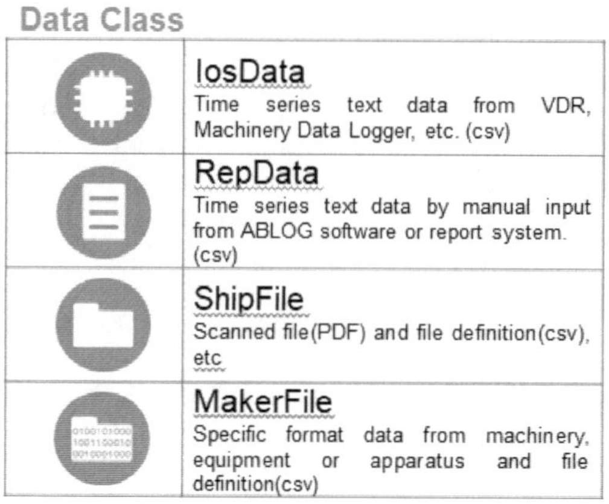

Fig.7 Data Class

'Data Type' is for specifying the data definition in "IoSData" and "RepData".

For data access control ShipDC provides sub-user setting function to grant access authority based on data ownership and schemes to detect/correct error data. ShipDC issues and distributes data keys with an access range set by ShipDC based on the request of data ownership holder. Users access data using the data key via software/application authorized by ShipDC. For data quality improvement, it also provides an error data detection/notification scheme to mechanically detect/notify errors including email non-delivery, data file damage, numerical data blanks and notification.

REQUIREMENT OF SHIPBOARD DEVICES (PP)

From onboard to shore, the data is stored in the database automatically by the email from a shipboard device to ShipDC with an attached data file as seen in Fig. 8. The following processes are necessary with the shipboard device.

- *Data output*: output text and file data in designated format
- *ZIP*: compress data files in a ZIP file
- *Authentication key code*: generate authentication key code by each sending action to prevent manipulation and unauthorized update
- *Send mail*: send mails to ShipDC with ZIP file and authentication key code file for data storage

RESTFUL API SERVICE

For SP, utilizing the data for their service or solution, RESTful API is provided as the data access interface. These API are:

- *DataSet API*: GET available ships and data type list
- *DataClass API*: GET accessible data with data key
- *Weather API*: GET free marine weather information mapping to ship position/ date & time in time series data provided by Japan Weather Association
- *Maker API*: GET specific data sending/storing available

for maker confidential data

- *Other API*: GET supplementary data such as data type definition (meta information)

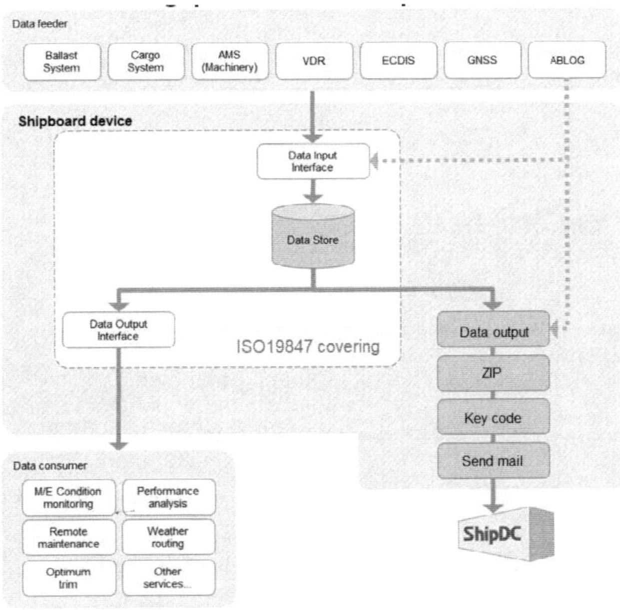

Fig.8 Requirement of Shipboard Devices (PP)

CONCLUSIONS

IoS-OP is the adaptive scheme to best utilize the big data in the maritime industry and consists of the data center with various functions to encourage the data use and "the common rules" to store, exchange, and utilize the data. Following its establishment in 2018, ShipDC data center and other operation for IoS-OP and partners had been working on the global launch of the service in 2019.

References

1) Bob Gill, Gartner Research Vice President. "Leading the IoT", https://www.gartner.com/imagesrv/books/iot/iotEbook_digital.pdf
2) ABI Research, "IoE VS. IoT VS. M2M WHAT'S THE DIFFERENCE?". https://www.abiresearch.com/pages/what-is-internet-things/
3) ClassNK Press Release, "Launch of IoS-OP & IoS-OP Consortium (Japanese only)", https://www.classnk.or.jp/hp/ja/hp_news.aspx?id=2880&type=press_release&layout=1
4) ShipDC Press Release, "Global launch of data utilization platform announced at Sea Asia Exhibition". https://www.classnk.or.jp/hp/en/hp_news.aspx?id=3782&type=press_release&layout=1
5) Smart Ship Association Platform, "SSAP had the presentation about our acitivities in Sea Japan 2018 at Tokyo, JAPAN", https://www.jsmea.or.jp/ssap/topics/20180413.html

Proceedings of the 6th International Conference on Ship and Offshore Technology
ICSOT 2019
November 7-8, 2019, IIT Kharagpur, India

ICSOT INDIA 2019-20

SPECTRAL FATIGUE ANALYSIS OF A SHIP PART 2: USING FEM BASED STRUCTURAL RESPONSES

Yogendra Parihar
Indian Register of Shipping
Mumbai, Maharashtra, 400072
India
yogendra.singh@irclass.org

Amresh Negi
Indian Register of Shipping
Mumbai, Maharashtra, 400072
India
amresh.negi@irclass.org

Suhas Vhanmane
Indian Register of Shipping
Mumbai, Maharashtra, 400072
India
suhas.vhanmane@irclass.org

ABSTRACT

The present paper discussed about the Spectral Fatigue Analysis (SFA) using FE based approach to determine Stress Transfer Functions STFs. Full stochastic spectral fatigue analysis is performed through the process of hydrodynamic response analysis, interaction of direct loads to finite element (FE) model, structural analysis and prediction of fatigue damage. As explained in Part-1, one of the most crucial aspect of SFA, is to obtain STFs. Beam theory based STFs are used in Part-1 while FE based structural responses (STFs) are used in PART-2. It is a computationally intense and expensive approach. An intermediate compute program is used to transfer the motions and hydrodynamic loads to FE model efficiently. Complexities involved in the load application on FE model are discussed in detail. Eventually a comparison of FE based SFA is made with other fatigue assessment approaches as discussed in Part-1 of this paper.

KEY WORDS

Spectral fatigue, hydrodynamic loads, direct load application, fatigue damage

INTRODUCTION

Fatigue is a failure of material due to cyclic loads. In case of ships, hull girder undergoes through various cyclic loads due wave actions which can induce the crack and fatigue failure. Prediction of fatigue life can be performed using S-N curve or fracture mechanics based approach. Present analysis used the S-N curves based approach. S-N curve based fatigue assessment can be performed using simplified method or spectral method. Simplified fatigue assessment is performed using the classification society's rule. These empirical load formulations presumes a wave environment (e.g. North Atlantic) which provide highest wave loads, would typically results conservative estimate of fatigue damage. On contrary, actual environmental condition can be accounted in case of spectral based fatigue analysis. The spectral analysis is a frequency domain based numerically intensive technique. The hydrodynamic loads are calculated using potential theory based sea-keeping programs. Various numerical techniques for computation of hydrodynamic loads can be found in [1], [2].

Spectral analysis is quite popular for reliable prediction of fatigue life. Many papers/literature discussed about the spectral fatigue analysis [3]–[6]. Also, classification societies [7]–[10] have published their detailed guidelines outlining spectral fatigue analysis procedures. Spectral fatigue assessment technique uses the STFs evaluated directly from structural analysis. STFs are the stress ranges defined for a set of discrete wave frequencies for a given wave heading, loading condition, structural location and unit wave amplitude. Stress range is normally expressed in terms of probability density functions for different short-term intervals corresponding to the individual cells of the wave scatter diagram. It is assumed to follow Rayleigh distribution with each short term response. The cycle count is according to zero crossing period of short term stress response. Using the Palmgren-Miner's Rule, linear summation of short term damages sustained over all the sea states gives the total fatigue damage for the structural detail. Total fatigue damage (T_D) accumulated over operational service life can be estimated by accounting for all sea states encountered with the different wave directions and all possible representative load cases.

Present papers discussed about the procedure of spectral fatigue analysis which is categorized into four parts namely sea-keeping analysis, structural analysis, short-term analysis and spectral analysis. Evaluation of direct hydrodynamic loads is performed using 3D panel methods. In present work the main focused is to obtain structural responses based on the finite element (FE) analysis. Green function based Panel method is used here to compute the hydrodynamic loads. Panel method is useful in capturing the hydrodynamic hull pressure compared to 2D strip method [11] and is more convenient for mapping the pressure from hydrodynamic model panels to FE mesh. For this purpose an in-house developed program has been used to transfer the load data and balancing of model. General FE package is used to record the structural response of ship. Finally, fatigue damage is calculated using the spectral method for given structural details.

Table 1. SUMMARY OF THE FATIGUE ASSESSMENT APPROACHES

Method – IDs	Load Evaluation Method	Fatigue Damage Approach
FAM-1	Semi analytical formulation	Closed form approach based on long-term response
FAM-2	2D Strip theory	Closed form approach based on long-term response
FAM-3	Semi analytical formulation	*Spectral approach Based on short term response
FAM-4	2D Strip theory	*Spectral approach Based on short term response
FAM-5	3D Panel method	*Spectral approach Based on short term response

*long term stress range distribution is defined through a short-term Rayleigh distribution within each short term sea state

Table 1 summarizes the fatigue damage prediction approaches which have been followed in the present papers (Part-1 and Part-2). These approaches are the combination of the different load evaluation methods and damage prediction approaches. Hence forth the method IDs are used when referring the various methods as mentioned in Table 1. Description of methods FAM-1 to FAM-4 has already been discussed in Part-1 of paper. This paper is focused on FAM-5 method. FAM-5 consists of the direct application of hydrodynamic loads (pressure and motions) to FE model of ship. In present paper, methodology and complexity involved in using FAM-5 for the fatigue damage have been described in detail.

METHODOLOGY

The overview of spectral fatigue analysis is shown in Figure 1.

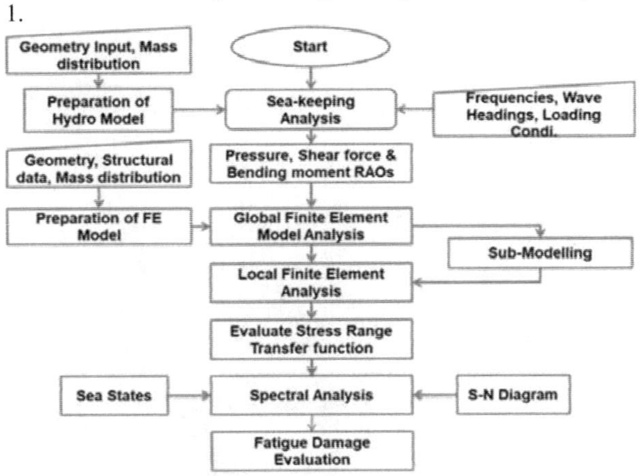

Figure 1: PROCEDURE FOR FATIGUE ANALYSIS

HYDRODYNAMIC LOAD COMPUTATION

Potential theory based sea-keeping programs are usually preferred for the computation of motions and loads. These potential theory based methods can be distinguished as zero forward speed Green function method and Rankine panel method. Forward speed problem in ship sea-keeping is constitute of total potential that can be defined as the sum of steady contribution due to ship's forward speed and unsteady potential due to ship motion. It is required to have solution for steady potential beforehand and then solving the unsteady potential around this potential which is very complicated mathematically and intense numerically. However, conventional cargo ships such as bulk carriers and oil tankers are having relatively low Froude number. Therefore, zero forward speed green function based panel method with higher order forward speed correction can be used to compute the loads/pressure for ships. In the present paper 3D Green function based panel method has been used to compute the hydrodynamic loads. The ship forward speed has been accounted using the speed correction factors.

Figure 2: PANEL DISTRIBUTION SHIP GEOMETRY OF BULK CARRIER IN HOMOGENEOUS LOADING CONDITION

INTERACTION OF LOADS

Sea-keeping programs provide wetted hull pressures, motions and hull girder loads such as shear forces (SF) and bending moments (BM) as outputs. But, application of these direct loads on finite element (FE) structural model is not a straight forward. In regards to panel pressure, Zhao [12] has demonstrated the utility of panel based methods. It is more convenient to transfer hydrodynamic panel pressures to 3D FE structural models therefore, panel based method is preferred. Apart from this approach, application of direct shear force and bending moments have been discussed in [13], [14]. The pro and cons of alternative method compare to panel pressure, and complexities of hydro-structure interaction have been discussed in [15]–[17]. Present study is using the panel pressure for structural analysis. Procedure for same is discussed below.

Considering the hydro-structural interaction as most critical task, the following procedure needs to be performed to transfer the direct hydrodynamic loads on FE model:

- Weight distribution of light weight, cargo weight, ballast water, bunkering and other appendages etc.
- Consideration of inertial loads (acceleration to FE model as shown in Figure 8)
- Mapping of panel pressure onto FE mesh
- Balancing of the structural model (inertia relief method)
- Solution of the numerical problem

Weight Distribution

The lightship weights are simulated by providing the density of material or using nodal mass elements. Minor differences in the lightship weight of ship can be adjusted by providing lumped mass elements or by changing the steel density (special cases). The cargo weights are modeled considering the loading condition as provided in the trim and stability booklet. The bulk cargo masses are simulated using point/lumped mass element at the CG of respective cargo holds. The rigid body elements are used to connect the mass element to relevant hold surfaces as shown in Figure 3 and Figure 4. Ballast and bunker weights are simulated as tank pressures. The appropriate pressures acting on internal surfaces of liquid tanks are calculated and applied to selected entities in FE model as shown in Figure 5 and Figure 6. With above discussed task, procedure of weight distribution is to be performed such a way that CG of structural model (vertical, transverse and longitudinal centers of gravity) remains as given in loading manual/stability booklet. The external fluid pressure is applied on FE model using the suitable pressure mapping technique as shown in Figure 7.

Figure 3: DISTRIBUTION OF CARGO HOLD The bulk cargo **masses** IN HOLD NO. 3

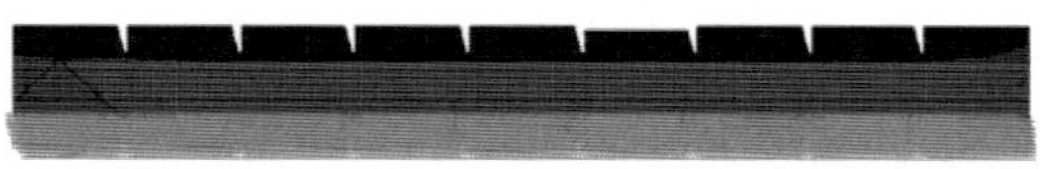

Figure 4: DISTRIBUTION CARGO HOLD MASS IN ALL HOLDS

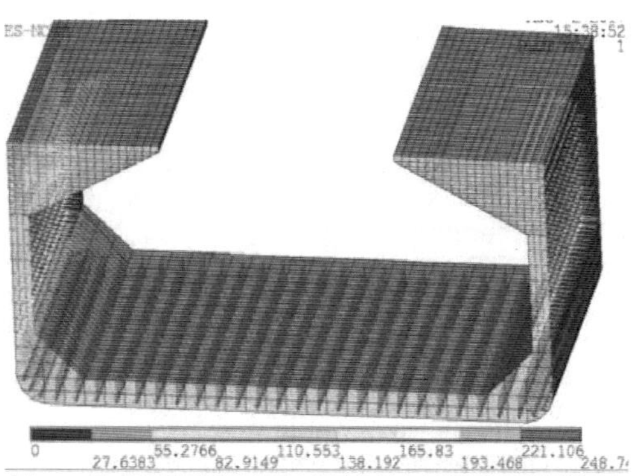

Figure 5: DISTRIBUTION OF BALLAST PRESSURE (NORMAL BALLAST CONDITION)

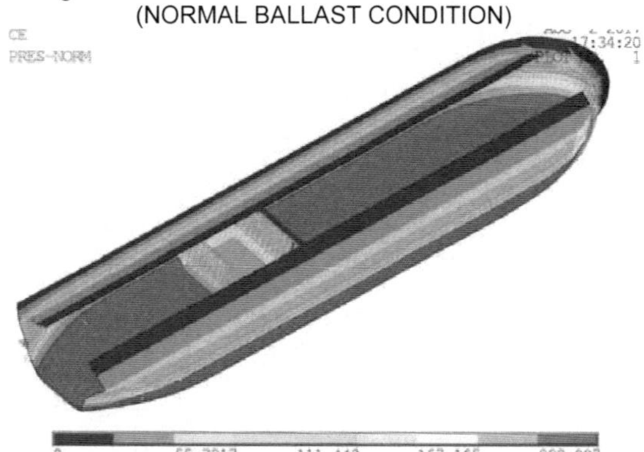

Figure 6: DISTRIBUTION OF BALLAST PRESSURE IN STARBOARD, PORT SIDE TANKS AND ONE CARGO HOLD (HEAVY BALLAST CONDITION)

Inertial Loads

Each type of loads discussed above consists the static and dynamic load components. The dynamic loads can be decomposed into quasi-static and inertial loads. The quasi-static components are induced due to ship's roll and pitch inclination. The directions of resultant gravity loads in ship's fixed coordinate system are shown in Figure 8 with the variation of roll and pitch motion respectively. The inertial loads in FE model are to be simulated by using the Eqns. (1)-(4). In case of liquid tank pressure, inertia loads are calculated using the following (5)-(6). All the relevant inputs (ship motions and accelerations) for Eqn (1) are to be obtained from sea-keeping analysis. IRS guideline for direct application of loads [18] can be referred for detailed procedure of weight distribution, pressure mapping, inertial and quasi-static loads distribution.

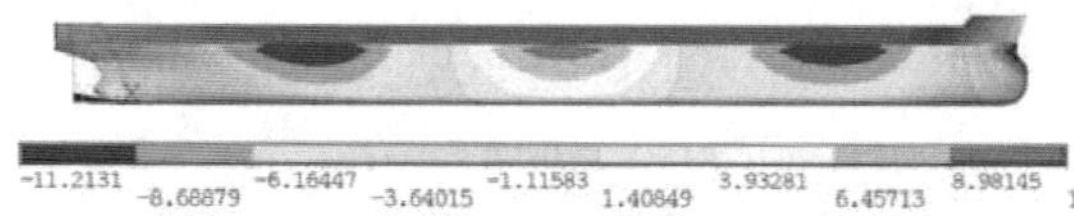

Figure 7: DISTRIBUTION OF REAL COMPONENT OF COMPLEX PRESSURE IN FREQUENCY DOMAIN (HOMOGENEOUS CONDITION, HEAD.=180°)

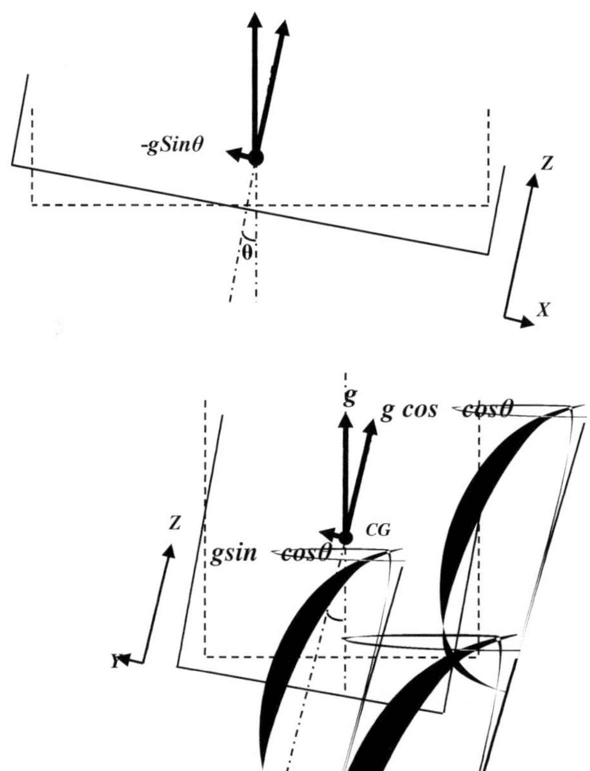

Figure 8: INERTIAL LOAD COMPUTATION

$$\overrightarrow{A_e} = \vec{A} + \vec{\Theta} \times \vec{R} \quad (1)$$

where:

$\vec{A}$: Acceleration vector at ship's CG can be estimated as follows in respective directions using Eqns (2), (3), and (4):

$$A_x = -g \sin\theta + a_x \quad (2)$$

$$A_y = g \sin\phi \cos\theta + a_y \quad (3)$$

$$A_z = g \cos\phi \cos\theta + a_z \quad (4)$$

$\vec{\Theta}$: Roll, pitch and yaw acceleration vector

$\vec{R}$: Distance vector from ship's CG to calculation point

a_x, a_y, a_z : Surge, sway, heave accelerations due to ship motions

θ, ϕ : Pitch and roll angle (by sea-keeping program)

$-g\sin\theta$, $g\sin\phi\cos\theta$, $g\cos\phi\cos\theta$ are the quasi-static components in x, y and z direction respectively with reference to ship's fixed coordination system due to combined motions of roll and pitch.

Liquid pressure in tank is calculated using following equations:

$$P = P_0 + \rho_l h_i g_{eff} \quad (5)$$

where (6)

$$g_{eff} = \sqrt{(A_{el}{}^2 + A_{et}{}^2 + A_{ev}{}^2)}$$

ρ_l : density of liquid

h_i : internal pressure head at CG of element measured from the top of tank to the load point

A_{el}, A_{et}, A_{ev}: Accelerations in longitudinal, transverse and vertical direction respectively are to be calculated using Eqn. (1).

CASE STUDY

A bulk carrier is taken for the analysis purpose. The main particulars of ships are as follows:

Table 2: MAIN PARTICULARS OF SHIPS

Ship Particulars	Bulk carrier
Length overall [m]	287.50
LBP [m]	279.00
Breadth (moulded) [m]	45.00
Depth (moulded) [m]	24.10
Scantling Draught [m]	18.49
Max Service speed [knots]	14.60

Motion and Loads

The convention used in sea-keeping and throughout in analysis is as follows:

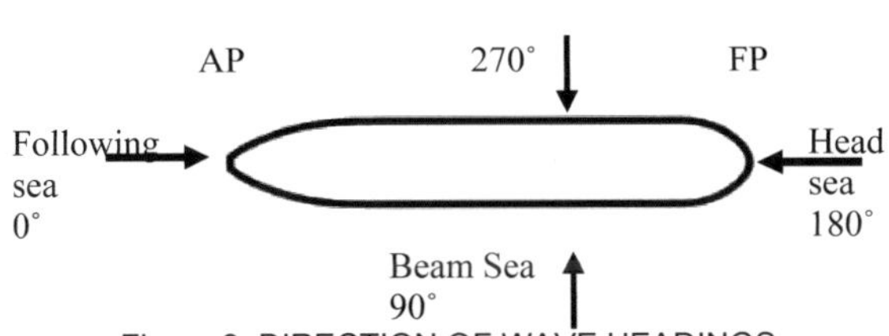

Figure 9: DIRECTION OF WAVE HEADINGS CONSIDERED FOR SEA-KEEPING ANALYSIS

As loads interact from hydro model to FE model, therefore, a consistency needs to be maintained between these two models. The two models should be geometrically similar. To maintain the same, one IGES file was used to create the hydro and FE models. The considered hydrodynamic model is represented by the total 1992 panels till mean waterline in homogeneous loading condition as shown in Figure 2. The dimension of panel can be decided by the considered wave lengths for analysis, atleast 6 to 10 panel points should present over one wave length. Total 27 frequencies (λ/L ranges from 0.1 to 5) and 12 wave headings from 0 to 330 with an interval of 30 are considered in present analysis.

Another important aspect in sea-keeping analysis is mass distribution, as inertia matrices being used in the motion equation. This required information regarding the mass properties and can be derived from mass model. Therefore, mass model must reflect the actual weight distribution of hull, cargo, ballast, bunkering appendages etc. for considered loading conditions. Any inaccuracy in the mass matrix may result in an unbalanced hydro model and thus incorrect end shear forces and bending moments. There can be small difference in the calculated mass by hydrodynamic analysis and total mass given in loading manual. An iteration process for tuning the mass and CG locations of ship can be carried out till the difference tends to negligible quantity.

Structural Modelling

A full length structural model of bulk carrier is created using FEM software. All primary and most secondary structural members are modeled in order to simulate actual stiffness of the hull girder. All applicable weights are simulated in FE model as discussed above. Simulation of weights is to be done accurately. The consequent vertical, transverse and longitudinal centers of gravity (VCG, TCG and LCG) as obtained from the model are compared and verified with the corresponding values listed in stability booklet.

Type of Elements and Model Idealization: Based on the usage, the following elements are used in the analysis.

Table 3: STRUCTURAL IDEALIZATION

Element type	Idealization
Beam	Stiffeners
Shell	Girder, plates, bracket
Mass	Lumped mass e.g. machinery components, cargo weight
Rigid link	Connection of node-node, node-mass

Material Properties: Standard material properties are given in Table 4.

Table 4: MATERIAL PROPERTIES

Material	Young's Modulus (N/mm^2)	Poisson Ratio	Density (t/mm^3)
Steel	2.05E+05	0.3	7.8E-09

Boundary Conditions: The right-handed co-ordinate system, with the x-axis positive forward, y-axis positive to port and z-axis positive vertically from baseline to deck are taken for the analysis as shown in Figure 10. The origin was located at the intersection between aft perpendicular (AP), baseline and centerline. The boundary conditions are applied on the FE model as listed in Table 5 and shown in Figure 10.

Table 5: BOUNDARY CONDITIONS FOR GLOBAL MODEL

	Location	Direction
Engine Room Front Bulkhead	SB & PS	Z
	CL	Y
Collision Bulkhead	CL	X, Y, Z

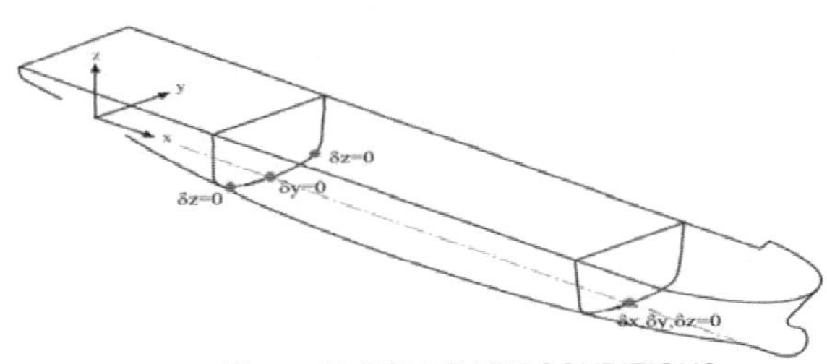

Figure 10: BOUNDARY CONDITIONS

Model Balancing

Hydrodynamic model is perfectly balanced as equilibrium is implicitly imposed in solution of equation of motion. The critical issue is balancing of the FE model when the loads are transferred from hydro model to FE model. Problem arises due to difference in mesh size between these two models. Hydro model is having the 1500 to 2000 panels in ship's geometry while outerhull geometry of FE model can have 20000 to 30000 elements. Mapping of pressure from coarser panels to finer elements can lead to unbalanced forces at artificial supports. Appropriate interpolation scheme is to be used for the pressure application. Present study used the 3D interpolation technique to map the pressure onto hull surface. Unbalanced non-zero reaction forces at the artificial supports are corrected using the inertia relief method. Inertia relief is a technique in which applied forces and moments are balanced by counter forces induced by accelerating the body. The application of these accelerations is performed such a way that it precisely cancels or balances the additional forces. However one should be cautious that application of inertia relief can alter the response profile of structure. Application of inertia relief method in ship structures is limited to cases where unbalanced forces are within engineering limits.

Structural Analysis

Static structural analyses are performed for given number of load cases corresponding to each loading condition. For each loading condition, 12 headings, 27 frequencies and two parts of frequency (real and imaginary) are taken for analysis at given speed. Total number of 648 ($1\times12\times27\times2\times1$) load cases are considered in single loading condition. Four loading conditions (homogeneous, alternate, heavy ballast and normal ballast) are analyzed.

Equilibrium check for each load case is mandatory. The applied hydrodynamic panel pressure must be in equilibrium with the counter loads induced due to motion of loaded (distributed weights) FE model of ship. To estimate the imbalance forces, all the forces and moments are to be summed up in global direction. In case of unbalanced forces, a suitable method to balance the FE model is employed as discussed above before performing the structural analysis.

Before preceding the structural response evaluation, analysis is verified with the following aspects as given in loading manual or stability booklet:

- Total weight and CG of each tank/hold
- Shear force distribution along the length of ship
- Moment distribution along the length of ship

Finally, hot-spot stress is evaluated using the recommended method as given in IIW recommendation [19] for fatigue design of welded joints. Fine mesh of size 50 mm × 50 mm is created at given locations as shown in Figure 11.

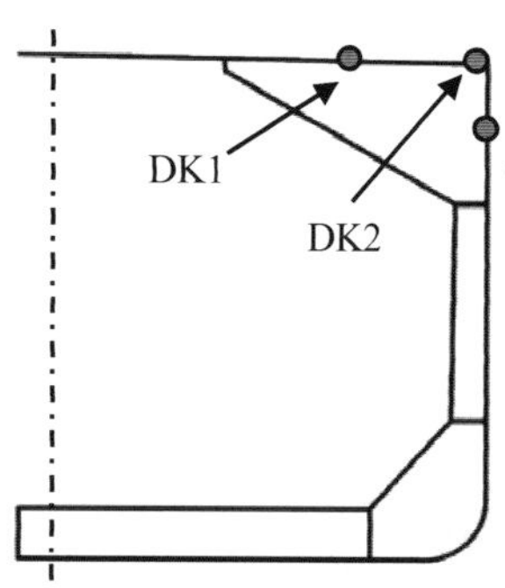

Figure 11: REPRESENTATIVE MID-SHIP SECTION SHOWING THE BUTT PLATE JOINTS (DK1, DK2 AND SS1)

SPECTRAL ANALYSIS

The fatigue damage at selected locations is predicted using a spectral approach. The spectral fatigue analysis is based on the Palmgren-Miner's linear damage summation rule and used the appropriate S-N curves. The procedure for spectral fatigue is well documented [7], [10], [20]. The detailed methodology can be referred from the literature [3], [6], [7], [13], [21], [22].

Spectral Approach Based on Short Term Response:

The closed form expression for total cumulative fatigue damage *D*, where the long term stress range distribution is defined through a short term Rayleigh distribution within each short term sea state, the fatigue damage for bi-linear S-N curve, is given as in Eqn. 7 [7].

$$D = \nu_0 T_d \sum_{i=1,j=1}^{\substack{All\ sea\ states \\ All\ headings}} r_{ij}\,[A+B] \tag{7}$$

$$A = \frac{\left(2\sqrt{2m_{0ij}}\right)^{m_1}}{\bar{a}_1}\Gamma\left(1+\frac{m_1}{2};\left(\frac{S_0}{2\sqrt{2m_{0ij}}}\right)^2\right)$$

$$B = \frac{\left(2\sqrt{2m_{0ij}}\right)^{m_2}}{\bar{a}_2}\gamma\left(1+\frac{m_2}{2};\left(\frac{S_0}{2\sqrt{2m_{0ij}}}\right)^2\right)$$

Where,

ν_0 = long-term average zero-up-crossing-frequency (Hz)

T_d = Design life in seconds.

$\bar{a}_1$, m_1 = S-N fatigue parameters for N < 10^7 cycles

$\bar{a}_2$, m_2 = S-N fatigue parameters for N > 10^7 cycles

γ = Incomplete Gamma function

Γ = Complementary Incomplete Gamma function

r_{ij} = the relative number of stress cycles in short-term condition *i, j*

S_0 = Stress range in *S-N* curve, where the change of slope occurs

$$m_n = \int_0^{\infty}\int_{\theta'=\theta-90}^{\theta'=\theta+90}\left(\frac{2}{\pi}\right)cos^2\theta'.(\omega_e^n S_\sigma).d\theta'.d\omega$$
$$\omega_e = \omega - \frac{V\omega^2}{g}cos\theta' \tag{8}$$

$$S_\sigma(\omega|H_S,T_Z,\theta) = |H_\sigma(\omega|\theta)|^2.S_\xi(\omega|H_S,T_Z) \tag{9}$$

Where, *i* and *j* indices denotes the number of sea states and direction respectively. $\sigma_{ij} = \sqrt{m_{0ij}}$ is the standard deviation of the stress response in the sea state and direction respectively Wirsching rainflow correction factor, thickness effect, mean stress effect have been considered in the stress range. The spectral moments are calculated using Eqn. 8. $S_\sigma(\omega|H_s,T_z,\theta)$ is stress spectrum generated by scaling the wave energy spectrum $S_\eta(\omega|H_s,T_z,\theta)$ and complex stress transfer function $H_\sigma(\omega|\theta)$ using Eqn. 3, where $H_\sigma(\omega|\theta)$ is the structural response for sinusoidal wave of unit amplitude computed. Total fatigue damage accumulated over operational service life is estimated by accounting for all sea states encountered with the different wave directions and given loading condition.

Result and Discussion

Spectral fatigue analysis is performed by following the procedure is given in Figure 1. D class of S-N curves [23], [24] is considered for selected weld location considering that hotspot stress range is being computed and used for evaluation of number of cycles. Wave data have been taken from the IACS recommendation 34 [25]. Spectral analysis is performed considering Pierson Moskovitz (P-M) wave spectrum. Stress responses were obtained for various sea states and respective directions in given loading conditions. The complex stress transfer functions are evaluated from FE analysis. Figure 12 to Figure 14 show the STFs in alternate loading condition. Total fatigue damage is evaluated using spectral method and STFs following Eqns. (7), (8) and (9). Mean stress effects are considered in the analysis.

Table 6: FRACTION OF TIME FOR CONSIDERED LOADING CONDITIONS

Loading conditions	Fraction of time
Homogeneous	0.25
Alternate	0.25
Normal ballast	0.20
Heavy ballast	0.30

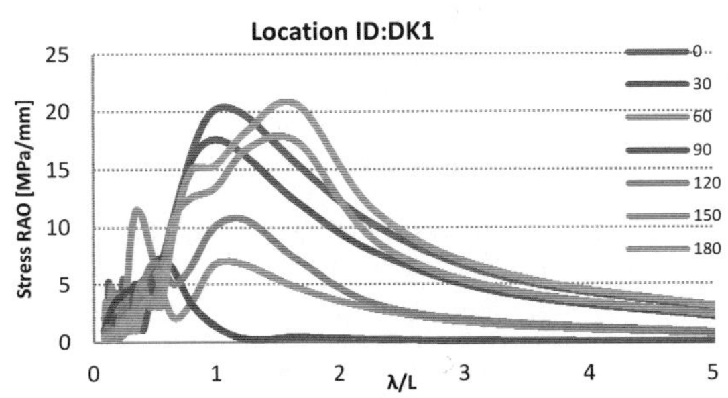

Figure 12: STFs AT DECK LOCATION DK1

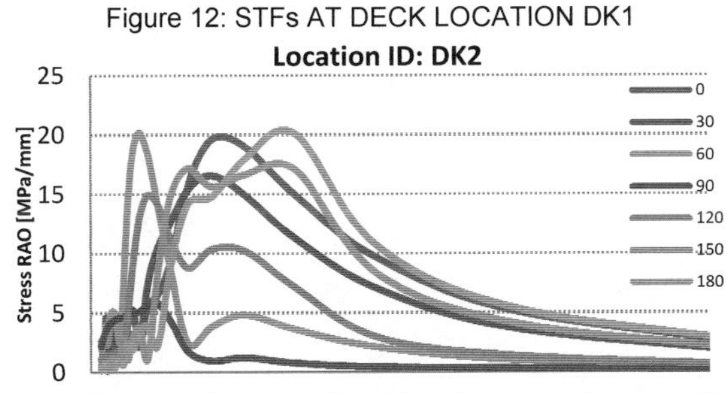

Figure 13: STFs AT DECK LOCATION DK2

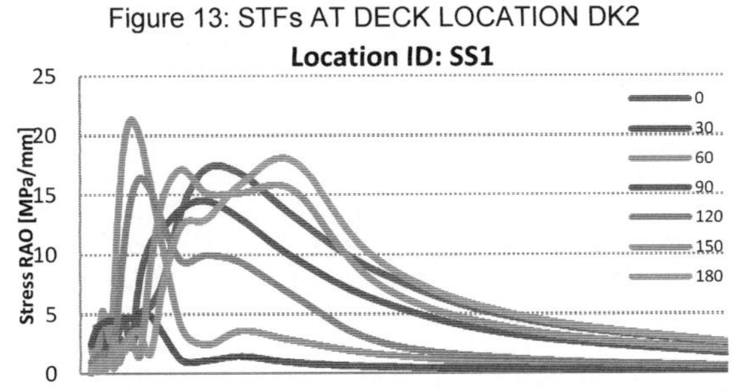

Figure 14: STFs AT SIDE SHELL LOCATION SS1

The fatigue damage for each loading condition is shown in Figure 15, where, homogeneous (max draught) and normal ballast (min draught) conditions make the larger contribution in fatigue damage. Although, the combined fatigue damage is calculated considering the fraction of each loading condition as shown in Table 6 and a factor of 0.85 considered for the ship sailing during its lifetime.

Table 1summarizes all the methods which have been followed for the load evaluation in Part 1 of the paper. Current approach is added as Fatigue Assessment Method-5 (FAM-5). The loads for FAM-1 and FAM-3 are calculated using the closed loop formulation based semi-analytical approach. Similarly, FAM-2 and FAM-4 have used the 2D strip theory based loads. The two load computational approaches (FAM 1, FAM 3 and FAM 2, 4) are differed by the way stress range responses are treated in fatigue damage assessment. Long distribution approach based on the Weibull Distribution is used for FAM-1 and FAM-2 However, short-term Rayleigh distribution within each sea state has been defined for the long term fatigue damage estimation in FAM-3 and FAM-4 methods. Current method FAM-5 is using the 3D panel based loads application to FE model to estimate the STFs. Fatigue damage is evaluated using the long term stress range distribution accumulated using short term Rayleigh distribution for each sea-state similar to FAM-3 and FAM-4.

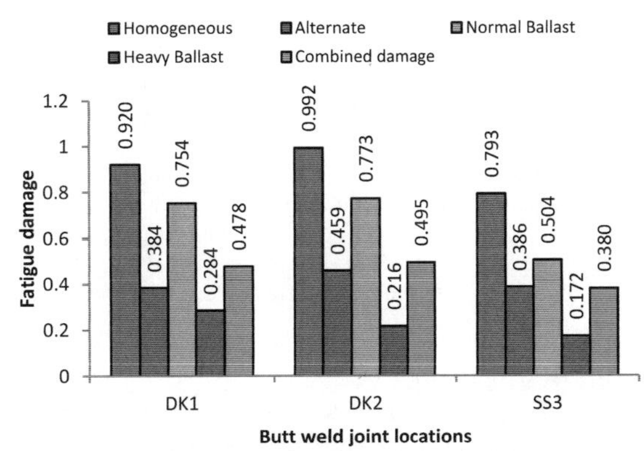

Figure 15: FATIGUE DAMAGE IN VARIOUS LOADING CONDITIONS AND COMBINED FOR CONSIDERED LOCATIONS

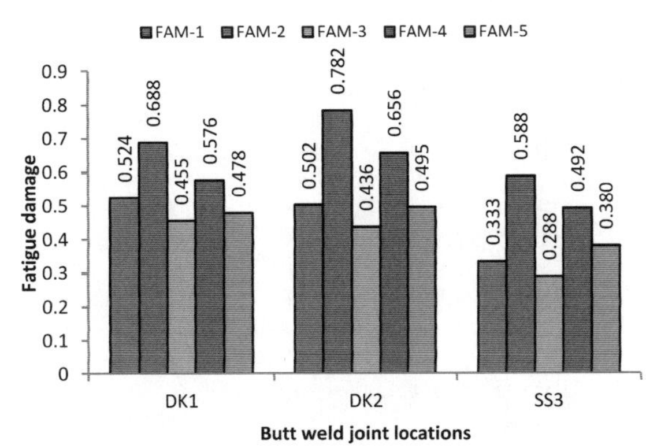

Figure 16: TOTAL FATIGUE DAMAGE AT GIVEN LOCATION USING SPECIFIED METHODS

To compare the above approaches, Figure 16 showed the total fatigue damage at specified locations estimated using all five approaches. FAM-5 is based on the comprehensive structural analysis for the evaluation of structural response. Therefore the following points are discussed and summarized with respect to FAM-5:

- Closed loop based semi-analytical approach is under predicting the fatigue damage but time and computational resources required by this method are very less.
- Strip theory based fatigue damage prediction is more conservative, although required time and computational resources are moderate compared to FE based approach.
- 3D panel and FE based fatigue damage values for considered location is coming in-between of above two methods. This method is considered to be more realistic. However, time and computational resources required to perform such analysis are very high.

CONCLUSION

Present paper showed the spectral fatigue analysis for the butt-weld joints for a bulk carrier. It detailed about the direct application of sea-keeping loads on FE model. Issues like the practical difficulties in implementation of direct loads and balancing of FE model were discussed. Processes, where the number of load cases are huge, can be automated as shown in present analysis.

Among three locations, location-2 (DK2) is more critical. Similar to 2D strip theory based damage assessment using FAM-2 and FAM-4, the detailed structural analysis based fatigue damage (FAM-5) shows the moderate influence of the HBM and torsion. Beam theory and FE based structural response approaches can be adopted for the fatigue assessment based on the design requirement as presented in detail. Semi analytical load evaluation method (FAM-1 and FAM-3) does not possess a feasible solution for the fatigue damage assessment in ship design. However, it may use for finding the critical locations for the comprehensive spectral fatigue analysis considering only the VBM.

ACKNOWLEDGEMENT

The authors express their sincere thanks to the Indian Register of Shipping (IRS) for providing the support to perform this study.

REFERENCES

[1] J. L. Hess and A. M. O. Smith, "Calculation of potential flow about arbitrary bodies," *Prog. Aeronaut. Sci.*, vol. 8, pp. 1–138, 1967.

[2] B. N. Salvesen, E. O. Tuck, and O. Faltinsen, "Ship Motions and Sea loads," in *Society of Naval Architects and Marine Engineers*, 1970, vol. 6, no. 6, pp. 250–287.

[3] Y. Wang, "Spectral fatigue analysis of a ship structural detail–A practical case study," *Int. J. Fatigue*, vol. 32, no. 2, pp. 310–317, Feb. 2010.

[4] T. Kukkanen and T. P. J. Mikkola, "Fatigue assessment by spectral approach for the ISSC comparative study of the hatch cover bearing pad," *Mar. Struct.*, vol. 17, no. 1, pp. 75–90, Jan. 2004.

[5] Y. Kim, S. Park, S. Jun, and J. Heo, "Spectral Fatigue Analysis for Side Longitudinals and Hopper Knuckle Connection in VLCCs and FPSOs," vol. 3, pp. 157–162, 2002.

[6] A. Negi and S. S. Dhavalikar, "Spectral Fatigue Analysis of a Transportation Barge," in *International Offshore and Polar Engineering Conference*, 2011, vol. 8, pp. 936–942.

[7] ABS, "Guidance Notes on Spectral-based Fatigue Analysis for Vessels," Houston, USA, 2012.

[8] DNV, "Fatigue Assessment of Ship Structures," Det Norske Veritas, 2014.

[9] BV, "Guidelines for Fatigue Assessment of Steel Ships and Offshore Units Guidance Note," Neuilly sur Seine Cedex – France, 2016.

[10] IRS, "Guidelines for spectral fatigue assessment for ship structures," Mumbai, 2018.

[11] Z. Li, W. Mao, J. W. Ringsberg, E. Johnson, and G. Storhaug, "A comparative study of fatigue assessments of container ship structures using various direct calculation approaches," *Ocean Eng.*, vol. 82, pp. 65–74, May 2014.

[12] C. Zhao, M. Ma, and O. Hughes, "Applying Strip Theory Based Linear Seakeeping Loads to 3D Full Ship Finite Element Models," in *32nd International Conference on Ocean, Offshore and Arctic Engineering*, 2013, pp. 1–9.

[13] Y. Parihar, "Fatigue Assessment of Container Ship," Indian Institute of Technology Kahargpur, 2014.

[14] Y. Parihar, S. K. Satsangi, and A. R. Kar, "Application of direct hydrodynamic loads in structural analysis," in *International Conference on Computational and Experimental Marine Hydrodynamics*, 2014, no. December.

[15] Y. Parihar, S. Dan, K. Doshi, and S. G. T, "Application of Direct Hydrodynamic Loads in Spectral Fatigue Analysis," in *Proceedings of the ASME 2017 36th International Conference on Ocean, Offshore & Arctic Engineering*, 2017, pp. 1–10.

[16] S. Malenica and Q. Derbanne, "Hydro-structural issues in the design of ultra large container ships," *Int. J. Nav. Arch. Ocean Eng*, vol. 6, pp. 2092–6782, 2014.

[17] M. Ma, O. Hughes, and C. Zhao, "Applying Sectional Seakeeping Loads to Full Ship Structural Models using Quadratic Programming," in *International Conference on Marine Research and Transportation*, 2012, no. June, pp. 25–28.

[18] IRS, "Guidelines for application of direct sea-keeping loads in structural analysis of ships," Mumbai, 2018.

[19] IIW, "Recommendation of fatigue design for welded joints and components - IIW-1823-07," Paris, France., 2008.

[20] BV, "Spectral Fatigue Analysis Methodology for Ships and Offshore," Paris, France., 2008.

[21] L. Xiang-chun, Guo- qing FENG, and Hui-long REN, "Study on the application of spectral fatigue analysis.pdf," *J. Mar. Sci. Appl.*, vol. 5, no. 2, pp. 42–46, 2006.

[22] K. T. Nguyen, Y. Garbatov, and C. G. Soares, "Spectral fatigue damage assessment of tanker deck structural detail subjected to time-dependent corrosion," *Int. J. Fatigue*, vol. 48, pp. 147–155, 2013.

[23] IACS, "Recommendation No 56 - Fatigue Assessment of Ship," 1999.

[24] UK HSE, "Offshore installations: guidance on design and construction," London, 1990.

[25] IACS, "Standard Wave Data - Recommendation 34," 2001.

Proceedings of the 6th International Conference on Ship and Offshore Technology
ICSOT 2019
November 7-8, 2019, IIT Kharagpur, India

ICSOT INDIA 2019-21

ANALYSIS OF LONGITUDINAL STRENGTH OF A BULK CARRIER IN WAVES

Sree Krishna Prabu Chelladurai
Department of Ocean Engineering and Naval Architecture, Indian Institute of Technology Kharagpur
skprabu@iitkgp.ac.in

Vishwanath Nagarajan
Department of Ocean Engineering and Naval Architecture, Indian Institute of Technology Kharagpur
vishwanath_n@naval.iitkgp.ernet.in

Om Prakash Sha
Department of Ocean Engineering and Naval Architecture, Indian Institute of Technology Kharagpur
ops@naval.iitkgp.ac.in

ABSTRACT
Ship's hull girder undergoes wave induced vertical and horizontal bending moment while navigating in seas. Both the bending moments are considered for ship strength. The horizontal bending moment in some conditions may influence ship maneuvering. This may occur due to rudder developing angle of attack because of hull girder deformation even though rudder angle is kept at 0°. In this study, the behavior of a bulk carrier ship in the wave for probabilistic longitudinal strength calculation is investigated. For large ships operating in random seas, it is necessary to investigate the longitudinal strength in waves. The bending moment of a bulk carrier for different wave conditions in fully loaded condition are investigated. For the probabilistic calculation of bending moment, North Atlantic wave data is used. The maximum bending moment usually occurs when the effective wavelength is of the order of ship's length. The computations are carried out in the head sea condition at zero forward speed of the vessel. The computation considers the sinkage and trim in wave. The bending moment in wave is compared with still water bending moment for fully loaded departure condition. The outcome of this investigation will suggest sufficient margin in the longitudinal strength during dynamic loading condition in heavy weather.

1. INTRODUCTION
Structural failure of a ship may result in loss of human life, environmental damage, and economic loss. Therefore, ships must be designed with adequate safety and reliability, and their designs must be acceptable from an environmental and economic point of view. Environmental data and models represent a necessary and important input to load and response calculations of the ship [1]. The ships during sailing are exposed to varying wave environment. This fact needs to be taken into consideration when specifying design and operational criteria. The wave load analysis is performed for head/ following sea, because the possibility of maximum hogging/ sagging bending moment is more compared to beam and quartering sea. The present analysis is carried out for zero forward speed condition. It is well known that the wave bending moment and shear force contributes more to the structural design than the still water bending moment. The wave bending moment can be approximately six times the still water bending moment [2]. In the present paper, the bending moment for a bulk carrier are computed in head/ following sea for various wave conditions. The analysis is carried out for fully grain loaded condition. For both the bending moment and shear force, there is a strong influence of the loading condition. For ships operating in waves, it is necessary to investigate the wave bending moment, which will occur with the ship at effective wavelength of the same order as the length of ship. It is evident that maximum vertical wave bending in a head sea occurs with waves slightly shorter than the length of the ship [3]. However, some larger ships may, under adverse conditions, be subjected to bending moments larger than those encountered in head seas. The return period for ship structures is 20 years for Ultimate Limit State (ULS) design [4]. Recently, IMO has increased the design life to 25 years in Goal Based Standards (GBS) for bulk carriers and tankers with length over 150 m [5]. Uncertainties involved in the suggested analysis and their consequences on bending moment are presented and the results in the extreme wave condition are discussed.

2. DESIGN MODEL
The methodology adopted to determine the bending moment is described. For the purpose of this study, wave conditions in North Atlantic Ocean are considered. The principal dimensions of the ship are given in Table 1. Once the hull form of the bulk carrier had been modelled using design software, their hydrodynamic performances in terms of sea keeping, and safety is assessed. To ensure that the designs are safe and reliable, rules have been developed by classification societies. Figure 1 shows the hull form and general arrangement of the ship. For cargo ships, the lightship weight is calculated by the use of semi-empirical formulas and design graphs, which depends on cube number, type of ship and cargo capacity. In the current analysis, the light ship weight is distributed as described in the Fig. 2 [6]. It may be noted that light ship weight is distributed irregularly along the ship's length. This depends on the arrangement of engine room, accommodation and outfit on the ship. For section modulus calculation, hull structure which are continuous along the ship's length as per classification rules are considered. Usually, the section modulus distribution is uniform along the

ship's length. Typical distribution of bending moment variation along ship's length assumed by the classification society is also shown in Fig. 2. The hull weight, superstructure, machinery weight, and outfit weight are estimated in the initial period of design process. The above items on board the ship correspond to lightship weight. The general arrangement provides the dimension of each hold, extent of the engine room, the position of components with their weight information, the LCG and VCG position. This information is used to calculate the stability of ship at a later stage. The modelling of bulk carrier using ship design software to determine the weight distribution, as well as calculations of bending moment in different wave height and wave period is the main objective of this work. The bulk carrier has five numbers of cargo holds, and top and bottom wing tanks. The wing tanks are along the cargo hold length, from fore peak tank to engine room. In the present analysis, the wave bending moments are determined and probability of different wave condition is computed. The wave crest position is varied along the bulk carrier for different wave height and wave period from forward to aft of ship. The wave height and wave period are varied between 0.5 to 13.5 m and 3.5 to 18.5 s respectively. The wave bending moments are computed with the vessel poised on a wave with its crest varying along the length of the ship. It is observed that the maximum bending moment occurs around the amidships region (0.2Lbp aft and 0.2Lbp forward of amidships). It is observed that when wave travels along the ship, the instant when the crest or trough comes to mid ship region, the ship usually experiences maximum bending moment. In addition, the magnitude of bending moment depends on the wave height [7]. The wave loads were analyzed in fully loaded departure condition considering the extreme response in different wave heights. The still water bending moment variation along the ship's length in design draft condition is shown in Fig. 3. The maximum bending moment is $-29.08 * 10^3$ t - m.

3. RESULTS

It was observed that the operational factors influencing the magnitude and distribution of the wave bending moment are the wave height, wave period and phase offset. The wave bending moments are analyzed with the vessel poised on a wave with its crest varying along the length of the ship. The maximum bending moment occurs around the amidships region. When the wave crest is between 0.15Lpp to 0.30Lpp, the bulk carrier experiences maximum sagging moment at mid ship region. When the wave crest is between 0.45Lpp to 0.65Lpp, the bulk carrier experiences maximum hogging moment at mid ship region. Likewise, when the wave crest is between 0.75Lpp to 0.95Lpp, the bulk carrier experiences maximum sagging moment. From Fig. 4a it is observed that the bending moment is maximum at wave period between 9.5 s to 11.5 s. Here the bending moment becomes maximum for wave height more than 11.5 m. When the wave period is between 8.5 s to 14.5 s, the bending moment undergoes a cycle of hogging and sagging as the wave crest travels along the length of the ship. When the wave period is less than 7.5 s, the bending moment variation is present mainly in sagging mode, and the hogging moment is very less as shown in Fig. 4a and Fig. 5. It is observed that when wave travels along the ship, the instant when the crest or trough comes around to mid ship region, the ship experiences maximum bending moment. Also the magnitude of bending moment depends on the wave height. When the ship encounters a wave of height 11.5 m and period 10.5 s, it experiences the maximum bending moment of $-321.01* 10^3$ t – m and it is named as case I. The probability of occurrence of this wave condition is 0.0314 % as shown in Fig. 5. In case of bulk carrier, it is usual to apply alternate hold loading condition while carrying heavy cargo like ores. In Fig. 4b the maximum observed bending moment along the ship's length for different wave conditions is shown. The corresponding loading condition details are shown in Table 2. It can be seen that the bending moment variation is different in case of uniform loading and alternate hold loading conditions. The hogging moment is maximum, in case of alternate hold loading condition. However, in case of uniform loading, the sagging moment is maximum. This has influence while calculating the fatigue loading of the hull structure in wave conditions. The magnitude and number of cycles of peak to peak stress variation affects the fatigue loading.

Figure 5 shows probability of occurrence of bending moment for different wave conditions. From Fig. 5, it can be observed that hogging moment is negligible till wave period of 8.5 sec. For wave condition of 1.5 m wave height and 7.5 sec wave period, the occurrence probability is 7.74%. The ship will experience a bending moment of $-37.75 * 10^3$ t – m for this probability, it is named as case - II. Likewise, when the ship encounters the wave condition 18.5 sec wave period, it experiences the bending moment of $-86.082 * 10^3$ t - m and the occurrence probability is 0.0001% as shown in Fig. 5. It is named as case - III. Here the bending moment is low as compared to case – I. When the wave period is large (18.5 s), the ship would rise and fall with the wave train without any appreciable variation of buoyancy along its length. In Fig. 6 the variation of wave elevation for a random wave is shown. It can be observed that the buoyancy distribution along the length of the ship will significantly change in random sea as compared to a regular wave. When this is combined with the actual flexure of the structure, the variation could be significant especially with regard to probabilistic fatigue assessment. In Table 3, the horizontal and vertical wave bending moment computed for the subject ship is shown. The values correspond to the probability level of 10^{-8}. It corresponds well with the values estimated in this paper and shown in Fig. 4a. In Fig. 7, typical horizontal mode shapes for a ship is shown [8]. The magnitude of horizontal wave bending moment is less as compared to vertical wave bending moment. The IACS rules for hull scantlings and structural details applicable to all types of bulk carriers consider the following ship characteristics: $L < 350$ m, $L / B > 5$, $B / D < 2.5$, $CB \geq 0.6$. This shows that the ships are structurally stiff for horizontal bending as compared to vertical bending. For bulk carrier and container ships, usually large deck opening are provided for accommodating cargo hatches. This will decrease stiffness for horizontal bending. The corresponding increase in horizontal hull flexure may induce an effective angle of attack on the rudder. This will cause coupling of horizontal hull bending with ship maneuvering dynamics and result in a variety of hydroelastic phenomena. This type of coupling has been extensively investigated for aircrafts [9]. This requires further investigation for the large bulk carriers and container ships. For lifetime maximum extreme values, there are two estimation procedures which are described as follows. In the short-term approach, the lifetime maximum extreme values are estimated by finding the maximum of the short-term extreme values for three hour duration from the 25 year return sea states. This estimation is based on the assumption that the 25 year return sea states have approximately 10^{-5} probability of occurrence, and the most probable short-term extreme values for three-hour duration have about 10^{-3} probability of exceedance. Combining these two probabilities, the extreme values in the 25-year return sea states are representative of a 10^{-8} probability level. In the long-term approach, the joint probability of short-term extreme values and occurrence of sea states in the wave scatter diagram is considered. Simulations are carried out for generating random wave for each wave height/ period combination using desired sea spectrum. Corresponding to the random wave, random hull bending stress time series is determined. From the time series, the probability of exceeding permissible stress limit is determined. Combining the two probabilities and adding probabilities for all the possible combination of wave height and wave period, the probability of exceeding safe stress limit is determined. If probability of exceeding permissible stress is less than 10^{-8}, the structural design is acceptable [10]. We also need to check for fatigue cycles induced by wave loads. For this some robust numerical computational technique like rain flow counting method need to be adopted.

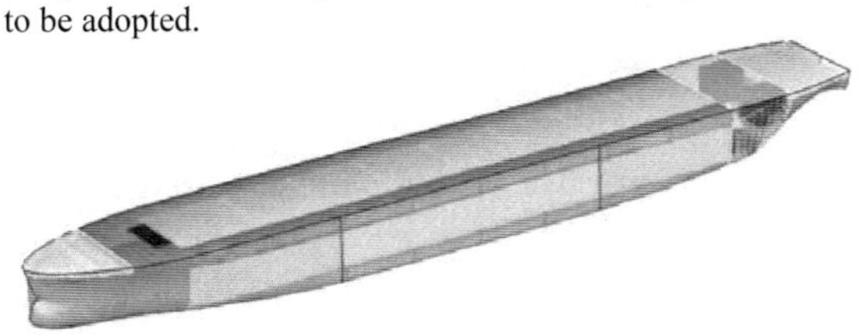

Figure 1. General arrangement of bulk carrier.

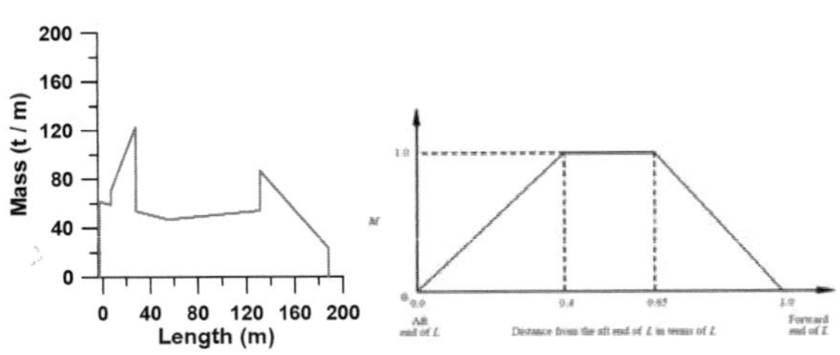

Figure 2. (L.H.S.) Typical distribution of lightship weight of a bulk carrier. (R.H.S.) Typical assumed distribution of bending moment along the ship's length.

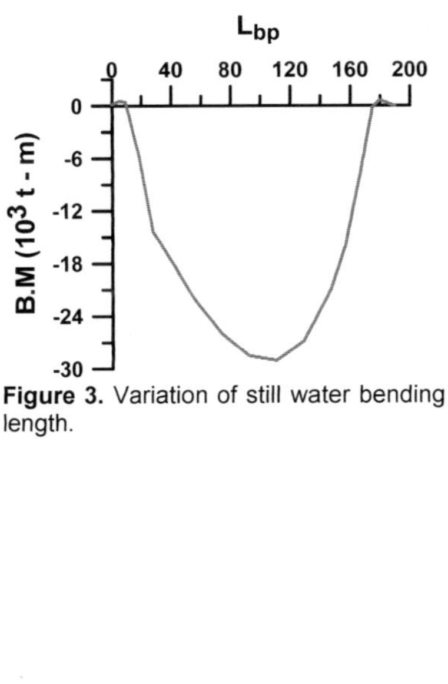

Figure 3. Variation of still water bending moment along ship's length.

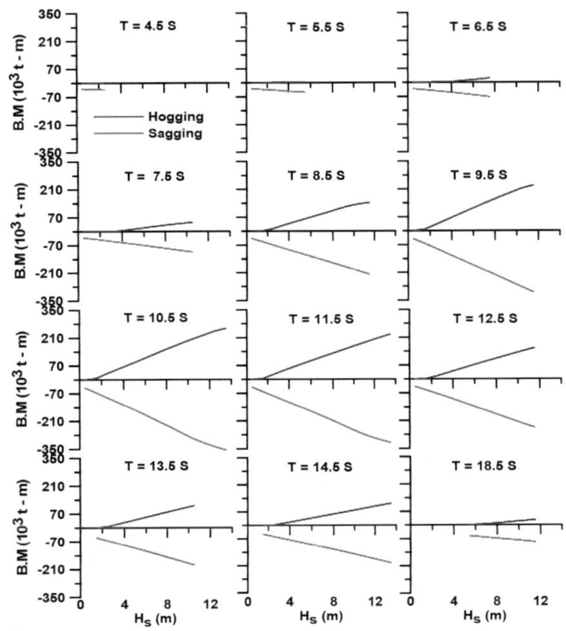

Figure 4a. Bending moment of fully grain loaded bulk carrier around amidships region in different wave heights.

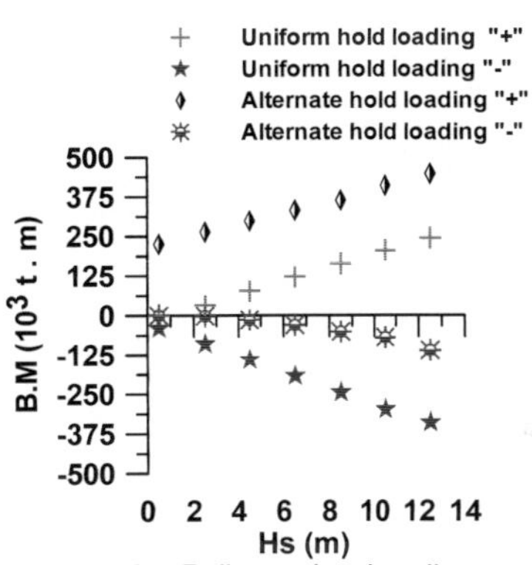

Figure 4b. Bulk carrier bending moment at two different loading conditions (Uniform hold loading and alternate hold loading) in regular wave.

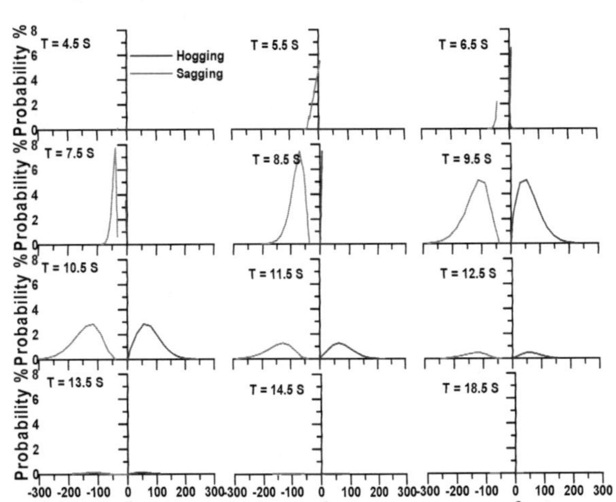

Figure 5. Probability of bending moment of fully grain loaded bulk carrier at different wave periods.

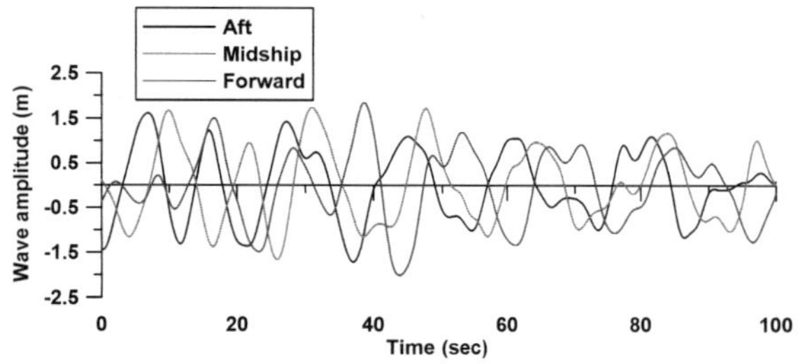

Figure 6. Variation of instantaneous wave elevation along a ship's length, 190.0 m. The wave elevations are taken at aft, midship and forward part at the same time instant. Significant wave height = 3.0 m, Mean wave period = 12.50 sec. Variation computed as per ITTC wave spectrum.

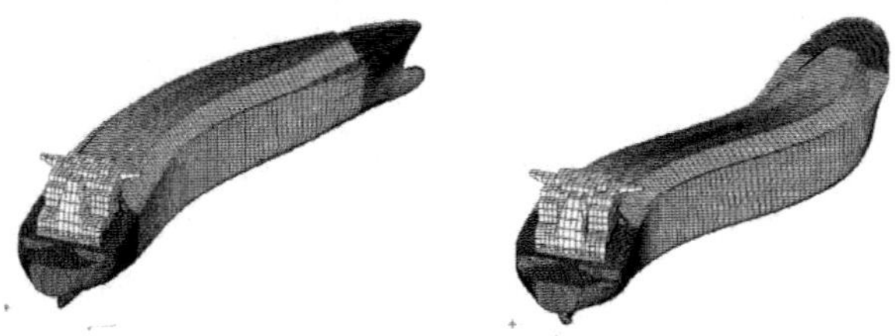

Figure 7. First two horizontal mode shapes for a ship [8].

Table 1. Main particulars bulk carrier.

Sr. No.	Dimension	Value
1	$Length_{oa}$ (m)	193
2	$Length_{bp}$ (m)	184
3	Breadth (m)	32.25
4	Depth (m)	17.15
5	Draft (m)	11.20
6	Dead weight (t)	45000
7	Speed (Knots)	14.6

Table 2. Detail of bulk carrier's two different loading conditions.

Sr. No.	Item	Uniform hold loading	Alternate hold loading
1	Mean draft	10.7 m	10.7 m
2	Cargo density	0.76 (t / m^3)	2.5 (t / m^3)
3	Loading Quantity	95 %	48 %
4	Loaded Holds	1 to 5	1, 3, 5

Table 3. Bending moments computed as per ABS class rule [10].

Sr. No.	Particulars	$x10^5$
1	Horizontal wave bending moment (t-m)	1.16722
2	Vertical wave bending moment (hogging) (t-m)	1.62904
3	Vertical wave bending moment (sagging) (t-m)	1.75817

4. CONCLUSION

The results of ultimate wave bending moment and the probability analysis give the safety requirements of margin for dynamics load in the form of an allowable percentage of still water bending moment. The important outcome of relevant studies on the prediction of probability of bending moment in waves has an important role in the evaluation of safety of ship. The results of the analysis are presented in the graphs, which can provide valuable guidance to a designer when selecting scantling of structural members at the early stages of design.

Acknowledgments

This work was financially supported by IIT Kharagpur research fund.

REFERENCES

[1] Vecleler A. (1965), Longitudinal Strength of Great Lakes Bulk Carriers, Marine Technology.

[2] Bitner-Gregersen E. M., Hovem L., Skjong R. (2002), Implicit reliability of ship structures. Proceedings OMAE 2002, Oslo, 23 – 28 June 2002.

[3] Paik J. K., Thayamballi A. K., Che J. S. (1996), Ultimate strength of ship hulls under combined vertical bending, horizontal bending and shearing forces. SNAME Transactions, Vol. 104, pp. 31 - 59.

[4] Hu Y., Zhang A., Sun J. (2001), Analysis on the ultimate longitudinal strength of a bulk carrier by using a simplified method, Marine Structures 14 (3): 311 - 330.

[5] Bitner-Gregersen E. M., Eide L. I., Hørte T., Skjong R. (2010), Ship and offshore structure design in climate change perspective. Springer Briefs in Climate Studies. DOI 10.1007/978-3-642-34138-0.

[6] Jensen J. J., Pedersen P. T. (1981), Bending moments and shear forces in ships sailing in irregular waves, Journal of Ship Research, Vol. 25, No. 4, pp. 243 - 251.

[7] Gordo J. M., Soares C. G. (1997), Interaction equation for the collapse of tankers and containerships under combined bending moments, Journal of Ship Research, Vol. 41, Issue 3, pp. 230 - 240.

[8] Guidance notes on ship vibration (2006), American Bureau of Shipping.

[9] Bisplinghoff R. L., Ashley H., Halfman H. (1996), Aeroelasticity. Dover Science, ISBN 0-486-69189-6..

[10] Rules for building and classing steel vessels (2009), Part 5B Specific vessel types, Common Structural Rules for Bulk Carriers, American Bureau of Shipping.

Proceedings of the 6th International Conference on Ship and Offshore Technology
ICSOT 2019
November 7-8, 2019, IIT Kharagpur, India

ICSOT INDIA 2019-22

HYDROFOIL WITH DIFFERENT CONFIGURATIONS OF TWO LEADING-EDGE PROTUBERANCES

Rajni Kant
Indian Institute of Technology
Kharagpur, West Bengal, 721 302
India
rajnikant.mech012@iitkgp.ac.in

Anirban Bhattacharyya
Indian Institute of Technology
Kharagpur, West Bengal, 721 302
India
ab@naval.iitkgp.ac.in

ABSTRACT

Leading edge protuberances on the pectoral fins of humpback whales have been widely adopted to the designs of foils to provide superior lifting characteristics in the post-stall regimes. The objective of this paper is to study the lift and drag characteristics of a rectangular hydrofoil having different configuration of two protuberances on the leading edge using numerical and experimental methods.

A base hydrofoil with NACA 63_4-021 section and aspect ratio of 2 was chosen for this study. The results indicate that, in general, a foil with larger protuberances have better post-stall lift characteristics whereas that with smaller protuberances have comparatively higher pre-stall lift. The lift and drag forces obtained from CFD calculations are in good agreement with those from experiments with hydrofoil models carried out in the towing tank facility.

In the numerical study, streamwise vortices are observed due to protuberances on the leading edge. An interesting observation is that the two protuberances can restrict the zone of separation between them at high angles of attack. This study reveals that the use of twin leading edge protuberances on foils can help in generation of higher post-stall lift, and the size of protuberances will influence the lift depending on the operation profile of the hydrofoil.

***Keywords**: Protuberance, Hydrofoil, Lift, Flow separation.*

NOMENCLATURE

α	Angle of attack
c	Chord Length
C_D	Drag Coefficient
C_L	Lift Coefficient
D	Drag Force
L	Lift Force
n	Number of tubercles
Re	Reynold's number with respect to the chord
S	Span
U_∞	Freestream Velocity
x	x-coordinate
z	z-coordinate
v	Dynamic viscosity
ρ	Density
λ	Wavelength of tubercle
A	Amplitude of tubercle

1. INTRODUCTION

The humpback whales are the most agile among their species. Their excellent maneuverability has been attributed to the tubercles over the leading edge of their flippers [1]. These tubercles or humps alter the flow hydrodynamics around the flippers and work as lift enhancement devices [2,3] They help the flow to remain attached up to a larger angle of attack thus increasing the stall angle [4].

Several research works have been done investigating the effects of tubercles over the leading edge of foil using numerical and experimental methods. A wind tunnel experiment was done by [3] showing a 40% delay in the stall angle with an increase in maximum lift and decrease in drag. A numerical study performed by [5] over NACA 63_4-021 baseline foil and other two modified foils with leading-edge protuberances having different amplitudes reported an increase in post-stall lift. The modified foil with a smaller amplitude of protuberance performs similar to the baseline foil, while lift increases in the post-stall region for the modified foil with higher amplitude protuberances. A number of studies have been done on the effect of protuberance amplitude over lift characteristics. An experimental study with varying tubercle A (0.025, 0.05c and 0.12c) and λ (0.25c and 0.5c) was presented in [6]. It has been found that the wavelengths of the protuberances play a minor role in the lift enhancement. On the other hand, amplitude had a significant effect i.e smaller amplitude performs better in the pre-stall regime and larger amplitude in the post-stall. A similar effect has been found from many other numerical [5] and experimental [7,8] literature.

Considering the available literature on leading-edge modified foils presented till now, the modifications were restricted to uniformly distributed humps/protuberances over the span of the foil. However, in some published works, the distribution of leading edge protuberance is not uniform. A numerical study performed by [9] having protuberances on the leading edge separated at a distance and found that it is as effective as that of continuous protuberances. Numerical and experimental study done by [10] on an airfoil having a single leading-edge protuberance showed that it can generate streamwise vortices like continuous protuberances. An experimental and numerical study performed by [11] over a rudder had found that the leading-edge tubercle similar to that of a humpback whale is able to restrict the flow separation between two large tubercles which led to better post-stall lift characteristics as compared to continuous protuberances.

In the present study, the two prominent protuberances which are mimicked from the first and fourth tubercle of the humpback whale are added over the leading edge of a hydrofoil to obtain a better post-stall lift. The location of twin protuberance is compared to the bio-mimicked configuration of [11] and [12] is shown in Figure 1.

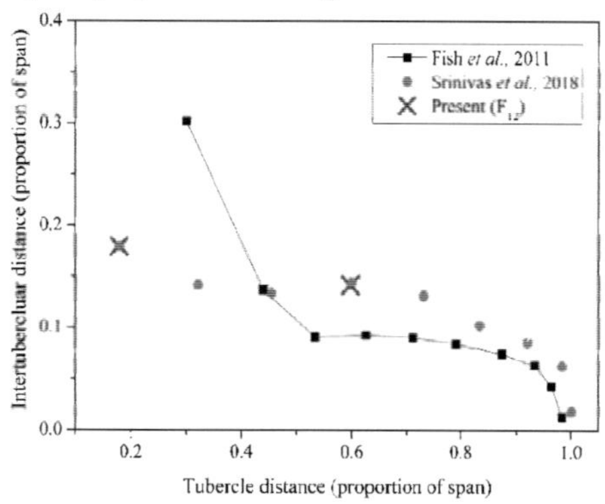

Figure 1. POSITION OF LEADING-EDGE PROTUBERANCES ALONG THE SPAN.

2. GEOMETRY

A symmetrical hydrofoil (NACA 63_4-021) was taken as baseline design having span (s = 200 mm) and chord length (c = 100 mm). The protuberance over the leading edge is defined geometrically using equation (1). The baseline foil along with the design of a leading-edge protuberance is shown in fig 2. $\bar{A}$ and λ are the amplitude and wavelength of the protuberances. z_i and z_f are the initial and final z coordinates respectively of the protuberances over the leading edge.

$$\Delta c = \bar{A}\sin\left[2\pi\left(\frac{z - z_i}{\lambda}\right)\right] \quad \forall \quad z_i \le z \le z_f \qquad (1)$$

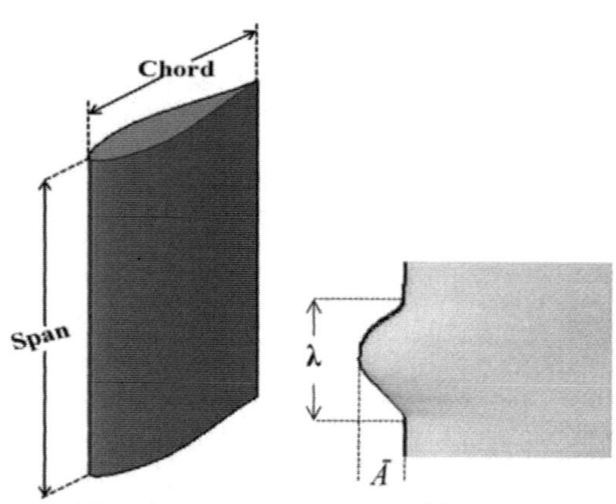

(a) (b)

Figure 2. (a) ISOMETRIC VIEW OF THE BASE HYDROFOIL, (b) SPECIFICATION OF THE PROTUBERANCE.

Three modified hydrofoils having twin protuberance over the leading-edge with varying the amplitude are first used foils with twin protuberance of amplitude (0.15c), (0.10c) and a foil with the first protuberance of 0.10c and second of amplitude 0.15c as shown in figure 3.

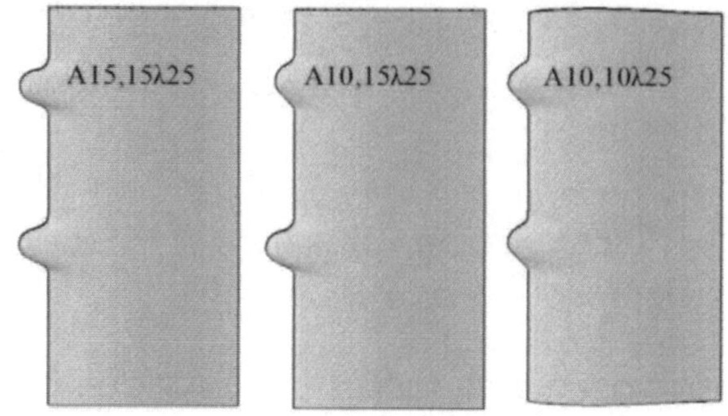

Figure 3. MODIFIED FOIL HAVING BIO-MIMICKED LEADING EDGES USED FOR THE COMPARISON.

The details of the amplitude and wavelength for the modified foils and the baseline foil are listed in table 1.

Table 1. DESIGN CONFIGURATION OF MODIFIED FOIL.

SL	Configuration	Label	A/λ ratio
1	Baseline	F_B	-
2	A = 0.15c, λ = 0.25c; A = 0.15c, λ = 0.25c;	A15,15 λ25	0.6; 0.6
3	A = 0.10c, λ = 0.25c; A = 0.15c, λ = 0.25c;	A10,15 λ25	0.4; 0.6
4	A = 0.10c, λ = 0.25c; A = 0.10c, λ = 0.25c;	A10,10 λ25	0.4; 0.4

3. METHODOLOGY

A comparative analysis of lift and drag characteristics for the modified foils is done using experimental and numerical methods.

3.1 Computational Set-up

The Reynolds- averaged Navier-Stokes (RANS) solver of STAR CCM+ is used with SST k-ω turbulence model because it performs better for flows with adverse pressure gradient [13]. The governing equations used for CFD simulations are as follows.

$$\frac{\partial \overline{U_i}}{\partial x_i} = 0 \qquad (2)$$

$$\rho\frac{\partial k}{\partial t} + \rho U_j\frac{\partial k}{\partial x_j} = \tau_{ij}\frac{\partial U_i}{\partial x_j} - \beta^* \rho k\omega + \frac{\partial}{\partial x_j}\left[\left(\mu + \sigma^* \mu_T\right)\frac{\partial k}{\partial x_j}\right] \qquad (3)$$

$$\rho\frac{\partial \omega}{\partial t} + \rho U_j\frac{\partial \omega}{\partial x_j} = \alpha\frac{\omega}{k}\tau_{ij}\frac{\partial U_i}{\partial x_j} - \beta^* \rho\omega^2 + \frac{\partial}{\partial x_j}\left[\left(\mu + \sigma\mu_T\right)\frac{\partial \omega}{\partial x_j}\right] \qquad (4)$$

In the above equations U_i represents the flow velocity in *i*-direction, ρ-the density, *k*- turbulence kinetic energy, *ω- the* rate of dissipation of the turbulence kinetic energy and μ- the dynamic viscosity of the fluid.

A cylindrical computational domain was constructed having polyhedral elements around the hydrofoil as shown in figure 4. The inlet and outlet boundary are set at 5c and 13c; whereas, the far-field radius is 5c. To capture the velocity gradient two volumetric cylinders are placed at the leading and trailing edge of the hydrofoil. The total grid size of the computational domain is around 2 million. The wall y+ value close to 1 is maintained near the foil boundary. A grid dependency study is performed at 15° angle of attack as shown in figure 5, where it has been found that the C_L and C_D converge satisfactorily for the chosen grid resolution.

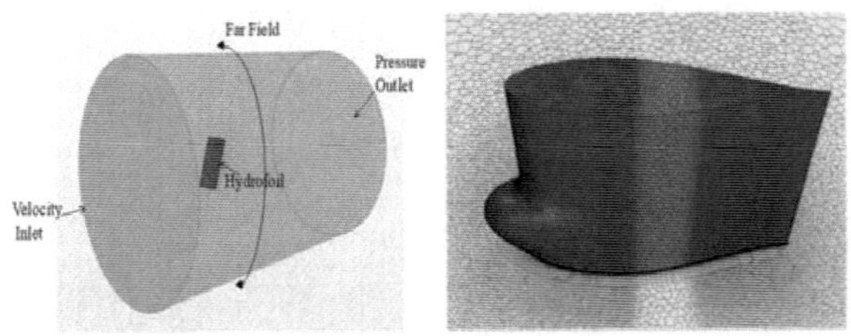

Figure 4. COMPUTATIONAL DOMAIN AND MESH.

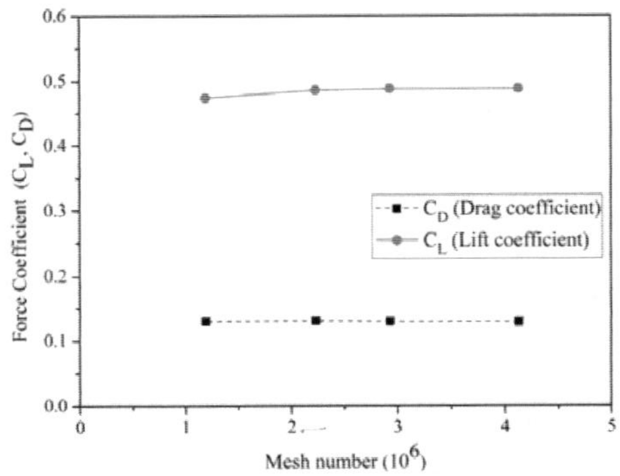

Figure 5. FORCE COEFFICIENT VARIATION WITH MESH NUMBER.

3.2 Experimental Set-up

Force measurement were conducted in the towing tank (figure 6) of the Department of Ocean Engineering and Naval Architecture at IIT Kharagpur. The foil is connected to the towing carriage using a structural mount(figure 7). The structural mount is equipped with an encoder, a stepper motor, and a load cell. The stepper motor is used to rotate the hydrofoil and it is attached with an encoder (3600 pulses per revolution) which gives feedback of actual angle turned. Different hydrofoils are alternately attached to the multi-axis load cell which is connected to the data acquisition system. Force measurements averaged over 20 sec were collected via National Instruments USB-9188 data acquisition system. The required flow velocity of 2 m/s is obtained by moving the carriage on rails installed along the length of the tank.

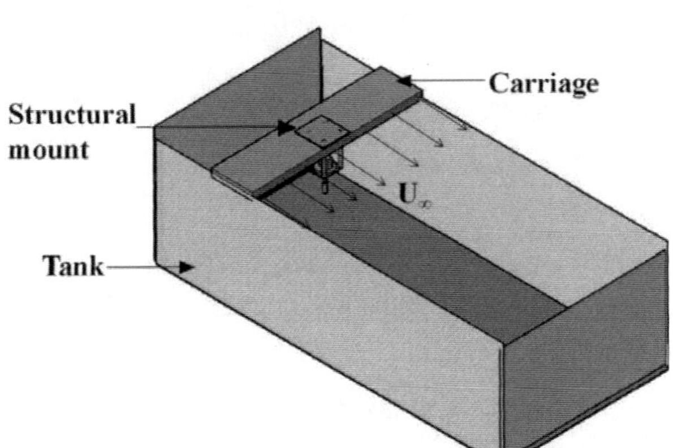

Figure 6. TOWING TANK WITH THE CARRIAGE.

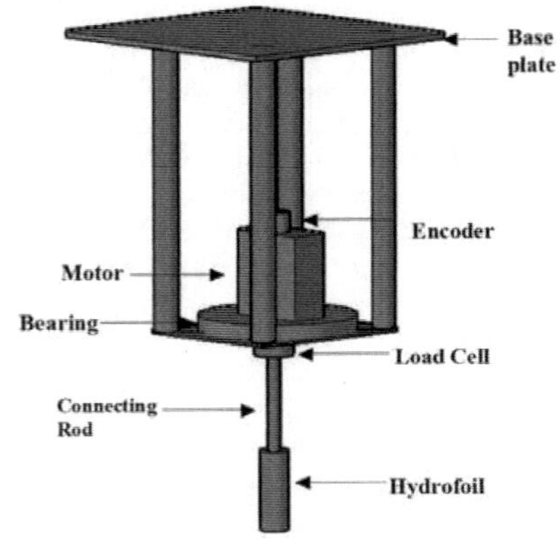

Figure 7. SKETCH OF LOAD CELL ARRANGEMENTS.

4. RESULTS AND DISCUSSIONS

The lift and drag coefficient of the three twin protuberance hydrofoils are compared with the base hydrofoil using CFD and experimental approaches. Thereafter, comparative flow analysis at different angles of attacks is presented.

4.1 C_L and C_D characteristics.

Numerical and experimental investigations of lift and drag are carried out at Re = 2 × 10^5. which are presented in this section.

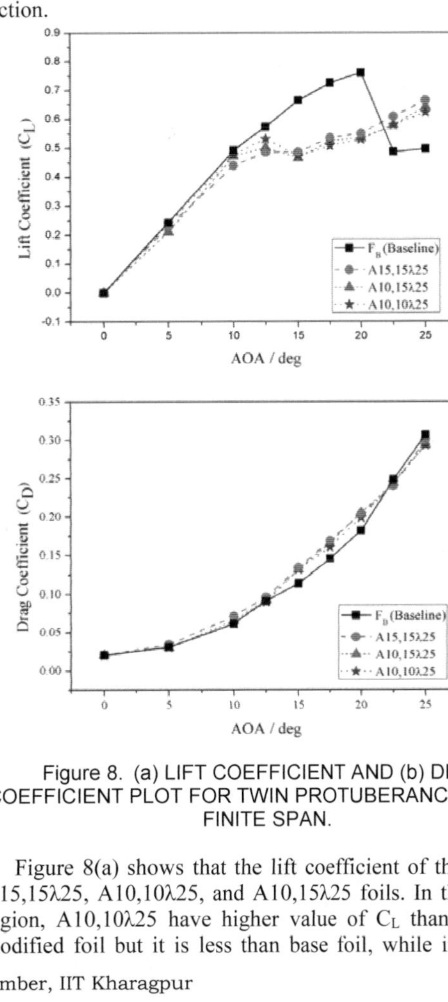

(a)

(b)

Figure 8. (a) LIFT COEFFICIENT AND (b) DRAG COEFFICIENT PLOT FOR TWIN PROTUBERANCE FOIL AT FINITE SPAN.

Figure 8(a) shows that the lift coefficient of the baseline, A15,15λ25, A10,10λ25, and A10,15λ25 foils. In the pre-stall region, A10,10λ25 have higher value of C_L than other two modified foil but it is less than base foil, while in post-stall region A15,15λ25 have higher C_L value than others. Similarly, in figure 8(b) the calculated drag coefficients are shown for all the hydrofoils, and it can be seen that all the hydrofoils are having almost similar drag coefficient curve. In figure 9, the C_L/C_D of all the foils are shown. The C_L/C_D for A10,10λ25 is 17%, 33% higher than A15,15λ25 at 5 and 10 deg respectively and similarly for A10,15λ25 it is 14%, 7% higher. In the post-stall region the C_L/C_D for all three modified foils is almost similar, and it is higher than the base foil only at $\alpha \geq 22.5$ deg.

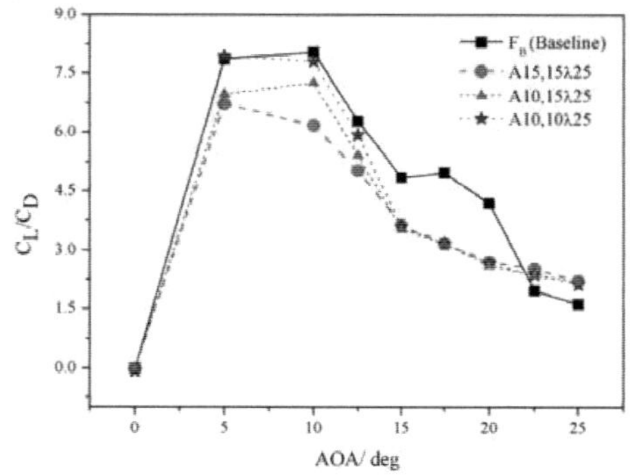

Figure 9. THE RATIO OF C_L/C_D FOR ALL TWIN PROTUBERANCES.

The lift and drag coefficients for the baseline, A15,15λ25 and A10,10λ25 hydrofoils obtained from experiments are compared with CFD in figures 10-12. The compared experimental results at different angles of attack are in good agreement with CFD results. The difference in the nature of C_L between the basic and modified foils is clearly visible. Uncertainty study of the experimental results is performed by taking five repeated measurements of drag and lift values for A10,10λ25 at α = 15 deg. The calculated uncertainties using Students 't' distribution at 95% confidence interval are less than 2% for drag and lift coefficients.

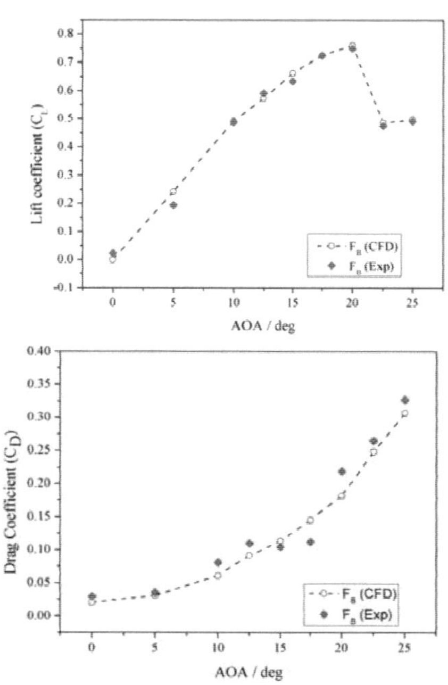

Figure 10. LIFT AND DRAG COEFFICIENT FOR F_B (EXPERIMENTAL AND CFD).

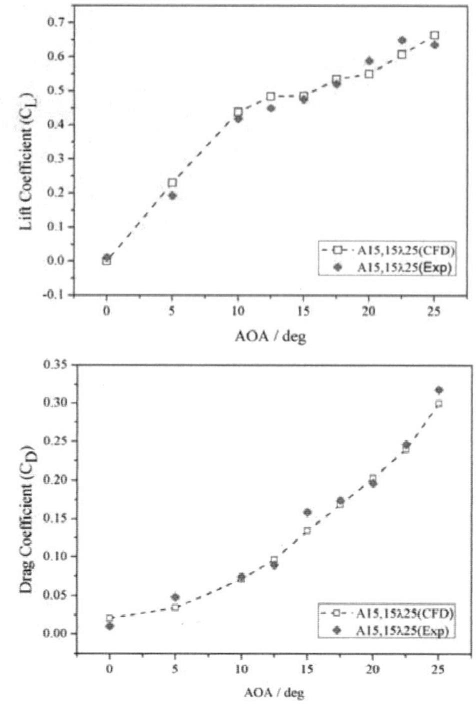

Figure 11. LIFT AND DRAG COEFFICIENT FOR N2A15,15λ25 (EXPERIMENTAL AND CFD).

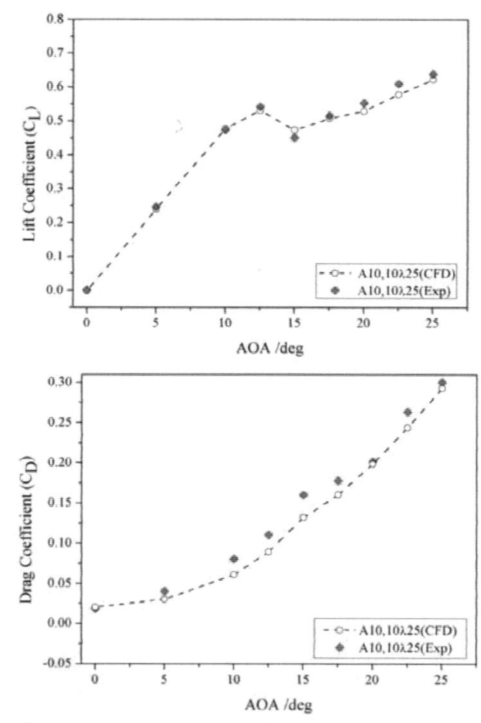

Figure 12. LIFT AND DRAG COEFFICIENT FOR N2A10,10λ25 (EXPERIMENTAL AND CFD).

4.2 Flow comparision using CFD

The lift and drag characteristics are mainly depended on the surface flow patterns and the stall angles are related to the flow separation at higher angles of attack. In order to study flow separation, a 3D volume plot of negative velocity over the hydrofoils at different angles of attack is shown in figure 13.

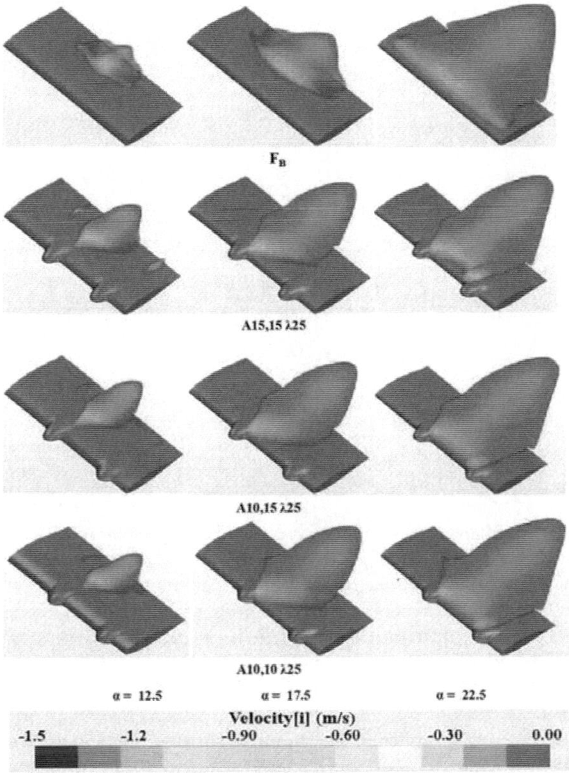

Figure 13. NEGATIVE VELOCITY SHOWING FLOW SEPARATION OVER THE SUCTION SIDE AT DIFFERENT A (NUMERICAL RESULTS).

It is observed that the domain of flow separation over the suction side of all the foils increases sharply with the increase of α. At α = 17.5, the separation is almost equal for all three modified hydrofoils. At α = 22.5, the flow separation is more for A10,10λ25 as the bigger protuberances are able to restrict the flow separation in between them, while the domain of flow separation is very large for the base hydrofoil.

4.3 Effect of Amplitude on Twin Protuberances

In this section, NACA 63_4021 foils with symmetrical twin leading-edge protuberance having three different amplitude (0.15c, 0.10c and 0.075c) are compared. The results of the lift coefficient and lift/drag ratio are presented in figure 14. The foil with the lowest amplitude has the least lift coefficient and lift/drag ratio in the post-stall region, while in the pre-stall region it is equal to the N2A10,10λ25. The protuberance of amplitude(0.15c) has a lower and higher lift coefficient in pre and post-stall region respectively, while the lift/drag ratio shows that the amplitude(0.10c) performs better the entire range of angle of attacks(α≤20 deg). Therefore for further study twin protuberances of amplitude(0.10c) is chosen.

The difference in C_L is more prominent in the post-stall region. Hence, to understand the streamwise vortex shedding for the twin protuberances a vorticity magnitude plot is shown in figure 15 for α=25 deg. It can be seen that as the amplitude of twin protuberances decreases the magnitude of vorticity also decreases, because of which transfer of energy in the flow separation decreases which can be seen in figure 16.

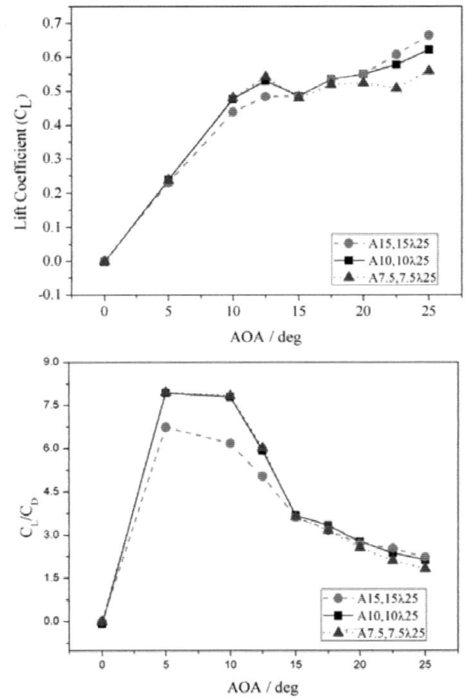

(a)

(b)

Figure 14. (a) LIFT COEFFICIENT AND (b) LIFT/DRAG RATIO FOR TWIN PROTUBERANCES HAVING DIFFERENT AMPLITUDE.

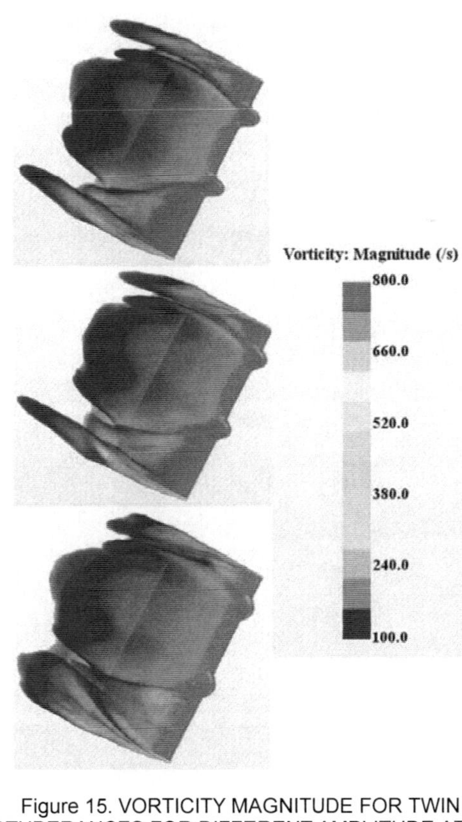

Figure 15. VORTICITY MAGNITUDE FOR TWIN PROTUBERANCES FOR DIFFERENT AMPLITUDE AT A=25 DEG.

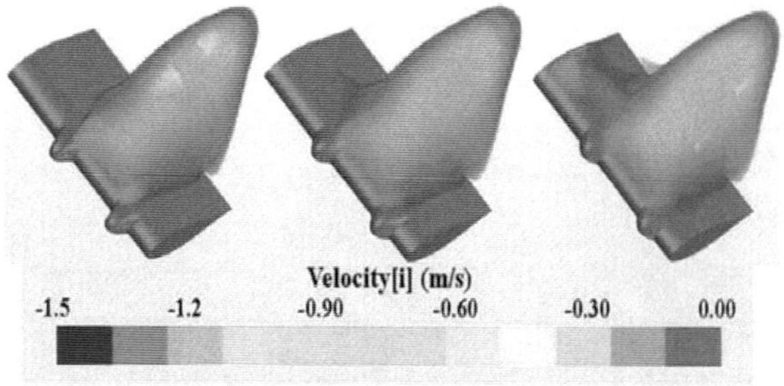

Figure 15. NEGATIVE VELOCITY FOR TWIN PROTUBERANCES AT THREE DIFFERENT AMPLITUDE(0.15C, 0.10C, 0.075C).

4.4 Position study for twin protuberances.

From the above studies of modified hydrofoils, it has been found that the twin protuberances with amplitude (0.10c) perform in general better as compared to the other foils. Therefore a position study is performed for the twin protuberance considering three cases i.e (L=0, 0.1, 0.2 of chord length), where L is the distance of protuberance base from the centerline of the span as shown in figure 17.

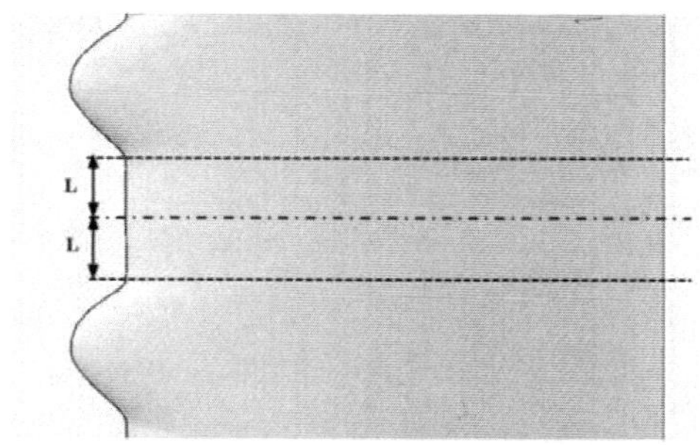

Figure 17. TWIN PROTUBERANCE POSITION FROM CENTERLINE.

The C_L and C_L/C_D comparison of foils with twin protuberance at different position from the center of the span is shown in figure 18. The lift coefficient is found to increase as L increases in the pre- stall region, while in the post-stall region the lift coefficient is maximum when L is minimum. From the C_L/C_D, it can be seen that closer protuberance performs better in the post-stall region, while the protuberance at larger distance performs better in the pre-stall region.

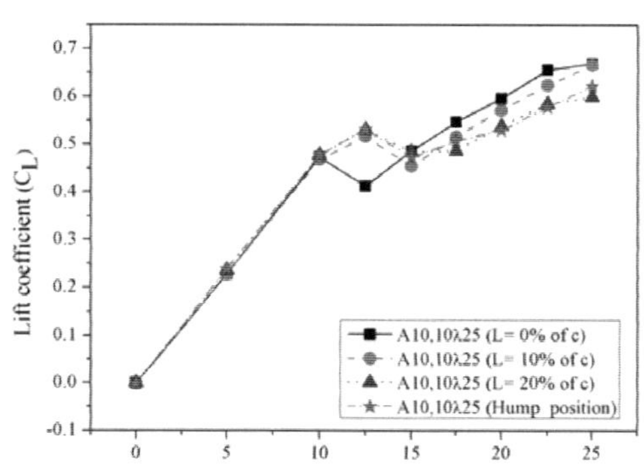

(a)

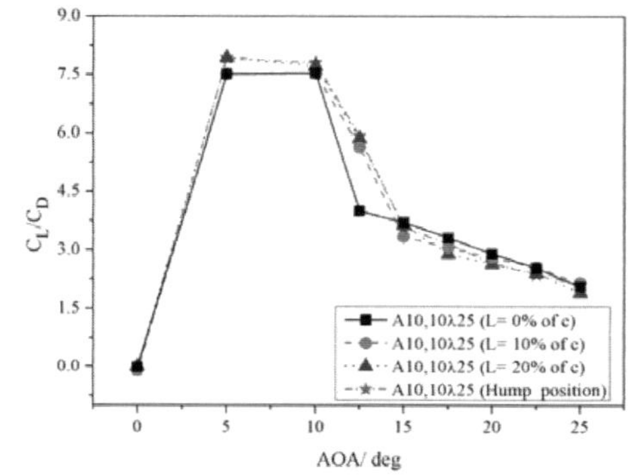

(b)

Figure 18. (A) LIFT COEFFICIENT AND, (B) LIFT/DRAG RATIO FOR TWIN PROTUBERANCES HAVING DIFFERENT POSITION.

5. CONCLUSIONS

The lift and drag characteristics of a hydrofoil of NACA 63_4-021 section with twin protuberances over the leading edge are investigated using experiments and CFD methods. The twin protuberances over the leading edges enhance the lift in the post-stall region due to the chordwise vortices generated by each protuberance, which reduce the flow separation at high angles of attack. The larger protuberances perform better in the post-stall region due to their high strength of vortices. The lift characteristics depend largely on the position of twin protuberance. The protuberances closer to the centerline of span perform better in the post-stall region, while as the distance of protuberances increases its lift characteristics improves in the pre-stall region.

REFERENCES

[1] Fish, F. E., and Battle, J. M., 1995, "Hydrodynamic Design of the Humpback Whale Flipper," J. Morphol., **225**(1), pp. 51–60.

[2] Fish, F. E., and Lauder, G. V, 2006, "Passive and Active Flow Control by Swimming Fishes and Mammals," Annu. Rev. Fluid Mech., **38**, pp. 193–224.

[3] Miklosovic, D. S., Murray, M. M., Howle, L. E., and Fish, F. E., 2004, "Leading-Edge Tubercles Delay Stall on Humpback Whale (Megaptera Novaeangliae) Flippers," Phys. fluids, **16**(5), pp. L39--L42.

[4] Fish, F. E., Weber, P. W., Murray, M. M., and Howle, L. E., 2011, "Marine Applications of the Biomimetic Humpback Whale Flipper," Mar. Technol. Soc. J., **45**(4), pp. 198–207.

[5] Cai, C., Zuo, Z., Liu, S., and Wu, Y., 2015, "Numerical Investigations of Hydrodynamic Performance of Hydrofoils with Leading-Edge Protuberances," Adv. Mech. Eng., **7**(7), p. 1687814015592088.

[6] Johari, H., Henoch, C. W., Custodio, D., and Levshin, A., 2007, "Effects of Leading-Edge Protuberances on Airfoil Performance," AIAA J., **45**(11), pp. 2634–2642.

[7] Chen, J. H., Li, S. S., and Nguyen, V. T., 2012, "The Effect of Leading Edge Protuberances on the Performance of Small Aspect Ratio Foils," *15th International Symposium of Flow Visualization ISFV15-040-S21*.

[8] Wei, Z., Toh, J. W. A., Ibrahim, I. H., and Zhang, Y., 2019, "Aerodynamic Characteristics and Surface Flow Structures of Moderate Aspect-Ratio Leading-Edge Tubercled Wings," Eur. J. Mech.

[9] Arai, H., Doi, Y., Nakashima, T., and Mutsuda, H.,

2010, “A Study on Stall Delay by Various Wavy Leading Edges,” J. aero aqua bio-mechanisms, **1**(1), pp. 18–23.

[10] Cai, C., Zuo, Z., Liu, S., and Maeda, T., 2018, “Effect of a Single Leading-Edge Protuberance on NACA 634-021 Airfoil Performance,” J. Fluids Eng., **140**(2), p. 21108.

[11] Srinivas, K. S., Datta, A., Bhattacharyya, A., and Kumar, S., 2018, “Free-Stream Characteristics of Bio-Inspired Marine Rudders with Different Leading-Edge Configurations,” Ocean Eng., **170**, pp. 148–159.

[12] Fish, F. E., and Kocak, D. M., 2011, “Biomimetics and Marine Technology: An Introduction,” Mar. Technol. Soc. J., **45**(4), pp. 8–13.

[13] Menter, F. R., Kuntz, M., and Langtry, R., 2003, “Ten Years of Industrial Experience with the SST Turbulence Model,” Turbul. heat mass Transf., **4**(1), pp. 625–632.

Proceedings of the 6th International Conference on Ship and Offshore Technology
ICSOT 2019
November 7-8, 2019, IIT Kharagpur, India

ICSOT INDIA 2019-23

WATER COLUMN RESPONSE WITHIN A BARGE HAVING BOTTOM OPENINGS

Suraj Garad
Research Scholar
Indian Institute of Technology, Kharagpur
West Bengal, India
surajk.garad101@gmail.com

A. Bhattacharyya
Asst. Professor
Indian Institute of Technology, Kharagpur
West Bengal, India
ab@naval.iitkgp.ac.in

R. Datta
Asst. Professor
Indian Institute of Technology, Kharagpur
West Bengal, India
ranadev@naval.iitkgp.ac.in

ABSTRACT

The motion characteristics of a barge with an opening or moonpool at the center and alternately at two different locations are investigated numerically in a series of regular head waves. The motion response analysis is done using the linear potential flow BEM solver WAMIT – alternately for a fixed barge case, and a barge with heave and pitch motions. It is observed that the motions of the water column within the opening and the resulting free surface elevations depend on the location of the opening and the incident wave frequency.
The nature and occurrence of vertical piston mode and longitudinal sloshing modes for the free surface of the opening is studied under the assumption of infinite water depth. The results indicate that the resonant frequency is almost independent of the moonpool location for the fixed barge. However, the resonance behaviour changes for the free cases. Investigations at different drafts show that for all the cases the resonant frequency decrease with the increase of the vessel draft. The main objective of this study is to highlight the resonance characteristics of water column for different positions of the bottom opening within a barge.

KEYWORDS

Moonpool, Resonant water motions, Piston mode, Sloshing mode, Free surface elevation

INTRODUCTION

Moonpools are vertical openings through the deck and hull structure to support underwater operations on offshore platforms and marine vessels. The water inside the moonpool is required to provide a moderate environment for vessel operations. However, large water motions may occur inside the moonpool under resonant conditions which includes vertical piston motions and longitudinal sloshing motions. The majority of published research works in this domain are focused on the behavior of water column within centrally located moonpools, which is typically the case for drilling vessels.
Fakuda (1974) performed experiments for investigation of water behavior inside the moonpool and its effect on the vessel motions. He obtained an empirical formulation which could be used for calculation of resonant frequency of water motions. Aalbers (1984) developed a mathematical model to describe behaviour of water inside the moonpool, and also performed model tests for to understand the damping mechanisms and motion behaviours. Molin (2001) carried out formulation for moonpool natural frequency (piston and sloshing modes) under the assumption of fixed barge of infinite length and beam and considering linear potential theory for 2D and 3D cases.
Nishimoto et al. (2001) performed numerical investigations on the hydrodynamic performance of a hull with moonpool by considering different moonpool sizes. Wei et al. (2011) had also presented a similar investigation using numerical as well as experimental methods. A numerical study on concentric cylindrical moonpool case was performed in Malta et al. (2006). Kristiansen et al. (2008) found that the water motions response inside the moonpool is overestimated compared to experimental

results due to neglecting fluid viscosity. They developed fully nonlinear NWT coupled with an inviscid vortex tracking method to investigate the impacts of fluid viscosity and nonlinear effects associated with the free surface.

Bull (2015) investigated the application of moonpools within two different rigid body profiles- a duct buoy and an axisymmetric point absorber, for energy extraction using oscillating water column mechanism.

Guo et al. (2016, 2017) investigated the hydrodynamic performance of a deep water drillship and water motions inside the rectangular moonpool using experimental and numerical study, where multiple peaks were identified from the vessel RAOs. The maximum water surface elevation was found to reach upto two times the external encountered wave height under resonant conditions. For a barge with dimensions similar to the drillship mentioned above, Molin (2017) and Newman (2018) presented theoretical and numerical studies respectively with rectangular moonpool having recess.

The present work aims to investigate the vessel responses as well as the water column behavior within a bottom opening when placed at different longitudinal locations of a rectangular barge.

1 THEORETICAL BACKGROUND

Boundary value problem

The numerical analysis is done using WAMIT software. The free surface conditions and the body boundary conditions are linearized. Under the assumption of irrotational motion and incompressible fluid, the fluid motion is governed by velocity potential $\Phi(x,y,z,t)$ which satisfies Laplace equation in the fluid domain,

$$\nabla^2\Phi(x,y,z,t)=0 \quad (1)$$

Here (x,y,z) represents a Cartesian coordinate system fixed relative to the undisturbed positions of free surface and body with the z-axis positive upwards.

$$\Phi(x,y,z,t)=\mathrm{Re}(\phi(x,y,z)e^{i\omega t}) \quad (2)$$

The free surface condition in linearized form is,

$$\phi_z-\frac{\omega^2}{g}\phi=0 \quad \text{on } z=0 \quad (3)$$

The velocity potential of incident wave is,

$$\phi_I=\frac{igA}{\omega}\frac{\cosh[k(z+h)]}{\cosh kh}e^{-ikx\cos\beta-iky\sin\beta} \quad (4)$$

Dispersion relation is used to calculate the wave number (k),

$$\omega^2=gk\tanh kh \quad (5)$$

A: Incident wave amplitude

g: Acceleration due to gravity = 9.81 m/s^2,
h: Water depth

ω: Wave frequency

β : Angle between direction of propagation of incident wave and the positive x- axis.

Due to assumption of linearity, the velocity potential (ϕ) can be decomposed into radiation potential (ϕ_R) and diffraction potential (ϕ_D),

$$\phi=\phi_R+\phi_D \quad (6)$$

$$\phi_R=i\omega\sum_{j=1}^{6}\xi_j\phi_j \quad (7)$$

$$\phi_D=\phi_I+\phi_S \quad (8)$$

ξ_j are the complex amplitudes of the body oscillatory motions in six rigid body degrees of freedom and ϕ_j is the corresponding unit amplitude radiation potentials. The velocity potential ϕ_s is the scattered disturbance of the incident wave by the body fixed at its undisturbed position.

On the undisturbed position of the body boundary, the radiation and diffraction potentials are subjected to,

$$\phi_{jn}=n_j \quad (9)$$

$$\phi_{Dn}=0 \quad (10)$$

Green's theorem is used to derive integral equations for radiation and diffraction velocity potentials on the body boundary.

Radiation potential evaluation

$$2\pi\phi_j(x)+\iint_{S_b}\phi_j(\xi)\frac{\partial G(\xi;x)}{\partial n_\xi}d\xi=\iint_{S_b}n_jG(\xi;x)d\xi \quad (11)$$

Diffraction potential evaluation

$$2\pi\phi_D(x)+\iint_{S_b}\phi_D(\xi)\frac{\partial G(\xi;x)}{\partial n_\xi}d\xi=4\pi\phi_0(x) \quad (12)$$

Where S_b is the wetted surface area of the body below the plane $z = 0$, in its fixed mean position. $\partial G(\xi; x)$ represents the Green function referred to as wave source potential. It is the velocity potential at the point x due to a point source of strength -4π located at the point ξ and satisfies the free surface and radiation conditions.

The added mass and damping coefficients can be obtained by,

$$A_{ij} - \frac{i}{\omega} B_{ij} = \rho \iint_{S_b} n_i \phi_j dS \qquad (13)$$

The elevation of water inside the moonpool is evaluated by,

$$\eta = -\frac{1}{g}\left(\frac{\partial \phi}{\partial t}\right)_{z=0} \qquad (14)$$

2 VALIDATION STUDY

2.1 Numerical setup

The numerical setup for validation is taken from Newman (2018). The resonant modes of motion are analysed for the moonpool which is centered at midship with a recess at the aft end of moonpool. The recess part has two main effects. It makes possible for the piston motion to get transformed into a longitudinal sloshing motion, leading to more complex water motions and also provides the obstacle effect or shallow water effect. A similar study was previously done experimentally and numerically by Guo et al. (2016, 2017) and Molin (2017) respectively. The complete structure consisting of barge and moonpool is analysed by solving the diffraction problem with incident waves and the resonant frequencies are found which maximises the free surface elevation inside the moonpool. The computations are made with a fixed barge (dimensions prescribed in Table (1)) in head waves to get free surface elevations inside the moonpool. The frequencies at which the free surface elevation inside the moonpool gets maximised are called as resonant frequencies. The free surface elevation is computed along the centre line of the moonpool at different field points to find the point of maximum elevation. The geometry of the barge is shown in Fig (1), and the moonpool dimensions are provided in Fig (2).

The surface of barge and moonpool are represented by panels. The analysis is done by using lower order option, where the potential in each panel is considered as uniform. Field points are placed 1 m apart along the length on the free surface along the centreline of moonpool to measure the free surface elevation. For validation purpose, frequency increment of 0.001 rad/sec is considered.

Table 1: BARGE DETAILS

Barge parameters	Dimensions (m)
Length	160
Breadth	32
Depth	17
Draft	11

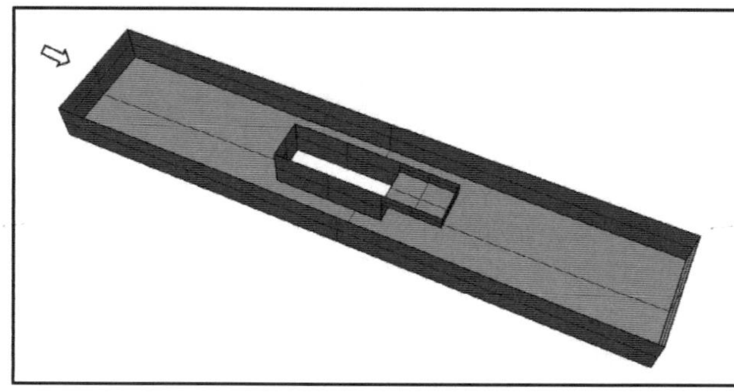

Figure (1): PERSPECTIVE VIEW OF RECTANGULAR BARGE WITH MOONPOOL

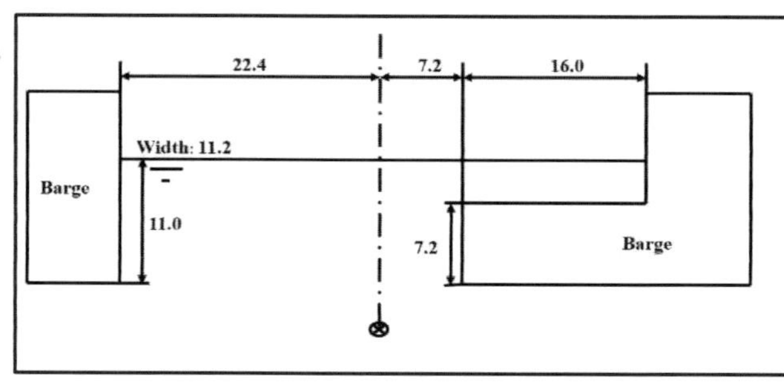

Figure (2): MOONPOOL WITH RECESS (WITH DIMENSIONS (m))

2.2 Discussion

Through analytical study, Molin (2001) have given the piston mode and longitudinal sloshing mode frequencies for two and three dimensional cases. The resonant frequencies for three dimensional case are,

Piston mode

$$\omega_{00} \simeq \sqrt{\frac{g}{h + b f_3(b/l)}} \qquad (15)$$

Where,

$$f_3 = \frac{1}{\pi}\left\{ sinh^{-1}\left(\frac{l}{b}\right) + \frac{l}{b} sinh^{-1}\left(\frac{b}{l}\right) + \frac{1}{3}\left(\frac{b}{l}+\frac{l^2}{b^2}\right) - \frac{1}{3}\left(1+\frac{l^2}{b^2}\right)\sqrt{\frac{b^2}{l^2}+1}\right\}$$

Longitudinal sloshing mode

$$\omega_n^2 \simeq g\lambda_n \frac{1+J_{n0} tanh\lambda_n h}{J_{n0}+tanh\lambda_n h} \tag{16}$$

Where,

$$J_{n0} = \frac{2}{n\pi^2 r}\left\{\int_0^1 \frac{r^2}{u^2\sqrt{u^2+r^2}}\left[1+(u-1)cos(n\pi u) - \frac{sin(n\pi u)}{n\pi}\right]du + \frac{1}{sin\theta_0} - 1\right\}$$

r = b/l and $tan\theta_0 = r^{-1}$

b = width of moonpool

l = length of moonpool

λ_n = incident wavelength

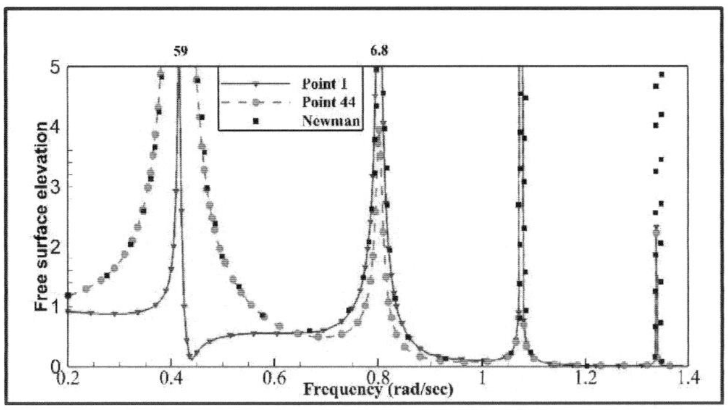

Figure (3): COMPARISON OF FREE SURFACE ELEVATION WITH NEWMAN (2018)

From Fig (3), it can be clearly observed that the free surface elevation is maximum at 0.416, 0.802, 1.075 and 1.338 rad/sec which can be termed as resonant frequencies which are same as from the Molin (2001) formulation as represented by Eq (15) and (16). The locations (field points: first or last field point, adjacent to end) and the frequency at which resonance occurs are similar to that presented in Newman (2018). For piston mode, the point of maximum elevation is at the shallow end of the recess while for first two sloshing modes it is at the deep end of the moonpool, which can be seen in Fig (3). The concept of piston mode frequency comes from the wave with longer wavelength for comparatively small moonpool dimensions. Here the free surface moves simply up and down inside the moonpool giving the piston mode. So, it is clear that piston mode occurs at 0.416 rad/sec while the first sloshing mode occurs at 0.802 rad/sec.

3 BARGE WITH OPENINGS AT DIFFERENT LOCATIONS

The goal of the present work is to investigate how the responses of the water column and the vessel changes, when the bottom opening is kept at different locations away from the center of the vessel. To extract wave energy by making use of water oscillations within the moonpool of the barge, it is similar to offshore oscillating water column (OWC) type wave energy convertor. A detailed study regarding the energy capture possibilities of an OWC device linked to moonpools within floating rigid bodies can be found in Bull (2015).

If a similar concept is applied to a barge, it is imperative to get a suitable opening location on the barge to obtain maximum conversion. A comparative study is done for threee different locations of the moonpool, and the wave frequencies at which resonance occurred. Considering energy conversion from water column motion in such an opening, the piston mode would be more suitable. One of the aspects of the present work is to find the locations and incident wave frequency at which free surface elevation is maximum

Table 2: BARGE DETAILS

Parameters	Dimensions (m)
Length	2
Breadth	0.5
Depth	0.4
Draft	0.2, 0.25, 0.3
Moonpool diameter	0.19

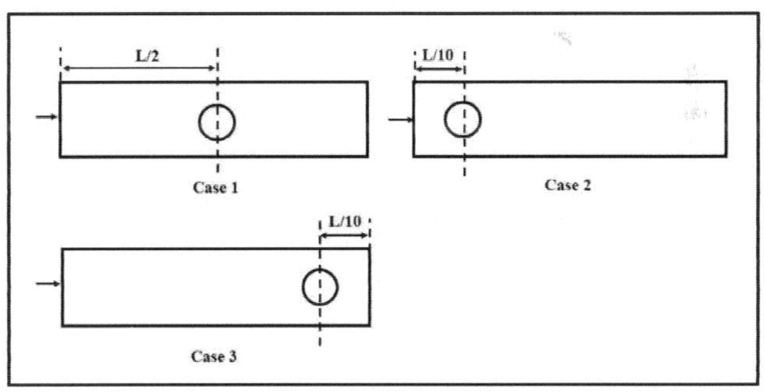

Figure (4): LOCATION OF OPENING WITHIN THE BARGE (1) OPENING AT CENTER (2) OPENING AT DISTANCE OF L/10 FROM THE FORWARD END (3) OPENING AT DISTANCE OF L/10 FROM THE BACK END

The barge details are selected as per possibility of performing model tests in the towing tank facility at IIT Kharagpur. It is observed that with this model the resonant frequencies fall within the range which can be generated by the wavemaker installed in the towing tank. The existing wavemaker can produce waves of maximum wave height upto 30 cm and time period from 1 to 4 sec.

The details of the barge are mentioned in Table (2). Fig (4) represents the locations of the openings which have been analysed. Keeping the same opening dimensions, three different cases are studied:

Case (1): Opening at the center
Case (2): Opening at distance of L/10 from the forward end
Case (3): Opening at distance of L/10 from the rear end

An analysis is carried out for three draft cases i.e. 0.2, 0.25 and 0.3 m. Free surface elevation is calculated at different field points on the free surface located along the centerline in the longitudinal direction. The respective locations are same for all three cases, which is shown in Fig. (5)

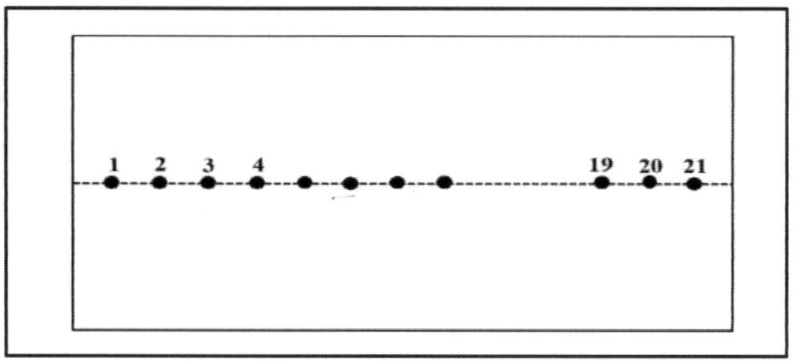

Figure (5): LOCATION OF FIELD POINTS ALONG THE CENTERLINE WITHIN THE OPENING

3.1 Numerical setup

The computational setup is kept similar to that which had been used for validation study. Analysis is done by making use lower order panels. The analysis is done for regular head waves under deep water consideration for all three cases. The distance between consecutive field points is taken as 0.01m. The frequency increment of 0.001 rad/sec is taken for computations. The computations are made for two different cases: one is similar to the validation study keeping the barge fixed, while in second case the barge is free to move in heave and pitch mode.

3.2 Results and discussions

The free surface elevation comparison is made separately for fixed and freely floating barge cases. Out of various field points, the free surface elevation is plotted at the point where maximum free surface elevation is obtained. Fig (6) represents the free surface elevation for fixed barge case having opening at central position, while Fig (7) is a freely floating barge case with central opening. An interesting observation is that the field points where free surface elevation is maximum are similar for freely floating and fixed barge case. The frequencies at which resonance is obtained are quite similar for both free and fixed barge cases. Applying the similar concept of piston and sloshing modes as defined for a fixed barge case to the freely floating barge, the piston mode and first sloshing mode are obtained at the lowest resonating frequency and the next one to the next resonating frequency respectively. From the formulation of piston mode and sloshing frequency by Fakuda (1977),

$$\omega_0 = \sqrt{\frac{g}{h+0.41\sqrt{S}}} \tag{17}$$

$$\omega_n = \sqrt{\frac{n\pi g}{l}} \quad n = 1, 2, 3.... \tag{18}$$

Where,
h = Draft of the barge,
S = Cross-sectional area of opening,
l = Length of moonpool

From Eq. (17), the resonant frequency obtained is 6.0385 rad/sec for 0.2 m draft, 5.5452 rad/sec for 0.25 m draft while for 0.3 m draft it is 5.1558 rad/sec.

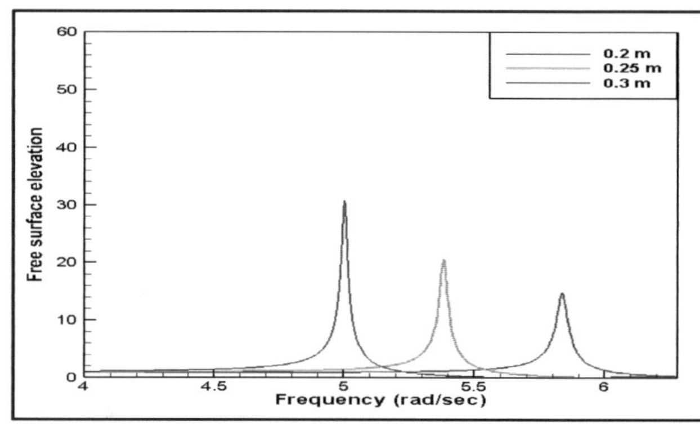

Figure (6): FREE SURFACE ELEVATION FOR FIXED BARGE HAVING MOONPOOL AT CENTER

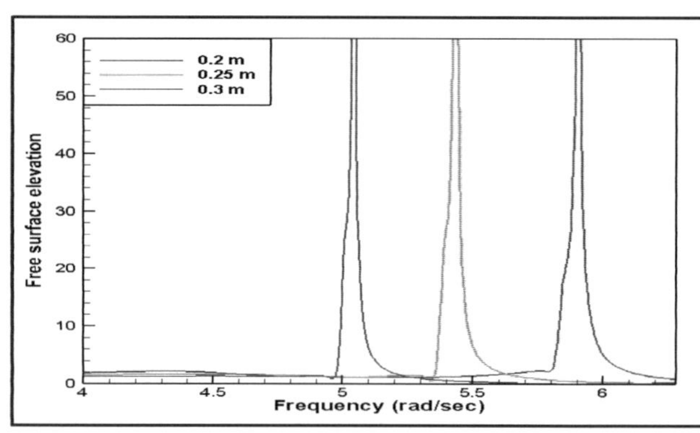

Figure (7): FREE SURFACE ELEVATION FOR FREE BARGE HAVING MOONPOOL AT CENTER

From Fig (6) and Fig (7), it can be observed that, with increase in draft the resonant frequencies decrease for fixed and freely floating barge, which follows Fakuda's formulation. For the central opening case, the resonant motions are only due to heave motion. Comparing to Fakuda's formulation the resonant frequencies are slightly lower, and within the range of incident frequencies only one resonant frequency is present. The free surface elevation obtained is unrealistically high as the viscous damping is ignored by the WAMIT. In this study, the main objective is to find out the resonant frequencies.

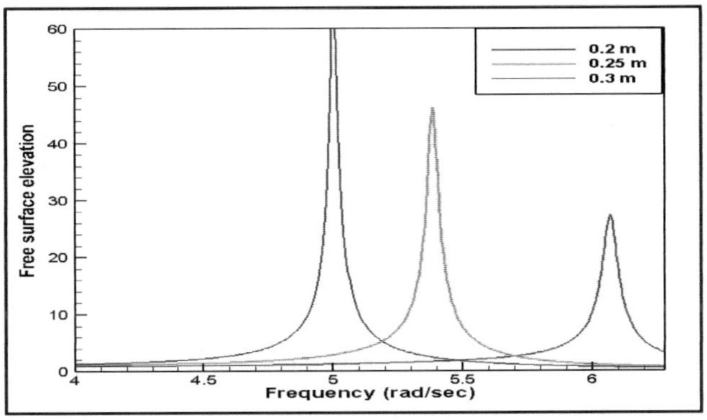

Figure (8): FREE SURFACE ELEVATION FOR FIXED BARGE HAVING MOONPOOL AT FORWARD POSITION

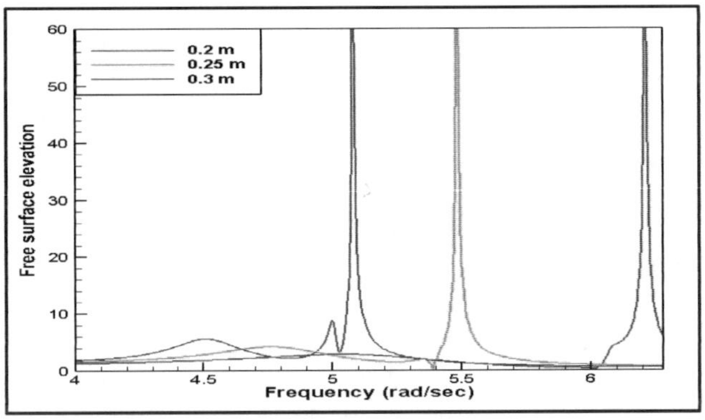

Figure (9): FREE SURFACE ELEVATION FOR FREE BARGE HAVING MOONPOOL AT FORWARD POSITION

For the next case, the moonpool is at a distance of L/10 from the forward end. Fig (8) represents the free surface elevations for fixed barge case while Fig (9) for the freely floating barge case having moonpool at forward location. Comparing with the fixed barge having moonpool at center, in this case resonant frequency is higher for 0.2 m draft but it is similar for 0.25 m and 0.3 m draft.

In moonpool at forward end for fixed condition, there is only one resonating frequency as shown in Fig (8) whereas for the freely floating barge case, multiple resonant frequencies are observed in Fig (9). However due to combined effect of heave and pitch motions, by changing the opening position, the points at which free surface elevation is maximum do not get altered.

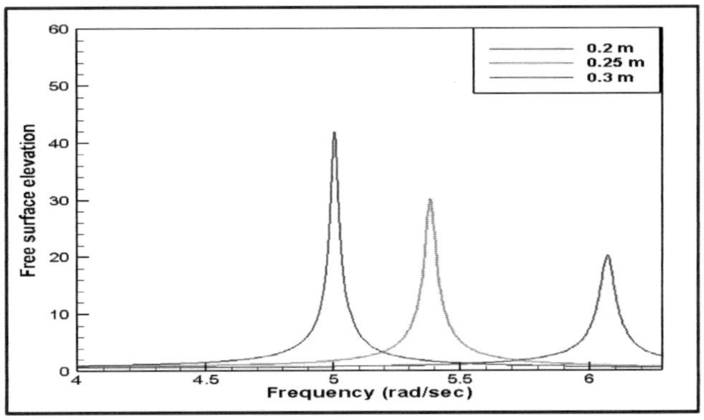

Figure (10): FREE SURFACE ELEVATION FOR FIXED BARGE HAVING MOONPOOL AT BACK POSITION

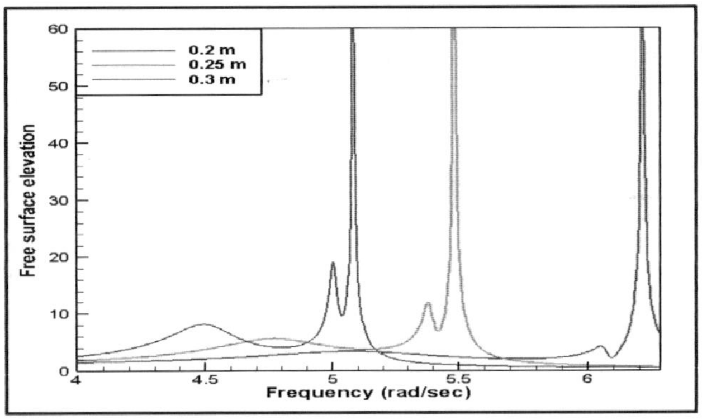

Figure (11): FREE SURFACE ELEVATION FOR FREE BARGE HAVING MOONPOOL AT BACK POSITION

Next, the resonant frequency analysis is done for moonpool at rear end having L/10 distance from the end. Fig (10) represents the free surface elevations for fixed barge case while Fig (11) for freely floating barge case having moonpool at rear end of barge. Similar like moonpool at forward location, after comparing with the fixed barge having moonpool at center, in this case resonant frequency is higher for 0.2 m draft but it is similar for 0.25 m and 0.3 m draft.

In moonpool at rear end for fixed condition, there is only one resonating frequency as shown in Fig (10) whereas for freely floating barge case, multiple resonant frequencies are observed as Fig (11). Similar to the forward moonpool case, due to combined effect of heave and pitch motions. Also, maximum free surface elevation occurs at the same points as before.

The free surface elevation comparison is done for the moonpool at forward and rear end for 0.2, 0.25 and 0.3 m drafts as shown in Fig (12), Fig (13) and Fig (14) respectively for fixed and freely floating barge conditions. From the Fig (12), Fig (13) and Fig (14) it can be clearly seen that resonant frequencies are same for fixed case at rear and forward end moonpool and also freely floating barge case at rear and forward end. This may be due to the symmetricity of the model.

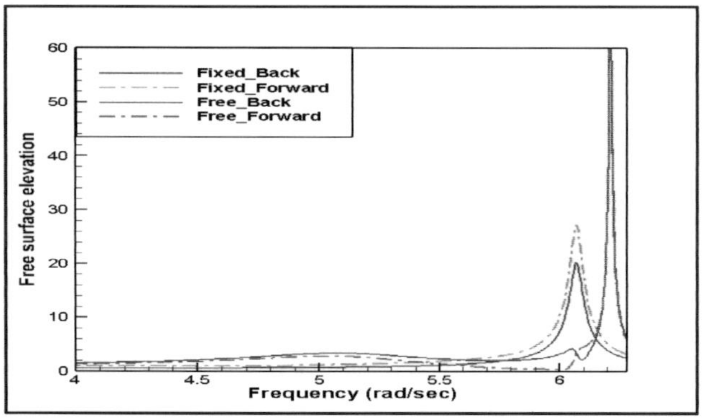

Figure (12): FREE SURFACE ELEVATION FOR FIXED AND FREE BARGE HAVING MOONPOOL WITH 0.2 M DRAFT

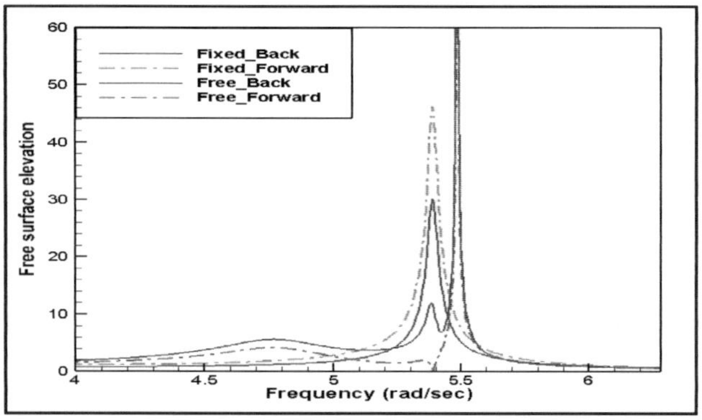

Figure (13): FREE SURFACE ELEVATION FOR FIXED AND FREE BARGE HAVING MOONPOOL WITH 0.25 M DRAFT

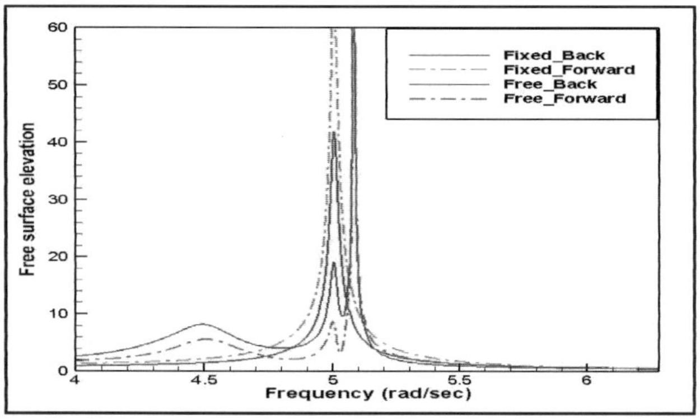

Figure (14): FREE SURFACE ELEVATION FOR FIXED AND FREE BARGE HAVING MOONPOOL WITH 0.3 M DRAFT

4. CONCLUSION

The response of water column for different locations of a bottom opening within a rectangular barge is investigated in a range of wave frequencies. Both fixed vessel case and vessel free to heave and pitch are studied. The barge motion is found to have significant effect on the occurrence of number of resonance frequencies. For all the three cases and for higher drafts (0.25 and 0.3 m) in fixed condition, the resonant frequencies are similar. These water column responses for different locations may provide some insight into the energy extraction potential if a suitable device is installed in the vessel. In order to investigate the practical possibilities, further studies need to be performed, both numerical and experimental, involving different opening configurations.

REFERENCES

[1] Fukuda, K., 1977. "Behaviour of Water in Vertical Well with Bottom Opening of Ship, and its Effects on Ship-Motion". J. of the Soc. Of Naval Architects of Japan 141, 107–122.

[2] Aalbers, A. B., 1984. "The water motions in a moonpool". Ocean Eng., 11(6), 557–579.

[3] Molin, B., 2001 "On the piston and sloshing modes in moonpools". J. Fluid Mech, 430, 27–50.

[4] Nishimoto, K., Videiros, P; M., Fucatu, C.H., Matos, V., Cueva, D., Cueva, M., 2001. "A Study of Motion Minimization Devices of FPDSOs". OMAE 2001-1131.

[5] Malta, E. B. Cueva, M., Nishimoto, K. and Masetti, I., 2006. "Numerical moonpool modelling". Proc 25th Int Conf on Offshore, Mech and Artic Eng, Hamburg, OMAE 2006-92456.

[6] Kristiansen, T. and Faltinsen, O. M., 2008. "Application of a vortex tracking method to the piston-like behaviour in a semi-entrained vertical gap". Applied Ocean Research, 30(1), 1–16.

[7] WAMIT User Manual Version 7.0 (2013).

[8] Bull D., 2015. "An improved understanding of the natural resonances of moonpools contained within floating rigid-bodies: Theory and application to oscillating water column devices". Ocean Engineering 108, 799-812.

[9] X. Guo, H. Lu, J. Yang, T. Peng., 2016. "Study on hydrodynamic performances of a deep water drillship and water motions inside its rectangular moonpool". Proc. 26th Int. Ocean and Polar Engineering Conference, ISOPE, Rhodes, 2016.

[10] X. Guo, H. Lu, J. Yang, T. Peng., 2017. "Resonant water motions within a recessing type moonpool in drilling vessel". Ocean Eng. 129, 228-239.

[11] B. Molin., 2017. "On natural modes in moonpools with recess". Applied Ocean Research 67, 1-8.

[12] Newman J. N., 2018. "Resonant response of a moonpool with a recess". Applied Ocean Research 76, 98-109.

Proceedings of the 6th International Conference on Ship and Offshore Technology
ICSOT 2019
November 7-8, 2019, IIT Kharagpur, India

ICSOT INDIA 2019-24

VIBRATIONAL ANALYSIS OF CUTTER SUCTION DREDGER

Chinmaya Ranjan Barik
Indian Institute of Technology
Kharagpur
India
chinmaya.barik@iitkgp.ac.in

Kiran Vijayan
Indian Institute of Technology
Kharagpur
India
kiran.vijayan@naval.iitkgp.ac.in

Om Prakash Sha
Indian Institute of Technology
Kharagpur
India
ops@naval.iitkgp.ac.in

ABSTRACT

Understanding the complex dynamic interaction during the dredging operation using a Cutter Suction Dredger can improve the production. Some of the parts of the CSD are the dredge hull, spud and the cutter shaft. The dredge hull is analysed for the ocean wave by the scaled down model to determine the motions in all six directions. By understanding the effect of the dredge hull over the spud, the spud is designed to the scaled down model and analysed for the motions due to the dredge hull as an input. The transient analysis of the spud-soil system has been done for the time variant water waves based on Morison equation below the water surface length of the spud. The transmission shaft is analysed for the modal characteristics which consists of two inclined shaft coupled using universal joints. A finite element model of the transmission shaft is developed. The shafts were modelled as a Euler-Bernoulli beam with five degrees of freedom per node and the universal joint were modelled as spring elements. The eigenvalue problem was analysed to determine the natural frequency and mode shapes for a straight and shafts inclined at a specific configuration. External excitation in the form of wave loading was calculated based on Morison equation. Considering the cutter motion at the sea bed as an impulse to the free end the temporal response due to these excitations were calculated. A parametric study was carried out to understand the influence on the natural frequency and mode shapes by varying the spring stiffness and damping of universal joint. With the different spring stiffness at the joint the natural frequency and mode shapes along with the time response were determined. The mode shape and temporal response showed sensitivity to change in damping and stiffness of the universal joint. Therefore, by careful tuning of the stiffness and damping of the universal joint the vibration transmission along the length of the shaft can be controlled.

NOMENCLATURE

ρ	Density of water (kg m^{-3})
a	Amplitude of wave (m)
A	Area of the cylindrical shaft (m^2)
[C]	Global damping matrix
C_d	Drag coefficient
C_m	Inertia coefficient
D	Diameter of the shaft (m)
E	Modulus of Elasticity (N m^{-2})
F	Wave force (N)
g	Acceleration due to gravity (m s^{-2})
G	Modulus of Rigidity (N m^{-2})
H	Depth of water (m)
I	Moment of inertia (m^4)
J	Polar moment of inertia (m^4)
K	Elemental stiffness matrix
[K]	Global stiffness matrix
k	Wave number
L	Length (m)
M	Elemental mass matrix
[M]	Global mass matrix
Ma	Added mass (kg)
t	Time period (s)
U	Water particle velocity (m s^{-1})
V	Velocity of the structure (m s^{-1})
ω	Water wave frequency (Hz)

INTRODUCTION

Increase of world population is a major issue for all human being. More and more lands are required for human activity. Dredging is a relevant process to create new lands. That is why the world dredging market is growing significantly during few past years. It involves excavation of waterways as well as discharging or filling over new areas to create new lands. A typical example of this is the Kolkata waste land development in India.

The purpose of dredging is not only for creating new lands it is also important to maintain or increase the depth of navigation channels, anchorages, or berthing areas for safe passage of boats and ships. So dredging operation is frequently carried out at the port to maintain the draft condition for the entering vessel. This involves in the removal of sediments and debris from the bottom of lakes, rivers, harbors, and the other water bodies. One of the equipment used for the dredging operation is the cutter suction dredger. The layout of the cutter suction dredger system is shown in the figure below.

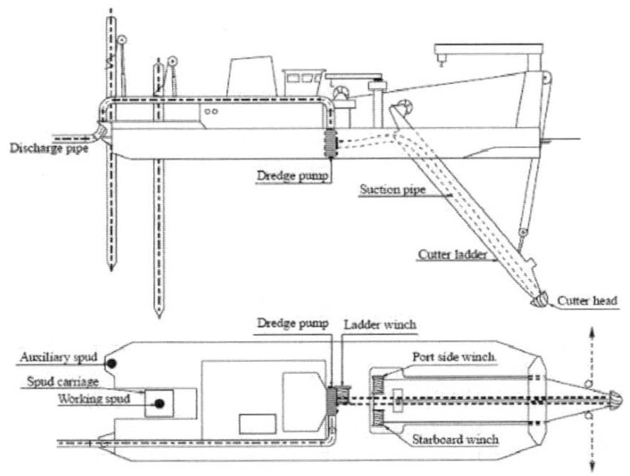

Fig 1: Schematic overview of the cutter suction dredger. [1]

For smooth and economical operation of a dredger, it is important to understand the vibration characteristics. The major parameters responsible for the performance and workability are hull characteristics, restraint imposed due to the spud and the dynamics of the spud. Many researchers have been working on these areas to increase the efficiency of the dredger. An estimation of response due to restraint imposed on the dredger hull from the spud can be helpful for quantifying the overall contribution to the vibrational effect from response to impulsive load due to cutting action of the cutter head. This paper is primarily focused on the influence over the linear vibration of the cutter suction dredger subjected to wave loading.

The spud and the dredge hull are also the important part of the CSD. The spud provides restraint to the system during the dredging operation. [1]analyzed pile-soil interaction subjected to both axial and lateral load using Finite Difference method by developing a three-dimensional Finite Element model. The result shows that there are reductions in induced bending moment and lateral displacement by increasing the amount of clay surrounding the piles resulting increased capacity for the resistance to lateral loading. Zachert et.al. [3] demonstrated the effect of soil structure interaction due to cyclic loading on foundations for offshore wind turbines. The soil is modelled by high cycle accumulation (HCA) model for the prediction of the deformation and soil-structure interaction and the effect on bending moment of the model test pile due to cyclic loading. Lopez-Querol et.al. [4] numerically analyzed the soil structure interaction of a mono pile by using API method, DEM and FEM. The FEM analysis is carried out using ANSYS software and the complete analysis is carried out for both static and dynamic loading. From the analysis it is found that the API method is a pseudo static approach and the cyclic bearing capacity is significantly lower than the static bearing capacity.

The force calculation on the cutter is the primary factor for dredging which is responsible for dredging performance. The dredging process during excavating material from sea bed surface using cutter suction dredger, it is very important to know about the forces exerted on the cutter. Young et al. [5] developed an experimental set up to calculate the forces in horizontal, vertical and axial direction. To know these forces a static analysis was done on carriage by the application of static loads to the cutter head in horizontal vertical and axial directions. The static equations were validated by the static finite element analysis through SolidWorks, a modeling software. The experimental and theoretical calculations were compared over the dredge cutter head. This is also important to make the optimal design of the cutter to reduce the undesired forces resulting less damage to the cutter and increase the performance which can ultimately reduce the cost of the dredging project. Ma Yasheng et al. [6] explored the 3-d cutting forces calculation of the cutter head. The emphasis was given on blade cutting forces with the small cutting angle and calculated for the saturated sand. The pressure distribution also calculated using MATLAB to ensure the similarity of the results with their experimental work.

The performance of the dredge is also affected by different wave condition. Wave height and frequencies are important parameters to be considered for the dredging operation. Koning et al. [7] experimented in laboratory for the soil-cutter head interaction under wave condition. It was found that the wave frequency influences the interaction between soil and cutter head. It was observed that the load on the cutter head can be reduced by low swing velocities and higher cutter revolution. Keuning et al. [8] worked in irregular waves to develop mathematical model for computation of behavior of cutter suction dredger. The work shows the importance of the soil behavior in the mathematical model. The behavior of dredge is heavily influenced by nonlinear characteristics of soil. This dynamic behavior depends on the type of soil. Hence it is important to get the knowledge about the soil reaction forces on the cutter.

The paper focuses on the dynamic behavior and response of the cutter suction dredger shaft. The shaft is subjected to wave loading. An impulse is provided at the free end to emulate the condition of a cutter in a linear manner.

MODELLING OF THE DREDGE CUTTER SHAFT

The cutter suction dredger consists of Hull, spud and a cutter shaft. A schematic sketch of the system is shown in the Fig. 2.

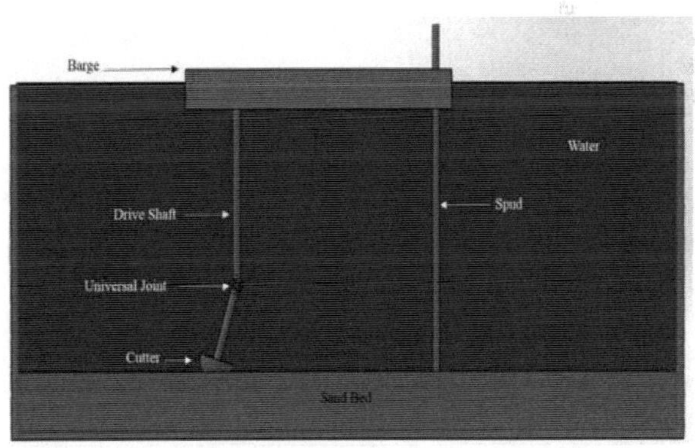

Fig 2: Equivalent model for the dredging operation

The cutter shaft consists of two shafts connected using a universal joint. The purpose of the universal joint is to transmit the torque from the straight shaft connected to the motor. The straight shaft which is connected to the prime mover is longer in comparison to the inclined shaft.

The scaled down model has been considered for the modelling of the cutter shaft. The scale factor (λ) is taken as four for the whole system. The cutter shaft is scaled down as per the Cauchy scaling law. The details of the scaled model are provided in Table 1.

Table 1: Material properties and shaft dimensions

Total length of the shaft	1.687 m
Length of larger shaft	1.25 m
Length of smaller shaft	0.437 m
Diameter of the both shaft	0.05 m
Material of the shaft and universal joint	Magnesium
Modulus of elasticity	45 GPa
Density of the structure	1800 kg/m3

A theoretical model of the individual shafts of the cutter suction dredger was modelled using finite element analysis. The shaft was considered as a Euler-Bernoulli beam. The effect of bending was considered in two planes. Each node has five degrees of freedoms shown in Fig 3.

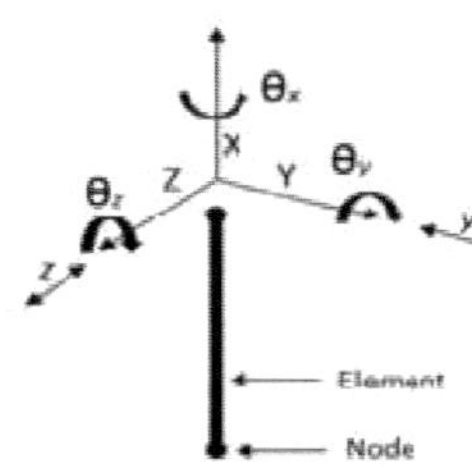

Fig 3: 2-Noded element with all 5 dof

The elemental mass matrix of the system is [9]

$$M=\rho AL/420\begin{bmatrix} 156 & 0 & 0 & 0 & 22L & 54 & 0 & 0 & 0 & -13L \\ 0 & 156 & 0 & -22L & 0 & 0 & 54 & 0 & 13L & 0 \\ 0 & 0 & 140J/A & 0 & 0 & 0 & 0 & -70J/A & 0 & 0 \\ 0 & -22L & 0 & 4L^2 & 0 & 0 & -13L & 0 & -3L^2 & 0 \\ 22L & 0 & 0 & 0 & 4L^2 & 13L & 0 & 0 & 0 & -3L^2 \\ 54 & 0 & 0 & 0 & 13L & 156 & 0 & 0 & 0 & -22L \\ 0 & 54 & 0 & -13L & 0 & 0 & 156 & 0 & 22L & 0 \\ 0 & 0 & -70J/A & 0 & 0 & 0 & 0 & 140J/A & 0 & 0 \\ 0 & 13L & 0 & -3L^2 & 0 & 0 & 22L & 0 & 4L^2 & 0 \\ -13L & 0 & 0 & 0 & -3L^2 & -22L & 0 & 0 & 0 & 4L^2 \end{bmatrix}$$

The elemental stiffness matrix of the system is [9]

$$K=\begin{bmatrix} 12EI/L^3 & 0 & 0 & 0 & 6EI/L^2 & -12EI/L^3 & 0 & 0 & 0 & 6EI/L^2 \\ 0 & 12EI/L^3 & 0 & -6EI/L^2 & 0 & 0 & -12EI/L^3 & 0 & -6EI/L^2 & 0 \\ 0 & 0 & GJ/L & 0 & 0 & 0 & 0 & -GJ/L & 0 & 0 \\ 0 & -6EI/L^2 & 0 & 4EI/L & 0 & 0 & 6EI/L^2 & 0 & 2EI/L & 0 \\ 6EI/L^2 & 0 & 0 & 0 & 4EI/L & -6EI/L^2 & 0 & 0 & 0 & 2EI/L \\ -12EI/L^3 & 0 & 0 & 0 & -6EI/L^2 & 12EI/L^3 & 0 & 0 & 0 & -6EI/L^2 \\ 0 & -12EI/L^3 & 0 & 6EI/L^2 & 0 & 0 & 12EI/L^3 & 0 & 6EI/L^2 & 0 \\ 0 & 0 & -GJ/L & 0 & 0 & 0 & 0 & GJ/L & 0 & 0 \\ 0 & -6EI/L^2 & 0 & 2EI/L & 0 & 0 & 6EI/L^2 & 0 & 4EI/L & 0 \\ 6EI/L^2 & 0 & 0 & 0 & 2EI/L & -6EI/L^2 & 0 & 0 & 0 & 4EI/L \end{bmatrix}$$

The global equation of motion of the system can be obtained as:

$$[M]\ddot{X}(t)+[C]\dot{X}(t)+[K]X(t)=0 \qquad (1)$$

C is the damping matrix obtained by using Rayleigh damping assumption.

The first three natural frequency of the individual shafts are provided in the table below.

Table 2: Natural frequency of Longer and shorter shaft

Sl. No.	Longer Shaft Natural Frequency (Hz)	Shorter Shaft Natural Frequency (Hz)
1	3.67	30.08
2	23.09	188.52
3	64.66	290.47

MODELLING OF THE DREDGE HULL

Dredge hull is the major part of the cutter suction dredger. A typical sketch of the dredger hull is given in Fig 4. Due to the ocean wave the hull will be subject to rigid body motion in the six degrees of freedom.

For the modelling of the hull of the cutter suction dredger the scaled down model has been considered. The scale factor (λ) was taken as four and the hull is scaled down using Froude scaling law. The material of the hull is taken as Acrylic. The length of the hull is 2.7 m and the beam is 1 m. The draft of the hull is 0.2 m and the depth is 0.305 m.

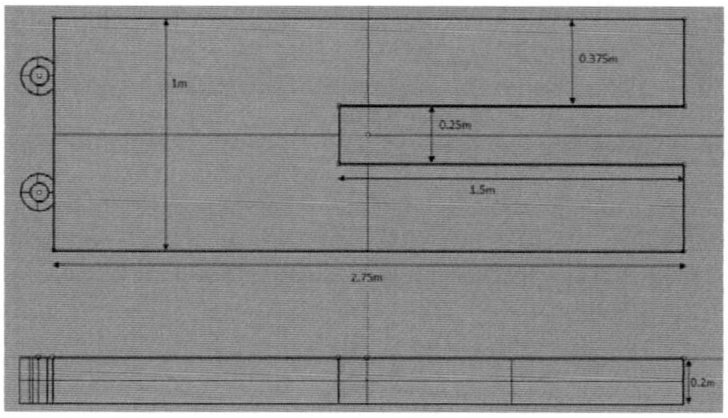

Fig 4: Geometry of Dredge Hull

The hull of the cutter suction dredger is modelled in WAMIT [10]. The WAMIT software computed the dredge hull motion in six degrees of freedom. As the scaling factor (λ) for the whole system is four, the scaled down wave input parameter for the computation is considered to be the wave height as 20 cm with different periods from 0.5s to 5s with head sea condition. The first input of the model is the geometric discretisation for the submerged part of the dredge hull which is created using Rhinoceros 4.0 [11].

The inertia properties for WAMIT analysis is obtained from Rhinoceros 4.0 and provided as shown in equation (2).

$$I = \begin{bmatrix} \rho_w \nabla & 0 & 0 & 0 & 0 & 0 \\ 0 & \rho_w \nabla & 0 & 0 & 0 & 0 \\ 0 & 0 & \rho_w \nabla & 0 & 0 & 0 \\ 0 & 0 & 0 & I_{xx} & 0 & 0 \\ 0 & 0 & 0 & 0 & I_{yy} & 0 \\ 0 & 0 & 0 & 0 & 0 & I_{zz} \end{bmatrix} \quad (2)$$

$$I_{xx} = K_{xx}^2 \rho_w \nabla \quad (3)$$

$$I_{yy} = K_{yy}^2 \rho_w \nabla \quad (4)$$

$$I_{zz} = K_{zz}^2 \rho_w \nabla \quad (5)$$

where ρ_w is the density of the water taken as 1000 kg/m3, '∇' is the total displacement, K_{xx}, K_{yy} and K_{zz} are the radius of gyrations in the direction of x, y and z-axis respectively.

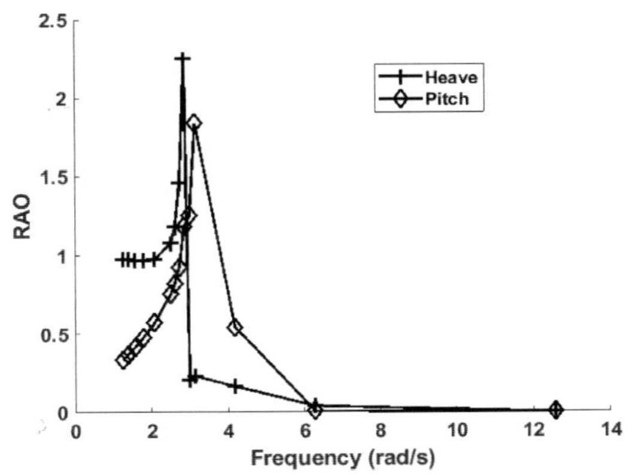

Fig 5: Heave and Pitch transfer function

Fig 5 shows the RAO computed from WAMIT for the heave and pitch motion for different wave encounter frequencies. Mainly heave and pitch is considered since the restraint provided by the spud restricts these two motions. As expected the encountering frequency corresponding to maximum heave and pitch is different. From the result it is found that the both the heave and pitch RAOs are comparatively higher at the frequency of 2.855 rad/s i.e. at the wave period of 2.2s. Therefore, to analyse the effectiveness of the restraint the above wave parameter was considered as excitation frequency.

MODELLING OF THE SPUD WITH SOIL

The spud restraints the motion of the hull by anchoring into the soil. The spud is scaled down by Cauchy scaling law. By fulfilling the scaling laws, the spud material is taken as Magnesium material.

The spud is embedded up to 0.5m deep inside of a cylindrical soil layer of 1 m deep and 2m wide. The soil is considered to be as clayey sand of specific gravity 2.5 and the water content level is 10%. The spud is hollow type with 5.5 mm thickness and the outside diameter is 8.1 cm. The spud height is 2.25m and the water depth is 1.75m above the soil surface. The software ANSYS 18.1 [12] was used to create the 3D model of the spud and soil foundation. The Transient Structural Analysis was used to analyse the dynamic wave loading condition.

As the spud is undergoing through wave, the force acting on it can also be derived from Morison's equation.

$$F(t)=F_I(t)+F_D(t) \quad (6)$$

Where $F_I(t)$ is the inertial force term and $F_D(t)$ is the drag force term. The inertial and drag force can be determined by integrating over the height of the structure.

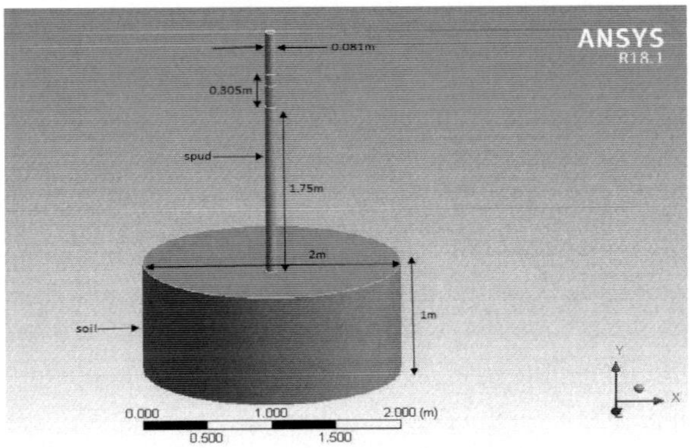

Fig 6: Model of spud with soil

For a structure fixed to the seafloor the inertial force due to waves impinging on a circular cylinder is

$$F_I(t) = -\int_{-H}^{0} \rho C_M \frac{\pi}{4} d^2 \frac{h}{2} \omega^2 \frac{\cosh[k(z+H)]}{\sinh(kH)} \sin(\omega t) dz \quad (7)$$

Similarly, the drag force is

$$F_D(t) = \int_{-H}^{0} \frac{1}{2} \rho C_D d \frac{h^2}{4} \omega^2 \frac{\cosh^2[k(z+H)]}{\sinh^2(kH)} \cos(\omega t) |\cos(\omega t)| dz \quad (8)$$

The heave response from the dredge hull is applied as acceleration on the spud in vertical direction and the pitch response is applied as the angular acceleration. The time variant Morison force is varying over the length of the spud below the water surface. The spud was represented as a linear-elastic, isotropic, homogeneous material of Magnesium. The soil is taken as clayey sand and Kinematic Hardening Law [4] have been employed for defining the properties of the soil. For the transient analysis of the soil- spud model the time step was taken as 0.01s over a period of 30s.

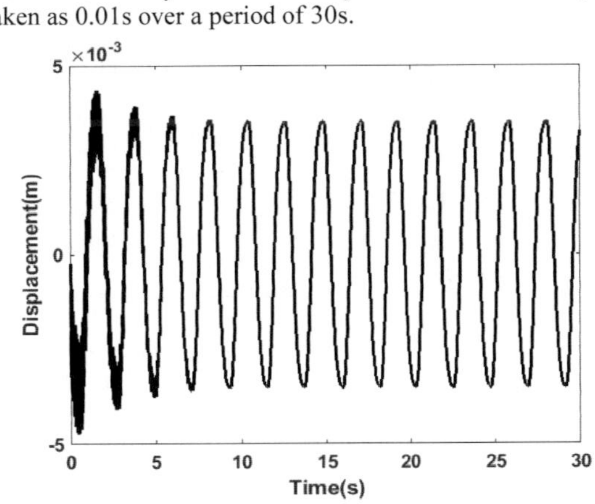

Fig 7a: Temporal response of the spud displacement at the connection location with the dredge hull.

The temporal response of the spud is shown in Fig 7a. The initial transients are due to the structural modes and as time progresses the system attains steady state as the wave excitation frequency. This also observed in the frequency response of the system shown in Fig 7b. From the frequency response it is clear that the response of the system is high at the natural frequency of the cutter shaft. However, for the further analysis of the assembled system the dynamics of the hull spud assembly is ignored and the cutter shaft is assumed to be fixed at one end.

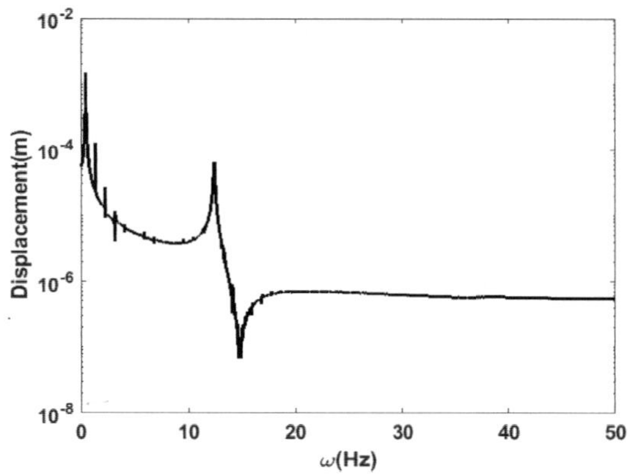

Fig 7b: Frequency response of the spud at the connection location to the dredge hull.

MODELLING OF CUTTER SHAFT ASSEMBLY

Next the analysis on the assembly of shorter and the longer shaft connected using universal joint is carried out in MATLAB. The connection with the prime mover is assumed fixed since the frequency of the hull spud assembly is low compared to individual shafts. The universal joint is modelled as a spring element. The inclined portion of the shaft includes a transformation matrix. The universal joint was modelled using a spring element of high stiffness.

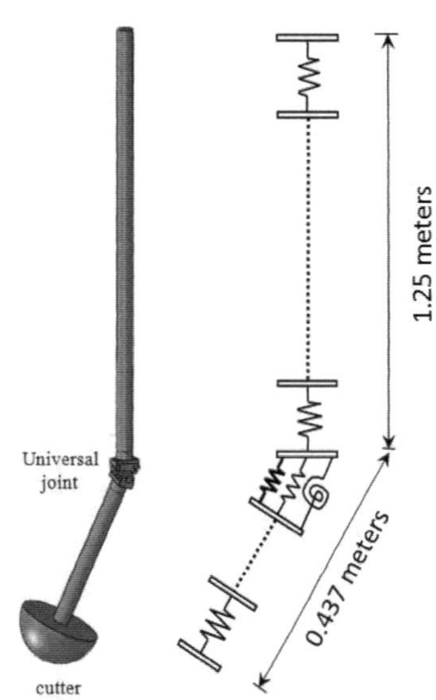

Fig 8 Diagram of cutter drive with cutter and its spring equivalent system

The addition of the universal joint would constraint the mode at the location of the Hooke's joint. This is observed in the mode shapes obtained for the shaft system by considering high stiffness at the connection point shown in Fig 9. The natural frequencies of the system with the addition of universal joint considering both high stiff and low stiff material are shown in table below.

Table 3: Natural frequencies of the system with high and low stiffness consideration of universal joint

Sl. No.	High Stiffness Natural Frequency (Hz)	Low Stiffness Natural Frequency (Hz)
1	23.44	23.04
2	40.92	38.68
3	64.62	63.46

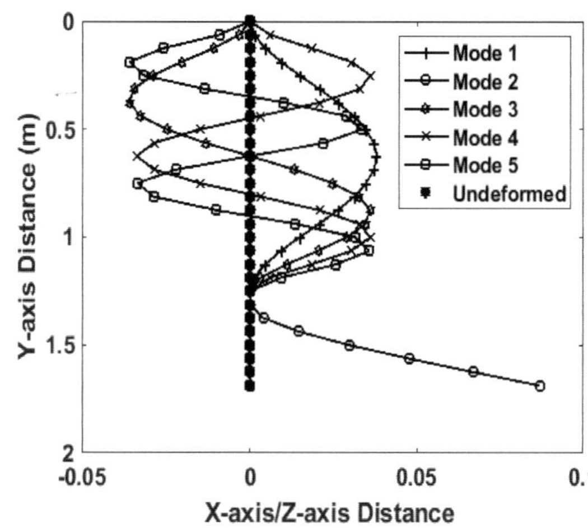

Fig 9: Modeshapes of the system with high stiffness at the joint

WAVE MODEL

The Morison's equation for the forces acting on an oscillating structure is given as:

$$F(\mathrm{t}) = \tfrac{1}{2}\rho C_{\mathrm{d}} DL(U - V)\left|U - V\right| + \rho C_{\mathrm{m}} AL(\dot{U} - \dot{V}) + \rho AL\dot{V} \tag{9}$$

where the water particle velocity is

$$U = \frac{agk}{\omega}\frac{\cosh k(H + z)}{\cosh kH}\cos(kx - \omega t) \tag{10}$$

and the water particle acceleration

$$\dot{U} = agk\frac{\cosh k(H + z)}{\cosh kH}\sin(kx - \omega t) \tag{11}$$

The equation of motion of the system can be derived as

$$[M]\ddot{X}(\mathrm{t}) + [C]\dot{X}(\mathrm{t}) + [K]X(\mathrm{t}) = F(\mathrm{t}) \tag{12}$$

The external force includes an extra inertia term due to added mass effect. The modified equation of motion is obtained as:

$$[M]\ddot{X}(t)+[C]\dot{X}(t)+[K]X(t)=F(t)-[M_a]\ddot{X}(t) \tag{13}$$

$$[M+M_a]\ddot{X}(t)+[C]\dot{X}(t)+[K]X(t)=F(t) \tag{14}$$

where $M_a = \rho\pi D^2 L/4$

TRANSIENT ANALYSIS

The rotation of the cutter over the bed surface is considered to be producing the impulsive force which is applied at the end of the cutter-drive.

The impulsive force in terms of initial condition is provided as an initial displacement and initial velocity at the free end. The undamped and damped natural frequencies are calculated considering the effect of both universal joint and added mass. Only proportional damping of the individual shaft is included. Table 4 shows the first 4 damped natural frequencies of the complete system.

Table 4: Damped natural frequencies of the structure

Sl. No.	Damped Natural Frequency (Hz)
1	19.13
2	23.43
3	31.55
4	40.89

The shaft system is considered as straight shaft and the initial displacement and velocity are taken as 0.01 m and 100 m/s respectively. The time series response over the entire length of the system is calculated. Universal joint was modelled by an increased stiffness at the connection of the two shafts. Two case studies were carried with a stiffness of 10 and 1.5 times elemental stiffness at the connecting node. The time series responses of the cutter shaft at different instant have been plotted below.

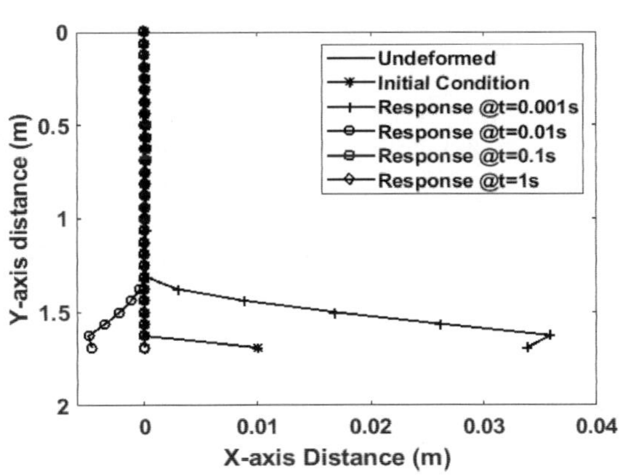

Fig 10a: Time series response at different times at t=0.001s, t=0.01s, t=0.1s and t=1s considering low stiffness at the universal joint.

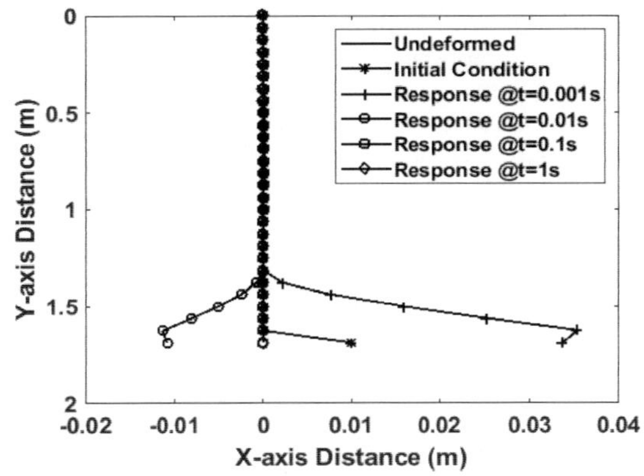

Fig 10b: Time series response at different times at t=0.001s, t=0.01s, t=0.1s and t=1s considering high stiffness at the universal joint

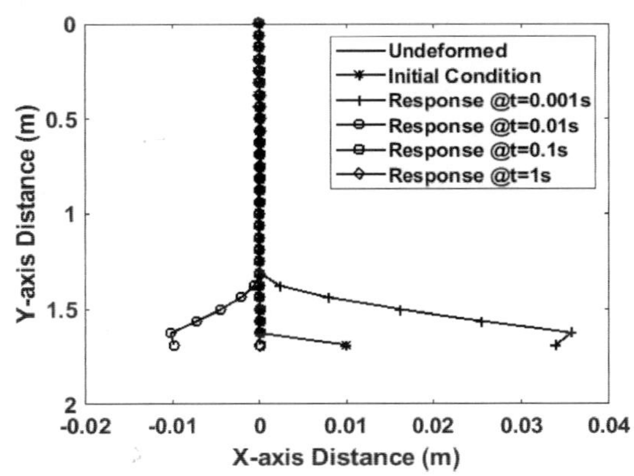

Fig 10c: Time series response at different times at t=0.001s, t=0.01s, t=0.1s and t=1s considering high damping and low stiffness

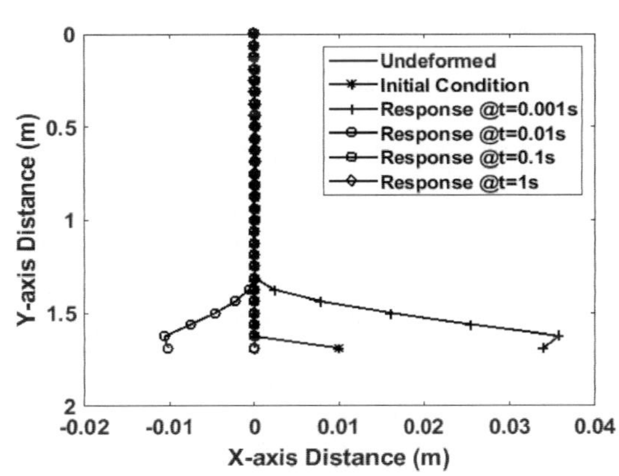

Fig 10d: Time series response at different times at t=0.001s, t=0.01s, t=0.1s and t=1s considering both high damping and high stiffness

From Fig 10a and Fig 10b it is found that due to the impulsive force applied at the free end, the response is higher at each instant in case of the universal joint with low stiffness in comparison to high stiffness universal joint. Also with a low stiffness, the vibration transmission occurred on the longer shaft.

Next a case study was carried out by providing damping at the universal joint. From the Fig 10c and Fig 10d it has been found that due to the impulsive force, the response is getting reduced. As expected with a higher damping the vibration response was reduced faster. Further in the low stiffness universal joint the vibration transmission to the longer shaft was also reduced.

CONCLUSIONS

For making the dredging operation smooth and economical an attempt has been taken to understand the vibration characteristics of cutter suction dredger. The dredge hull, spud and the cutter shaft were analysed for different environmental condition.

The dredge hull is designed in Rhinoceros and analysed for different wave condition in WAMIT. The responses were evaluated for heave and pitch. The optimum wave parameter was found where the heave and pitch RAOs are comparatively high which is the wave input parameter for the transient analysis of spud-soil system.

The soil-spud system dynamics is studied. The model is designed and analysed in ANSYS. The heave and pitch motions from the dredge hull are the input parameters in terms of acceleration and angular acceleration over the spud. The time variant wave force has been considered over the length of the spud is calculated by Morison equation inside the water. For the transient analysis of the soil spud system the spud is considered as linear elastic, isotropic homogeneous material and the soil is governed by kinematic hardening law.

The dynamics of the shaft system of the cutter drive is also studied and designed as two shafts connected by means of a universal joint. The dynamic equation is generated by considering the system as two nodded finite element beam element with five degrees of freedom per each node. The wave forces have been calculated over the beam segment using Morison equation. The time series responses have been determined over the entire length including the added mass effect for a cylindrical member.

An impulsive load was provided at the extreme end of the shaft system and response due to variation in universal joint configuration was determined. The results from the study indicated that the shaft vibration response was sensitive to the stiffness and damping of the universal joint. The time response analysis of the soil-spud system indicates that there is no increase in displacement of the spud with time. The equivalent stress generated in the spud is way below the design stress requirement which gives the information about the safe designing of system for the experimental purpose. The future work will look into the experimental analysis of the system.

ACKNOWLEDGEMENTS

The authors wish to acknowledge Kshitij Srivastava for his help and support during the development of this technical paper.

REFERENCES

[1] "Onslow Marine Support Base Stage 2: Dredging & Spoil Disposal Management Plan," 2018.

[2] Y. Khodair and A. Abdel-Mohti, "Numerical Analysis of Pile–Soil Interaction under Axial and Lateral Loads," Int. J. Concr. Struct. Mater., vol. 8, no. 3, pp. 239–249, 2014.

[3] H. Zachert, T. Wichtmann, and T. Triantafyllidis, "Soil structure interaction of foundations for offshore wind turbines," in Proceedings of the International Offshore and Polar Engineering Conference, 2016, vol. 2016, pp. 68–75.

[4] S. Lopez-Querol, L. Cui, and S. Bhattacharya, "Numerical Methods for SSI Analysis of Offshore Wind Turbine Foundations," in Wind Energy Engineering: A Handbook for Onshore and Offshore Wind Turbines, Elsevier Inc., 2017, pp. 275–297.

[5] D. R. Young and R. E. Randall, "Laboratory Measurement of Forces on a Dredge Cutter Head," pp. 37–53.

[6] M. Yasheng, N. Fusheng, and S. A. Miedema, "Calculation of the Blade Cutting Force for small Cutting Angles based on MATLAB," in The 2nd China Dredging Association International Conference & Exhibition, 2006.

[7] S. A. Miedema, J. de Koning, and A. Zwartbol, "Soil/Cutterhead Interaction Under Wave Conditions," in Proc. WODCON X, 1983.

[8] P. J. Keuning and J. M. . Journée, "Calculation Method for the Behaviour of a Cutter Suction Dredge Operating in Irregular Waves," World Dredg. Congr., pp. 79–96, 1983.

[9] M. Paz and W. Leigh, Structural Dynamics: Theory and Computation. Springer Science+Business Media, LLC, 2006.

[10] "WAMIT® User Manual, Version 7.0," WAMIT, Inc. .

[11] "Rhinoceros User's Guide," 2007. [Online]. Available: http://docs.mcneel.com/rhino/6/usersguide/en-us/index.htm. [Accessed: 20-Sep-2019].

[12] "ANSYS® Academic Research Mechanical, Release 18.1, Help System," ANSYS, Inc. .

Proceedings of the 6th International Conference on Ship and Offshore Technology
ICSOT 2019
November 7-8, 2019, IIT Kharagpur, India

ICSOT INDIA 2019-25

UNDERSTANDING POLAR ENVIRONMENT: PRELIMINARY RESULTS FROM DEEP-LEARNING-BASED SEGMENTATION OF OPTICAL ICE IMAGES

Nabil Panchi
IIT Kharagpur
Kharagpur, West Bengal, 721302
India
panchinabil@iitkgp.ac.in

Ekaterina Kim
NTNU
Trondheim, 7491
Norway
ekaterina.kim@ntnu.no

Anirban Bhattacharyya
IIT Kharagpur
Kharagpur, West Bengal, 721302
India
ab@naval.iitkgp.ac.in

ABSTRACT
Computer vision-aided scene understanding has drawn much attention, especially in autonomous vehicles and advanced driver assistance systems. Image segmentation forms the core of these systems, which is required for effective planning of maneuvers. Recently, with more navigational routes opening up in the Arctic ocean, computer-aided understanding of the ice conditions in front of a surface vessel has become very important for pilot assistance systems because understanding the severity of ice conditions is essential for safe ice navigation. The floating ice features need to be correctly detected, classified and accurately segmented, and the path needs to be carefully planned to avoid excessive ice loads on the ship hull and the ship besetting in ice. With the increasing popularity of deep learning in recent years, many deep architectures have been successfully proposed for solving a number of segmentation problems. This paper focuses on polar scene segmentation tasks. The goal is to find a model that can accurately classify and locate the following surface ice features: icebergs, deformed ice, level ice, broken ice, ice floes, floebergs, floebits, pancake ice, and brash ice. In this paper, it is shown that this segmentation task can be solved with a fully convolutional neural network (U-Net) model with a pretrained residual neural network architecture (ResNet) as the downsampling part. The performance of ResNet18, ResNet34, ResNet50 and ResNet101 on images collected online and from the cruise to the Fram Strait in 2012 was analyzed. In addition, the model performance coupled with convolutional conditional random fields (CRFs) and fully connected CRFs was analyzed and compared. Preliminary results indicate that the U-Net model with ResNet101 as the backbone in combination with the convolutional CRF-based postprocessing technique performs the best on available optical images of ice cover.

1. INTRODUCTION

As more industries begin to operate in polar areas, the number of vessels in ice-infested waters is increasing faster than the number of trained professionals who can safely and efficiently operate in these environments. Although considerable research has been devoted to autonomous shipping technology for aspects such as navigation support, equipment monitoring, and safety improvements, less focus has been on computer-aided scene assessment from ships that are traveling in ice. Currently, all navigation of surface vessels in ice-infested waters is based on the experience of ice pilots and captains. Figure 1 presents a typical ice transit decision process from [1]. To support this decision process, it is of great advantage to be able to automatically identify ice types within an overall ice regime. In this paper, we focus on the highlighted (with green color) sections of the diagram, i.e., on accurate localization and classification of various surface ice features from optical images taken aboard a ship with the help of deep learning methods.

Examples of relevant works dealing with the identification of ice conditions are given in [2-9] and include techniques and sensor systems for the analysis of ice concentrations, floe-size distributions, and drift speeds using shipborne images from optical cameras and/or marine radars or optical images from an unmanned autonomous underwater vehicles (AUV). To the best of our knowledge, no currently available model can discriminate between sea-ice types using optical images from a surface vessel. Attempts to segment images of river ice [10] are limited to a very few classes with the purpose of understanding the process of ice formation in rivers and thus lack the ice features that are important for navigation (e.g., icebergs and deformed ice). A recent attempt at classifying the ice object present in the optical image was made in [11], which addressed the classification of ice objects, as opposed to fine-grained segmentation of optical images, which is the objective of this paper.

The main objective of this work is to find a model that can analyze optical surface ice images and process pixels by pixel segmentation of these optical ice images to obtain the exact location and type of ice present in the ice regime. This paper presents a candidate model that can perform this task and demonstrates the ability of this model to segment ice surface features from close-range optical imagery gathered online and during the research cruise to the Fram Strait on the RV Lance in March 2012. The following categories of surface ice features

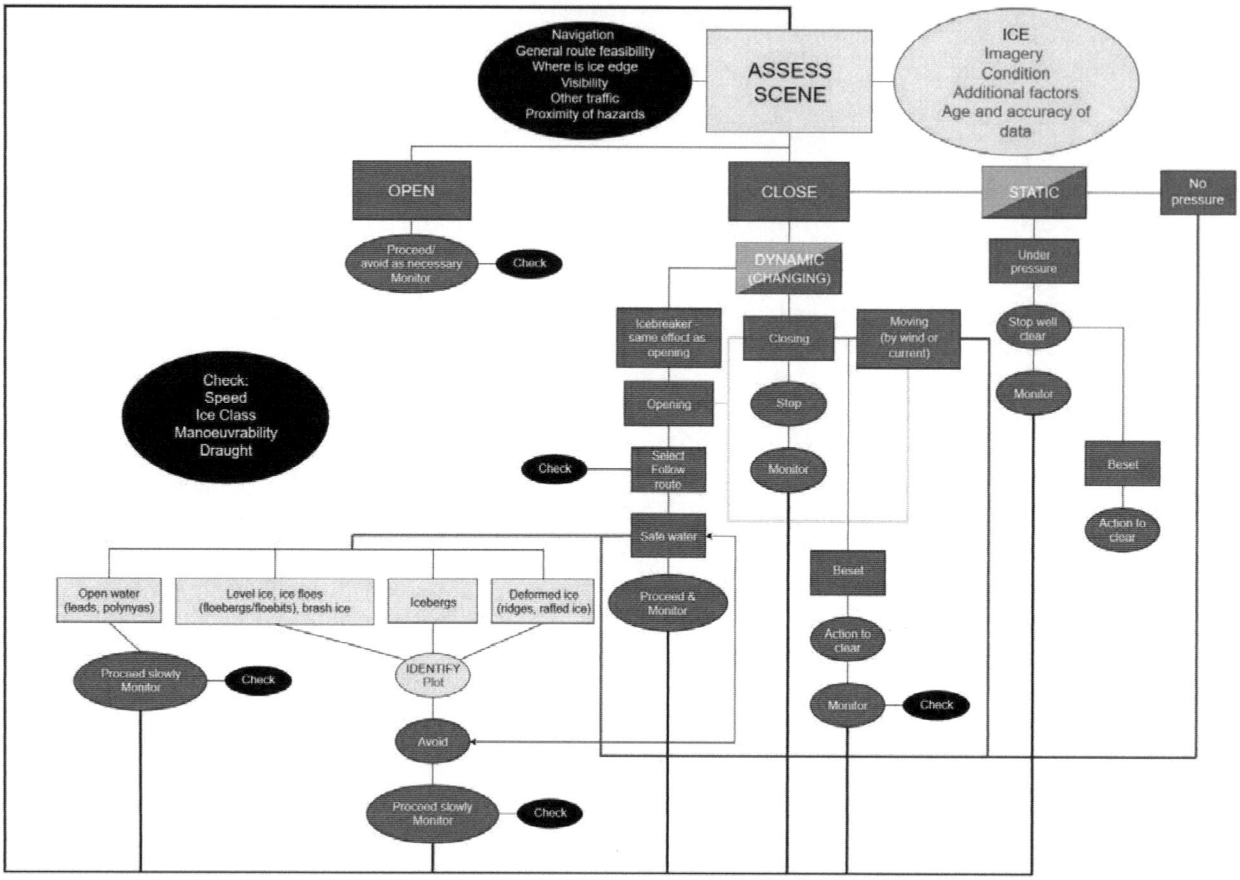

Figure 1. A TYPICAL ICE TRANSIT DECISION PROCESS (ADOPTED FROM [1] WITH SMALL MODIFICATIONS)

are considered: *level ice*, *deformed ice*, *broken ice*, *icebergs*, *floebergs*, *floebits*, *icefloes*, *pancake ice*, and *brash ice*. A variety of algorithms were validated on the collected ice imagery, and the results were compared from these algorithms.

2. METHODS

The deep learning methods used in this study are well established in research fields other than ice. For our ice image segmentation task, convolutional neural networks (CNNs) were chosen as they have been successfully applied to various classes of problems involving images such as image segmentation classification and object detection. See [12-18] for more details.

2.1 Dataset

The dataset consists of images containing the following categories of surface ice features: *icebergs, floebergs, floebits, level ice, deformed ice, broken ice, ice floes, pancake ice, brash ice,* along with five other categories, such as *underwater ice, melt ponds, open water, sky, and shore.* The images for the dataset were obtained from the data gathered during the research cruise to the Fram Strait on the RV Lance in March 2012 and from Google, Yandex, and Baidu search engines. The images in the dataset were mostly taken from onboard a vessel under good visibility conditions. These images were manually labeled according to the definitions provided in [19]. Additional features (leads, fractures, cracks and polynyas) were not considered in the labeling process; rather, categories such as open water, underwater ice, and melt ponds were used. Figure 2 contains sample images from the dataset along with their labels. Segmentation masks (ground truth mask, shown in Fig. 2) for all images were obtained by making precise polygons around the ice features using an open source tool from [20]. Details and the challenges of the labeling procedure are summarized in the appendix section of this paper.

The dataset consisted of 375 images. The dataset was split into a training set and a validation set with a ratio of 9:1, i.e., ten percent of the images (37 images) were chosen at random for the validation set.

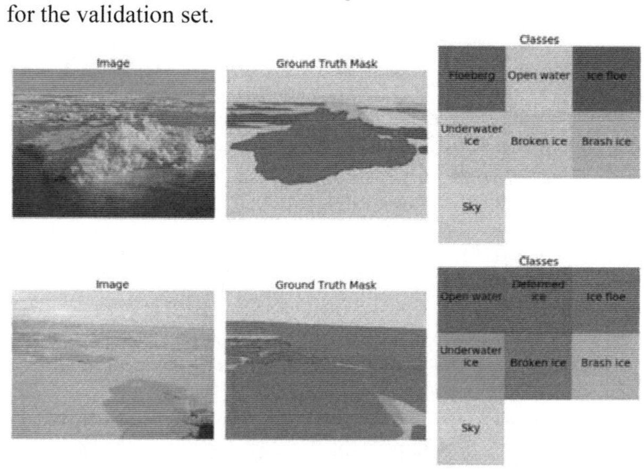

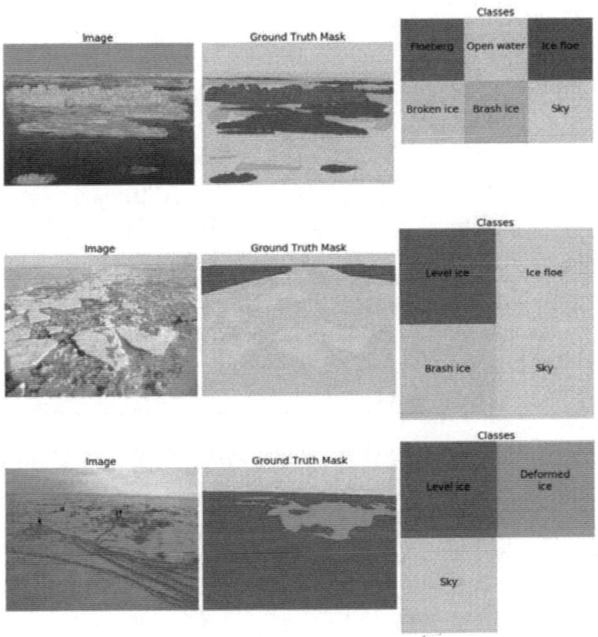

Figure 2. SAMPLES FROM THE DATASET ALONG WITH THEIR HAND LABELED SEGMENTATION MASK

2.2 Neural Network Architecture

We want a neural network that will take an image as the input and predict the masks as accurately as possible (for an example of the segmentation mask, refer to Fig. 2). The CNN model that is used in this study is a modified version of U-Net [21], which was initially proposed for the segmentation of medical images. There are two parts of a U-Net, the encoder (the downsampling part), which is used to obtain a high-dimensional representation of the image, and the decoder (the upsampling part), which is used to generate the segmentation masks from the extracted high-dimensional representations of the image. The encoder consists of a stack of convolutional layers in combination with max-pooling layers, and the decoder is a stack of transposed convolutional layers. This makes it an end-to-end convolutional neural network (also known as a fully convolutional neural network or FCNN). This end-to-end convolutional nature also allows this network to work with multiple image sizes without any adaptive average pooling layers. It also ensures that the number of learnable parameters in the network is less compared to the number of learnable parameters in networks with fully connected layers. In our implementation, the normal encoder part of the network was replaced with the more effective and efficient ResNet architectures [15]. We experimented with four types of ResNet architectures as the backbones, namely, ResNet18, ResNet34, ResNet50 and ResNet101.

2.3 Training

We used the ResNet models that were pretrained on ImageNet [22] as the encoders. This is known as transfer learning. The motivation for transfer learning comes from the fact that training a CNN from scratch with all the weights randomly initialized requires a large quantity of training data. However, obtaining such a large dataset is difficult because labeling images is challenging, time consuming, and expensive. In addition, many deep learning models are specific to a particular task or field. While they might be the state-of-the-art models in that field, the deep learning models suffer from considerable performance losses when used in a new task that might still be similar to the task that the model was trained on. Transfer learning goes beyond specific domains or tasks and enables us to utilize a part of the model across domains. In addition, it has been found that the earlier layers of a CNN learn more generic features such as edges or color blobs, but later layers of a CNN learn progressively complex features that are more specific to the dataset on which the CNN was trained. Refer to [23] for more details. The information learned by the ResNet model due to the training on ImageNet can be leveraged with the help of transfer learning. This is possible because edges, color blobs and other basic features of images can be found in all types of images. This helped us tackle the problem of limited training data to an extent.

We also applied data augmentation to the training data to address the problem of limited training samples. Data augmentation is needed because the deep learning model needs to generalize well on unseen data; for this to happen, the deep learning model must not become biased to very specific features of the dataset (a.k.a. overfitting). In contrast, it should learn to generalize. This means that the training process must ensure that the model learns the main features of the dataset and not the very specific ones. To ensure this, it is important that the feature space of the dataset also covers the spectrum of possibilities, i.e., it should contain a large number of representations of each feature of the dataset. We used augmentation techniques such as small rotations, horizontal flips, resizing, randomly cropping a part of the image, randomly adjusting the contrast and the brightness of the image, and perspective warping.

The selection of the learning rate for the training of the model was performed by looking at the loss versus the learning rate curve and choosing the highest learning rate at which the loss is still decreasing (see Fig. 3a). The learning rate is essentially a multiplier by which the gradients of the weights are multiplied before subtraction from the original weight; hence, it is important that the learning rate is carefully chosen. A very small learning rate will result in very slow model convergence, and an overly large rate might lead to divergence of the model and will result in an increasing loss until it becomes too large for the computer to handle as it causes memory overflow. Thus, one needs to find a learning rate at which the loss decreases as rapidly as possible without diverging. The graph of loss versus the learning rate is illustrated in Fig. 3a. Initially, the loss decreases as the learning rate increases. Then, the loss gradually stops decreasing and then starts rising quickly. This is because as the learning rate increases, the rate of decrease in loss continues to increase until the learning rate becomes too high. When the learning rate becomes too high, the loss starts diverging. The loss was plotted against the learning rate for all the models, and the learning rate at which the loss was still decreasing was chosen for training the respective model. For all the combinations in the experiments, the learning rate was chosen between $1e^{-6}$ and $1e^{-3}$. The learning rate was increased in the early phase of the training and decreased in the later phase of the training. Figure 3b shows how the value of the learning rate changes over the training iterations. Initially, the value of the learning rate is smaller than a certain maximum value; as the training progresses, the value of the learning rate increases to a certain maximum value and then starts decreasing, eventually reaching the initial value. The maximum value of the learning rate, in this case, was the one that we found by looking at the loss versus the learning rate graph, and the lower value was chosen to be ten times lower than the max value. The length of this cycle was set to a value slightly less than the total number of training iterations. This was suggested in [24], and this approach works better than normal annealing, where the learning rate only decreases over the period of training. Different learning rates were used for different groups of layers in the CNN. This was done to ensure that the information learned by the encoder of the network (due to the training on the ImageNet) was not destroyed. The learning rate used for the encoder was a magnitude smaller than the learning rate for the decoder.

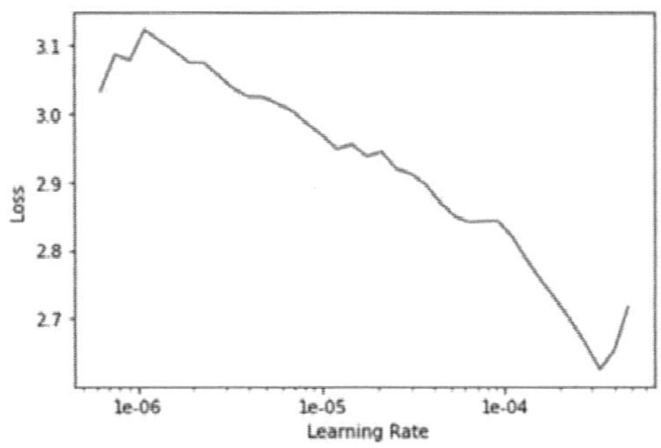

(a) Loss versus learning rate

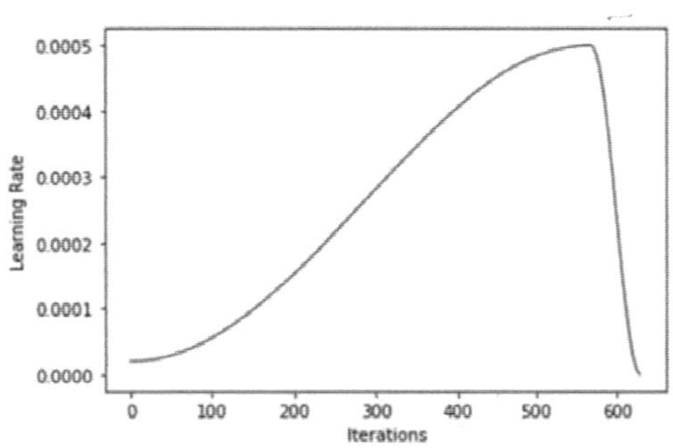

(b) Learning rate versus training iterations

Figure 3. LEARNING RATE GRAPHS

The CNN was trained in two phases; in the first phase, only the decoder was trained as it was initialized from scratch. In the second phase, all the weights were unfrozen, and the entire network was trained. The best weights from the first phase were chosen as the starting point for the training in the second phase. Furthermore, the model was trained on 256*256 images first, and then the best weights from this were used as a starting point for the training on 512*512 images. The idea is that while the model trained on smaller images learns fewer features than the model trained on large images because smaller images contain fewer details, the features that it learns are important. Thus, a model trained on smaller images also generalizes well on larger images. This is also known as progressive resizing in the fastai library [25]. In addition, smaller images are much easier to train on, which in turn leads to more experiments. The images in the training set were randomly cropped to obtain the desired resolution. For the validation set, a patch of desired resolution was cropped from the center of the image. Table 1 presents a summary of the parameters for the best performing model.

TABLE 1. PARAMETERS

Parameters	Value
Learning rate	$5e^{-4}$ (1st phase), $5e^{-4}/4$, $5e^{-4}/40$ (2nd phase)
Image size	256*256
Weight decay	$1e^{-2}$
Batch size	8 images per batch
Learning rate cycle parameters	Increase the learning rate for 90% of the training epochs, decrease for 10% of the epochs.
Total training epochs	65 (15 (1st phase) + 50 (2nd phase))
ResNet backbone	ResNet101
Postprocessing	Convolutional CRF (ConvCRF)

The models were trained on the Nvidia Tesla V100 GPU, located at UNINETT Sigma2 AS in Norway, which supports half-precision training as well. This reduces the GPU memory required for the model and hence allows for a larger batch size, which increases the speed of the training. Additionally, this becomes very important in the case of large models such as ResNet101.

All the deep learning models in this study were implemented with the help of FastAI [25] and PyTorch [26].

2.4 Postprocessing

The results from the CNN were noisy (contained small, random patches of misclassified pixels), as shown in Fig. 3. These predicted masks required cleaning to remove the noise. For that purpose, two different types of conditional random field (CRF)-based postprocessing techniques were used: fully connected CRF [27] (FullyConnCRF) and convolutional CRF [28] (ConvCRF). The convolutional CRF worked better and faster than the fully connected CRF. More details on the results and speed comparisons of the postprocessed predictions can be found in the 'Main results' section of the paper.

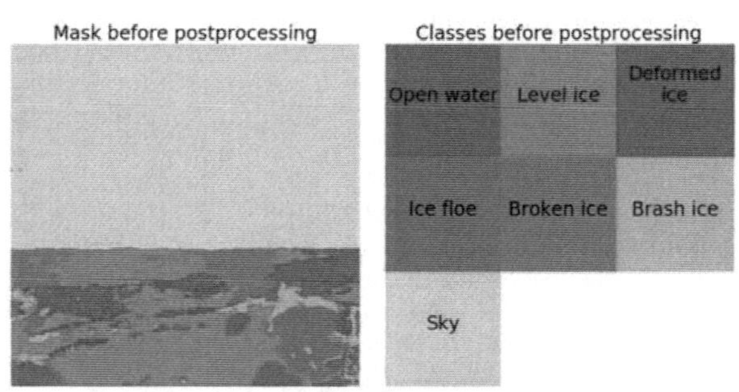

(a) Without postprocessing

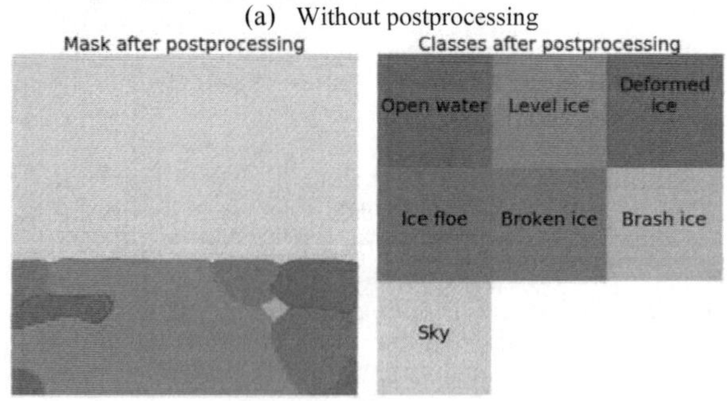

(a) With ConvCRF postprocessing

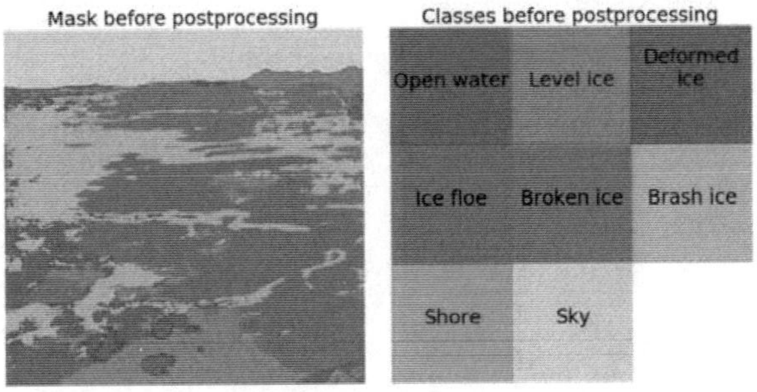

(b) Without postprocessing

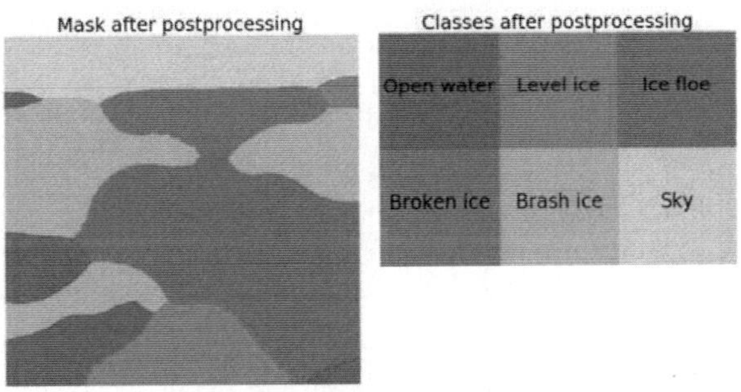

(b) With ConvCRF postprocessing

Figure 4. SAMPLES COMPARING LABELS WITHOUT ANY POSTPROCESSING AND WITH CONVOLUTIONAL CRF POSTPROCESSING

2.5 Evaluation

To evaluate the performance of the model and to compare the quality of predictions with different models, two metrics were used: the mean intersection over union (mean IOU) and the global accuracy. The mean IOU was calculated by averaging the IOU values for all classes. The IOU value for a class was calculated by dividing the intersection of the predicted and true label by their union. The mean IOU was treated as the main metric for evaluation. Global accuracy was calculated by obtaining the total number of correctly classified pixels and then dividing it by the total number of pixels. Global accuracy was not treated as the main metric for evaluation because the number of total pixels per class is highly imbalanced, and looking at the accuracy might not give us the true picture of how good the model is performing. Four different ResNets (ResNet18, ResNet34, ResNet50 and ResNet101) were used as the encoder (backbone) of the U-Net. All the models were trained on two different image sizes of 256*256 and 512*512. All of these models were then evaluated without any postprocessing and then with the convolutional CRF postprocessing technique.

3. MAIN RESULTS AND DISCUSSION

The following tables contain the metrics for all the conducted experiments. The highest scores are highlighted in bold.

Table 2. MEAN IOU FOR MODELS TRAINED ON IMAGE SIZE OF 256*256

Encoder	Mean IOU	Mean IOU-ConvCRF
ResNet18	0.684	0.773
ResNet34	0.678	0.792
ResNet50	**0.716**	0.792
ResNet101	0.712	**0.800**

Table 3. GLOBAL ACCURACY FOR MODELS TRAINED ON IMAGE SIZE OF 256*256

Encoder	Global Accuracy	Global Accuracy-ConvCRF
ResNet18	**0.849**	**0.853**
ResNet34	0.843	0.846
ResNet50	0.837	0.835
ResNet101	0.844	0.843

Table 4. MEAN IOU FOR MODELS TRAINED ON IMAGE SIZE OF 512*512

Encoder	Mean IOU	Mean IOU-ConvCRF
ResNet18	0.634	**0.735**
ResNet34	0.635	0.728
ResNet50	**0.654**	0.732
ResNet101	0.632	0.715

Table 5. GLOBAL ACCURACY FOR MODELS TRAINED ON IMAGE SIZE OF 512*512

Encoder	Global Accuracy	Global Accuracy-ConvCRF
ResNet18	**0.822**	**0.826**
ResNet34	0.809	0.813
ResNet50	0.803	0.804
ResNet101	0.815	0.82

If we compare in terms of the main evaluation metric, mean IOU, without any postprocessing, the model with ResNet50 as the backbone performs the best on both image sizes (i.e., 256*256 and 512*512) with mean IOU values of 0.716 and 0.654, respectively. Figure 5 shows the graph of mean IOU values for various image sizes coupled with ConvCRF postprocessing for the models with ResNet18, ResNet34, ResNet50 and ResNet101 as the encoders. In the case of images with size 256*256 coupled with ConvCRF postprocessing, a general increase in the mean IOU value can be seen as we move from the model with ResNet18 as the encoder to the models with deeper encoders such as ResNet101. However, the opposite trend is observed in the case of larger images (512*512), where the mean IOU value shows a general decrease as we move toward models with deeper encoders. For the case of images without any postprocessing, the model with ResNet50 as the encoder outperforms the other three variants in the case of both image sizes. If we look at the values of the accuracy, shallower models (models with ResNet18 and ResNet34 as the backbones) perform better than the deeper models. The performance varies by approximately ± (0.5 – 1) % over multiple runs of the experiments.

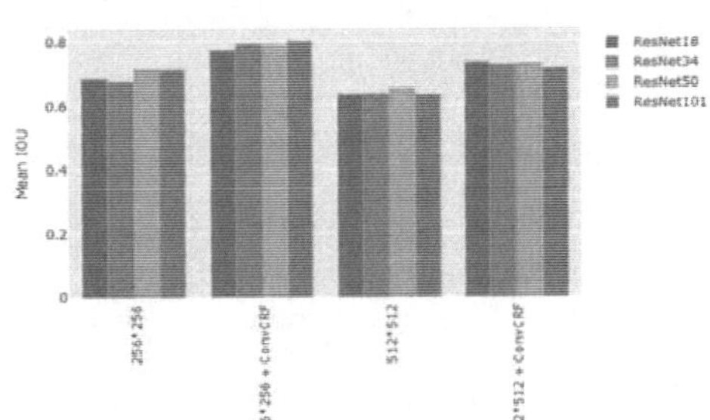

Figure 5. MEAN IOU VALUES FOR VARIOUS ENCODERS AND IMAGE SIZES

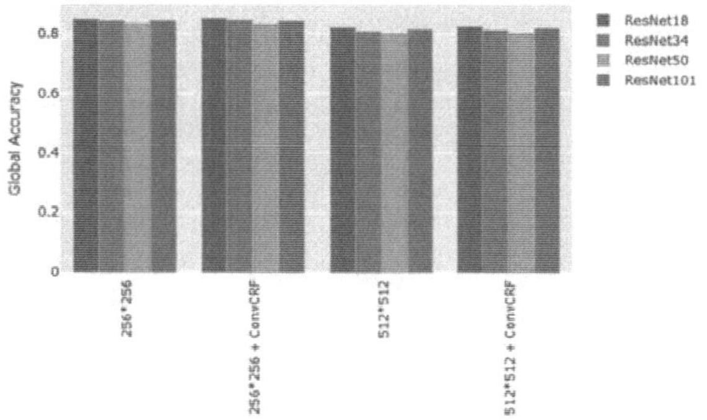

Figure 6. GLOBAL ACCURACY VALUES FOR VARIOUS ENCODERS AND IMAGE SIZES

The numbers in Table 6 show that there is a significant increase (gain) in the mean IOU values due to the CRF postprocessing. This increase depends on the image size and the backbone architecture. Figure 7 depicts the gain values for various encoders and image sizes for comparison. The gain was highest in ResNet34 as the encoder and lowest in ResNet101 as the encoder. This is true for both image sizes. The increase in the value of the mean IOU was higher in the case of smaller images (256*256) for all the backbones.

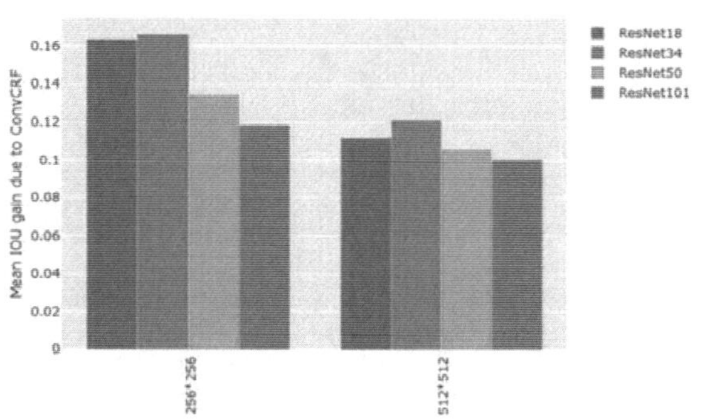

Figure 7. GAIN IN IN IN MEAN IOU VALUES DUE TO CONVCRF-BASED POSTPROCESSING

Table 6 compares the two postprocessing techniques. This comparison was performed with ResNet34 as the backbone and with an image size of 256*256. The gain values were calculated by averaging the gain over 40 epochs.

Table 6. GAIN IN METRICS DUE TO POSTPROCESSING

Metric-Postprocessing	Gain
Mean IOU-ConvCRF	0.186
Mean IOU-FullyConnCRF	0.107
Global Accuracy-ConvCRF	0.009
Global Accuracy-FullyConnCRF	0.009

We found that the convolutional CRF works better than the fully connected CRF. It also took approximately 40 seconds less than the fully connected CRF for 37 images of size 256*256. The gains in the global accuracy values were very small. This also highlights the importance of choosing mean IOU instead of global accuracy because, as shown in Fig. 4, the postprocessed predictions are cleaner and less noisy than the nonprocessed predictions, but the accuracy value does not vary substantially. Therefore, if we considered only global accuracy as the evaluation metric, it would not have been possible to quantify the effect of the CRF postprocessing. Figure 8 contains the values of the mean IOU over 40 epochs of training for the model with ResNet34 as the encoder. The mean IOU value for ConvCRF starts much higher compared to the other values and stays higher than the mean IOU value without postprocessing and the mean IOU value after postprocessing with FullyConvCRF.

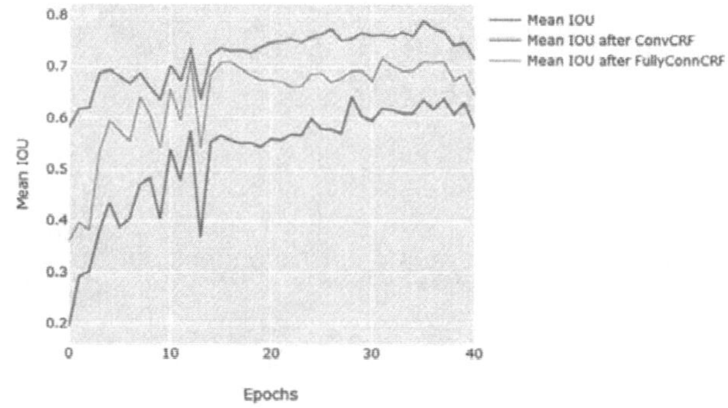

Figure 8. MEAN IOU VALUE FOR 30 EPOCHS OF TRAINING

The model with ResNet101 as the backbone in combination with convolutional CRF-based postprocessing performed the best, with a mean IOU value of 0.8. Figure 9 contains a few qualitative results from the best performing model.

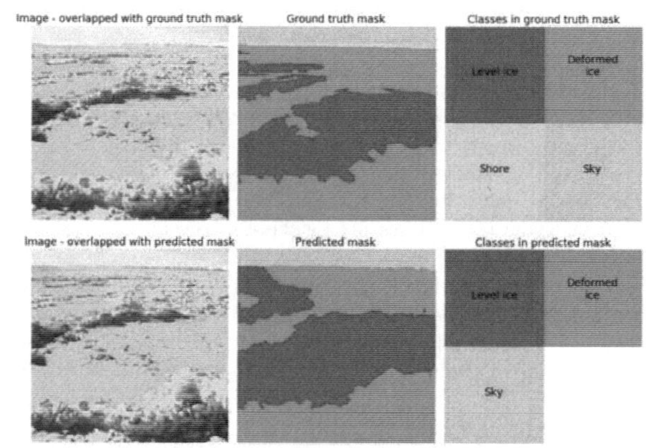

(a)

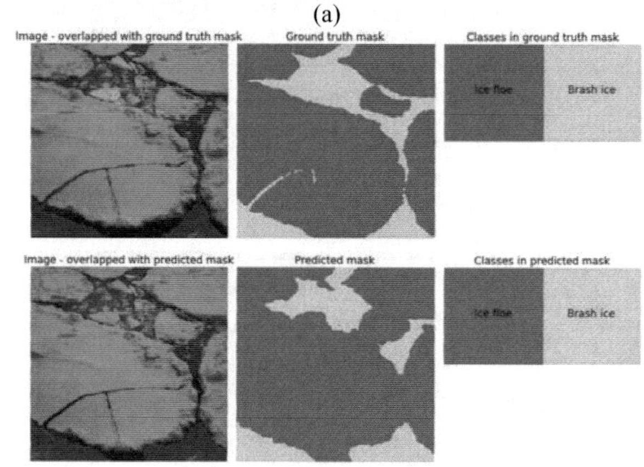

(b)

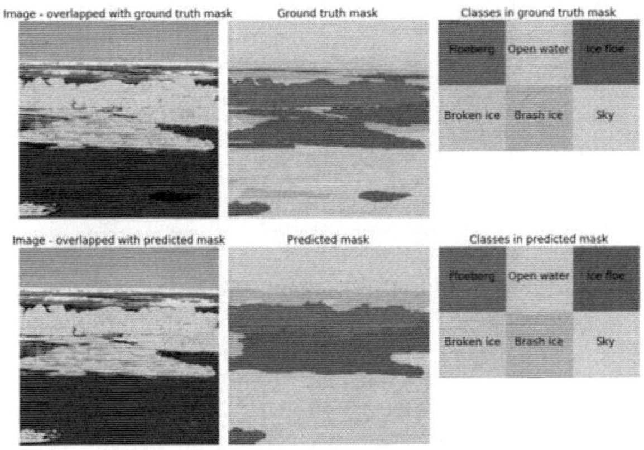

(c)

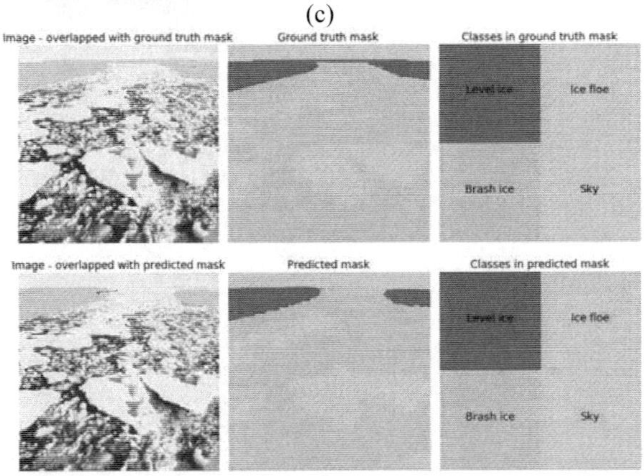

(d)

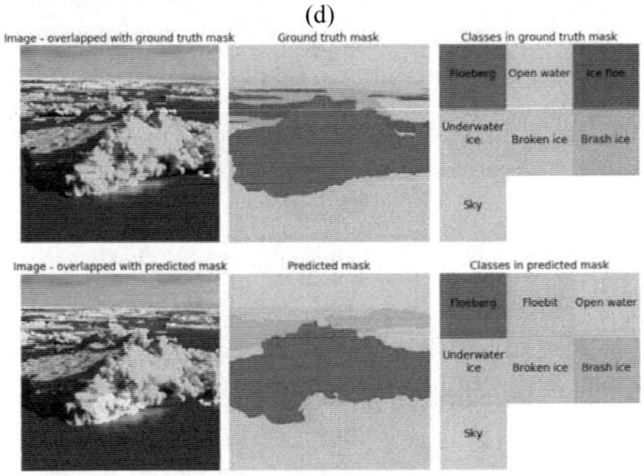

(e)

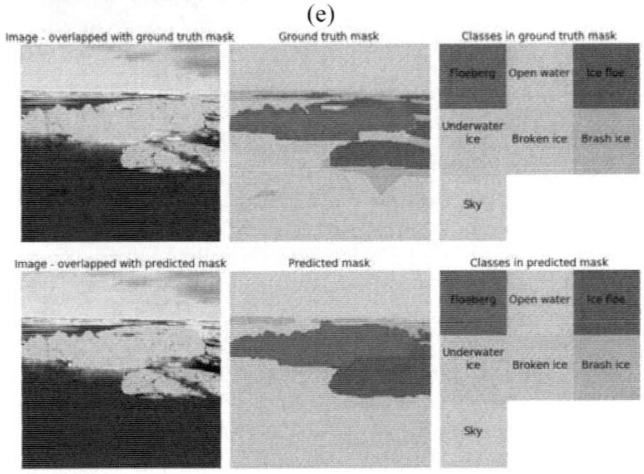

(f)

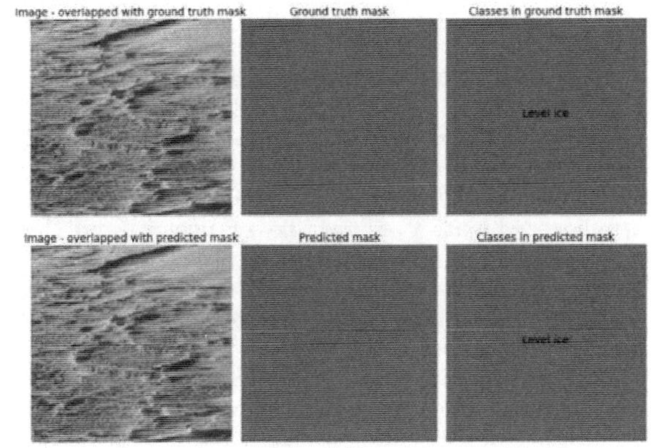

(g)

Figure 9. PREDICTIONS FROM THE MODEL

We presented the best and the worst results from the best performing model (ResNet101+ConvCRF image size - 256*256) in Fig. 10. The best result refers to a result where the number of misclassified pixels in the prediction is lowest among all the images in the validation set, and the worst result refers to a result where the number of misclassified pixels in the prediction is highest among all the images in the validation set. The image corresponding to the best prediction is a close-up shot of deformed ice, and the model performs well on that. The worst result is on the image that contains only pancake ice and sky, but the model predicts it as deformed ice, level ice, and brash ice.

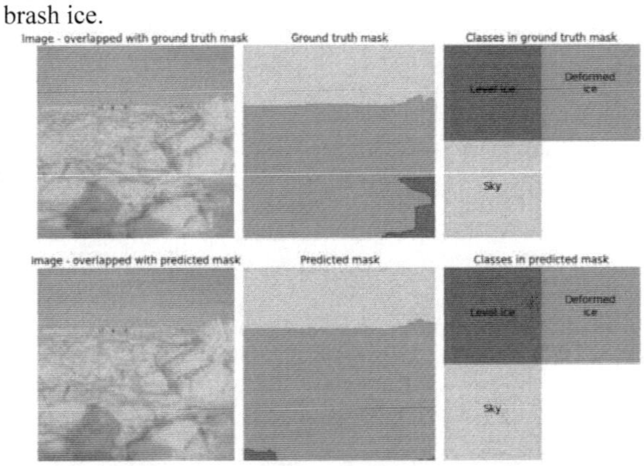

(a) Best

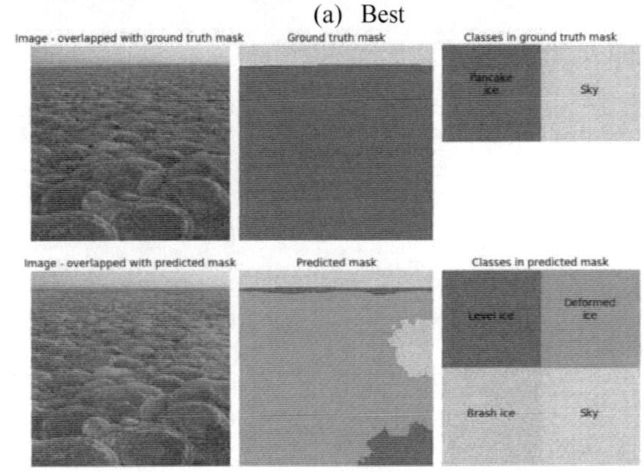

(b) Worst

Figure 10. BEST AND WORST PREDICTIONS

Table 7 contains the individual accuracy for all the classes. The individual accuracy for a class is calculated by dividing the total number of pixels that are correctly classified by the total number of pixels in that class. Sky is most accurately predicted. This is understandable because sky is present in more than 300 images out of 375 images (refer to Fig. 11 for the classwise distribution of images). Additionally, the sky appears in a fixed portion of the image (top part of the image), it is generally textureless, and it has a very light sky-blue color. All of these features are mostly consistent across the dataset, and hence, one would assume that sky must be the easiest to predict. The next five classes, level ice, deformed ice, ice floe, brash ice and open water, have an individual accuracy of more than 75%. All 5 classes are present in more than 100 images. The next 3 classes, iceberg, pancake ice and underwater ice, are predicted with decent accuracy. Although pancake ice is present only in approximately ~ 20 – 25 images, it is predicted with decent accuracy of 52%. We think that this may be due to the distinctive shape of the pancake ice in the available dataset. For underwater ice, the model predicts 32% (refer to the confusion matrix in Fig. 13 for the confusion scores) of the pixels pertaining to underwater ice as brash ice, which can be attributed to the fact that many images have underwater ice visible but covered by the brash ice on the surface. A significant number of pixels (31%) of shores are misclassified as sky. This may be because the shore and sky are very close and, in many cases, the shore is visible as a small strip on the horizon, so it is classified as part of the sky. The last two classes (sorted according to the individual accuracy) (floeberg and melt pond) have zero prediction accuracy. Most (70%) of the floeberg is classified as iceberg, which can be attributed to two possible reasons: floeberg lies in isolation, such as icebergs, and can be of similar dimensions (heightwise); thus, this might contribute to the confusion. The second reason could be due to the imbalance of training samples across classes. Figure 11 presents the number of images containing the corresponding classes. The images containing pancake ice, floebits floebergs, and melted ponds are very small in number. This imbalance leads to a deep learning model that is very good at recognizing features that are abundant in the dataset and poor at recognizing features that are less abundant in the dataset. This also impacts the choice of the evaluation metrics because accuracy can be easily swayed by the abundant class, and this might give the impression that the model is very good, while in reality, the model performs poorly on some of the classes. A more detailed analysis of the performance of the model is planned as a part of future work.

Table 7. INDIVIDUAL ACCURACY FOR ALL THE CLASSES

Class	Accuracy-ConvCRF
Sky	0.993
Level ice	0.896
Deformed ice	0.861
Ice floe	0.828
Brash ice	0.785
Open water	0.779
Iceberg	0.661
Pancake ice	0.622
Underwater ice	0.559
Shore	0.554
Broken ice	0.527
Floeberg	0.255
Floebit	0
Melt pond	0

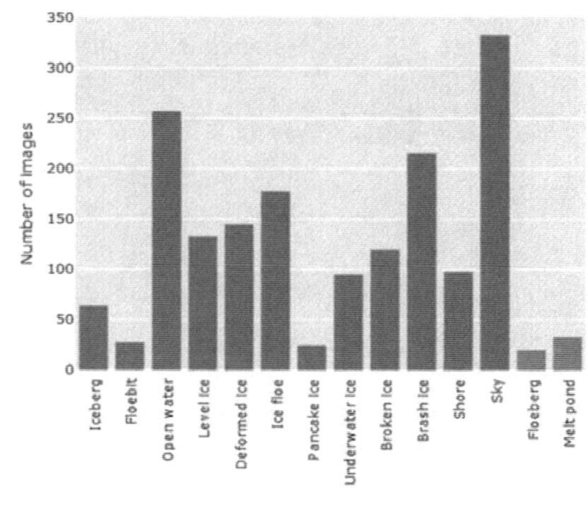

Figure 11. CLASSWISE DISTRIBUTION OF IMAGES; RED COLOR INDICATES THE CLASSES HAVING A PREDICTION ACCURACY OF ZERO

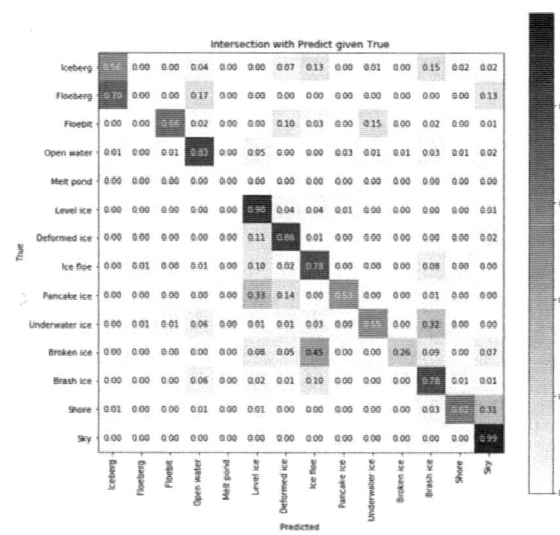

Figure 12. CONFUSION MATRIX

Figure 13 presents the classwise accuracy and the gain in classwise accuracy due to the postprocessing for all the encoders. It can be seen in Fig. 13a that the accuracies for the floeberg and the melt pond are zero for all the encoders. From Fig. 13b, we can see that the postprocessing affects half the classes negatively, especially the shore, where the postprocessing decreases the accuracy by as high as 10%. This is because in many images from the dataset, the shore is visible as a very small strip near the horizon, and the postprocessing considers that as a part of the sky. After postprocessing, the individual floes that are close together are combined to form a giant floe, and the brash ice in between the floes is misclassified as a part of the giant floe, causing the loss in accuracy due to the postprocessing in the case of brash ice. These shortcomings of postprocessing can be addressed to an extent by changing the parameters of the postprocessing techniques. An exhaustive search for the best parameters of the postprocessing techniques is planned as a part of future work.

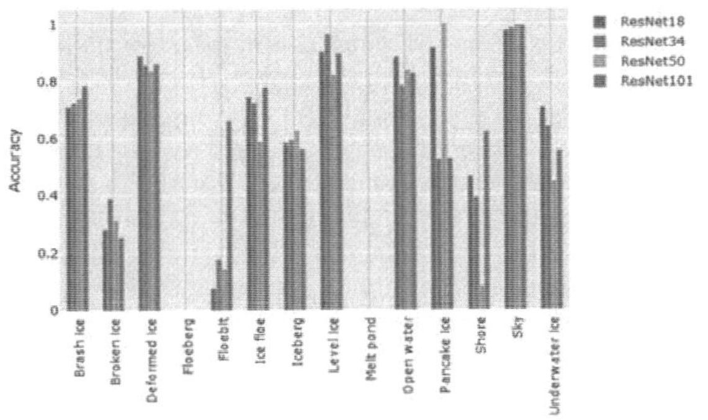

(a) Accuracy

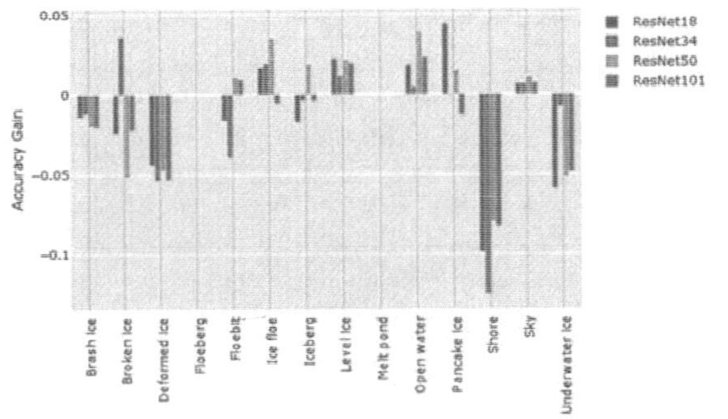

(b) Gain or loss in accuracy due to postprocessing using ConvCRF

Figure 13. INDIVIDUAL ACCURACY PLOTS FOR THE IMAGE SIZE 256*256

4. CONCLUSION

The objective of this study was to find a model that can accurately classify and locate the following surface ice features on optical images: *icebergs, deformed ice, level ice, broken ice, ice floes, floebergs, floebits, pancake ice*, and *brash ice*. First, we gathered and manually labeled optical ice images (a total of 375 images) that were later used to train and validate the model, i.e., the U-Net model with the ResNet architecture as the downsampling part. The model performance was evaluated in terms of IOU and accuracy. Different model configurations were analyzed and compared, including ResNet architectures (ResNet18, ResNet34, ResNet50, ResNet50), image sizes (256*256 and 512*512), and postprocessing techniques (fully conv CRF and conv CRF). All of the studied deep learning models were validated on a validation set of images containing 10% of the total labeled data. The best model was obtained through a comparison of the performance among several candidates. The preliminary results of this study can be summarized as follows and pertain specifically to the considered dataset:

- The model with ResNet101 as the backbone in combination with convolutional CRF-based postprocessing performed the best, with a mean IOU value of 0.8.
- For all of the tested models, both the IOU and the accuracy were slightly higher for images of size 256*256 than for 512*512.
- All of the tested models have limitations in detecting features that are scarce in the dataset, such as floebits, floebergs and melt ponds.

The latter limitations can be addressed by adding more image samples containing these ice features and thereby making a more balanced dataset. Other model limitations include the adverse effects of postprocessing on some of the classes, such as brash ice and shore. These can be addressed by exhaustively identifying the best parameters for the postprocessing methods, which is planned as a part of future work. Most of the images in the considered dataset were taken in broad daylight conditions with good visibility. More images in adverse weather conditions with heavy snow, rain, and poor lighting conditions are required to make the deep learning model more robust with respect to its performance in these conditions.

The presented study is a first step toward an automated ice feature detection system that can aid and complement the ice navigator tasks. The presented preliminary results serve as a foundation for automated identification of ice features using optical images. In the future, we plan to demonstrate the usefulness of the presented approach for understanding and characterizing polar environments. We hope that the presented findings can stimulate and support the development of ice navigation support systems.

ACKNOWLEDGMENT

The following table consists of the sources of the images:

Table 8. SOURCES OF IMAGES IN THIS PAPER

Image	Source
Fig. 10b	Igor Podgorny/LiveJournal
All other images	Sveinung Løset

We wish to acknowledge Professor Sveinung Løset for sharing his photographs from the cruise to the Fram Strait. We also wish to acknowledge UNINETT Sigma2 AS, the national infrastructure for computational science in Norway, for granting access to their data storage and processing resources.

REFERENCES

[1] Snider, D., 2018. *Polar ship operations: A practical guide*. Nautical Institute.

[2] Muramoto, K.I., Matsuura, K. and Endoh, T., 1993. "Measuring sea-ice concentration and floe-size distribution by image processing". Annals of Glaciology, 18, pp.33-38.

[3] Hall, R.J., Hughes, N. and Wadhams, P., 2002. "A systematic method of obtaining ice concentration measurements from ship-based observations". Cold Regions Science and Technology, 34(2), pp.97-102.

[4] Lu, P. and Li, Z., 2010. "A method of obtaining ice concentration and floe size from shipboard oblique sea ice images". IEEE Transactions on geoscience and remote sensing, 48(7), pp.2771-2780.

[5] Ji, S., Li, H., Wang, A. and Yue, Q., 2011. "Digital image techniques of sea ice field observation in the bohai sea". In Proceedings of the International Conference on Port and Ocean Engineering Under Arctic Conditions (No. POAC11-077).

[6] Zhang, Q. and Skjetne, R., 2014. "Image processing for identification of sea-ice floes and the floe size distributions". IEEE Transactions on geoscience and remote sensing, 53(5), pp.2913-2924.

[7] Lu, W., Zhang, Q., Lubbad, R., Løset, S. and Skjetne, R., 2016, October. "A shipborne measurement system to acquire sea ice thickness and concentration at engineering scale". In Arctic Technology Conference. Offshore Technology Conference.

[8] Heyn, H.M., Knoche, M., Zhang, Q. and Skjetne, R., 2017, September. "A system for automated vision-based sea-ice concentration detection and floe-size distribution indication from an icebreaker". In ASME 2017 36th International Conference on Ocean, Offshore and Arctic Engineering. American Society of Mechanical Engineers Digital Collection.

[9] Kjerstad, Ø.K., Løset, S., Skjetne, R. and Skarbø, R.A., 2018. "An Ice-Drift Estimation Algorithm Using Radar and Ship Motion Measurements". IEEE Transactions on Geoscience and Remote Sensing, 56(6), pp.3007-3019.

[10] Singh, A., Kalke, H., Ray, N. and Loewen, M., 2019. "River Ice Segmentation with Deep Learning". arXiv preprint arXiv:1901.04412.

[11] Kim, E., Dahiya, G.S., Løset, S. and Skjetne, R., 2019. "Can a computer see what an ice expert sees? Multilabel ice objects classification with convolutional neural networks". Results in Engineering, p.100036.

[12] Sun, K., Zhao, Y., Jiang, B., Cheng, T., Xiao, B., Liu, D., Mu, Y., Wang, X., Liu, W. and Wang, J., 2019. "High-Resolution Representations for Labeling Pixels and Regions". arXiv preprint arXiv:1904.04514.

[13] Zhu, Y., Sapra, K., Reda, F.A., Shih, K.J., Newsam, S., Tao, A. and Catanzaro, B., 2019. "Improving Semantic Segmentation via Video Propagation and Label Relaxation". In Proceedings of the IEEE Conference on Computer Vision and Pattern Recognition (pp. 8856-8865).

[14] Krizhevsky, A., Sutskever, I. and Hinton, G.E., 2012. "Imagenet classification with deep convolutional neural networks". In Advances in neural information processing systems (pp. 1097-1105).

[15] He, K., Zhang, X., Ren, S. and Sun, J., 2016. "Deep residual learning for image recognition". In Proceedings of the IEEE conference on computer vision and pattern recognition (pp. 770-778).

[16] Ren, S., He, K., Girshick, R. and Sun, J., 2015. "Faster r-cnn: Towards real-time object detection with region proposal networks". In Advances in neural information processing systems (pp. 91-99).

[17] Chen, L.C., Zhu, Y., Papandreou, G., Schroff, F. and Adam, H., 2018. "Encoder-decoder with atrous separable convolution for semantic image segmentation". In Proceedings of the European conference on computer vision (ECCV) (pp. 801-818).

[18] Touvron, H., Vedaldi, A., Douze, M. and Jégou, H., 2019. "Fixing the train-test resolution discrepancy". arXiv preprint arXiv:1906.06423.

[19] World Meteorological Organization, 1970. *WMO Sea-ice Nomenclature: Terminology, Codes, Illustrated Glossary and Symbols*. Secretariat of the World Meteorological Organization.

[20] S. Kim, 2019. label-tool, On the www.github.com. https://github.com/Slava/label-tool.

[21] Ronneberger, O., Fischer, P. and Brox, T., 2015, October. "U-net: Convolutional networks for biomedical image segmentation". In International Conference on Medical image computing and computer-assisted intervention (pp. 234-241). Springer, Cham.

[22] Deng, J., Dong, W., Socher, R., Li, L.J., Li, K. and Fei-Fei, L., 2009, June. "Imagenet: A large-scale hierarchical image database". In 2009 IEEE conference on computer vision and pattern recognition (pp. 248-255). IEEE.

[23] Zeiler, M.D. and Fergus, R., 2014, September. "Visualizing and understanding convolutional networks". In European conference on computer vision (pp. 818-833). Springer, Cham.

[24] Smith, L.N., 2018. "A disciplined approach to neural network hyper-parameters: Part 1--learning rate, batch size, momentum, and weight decay". arXiv preprint arXiv:1803.09820.

[25] Howard, Jeremy and others, 2018. Fastai v1, on the www.github.com. https://github.com/fastai/fastai.

[26] Paszke, A., Gross, S., Chintala, S., Chanan, G., Yang, E., DeVito, Z., Lin, Z., Desmaison, A., Antiga, L. and Lerer, A., 2017. "Automatic differentiation in pytorch".

[27] Krähenbühl, P. and Koltun, V., 2011. "Efficient inference in fully connected crfs with gaussian edge potentials". In Advances in neural information processing systems (pp. 109-117).

[28] Teichmann, M.T. and Cipolla, R., 2018. "Convolutional CRFs for semantic segmentation". arXiv preprint arXiv:1805.04777.

APPENDIX

Labeling

Hand-labeling of the dataset was challenging. The main challenges and the rules that were adopted to address them are summarized in Table A1 and pertain specifically to the considered dataset. Because a large portion of the dataset had these challenges, the rules in Table A1 were central to the labeling process and affected the model training and performance. Thus, further refinement of the labels is planned as a part of future work.

Table A1. LABELING CHALLENGES AND THEIR ADOPTED RULES/SOLUTIONS

Challenge	Solutions
1) It is hard to distinguish between a floebit, floeberg and an iceberg when these features are near the horizon as it is very hard to determine the size and the color of the feature.	1) If we were not absolutely certain of the ice feature on the horizon, it was labeled as a part of an iceberg. This was done considering that icebergs pose a more severe challenge than other ice features; thus, labeling it as an iceberg ensures that we err on the side of caution.
2) The images do not have any metainformation such as the height (from sea-level) and angle of the camera lens while taking the picture. This creates a problem in determining the size of the ice features, and size is an important factor in determining the class of the feature.	2) Other features in the image were analyzed to determine the size of the feature relatively.
3) Since most of the images were taken from onboard the vessels, some images have a part of the vessel in the frame.	3) These images were left out of the training dataset. It is understood that in a real-scenario, when a camera is mounted on the bridge of the vessel, there will be some part of the vessel visible in the image, but as this camera is going to be fixed, we would know exactly which part of the image is the ship, and that part can be cropped out before passing the image through the model for segmentation.
4) Labeling brash ice is difficult when it is truly scattered and exists in a very low concentration; thus, some part of open water is also labeled as brash ice.	4) The brash ice with lower concentration was very precisely labeled to ensure that open water was not labeled as a part of the brash ice.
5) While labeling an image with ice floes and broken ice, the floes that are present near the camera are separately labeled, and as we move further from the camera, it becomes hard to distinguish one floe from another.	5) The ice floes were labeled separately for the part of the image where the boundaries of the floes were visible. Beyond that, all floes were labeled together as broken ice.
6) Underwater parts of various ice features are also visible in the images; all of them are labeled as underwater ice.	6) Ideally, underwater parts of different ice features should have been labeled separately, but the dataset did not have enough images, and there would have been classes with only approximately 5-10 images if this type of distinction was followed.

Proceedings of the 6th International Conference on Ship and Offshore Technology
ICSOT 2019
November 7-8, 2019, IIT Kharagpur, India

ICSOT INDIA 2019-26

MITIGATION OF UNDERWATER EXPLOSION INDUCED SHOCK LOADING ON SANDWICH COMPOSITE WITH BIOMIMETIC CORE

Akurati Prabhakar
Department of Ocean Engineering
and Naval Architecture
Indian Institute of Technology Kharagpur
Kharagpur 721302, India
Email: prabhakar993@iitkgp.ac.in

Ritwik Ghoshal
Department of Ocean Engineering
and Naval Architecture
Indian Institute of Technology Kharagpur
Kharagpur 721302, India
Email: ritwik@iitkgp.ac.in

ABSTRACT

This paper investigates the dynamic response of light-weight bio-inspired sandwich composite panel subjected to underwater explosion induced shock loading. It is well known that bio-inspired systems balances the weight to strength ratio and also provides higher stiffness. Therefore, a sandwich composite core with branched stiffeners inspired from Victoria Amazonica (lily-pad) is proposed and investigated. The geometric pattern of cross-ribs on its underside acts as slender diagonal rods which provides lateral stiffness and high rigidity to the structure. Typically, the response of sandwich composite panel subjected to shock loading can be split into three different stages, stage I: fluid-structure interaction effect on face sheet; stage II: core material crushing; stage III: sandwich composite undergoes plastic bending and stretching. The prime focus of this work is to study the stage II, i.e., core crushing phenomenon with bio-inspired core. In this regard, a finite element based decoupled fluid-structure interaction model is developed using ABAQUS. Shock loading of the structure is estimated considering non-linear compressible water medium at both front and back side of the plate. A stress-strain curve idealised as rigid-perfectly-plastic-locking (R-P-P-L) is utilized to model the core compression. The initial response is assumed to be rigid, which is followed by propagation of plastic shock with in the core material that causes a sudden jump in compressive normal strain. Mitigation performance of the biomimetic core is investigated in term of arresting equivalent plastic strain using face sheet and core material as a sacrificial material. A comparative study on structural response of the back face-sheet, plastic shock arrest position is performed against the standard honeycomb core available in literature. These results can be useful in designing blast resistance sandwich composite panel for naval ships.

I INTRODUCTION

G.I.Taylor [1] demonstrated with closed form solution that on considering fluid-structure interaction effects reduces impulse transmission to the light plates subjected to shock loading due to underwater explosion. However, rigid plate with constant back pressure, low blast intensities, acoustic medium, and negligible compressibility effects assumptions has been restricted the applicability of Taylor's theory to submerged plates having varying back pressure due to hydrostatic pressure from the water. He also introduced a non-dimensional parameter which relates time scales of the duration of shock loading due to underwater explosion to fluid-structure interaction(FSI). Subsequently He showed that total reflection of blast wave occurs when the plate reaches infinite mass which acts as a fixed boundary. And, when the plate reaches negligible mass, the plate readily accelerates thus by decreasing reflected pressure wave.

Z.Liu and Y.Young(2008) [2] reformulated Taylor's theory to submerged free-standing plate with water back pressure and concluded that impulse transmission to plate has been reduced when compared with equivalent ABP within the acoustic range for both front and back side of the plate. Further, he explained that due to presence of varying back pressure and fluid on the back side of the plate results in formation of shock wave which relives the pressure on the front side plate results in receding movement of the plate thereby further reducing impulse transmission for light plates. Non-linear compressibility effects has not been considered in his study.

The particle velocity from an underwater explosion due to TNT (Trinitrotoluene)or RDX (Research Department Explosive) explosives are in the range of 3-7km/s. S.Ridah(1988) [3] proposed that non-linear compressibility effects in water has to be considered with particle velocity in the range of around 1 km/s.

Kambouchev *et al* [4] extended Taylor's theory to non-linear compressibility range for both very heavy and very light plates with help of numerical method based on Lagrangian formulation of the euler equations of compressible flow and shock capturing techniques. further, proposed modified non-dimensional parameter for Fluid-structure interaction(FSI) time-scale which depends on ratio of plate mass and product of peak over pressure, shock speed, and decay time. He concluded that the effect of non-linear compressibility enhances the mitigation of transmitted impulse for both exponential and uniform shock profile subjected to full range of plate masses.

W.Peng [5] develops an FSI model by considering resistance caused by the shock wave formation at the back of the free-standing plate subjected to blast wave in non-linear compressible air medium. Results from this FSI model shows that resistance increases for lighter plates and concluded that the effect of formation of a shock wave at the back side of the plate has to be considered while designing the blast resistance structures.

R.Ghoshal *et al.* [6] demonstrated the effect of non-linear compressible water medium for both front and back side of the plate. His analytical theory shows there is a significant reduction in impulse transmission to the free-standing plate and concludes that FSI effect is more predominant for VBP (varrying back pressure) case in comparison with CBP (constant back pressure). Further, impulse design transmission curves have been developed for both uniform and exponential shock wave profile for intermediate plate masses subjected to different ranges of shock intensities.

According to the literature found on the modelling of core crushing of sandwich panel researchers are using Rigid perfectly plastic locking (R-P-P-L) to idealise the stress stain curve for the crushable core material. Reid and Peng [7] studied the crushing dynamics of wood and commented that this idealization can be used to study the crushing of different material like aluminium foams [8] and lattice cores [9] [10] subjected to shock loading.

Deshpande and Fleck [11] studied the response of composite panel under shock loading using Taylor's FSI theory and R-P-P-L model for core-crushing and reported that plastic-locking model can be adopted to model compressible cores and suggested that weak sandwich cores enhances the underwater shock resistance.

Main and Gazonas [12] developed an analytical model for core crushing of sandwich panel under air-blast loading using kambouchev *et al* [4] for FSI theory and reported that alteration of face sheet mass distribution of sandwich composite panel can mitigate shock loading on the structure.

Wang *et al* [13] developed analytical model for FSI and conducted shock-tube experiments on E-glass Vinyl-Ester(EVE) face sheet composite with different foam cores and reported that their analytical model can predict accurate fluid structure interaction time between fluid and sandwich composites.

In the present study, design impulse load has been calculated using Ghoshal and Mitra FSI theory [6] which is applied as time dependent pressure load on the front surface of front face sheet of standard honeycomb sandwich composite ABAQUS model [14] with R-P-P-L model for core crushing phenomenon. The results are validated against existing one dimensional results [15]. A comparative study on structural response of the back face-sheet, plastic shock arrest position is performed against sandwich composite panel with square honeycomb core and proposed biomimetic core. Results obtained can be utilized in designing blast resistance sandwich composite panel for naval ships.

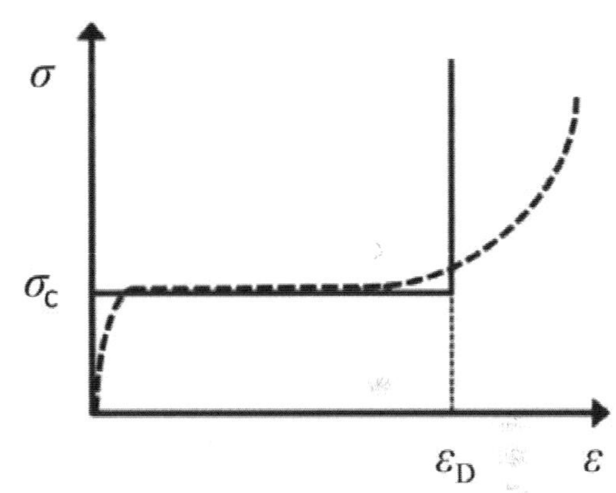

FIG. 1. Idealised stress strain curve

I.I OBJECTIVES AND APPROACH

The primary objective of this manuscript is to develop bioinspired underwater blast resistance sandwich composite panel. To achieve this, authors have proposed sandwich composite core inspired from lilypad(*Victoria Amazonica*) which is having double branch stiffeners to provide high stiffness and also balances the weight to strength ratio.

The outline of the paper is as follows: Summary of Ghoshal and mitra Fluid structure interaction model has been presented in sec. II. Coupled effect of FSI theory with R-P-P-L core crushing model has been summarized in sec. III. Section IV deals with details of FEM based ABAQUS model's and material modelling used in this manuscript. Validation of developed ABAQUS model against already existed results has been presented in sec V. Section VI provides results and discussions comparing the structural response of proposed biomemetic core with standard honeycomb core subjected to underwater explosion induced shock loading. Finally, in sec.VII conclusions are summarised.

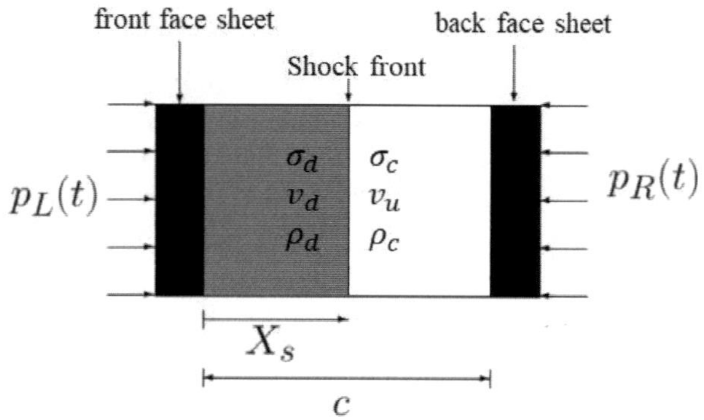

FIG. 2. schematic diagram of shock wave propagation through crushable core.

II FLUID STRUCTURE INTERACTION MODEL

The shock wave from underwater explosion was modeled as an exponentially decaying over-pressure given by

$$p(t) = p_s e^{(-t/t_i)}, \tag{1}$$

The magnitude of peak over-pressure (p_s) in MPa and decay time t_i in milliseconds.

For exact representation of non-linear compressible water medium, we consider equation of state (EOS) of form Mie-Gruneisen (MGEOS). Incident shock parameters i.e., particle velocity (u_p), shock velocity (U_s), and peak density (ρ_s^w) can be obtained by solving linear shock particle ($U_s - u_p$) velocity using Rankine-Hugoniot jump (RHJ) condition along with Mie-Gruneisen (MGEOS) which gives

$$u_p = \frac{-\rho_0^w c_0 + \sqrt{(\rho_0^w c_0)^2 + 4S_1 \rho_0^w p(t)}}{2S_1 \rho_0^w}, \tag{2}$$

$$\rho_s^w = \frac{\rho_0^w U_s}{U_s - u_p}, \tag{3}$$

and

$$U_s = c_0 + S_1 u_p. \tag{4}$$

where c_0 and S_1 are fitting coefficients and ρ_0^w is the initial density of water. Ghosal and Mitra FSI theory can capture shock irrespective of phase transition of water to ice VII at high pressure ranges. For practicality only phase-I, i.e., prior to phase transition, where fitting coefficients are $c_0 = 1450m/s$ and $S_1 = 2.166$ has been used in this manuscript.

Relative impulse transmission, i.e., ratio of transmitted (I_P) and incident impulse (I_i) of a free standing plate subjected to underwater explosion can be obtained by empirical relation [6] as

$$\frac{I_p}{I_i} = \gamma_R \left(\frac{C_R f_R}{\gamma_R}\right)^{\frac{\beta_s}{1+\beta_s}} \beta_s^{\frac{\beta_s}{1-\beta_s}} \tag{5}$$

where C_R is the reflected coefficient for fixed rigid flat plate subjected to constant shock i.e., ratio of reflected to incident shock and γ_R is the relative impulse transmission coefficient of a fixed rigid flat plate subjected to exponential shock(as mass $m_p \to \infty$).

$$\gamma_R = \lim_{m_p \to \infty} \frac{I_p}{I_i} = \frac{\int_0^\infty C_R(t) p(t) dt}{p_s t_i}. \tag{6}$$

The fluid structure interaction parameter (β_s) and non-dimensional parameter (f_R) are given as

$$\beta_s = \frac{\rho_s^w U_s t_i}{m_{FSI}}, \tag{7}$$

$$f_R = \sqrt{\frac{1}{p_S} \frac{(\rho_s^w - \rho_0^w))}{\rho_0^w (C_R - 1)} \rho_s^w U_s^2}, \tag{8}$$

where m_{FSI} is mass involved in fluid structure-interaction. According to available literature [16] pressure history obtained from relative impulse transmission has been applied on the front surface of the front face sheet

$$p_L(t) = C_R p_s e^{-\frac{t}{t_R}}, \tag{9}$$

$$\frac{t_R}{t_i} \equiv \frac{I_p}{C_R p_s t_i}. \tag{10}$$

where decay time $t_i \sim 0.1$ ms and total time $t \sim 0.2$ ms has been used for calculations in this manuscript wherein reflected pressure profile closely matches with pressure history obtained through numerical simulation by Ghoshal and Mitra.

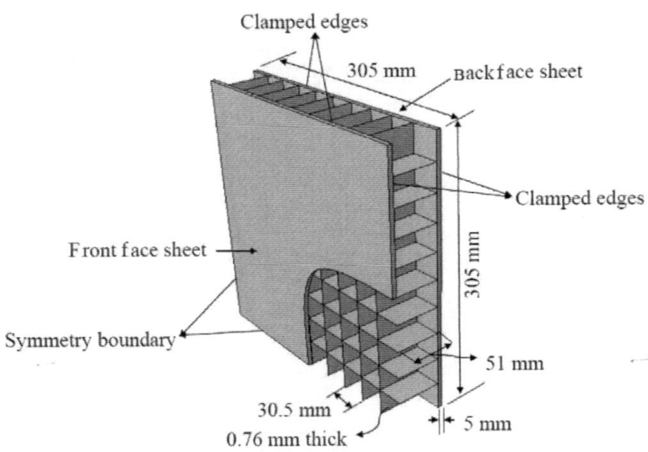

(a)

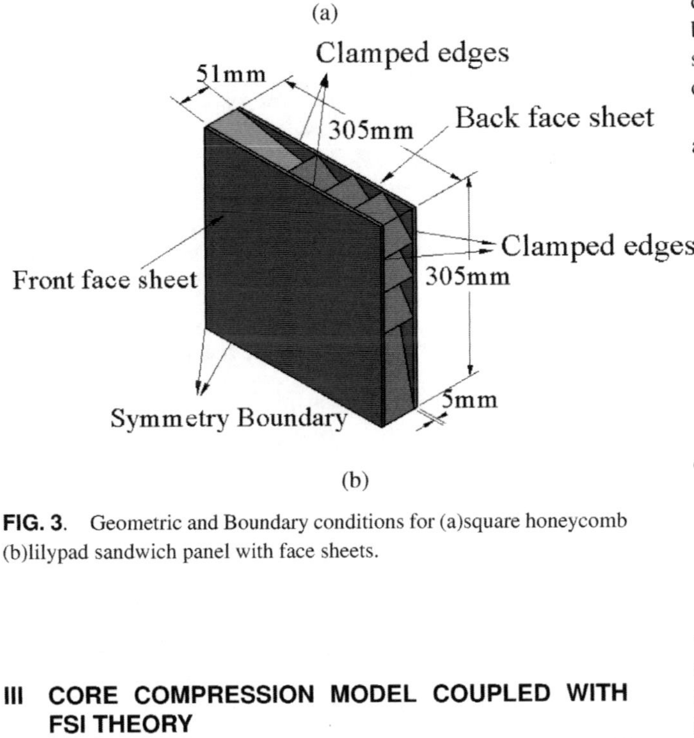

(b)

FIG. 3. Geometric and Boundary conditions for (a)square honeycomb (b)lilypad sandwich panel with face sheets.

III CORE COMPRESSION MODEL COUPLED WITH FSI THEORY

The sandwich panel considered for analysis has two rigid face sheets both front and back with mass m_f per unit area and the crushable core with mass m_c per unit area and thickness c. The behaviour of the core is treated as rigid, perfectly-plastic, locking like material having crushing strength σ_c and densification strain ε_D [11] with idealised stress-strain curve as shown in Fig. 1

$$\sigma_c = \overline{\rho}\sigma_Y, \tag{11}$$

$$\varepsilon_D = 0.8 - 1.75\overline{\rho}. \tag{12}$$

where $\overline{\rho}$ is the relative density of the core which is taken as 5% in this manuscript and σ_Y is the yield strength.

When the shock reaches the front face of the composite sandwich panel it accelerates until it reaches peak velocity while the core and back face sheet are initially at rest. Then a plastic shock wave was initiated in the core and travels through it with the speed of material shock velocity c_{pl}. Since the initial response of the core is rigid the elastic precursor wave has been developed then the stress immediately ahead of the shock front raises and reaches to σ_c. At any instant of time t, distance travelled by the plastic shock front was $X_s(t)$, measured in undeformed configuration as shown in fig. 2. At this instant, the back face sheet and undeformed core upstream from the plastic shock wave share a common velocity v_u and front face sheet and compressed core share a common velocity v_d.

Conservation of mass Eq.(13) and linear momentum Eq.(14) across the shock front gives [17] [18] [11]

$$\dot{X}_s = C_{pl} = \frac{v_d - v_u}{\varepsilon_D}, \tag{13}$$

$$\sigma_d - \sigma_c = \rho_c C_{pl}(v_d - v_u) = \frac{m_c}{C} C_{pl}(v_d - v_u). \tag{14}$$

The acceleration of the undeformed core and back face sheet can be obtained by using Lagrangian equilibrium relation

$$\frac{\partial \sigma}{\partial X} + \rho_i \dot{v} = 0. \tag{15}$$

where X is the position of the shock front, v is the material velocity, ρ_i is the undeformed/initial core density and σ is the cauchy stress which is compressive in nature. Integrating Eq. (15) over the two domains $(0 \leq X \leq X_s)$ contains deformed core, front face sheet gives and $(X_s \leq X \leq C)$ undeformed core, back face sheet separated by a plastic shock front gives

$$p_L(t) - \sigma_d = \left(m_f + m_c \frac{X_s}{C}\right) \dot{v_d}, \tag{16}$$

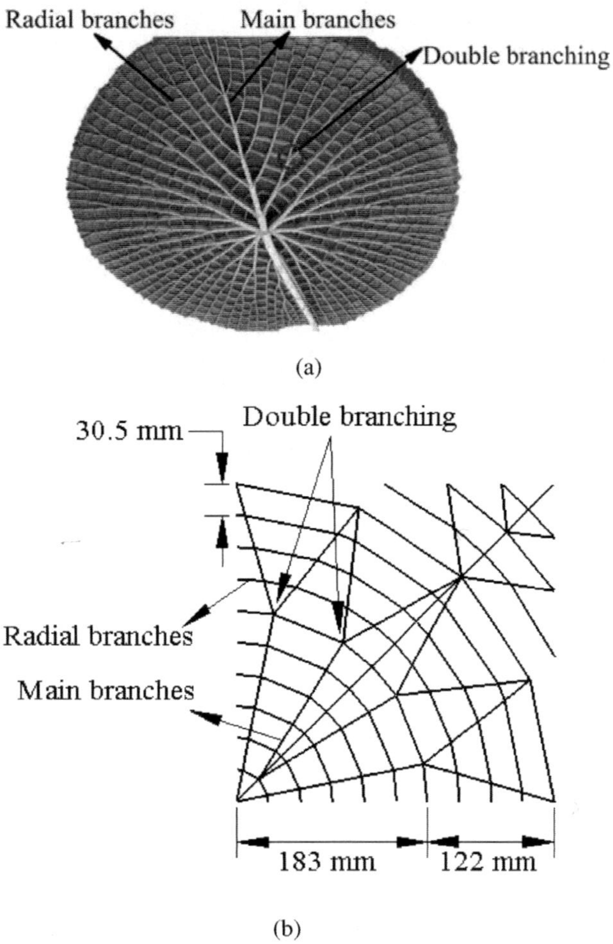

FIG. 4. (a)Underside of Amazon lilypad with double branching (b)Front view of biomimetic Lilypad core

$$\left[m_f + m_c\left(1 - \frac{X_s}{C}\right)\right] = \sigma_c - p_R(t). \tag{17}$$

where $p_L(t)$is the pressure calculated from Ghoshal mitra FSI theory [15] as stated in sec. II and p_R is the pressure acting on the back face sheet by the surrounding water medium. which can be obtained by solving linear shock particle $(U_b - u_b)$ velocity along with RHJ condition between upstream and downstream of the shock produced due to accelerating back face sheet which gives [5] [2]

$$p_R = \rho_0^w U_b u_b. \tag{18}$$

where U_b is shock speed produced due to accelerating back face sheet with a particle velocity (u_b) which is same as that of back face sheet velocity $(u_b = v_u)$. Substituting shock speed U_b and particle velocity (u_b) in Eq.(18) gives

$$p_R = \rho_0^w (c_0 + S_1 v_u) v_u. \tag{19}$$

Plastic shock wave within the core triggers only when relative core strength i.e., the ratio of crushing strength(σ_c) and peak pressure($C_R p_s$) is low. which gives us necessary condition for plastic shock wave to produce within the core as follows

$$\frac{\sigma_c}{C_R p_s} \leq \frac{1+\bar{m}}{2+\bar{m}}, \tag{20}$$

$$\bar{m} = \frac{m_c}{m_f}. \tag{21}$$

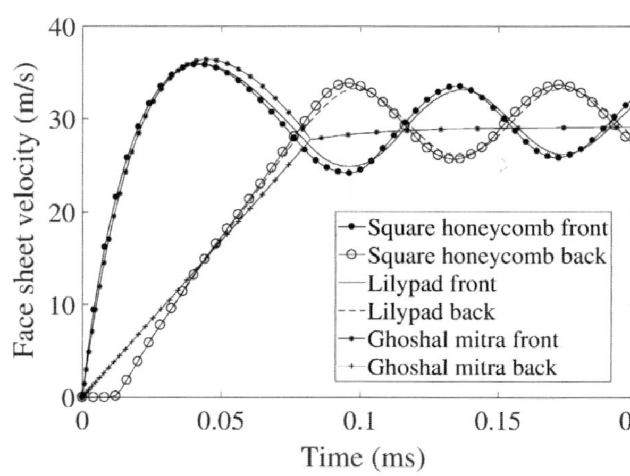

FIG. 5. Amazon lily-pad with double branching.

The differential equations(Eqs.(13),(16),(17)) has been solved simultaneously with the help of ordinary differential equation solver ($ode45$) in MATLAB using necessary condition as stated above along with initial conditions as $t = 0, \dot{v_d} > \dot{v_u}, v_d = v_u = X_s = 0$ and $P_R = 0$.

If the shock wave is arrested within the core i.e., $\dot{X_s} = 0$ then the pressure on the front face sheet changes to

$$p_L(t) = (2m_f + m_c)\dot{v} - p_R, \tag{22}$$

Then at the time of arrest ($t = t_a$) both front and the back face sheet travels with same velocity. Then initial velocity condition for solving Eq. (22) changes to $v = v_d = v_u$.

If the shock reaches the back face sheet i.e., $X_s/C = 1$ then the whole core is under plastic deformation which is undesirable.

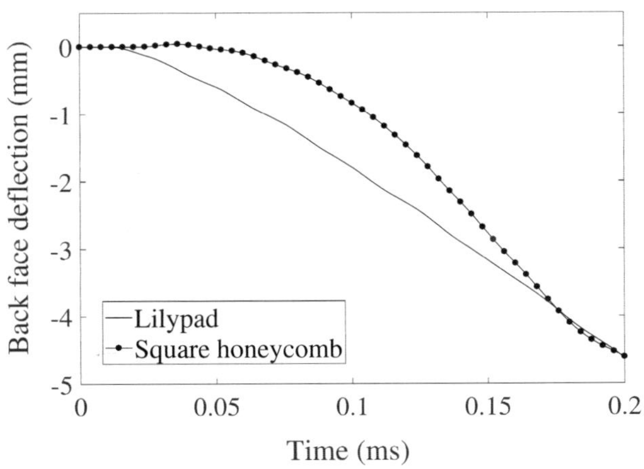

(a)

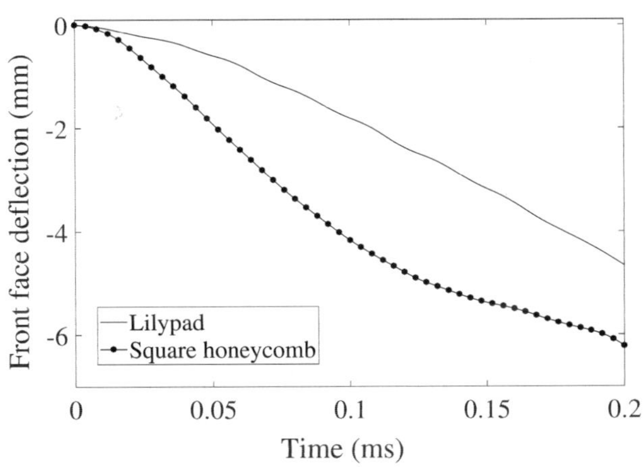

(b)

FIG. 6. Displacement history of (a) back face sheet (b) front face sheet for an incident peak overpressure ratio of 500

IV NUMERICAL SIMULATIONS

This section contains the description of the numerical model of a metallic sandwich with square honeycomb core to study the dynamic response of the sandwich composite plate subjected to underwater explosion induced shock loading coupled with fluid-structure interaction effect. Shock load i.e. the time dependent pressure history, calculated based on the theory described in the section II, is applied on the front surface of the front face sheet of the of the sandwich plate as shown in Fig.3. Assuming symmetry behaviour one quarter of both panels has been modelled using three dimensional continuum elements (C3D8R) for the face sheets and shell elements (S4R) for crushable core. The plate thickness for the both cores are taken as 0.76 *mm*. The spacing between radial branches are 30.5 *mm*. first stage of double branching of the main branches starts from single spacing i.e., 30.5 *mm*. Further second stage of the double branching starts from 183 *mm* ($\sim 6 \times 30.5$ mm). Then the third stage of branching strats from 122 *mm* ($\sim 4 \times 30.5$ mm). In both sandwich panels both front and back face sheets are tied to the core the core with help of tie-constraints.

In this research, the core is modelled with double branching as shown in Fig. 4(b) from the inspiration found in the Amazon lilypad (Fig.4(a)) which gives more strength to weight ratio and does not allow the plastic shock to reach the back face sheet.

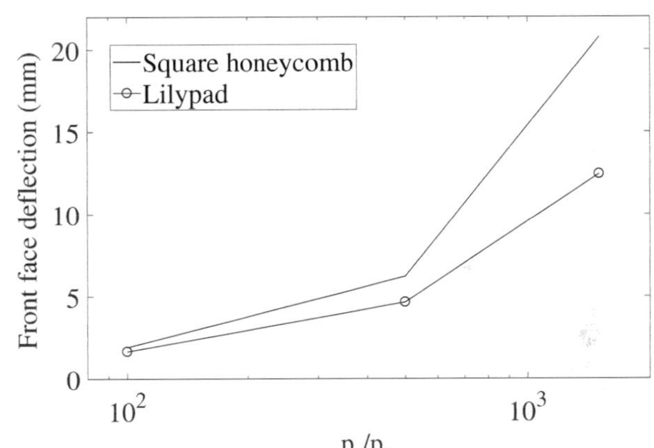

FIG. 7. Displacement at the center of front face sheet of sandwich composite panel

IV.I MATERIAL MODELLING

A RPPL model has been used to model AL6XN core of the both sandwich plates. AL6XN has high strength and toughness when compared to other marine grade stainless steels. The material properties of this material provided in Table. 1. To model behaviour of the plasticity in crushable core the plastic strain at the yield stress i.e., 400 MPa is taken as zero. Initial density of water is taken as 1000 kg/m^3 and atmospheric pressure as 1 bar.

TABLE 1. MATERIAL PROPERTIES OF AL6XN

Young's modulus	$1.61X10^5 MPa$
Poisson's ratio	0.3
Density	7850 kg/m^3
Yield stress	400 MPa

V VALIDATION

The ABAQUS models are validated against available results from the available literature [15]. R.Ghoshal *et al.* solved one dimensional underwater explosion induced shock loading on free standing sandwich composite panels with crushable core. poisson's ratio changed to zero to get one dimensional effect in three dimensional models. For validation peak overpressure ratio is taken as 500. Both face sheets of free standing sandwich composite panel are considered as rigid plates [15]. The face sheet velocities for both sandwich composite panels along with analytical simulation using Ghoshal mitra theory are plotted in Fig. V. Deviations from analytical velocity and time to achieve common velocity are caused due to three dimensional model effects.

VI RESULTS AND DISCUSSIONS

In this section, numerical results of both square honeycomb and lilypad sandwich composite panels subjected to underwater explosion induced shock loading has been provided. In order to study the dynamic responses of the both sandwich panels for peak overpressure ratios p_s/p_0 (ratio of peak over-pressure to atmospheric pressure) i.e., 100, 500 and 1500 the pressure history acting on the front face sheet plate has been calculated using the procedure provided in section II. The boundary conditions for these cases are explained in section. IV and are shown in Fig. 3.

In the current manuscript the C_R and γ_R values for above mentioned peak overpressure ratios are tabulated in table. 2 and the relative density of the core has been considered as 5% for the dimension of the plate provided in fig. 3. The dynamic response of the both Square honey comb and lilypad sandwich panel subjected to different shock intensities has bee compared in this analysis.

Spatial displacement history at the centre of both front and back face sheets for both square honeycomb and lilypad core sandwich composite panel are plotted in Fig. 6. It can be clearly seen that there is a significant reduction in displacement of front face sheet for lilypad core sandwich panel when compared to square honeycomb panel at the end of the time step i.e., 0.2 ms fig. 6(a). However, at the same time instance deflection of back face sheets for both the sandwich panels are same. 6(b)

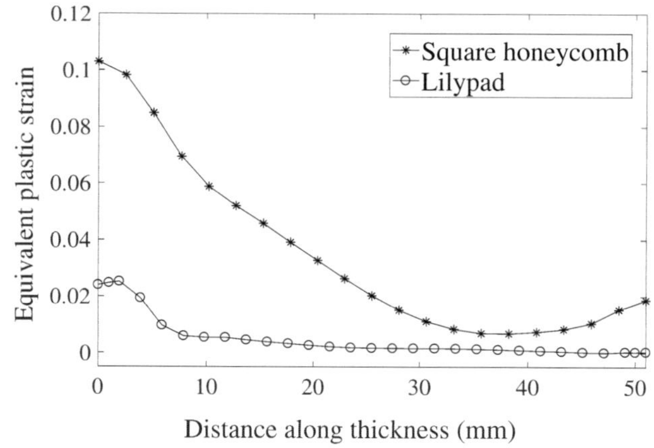

(a)

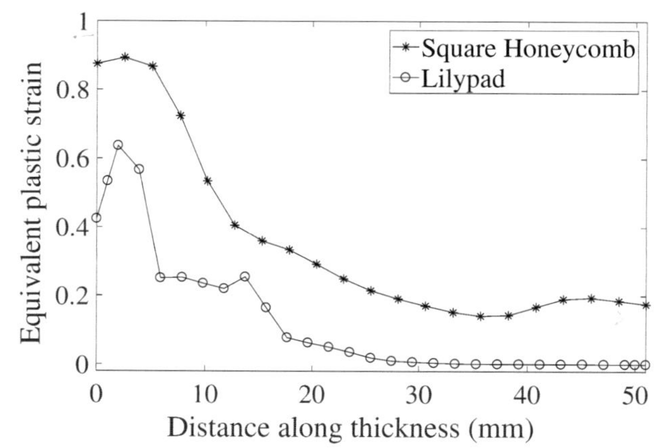

(b)

FIG. 8. Equivalent plastic strain along thickness of the core for peak overpressure ratio (a) 500 (b) 1500

The centre point deflection of front face sheet for both sandwich panels are plotted against peak overpressure ratio's in fig. 7. It clearly shows that at low shock intensities the deflection of front face sheet for both sandwich composite panels are same. But, under high shock intensities i.e., at p_s/p_0 500 front face sheet deflection for square honey comb is 33% more that lilypad core sandwich and at p_s/p_0 1500 it is more than 66% . so, the deflection at the centre point of the front face sheet for lily-

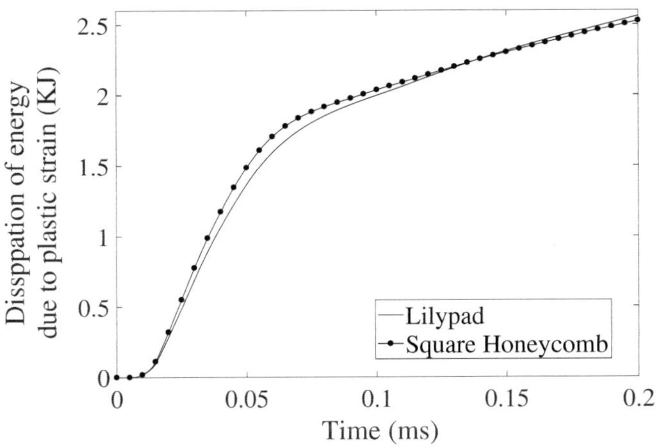

(a)

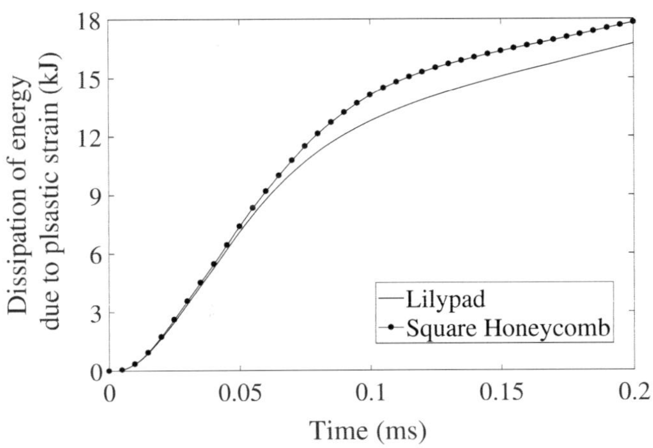

(b)

FIG. 9. Plastic dissipation energy time history for peak overpressure ratio (a) 500 (b) 1500

pad core sandwich panel is very less when compared with square honeycomb sandwich panel subjected to high shock intensities.

In Fig. 8 equivalent plastic strain along through thickness for the both the core are plotted. The core which is nearer to front face sheet experiences more plastic strain when compared to the core from it. This caused due to initiation and propogation of plastic shock wave within the core. lilypad core subjected to peak overpressure of 500 the plastic shock wave arrested around diatance of 10 mm (plastic strain is zero) from the front face sheet but in Square honeycomb sandwich panel significant plastic strain can be observed in the core near to back face sheet. It signifies that some amount of plastic shock wave has reached back face sheet in square honeycomb sandwich panel. Further, under high shock intensity i.e., peak over pressure ratio 1500 the plastic shock wave in lilypad is arrested around 28 mm distance from the front face sheet. But, in case of square honeycomb some amount of plastic shock wave reaches back face sheet. Maximum plastic strain is observed near to the front face sheet. Figure. 8(b). clearly shows that lilypad core is performing better in arresting plastic shock wave propagation within the core.

Plastic energy dissipation of the both sandwich panels against time are plotted in Fig. 9. For lilypad core sandwich panel the dissipation energy closely matches with square honeycomb at low shock intensities i.e., p_s/p_0 500 Fig. 9(a). But at high shock intensities i.e., p_s/p_0 1500 Fig. 9(b) the plastic energy is more for square honeycomb when compared with lilypad core sandwich panel.

TABLE 2. C_R and γ_R values

Overpressure ratio (p_s/p_0)	C_R	γ_R
100	2.023	1.741
500	2.11	1.784
1500	2.29	1.880

VII CONCLUSIONS

From the analysis of comparative study on structural response of both sandwich panels suggests that the biomemtic core sandwich panel performs better than standard square honeycomb sandwich panel. A more detailed study with different geometric parameters like effect of spacial distance between radial branches and stage I double branching distance from the centre of the plate and material parameters like plate thickness of radial and main branches has to perform to fully understand the capability of proposed biomimetic core. A finite element models with failure criteria has to be analysed, which is more important for blast resistance sandwich composite panel under high shock intensities.

REFERENCES

[1] Batchelor, G. K., and Street, R. E., 1961. "The scientific papers of sir geoffrey ingram taylor". *Physics Today*, ***14***(1), pp. 60–62.

[2] Liu, Z., and Young, Y. L., 2008. "Transient Response of Submerged Plates Subject to Underwater Shock Loading: An Analytical Perspective". *Journal of Applied Mechanics*, ***75***(4), 05. 044504.

[3] Ridah, S., 1988. "Shock waves in water". *Journal of Applied Physics*, ***64***(1), pp. 152–158.

[4] Kambouchev, N., Noels, L., and Radovitzky, R., 2006. "Nonlinear compressibility effects in fluid-structure interaction and their implications on the air-blast loading of structures". *Journal of Applied Physics,* ***100***(6), p. 063519.

[5] Peng, W., Zhang, Z., Gogos, G., and Gazonas, G., 2011. "Fluid Structure Interactions for Blast Wave Mitigation". *Journal of Applied Mechanics,* ***78***(3), 02. 031016.

[6] Ghoshal, R., and Mitra, N., 2012. "Non-contact near-field underwater explosion induced shock-wave loading of submerged rigid structures: Nonlinear compressibility effects in fluid structure interaction". *Journal of Applied Physics,* ***112***(2), p. 024911.

[7] Reid, S., and Peng, C., 1997. "Dynamic uniaxial crushing of wood". *International Journal of Impact Engineering,* ***19***(5), pp. 531 – 570.

[8] Hanssen, A., Enstock, L., and Langseth, M., 2002. "Close-range blast loading of aluminium foam panels". *International Journal of Impact Engineering,* ***27***(6), pp. 593 – 618.

[9] Harrigan, J., Reid, S., and Peng, C., 1999. "Inertia effects in impact energy absorbing materials and structures". *International Journal of Impact Engineering,* ***22***(9), pp. 955 – 979.

[10] Wadley, H. N., Fleck, N. A., and Evans, A. G., 2003. "Fabrication and structural performance of periodic cellular metal sandwich structures". *Composites Science and Technology,* ***63***(16), pp. 2331 – 2343. Porous Materials.

[11] Deshpande, V., and Fleck, N., 2005. "One-dimensional response of sandwich plates to underwater shock loading". *Journal of the Mechanics and Physics of Solids,* ***53***(11), pp. 2347 – 2383.

[12] Main, J. A., and Gazonas, G. A., 2008. "Uniaxial crushing of sandwich plates under air blast: Influence of mass distribution". *International Journal of Solids and Structures,* ***45***(7), pp. 2297 – 2321.

[13] Wang, E., Gardner, N., Gupta, S., and Shukla, A., 2016. "Fluid-structure interaction and its effect on the performance of composite structures under air-blast loading". *The International Journal of Multiphysics,* ***6***(3).

[14] Dharmasena, K., Wadley, H., Xue, Z., and Hutchinson, J., 2008. "Mechanical response of metallic honeycomb sandwich panel structures to high-intensity dynamic loading". *International Journal of Impact Engineering - INT J IMPACT ENG,* ***35***, 09, pp. 1063–1074.

[15] Ghoshal, R., and Mitra, N., 2014. "On core compressibility of sandwich composite panels subjected to intense underwater shock loads". *Journal of Applied Physics,* ***115***(2), p. 024905.

[16] Vaziri, A., and Hutchinson, J. W., 2007. "Metal sandwich plates subject to intense air shocks". *International Journal of Solids and Structures,* ***44***(6), pp. 2021 – 2035. Physics and Mechanics of Advanced Materials.

[17] Tan, P., Reid, S., Harrigan, J., Zou, Z., and Li, S., 2005. "Dynamic compressive strength properties of aluminium foams. part ii—'shock' theory and comparison with experimental data and numerical models". *Journal of the Mechanics and Physics of Solids,* ***53***(10), pp. 2206 – 2230.

[18] Gurtin, M. E., Fried, E., and Anand, L., 2010. *The Mechanics and Thermodynamics of Continua*. Cambridge University Press.

Proceedings of the 6th International Conference on Ship and Offshore Technology
ICSOT 2019
November 7-8, 2019, IIT Kharagpur, India

ICSOT INDIA 2019-27

AN AUTONOMOUS SPEED SETTING SYSTEM TO ENHANCE OPERATION OF UNMANNED PLANING CRAFTS IN A SEAWAY

Himabindu Allaka
ahimabin@campus.haifa.ac.il

Asaf Levy
asaf.levy89@gmail.com

Deborah Levy
dlrun14@gmail.com

Tali Treibitz
ttreibitz@univ.haifa.ac.il

Morel Groper
mgroper@univ.haifa.ac.il

The Hatter Department of Marine Technologies,
Leon H. Charney School of Marine Sciences,
University of Haifa
Haifa, 3498838
Israel

ABSTRACT

Traditional manned planing crafts may present a high risk to crew onboard when operated in harsh and hostile environments. Consequently, the use of unmanned, Autonomous high-speed Planing Crafts (APCs) is a promising alternative for fast operations in littoral, confined and dangerous waters. However, ensuring the seaworthiness of these crafts, presents a significant challenge, particularly in regard to large vertical motions developed in a seaway which pose a hazard to payload and to the craft's structural integrity. Therefore, a real time prediction and estimation of vertical accelerations in waves and a control system to set the craft's forward speed is considered as an essential requirement for safety operation of APCs in seaways. In the present study, a system to enhance the seaworthiness of APCs by employing an autonomous speed setting mechanism as response to the predicted APC behavior when operating in seaways is developed. The system employs an embedded analytical tool termed Motion Assessment of Planing Craft in a Seaway (MAPCS) for the prediction of vertical accelerations the crafts might encounter to the incoming waves. Information on the characteristics of those waves is provided by a dedicated vision system. The complete development of the vision-aided speed modulation system (VSMS) for the APCs' autonomous navigation and its test results are presented.

NOMENCLATURE

Hs	Significant wave height [m]
$H_{1/3}$	One third wave height [m]
H_{max}	Maximum wave height [m]
Tp	Peak period [sec]
W	Weight of the ship [N]
I	Mass moment of inertia in pitch [kg.m2]
$\theta_{cg}, \ddot{\theta}_{cg}$	Pitch angle and acceleration at CG
T	Thrust Force [N]
$\ddot{x}_{cg}, \ddot{z}_{cg}$	Surge and Heave accelerations at CG
D	Frictional Drag Force [N]
x_a, x_b, x_d, x_t	Moment arm of hydrodynamic, hydrostatic, drag and thrust force [m]

INTRODUCTION

The study of autonomous navigation of marine crafts in general and unmanned planing crafts in particular is a fast developing research field, but most ongoing investigations focus on navigation and path planing algorithms. Those are mainly based on automatic processing of nautical charts combined with some collision and obstacles avoidance systems [1–11]. Currently, to the best of our knowledge, the existing autonomous navigation systems cannot address the seaworthiness challenges posed to APCs when operating in real sea conditions. While in manned operation the skipper can, to some degree, limit the accelerations by adjusting the craft's speed and course in response to its behavior as he feels and expects, in an unmanned mode the speed and course are predefined in the mission plan and adjustment as a response to the craft's behavior to incoming waves is not provided. In this work it is considered that the prediction and estimation, in real time, of expected vertical accelerations in waves the APC might develop, and control algorithms embedded in the crafts' autonomous navigation systems to limit those are an essential need to ensure safe operation of APCs. Consequently, the ability for autonomous speed setting and course adjustment in response to the incoming waves is required [12]. Therefore, a computational model-based system, integrated into the APC navigation system, which can determine in real time the expected accelerations and motions of the craft based on knowledge of the characteristics of incoming waves, is developed. The system, based on a real-time motion computational model MAPCS and visual and inertial sensors, will assist the craft's autonomous navigation system by determining and then altering the craft's momentary speed and course in a seaway to limit movements and preserve its structure and payload safety and integrity. The incoming waves characteristics provided as an input to the MAPCS in real time by a dedicated vision system is also developed in the frame of this research. Figure 1 presents the general configuration of the above mentioned framework. This includes the following components:

a. The estimation of the characteristics (height and frequency) of incoming waves is essential for APC navigation. This data is obtained from the sea state estimation system installed onboard the craft and equipped with visual and inertial sensors.

b. Based on the incoming waves and craft's characteristics, the MAPCS model [13,14] predicts the vertical accelerations the APC might develop as a function of the forward speed.

c. The speed setting algorithm modulates the speed of the APC as a function of the predefined acceleration threshold and the MAPCS predicted acceleration values and sends a throttle command accordingly.

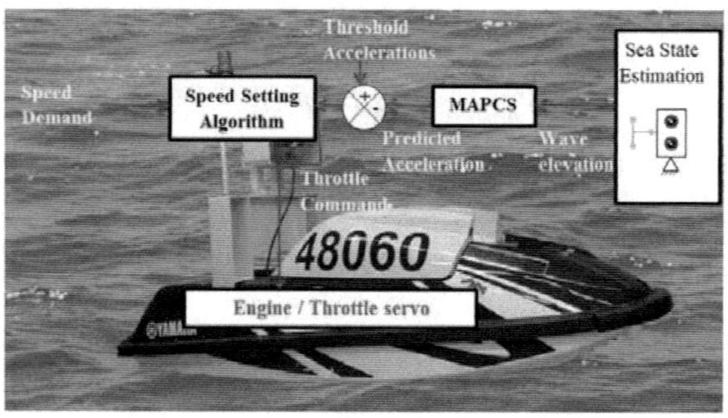

FIGURE 1. APC WITH VSMS FRAMEWORK

TYPICAL NAVIGATION SYSTEMS OF AUTONOMOUS SURFACE CRAFTS

Existing autonomous surface platforms that offer autonomous navigation capabilities developed for both research, commercial and military applications [1–11] are typically equipped with the following:

a) An auto pilot system, which allows operators to define a mission plan by demarcating lines, orbits or waypoints and speeds in the segments between waypoints using a dedicated GUI and existing, often system embedded, nautical charts. The auto pilot system controls the craft course and speed to follow this predefined plan.

b) An obstacle avoidance system (OAS), which makes deliberate, real time, navigational decisions to avoid collisions with other crafts or floating objects.

c) An over-the-horizon communication system, for remote control operations (takeover) and telemetry beyond line of sight distances.

Although several authors [2,3,8–11,15] developed obstacle avoidance systems for unmanned surface crafts using either radar or vision technologies or a combination of both, to the best of our knowledge, no research describing an autonomous navigation system for unmanned surface crafts that includes a sub-system to estimate incoming wave elevation and modulate vehicle speed accordingly to ensure safety has been published.

Sea state estimation

The routinely employed classical method to obtain sea state figures is to process motion measurements obtained from traditional wave rider moored buoys deployed at fixed locations to obtain the wave spectrum and, consequently, the sea state. In the context of this work, to enhance the navigation and safe operation of APCs in a seaway, accurate estimation of the sea state in the front of craft is essential. Thus, onboard sensors whose data is processed in real time to estimate sea state are investigated here.

There are only a limited number of works where onboard sensors such as Inertial Measuring Units (IMUs) [16–18], radars [19] and vision systems [20,21] were used to measure in situ sea state from sailing vessels.

Several authors [16–18] have worked on sea state estimation using vessel's motion measurements as input. A key assumption in these works is that in low to moderate sea states, the wave-induced six degrees-of-freedom motion of a vessel are linear with the incident waves. This linear assumption between waves and the associated vessel's responses facilitates the use of response amplitude operators (RAOs), which, once reversely used, express how the measured vessels response spectrum is transferred into wave spectrum.

Use of this approach in the context of this work is unsatisfactory because of two main limitations:

1) As the behavior of high-speed planing crafts in a seaway is highly non-linear, the use of RAOs in their classical linear sense may result in insufficiently accurate results.

2) As the intention of this work is to modulate the craft's speed in response to incoming waves, it is advantageous to estimate their height at a distance, in the undisturbed sea surface in front of the craft.

Cui [19] employed an onboard radar system to measure waves height. Specifically, the systems comprised two radar units, a long-range radar used to detect sea surface objects a few kilometers away from the craft and a short-range frequency modulated continuous wave radar to detect nearby obstacles. This short-range radar was employed in the estimation of the waves height in front of the craft and consequently enhance the safe navigation of the unmanned surface vehicle. Experimental results obtained from the short-range radar system from both wave tank and real sea tests were presented. The author concluded that the radar system was capable of estimating the sea state; however, no comparison with a validated wave estimation system was presented. Also, the author did not elaborate on how exactly the wave estimation as obtained from the radar system is utilized to enhance the safe navigation of crafts.

Few authors [20,21] employed stereo vision systems to estimate sea state. Corgnati et al. [21] employed a passive stereo vision system for real-time evaluation of sea wave characteristics, e.g., period and significant wave height, using a 3D model of the sea surface. The 3D reconstruction was based on solving correspondence problems (stereo matching) when creating a point cloud. The solution for wave estimation was based on spatial measurements and, therefore, a full 3D reconstruction was needed. The method relies on calculating wavelength; thus, the images must contain at least two waves and, accordingly, have to be taken from a significant height. For the goals of the present research, only wave elevation data in the time domain at a certain distance from the camera is required, eliminating the need for a complete 3D reconstruction of the captured images. Therefore, a computational fast and reliable methodology to obtain wave elevation at a certain distance in front the camera was developed and employed in this work and is described in the sections below.

VISION AIDED SPEED MODULATION SYSTEM (VSMS) OVERVIEW

The VSMS configuration schematically presented in Figure 2 comprises a vision system equipped with an, in house developed, stabilized stereo camera arrangement used to capture the state of the approaching sea in front of the craft. The stereo camera is stabilized in two axes (roll, and pitch). The vision system algorithm computes the elevation of the incoming waves at a constant, predefined distance ahead of the craft in the camera's reference frame by reconstruction using stereo imaging. Thereafter, the computed wave elevation data is heave-compensated using heave position data as recorded by an IMU installed on the stabilized stereo camera frame to obtain the wave elevation in the inertial frame. Although the obtained momentary wave elevation in front of the APC will not be exactly the momentary wave elevation that the craft will actually encounter as the waves propagates, in a statistical fashion and based on the high speed of advance of the planing craft, wave propagation is ignored. Consequently, the same vision system obtained wave elevation is considered as the one the craft will encounter, and the two momentary sea elevations, are treated as different realizations of the same sea state. Thus, the incoming seaway wave elevation data provided by the vision system algorithm and the actual velocity of the APC is fed as input to the MAPCS. Based on this input, the MAPCS predicts the vertical accelerations of the APC in response to the incoming waves. Then, the predicted acceleration values are compared with the threshold/permissible value. The difference between the predicted and threshold acceleration values are provided as input to the speed setting algorithm. If the predicted acceleration values exceed the threshold limit, the speed setting algorithm will momentarily and in real time reduce the actual speed demand based on a simplified linear algorithm. The updated speed demand is sent in real time to the autopilot system, which sends it as an updated value to the throttle command.

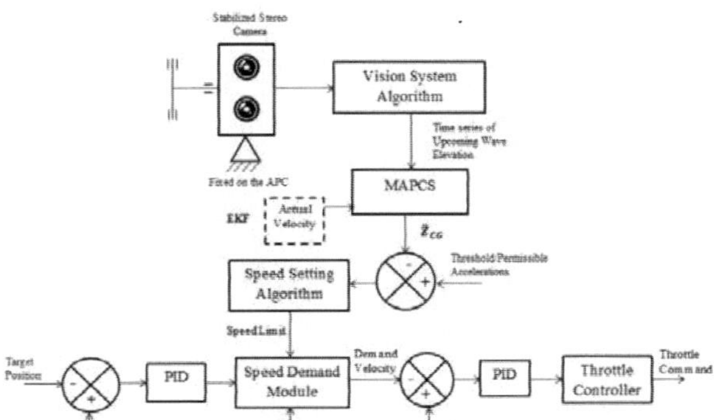

FIGURE 2. VISION AIDED SPEED MODULATION SYSTEM

MAPCS Model Overview

The MAPCS model used in the VSMS to estimate planing crafts' vertical accelerations was developed by the authors following the work of Zarnick [22] and by incorporating the pressure correction [23] and modified added-mass theories [24]. The craft is modelled as a series of strips or impacting wedges. MAPCS is capable to predict the craft's motions in the longitudinal plane (surge, heave and pitch) when operating in head or following seas.

The expressions for the surge x_{CG}, heave z_{CG}, and pitch θ_{CG} (where CG refers to the craft's center of gravity) directions are presented in expression (1):

$$\begin{aligned} M\ddot{x}_{CG} &= T\cos\theta_{CG} - F_{dyn}\sin\theta_{CG} - D\cos\theta_{CG} \\ M\ddot{z}_{CG} &= -T\sin\theta_{CG} - F_{dyn}\cos\theta_{CG} - F_{sta} + D\sin\theta_{CG} + W \\ I\ddot{\theta}_{CG} &= Tx_t + F_{dyn}x_a + F_{sta}x_b - Dx_d \end{aligned} \quad (1)$$

The total hydromechanical normal force acting on the planing craft is the vectorial summation of the hydrodynamic lift force F_{dyn} and the hydrostatic lift force F_{sta}. When a wave impinges on the hull, the immersion of hull volume changes along the length of the craft and there is a change in external forces due to hydrodynamic forces F_{dyn} and hydrostatic forces F_{sta}. The planing craft impinged by waves is shown in Figure 3.

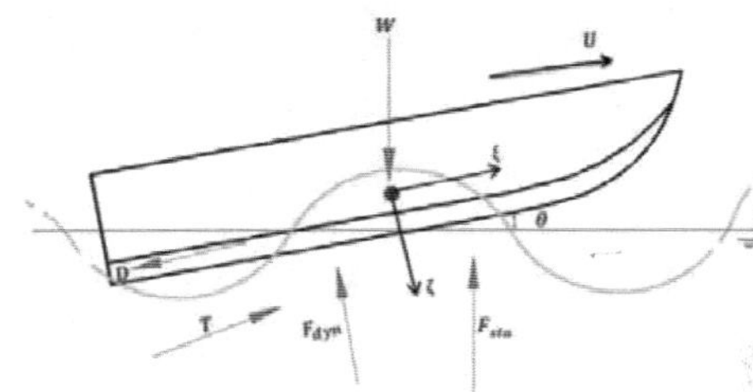

FIGURE 3. FORCES ACTING ON PLANING CRAFT IN WAVES

The solution of the nonlinear set of ODEs (1) allows the computation of the craft's motion in the longitudinal plane and in the time domain.

The detailed development of the MAPCS model, its validation and verification are detailed in [12–14].

APC CONFIGURATION WITH VISION SENSORS

To facilitate autonomous speed modulation in a seaway, a vision system was developed and installed on the APC, as shown in Figure 4. Currently, the vision system uses a stabilized ZED stereo camera https://www.stereolabs.com/zed/

FIGURE 4. APC PLATFORM WITH VISION SYSTEM

Stabilization of the vision system and its validation

The VSMS requires a stream of incoming wave elevation data in the time domain which is captured by the vision system. Ensuring accuracy of the wave elevation measurements is of significant importance to the correct operation of the system and optimal performance of the APC. To this end, it was decided to develop, in-house, a robust and stabilized mount for the vision system. To stabilize the ZED stereo camera in two axes (roll, and pitch), a gimbal frame made of carbon fiber plates and tubes was developed. Two brushless motors were installed onto the gimbal frame, one for stabilizing the roll axis and the other for stabilizing the pitch axis, as presented in Figure 5. A BGC 32-bit extended brushless gimbal controller was used to stabilize the system. The controller's proportional integral differential (PID) parameters were fine-tuned for faster response and perfect stabilization.

FIGURE 5. STABILIZED VISION SYSTEM SETUP

To get an estimation of how effective the in-house developed stabilization system performs when the APC operates at planing speeds in a real seaway, an IMU was mounted on the vision system camera to record its linear and angular motion, with an additional IMU installed inside the vision system cabinet on an unstabilized frame. A comparison was made between these two IMUs readings. Figure 6 (a) and (b) presents the time series of the roll and pitch angle of the unstabilized and stabilized frames, respectively, when the craft was advancing with a velocity of 7.5 m/sec. It can be observed that the stabilization system performs well but there are few events where it fails to stabilize. These are high impact slams the craft encountered when operating at high seas and at planing speeds. The effect of those events on the computed wave elevation data is numerically filtered.

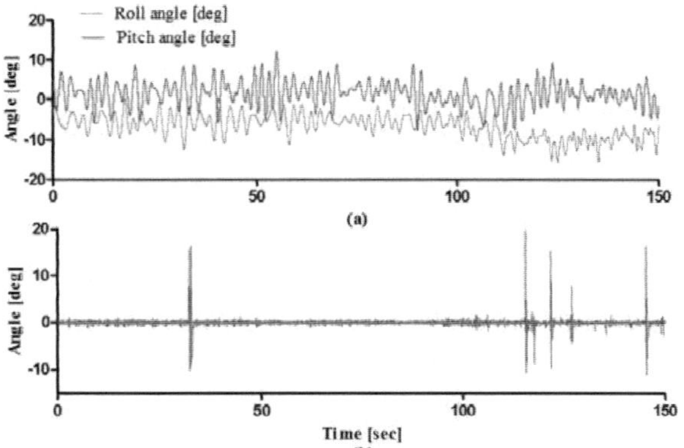

FIGURE 6. ROLL AND PITCH ANGLES AT 7.5 M/SEC CRAFT VELOCITY: (a) UNSTABILIZED IMU (b) STABILIZED IMU MEASUREMENTS

WAVE ELEVATION ESTIMATION FROM VISION AND INERTIAL SENSORS

The passive stereo vision system mounted on the craft is stabilized in roll and pitch axes (roll and tilt)). It will, however, be displaced in the heave direction (along the Z-axis) in response to the craft motion in a seaway; this non-compensated motion will cause a deviation in the computed waves' elevation. To take account of this motion, an additional IMU was installed on the stereo vision camera to estimate its exact heave position. The employed coordinate systems are shown in Figure 7: Stabilized, IMU coordinates $(O_{imu(s)}, x_{imu(s)}, y_{imu(s)}, z_{imu(s)})$, stereo system left $(O_{CL}, x_{CL}, y_{CL}, z_{CL})$, right $(O_{CR}, x_{CR}, y_{CR}, z_{CR})$ and the actual wave elevation coordinates (O_i, x_i, y_i, z_i). The wave elevation of the vision system is processed in the camera's left frame$(O_{CL}, x_{CL}, y_{CL}, z_{CL})$; thus, the stabilized IMU $(O_{imu(s)}, x_{imu(s)}, y_{imu(s)}, z_{imu(s)})$ was mounted in close proximity to the camera's left frame and the origins were considered to coincide.

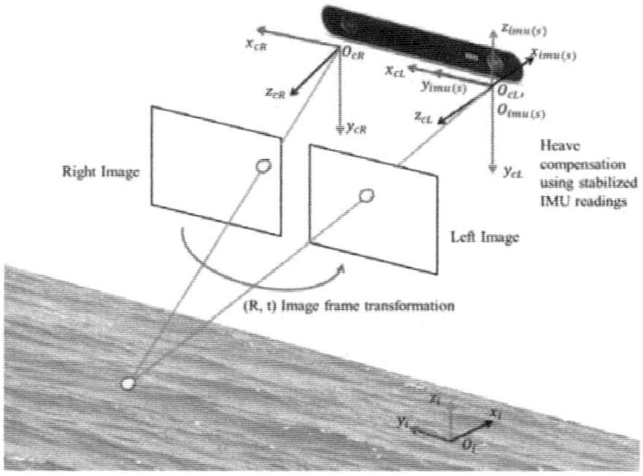

FIGURE 7. THE FOUR COORDINATE SYSTEMS INVOLVED IN THE VISION SYSTEM

Figure 8 presents a sample reconstruction of the wave elevation time history in the camera frame $(O_{CL}, x_{CL}, y_{CL}, z_{CL})$, red curve, the heave displacement of the stereo vision camera in the stabilized IMU frame $(O_{imu(s)}, x_{imu(s)}, y_{imu(s)}, z_{imu(s)})$, blue curve and the heave compensated wave elevation reconstruction, in the inertial frame(O_i, x_i, y_i, z_i), green curve.

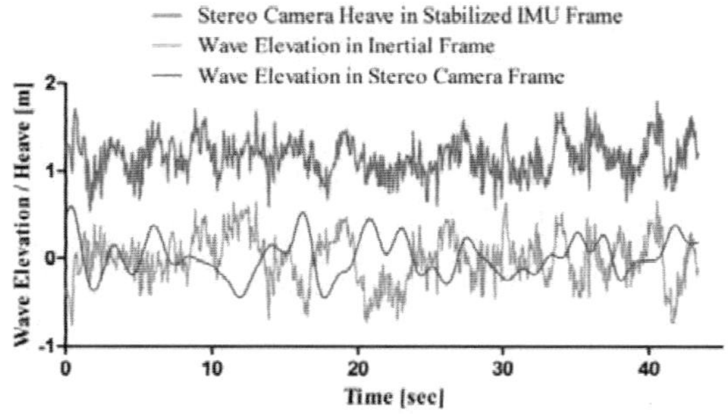

FIGURE 8. WAVE ELEVATION AND SYSTEM HEAVE DISPLACEMENT TIME HISTORY

Validation of wave elevation as obtained from the vision system in regular waves

To verify the accuracy of the vision system measured wave elevation data, an experiment was performed in a scientific wave tank which generates regular waves of constant amplitude. Wave height was first measured manually. Then, an in-house developed wave buoy equipped with an IMU was deployed in the wave pool to measure the wave height. The wave elevation data from the wave buoy is presented in Figure

9(a). Videos were also taken using the ZED stereo camera. Based on the stereo imaging, 3D reconstruction of the scene (except for occlusions) was performed in batch processing. Both the intrinsic and extrinsic parameters of the stereo cameras as supplied by the manufacturer were used. The next step was to find matching points in both images in order to triangulate the real-world data points. This task is quite difficult due to similarity of features in the images. To find dense matching, an epic flow algorithm [25] was employed and applied in a bi-directional manner: from the left image to the right and vice-versa. A match was considered to be valid only for pixels that had an end-point-error of 3 pixels or less for images with resolution of 1920x1080 pixels (HD). The point cloud was validated by measuring the size of the wave amplitude from the data, which is known from ground truth measurement to be 12 cm. The last step was to decide at which distance from the camera the water surface has to be observed. A distance of 4 m in front the camera was used. The amplitude values of all the pixels in that range were averaged to get the wave elevation as shown in Figure 9(b). The wave elevations from the wave buoy (Fig. 9(a)) and from the vision system (Fig. 9(b)) were observed to be equal, with a wave height of 12–12.5 cm corresponding to the manual measurement of 12 cm and frequency of 0.5 Hz.

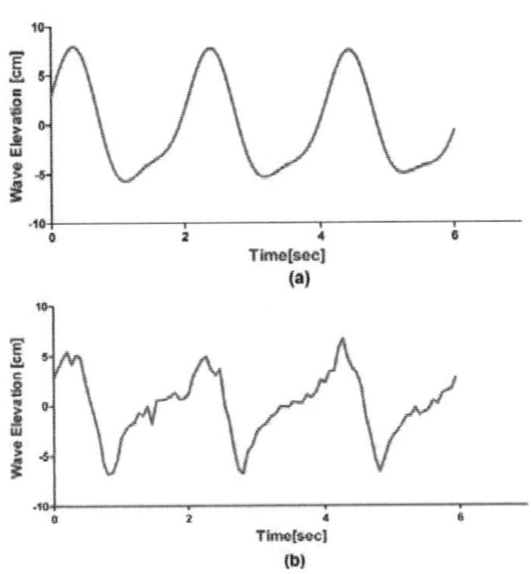

FIGURE 9. WAVE ELEVATION: (a) WAVE BUOY, (b) VISION SYSTEM

Validation of wave elevation as obtained from the vision system in real seas

To validate and confirm the performance of the vision system in a real operational scenario sea experiments were performed with the system installed onboard a manned 6.5m in length planing craft operated at planing speeds and in real seas. To verify the accuracy of the wave elevation data as obtained from the vision system, our in-house developed wave buoy [26] (Fig 10) was deployed in the region of the sea experiment.

The vision system captured recordings were batch processed to obtain wave elevation at three different distances in front of the craft. The wave elevation data compensated for heave and the spectral and time domain parameters of the wave records for these distances as obtained from the vision system were statistically compared against data collected by our in-house developed buoy and against a standard wave buoy operated by CAMERI [27] and located in a close vicinity to the experimental site and is presented in Figures 11 and 12.

FIGURE 10. IN HOUSE DEVELOPED WAVE BUOY (IMU INSIDE)

Figure 11 presents a comparison of the wave elevation spectral parameters, i.e. significant wave height (H_S) and peak wave period (T_P) as obtained from the stereo vision system vs. our wave buoy and vs. the standard CAMERI wave buoy. Figure 12 presents a similar comparison, this time in the time domain for $H_{1/3}$ and $H_{\max}$.

The abscissa represents the distance in front of the craft (where the sea state parameters were evaluated) and the results obtained from the employed wave buoys. The ordinate represents the value of the estimated parameters.

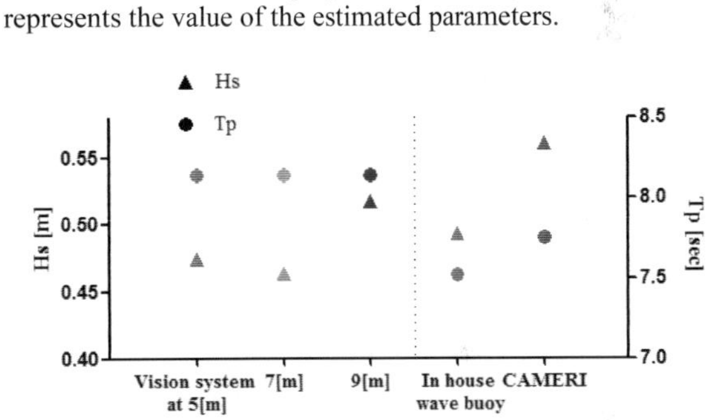

FIGURE 11. WAVE ELEVATION SPECTRAL PARAMETER COMPARISION

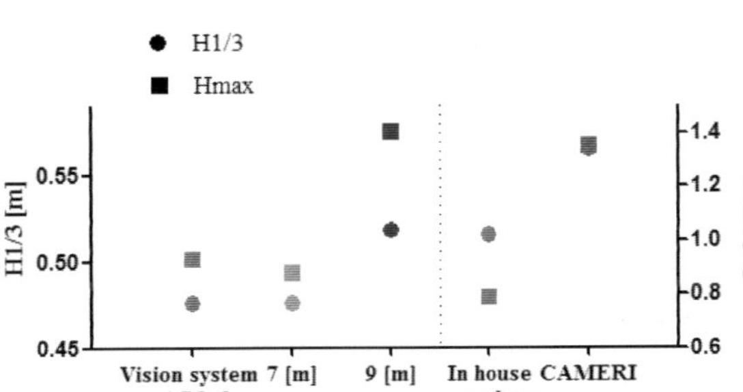

FIGURE 12. WAVE ELEVATION TIME DOMAIN PARAMETER COMPARISION

From figure 11, as can be observed at the day of the experiment the significant wave height as estimated using the vision system data was 0.47m, 0.46m and 0.52m at distances of 5m, 7m and 9m in front the craft accordingly. At the same sea realization the significant waves height as obtained using our wave buoy was 0.49m and the CAMERI buoy recorded a significant wave height of 0.55m. The peak period as estimated using the data provided by the vision system was

8.2s for all three distances where the data collected by our buoy provided a peak period equal to 7.7s and CAMERI 8.3s.

From figure 12 it can be observed that the $H_{1/3}$ estimated waves height using the vision system data was 0.47m, 0.47m and 0.51m at distances 5m, 7m and 9m accordingly. At the same sea realization the $H_{1/3}$ as obtained using our wave buoy was 0.51m and the CAMERI buoy recorded a $H_{1/3}$ of 0.56m. Similarly, the H_{max} estimated using the vision system data was 0.92m, 0.87m and 1.40m at distances 5m, 7m and 9m accordingly. At the same sea realization the H_{max} as obtained using our wave buoy was 0.80m and the CAMERI buoy recorded a H_{max} of 1.35m.

Based on this comparison it seems that the waves spectral and time domain sea state parameters as obtained using the stereo vision system operated in a real operational scenario shows a fair correlation with the values obtained from our self-developed buoy and with the data obtained from the standard wave buoy operated by CAMERI [27].

TESTING OF VSMS SYSTEM

The performance of the complete VSMS system was examined offline in a simulated MATLAB/SIMULINK environment using the experimental real seas' wave elevation data as captured by the vision sensors and the previously developed MAPCS model. The algorithm and conditions were set as follows:

1) As soon as the APC is operated the vision sensors begin accumulating wave data for the first 20 seconds while the VSMS remains idle, with the craft's speed of advance setting equal to the preset setting.
2) Once 20 seconds of wave data is accumulated, the VSMS system starts operating, with the MAPCS computing predicted vertical acceleration values each time the APC is operated at planing speeds, i.e. $F_r \geq 0.8$.
3) In planing speeds operation, the predicted vertical acceleration data is fed to the speed setting algorithm where the vertical acceleration peaks are compared against the user-defined threshold values and if the former exceed the threshold, through an iterative process, an appropriate speed limit is computed. For the simulation purpose it was decided to update the speed setting every 20s only. In the real sea experiments to follow, based on the real time computational time, the behavior of the APC propulsion plant and its reaction to speed changes, the update rate of the speed setting will be tuned.
4) This newly computed speed limit overrides the preset speed setting and the throttle is modulated accordingly through a PID controller.
5) As long as the APC is at planing speeds, the VSMS will continue its operation and compute speed limits to modulate the preset on-route speed as originally dictated in the mission planner.

Figure 13 presents the VSMS Simulink model where blocks of the craft geometry, the sea state and the state space initial conditions were fed as input to the MAPCS model. The sea state data provided as input was based on simulated data from the vision system at a distance of 20m in front of the APC.

Based on this input, the MAPCS model predicted the vertical acceleration peaks, which were fed to the iterative speed setting algorithm, and a speed limit was obtained. This new speed limit overrode the preset surge velocity in the state vector provided as input to the MAPCS model.

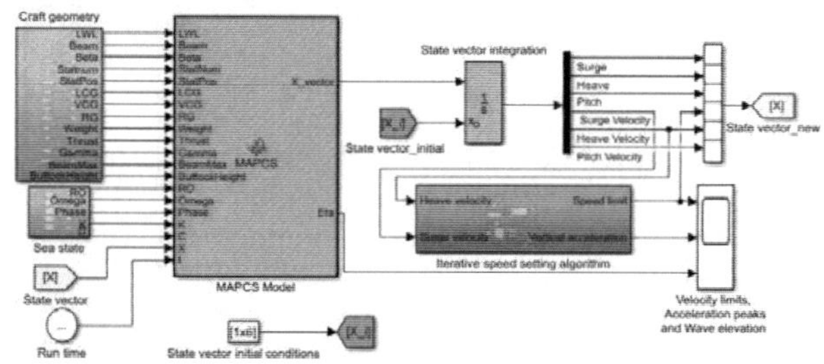

FIGURE 13. VSMS IN SIMULINK ENVIRONMENT

Figure 14 presents a comparison of vertical acceleration encountered by the APC 2.24m in length without an active VSMS (Fig. 14 (b)) and with the VSMS (Fig. 14(c)) active for the same wave elevation time history (Fig. 14(a)). The preset on-route speed was set to 12 m/sec. Figure 14(b) presents the encountered vertical acceleration (in green) without the intervention of the VSMS. The acceleration threshold is marked with a dotted red line (set at 60 m/s^2) and the actual speed of the craft in blue. As the VSMS is not active, the actual speed of the craft equals the present on-route speed.

Figure 14(c) presents the encountered vertical acceleration (in green) with the intervention of the VSMS, the acceleration threshold again marked with a dotted red line (set at 60 m/s^2) and the actual speed of the craft (in blue). It can be observed that when the VSMS in engaged, the craft's speed is modulated as dictated by the speed setting algorithm.

Comparing the vertical acceleration in Figures 14(b) and 14(c), when no VSMS system is employed (Fig. 14(b)), the vertical acceleration peaks cross the threshold acceleration limit often with a very high magnitude. In contrast, when the VSMS system is engaged (Fig. 14(c)), the speed modulation causes a drastic reduction in encountered vertical acceleration peaks and the vertical acceleration peaks rarely crossed the threshold acceleration limit.

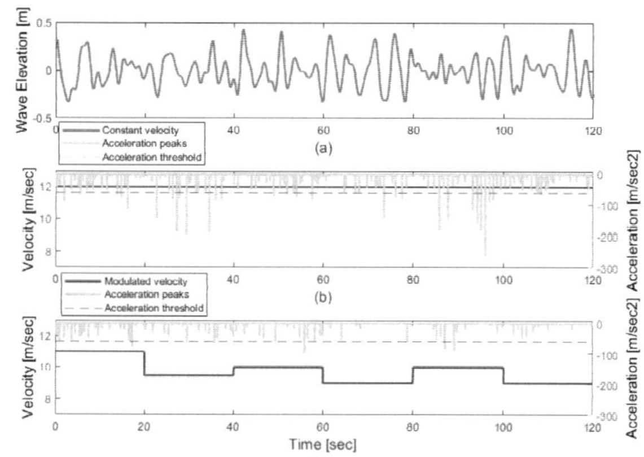

FIGURE 14. COMPARISION OF THE APC VERTICAL ACCELERATION RESPONSES WITH AND WITHOUT THE VSMS FOR THE SAME SEA STATE

CONCLUSIONS AND DISCUSSIONS

From the obtained results, it was confirmed that the vision system presents a feasible method for estimating the incoming waves in undisturbed waters in front of a planing craft. The stereo camera (ZED) used in this study, however, has a limited range of up to 20 meters for the purpose of reconstruction of the waves' elevation data and will be replaced in the future by two industrial cameras equipped with zoom lenses.

From the VSMS simulated results, the importance of a robust and reliable autonomous speed modulation system is clear.

Future works will include improvement of all modules of the VSMS and integration of all in the APC.

ACKNOWLEDGMENT

This study was supported by the Leona M. and Harry B. Helmsley Charitable Trust, the Maurice Hatter Foundation, and the Paul Amir Foundation. Research at the University of Haifa was conducted at the Hatter Department of Marine Technologies in the Sub Sea Engineering Lab. The authors are grateful to the Israel Shipyards Ltd. for providing an opportunity to conduct experiments on a Shaldag MK II class fast patrol boat. The authors also thank the Coastal and Marine Engineering Research Institute for providing real sea wave elevation data.

REFERENCES

[1] S. Campbell, M. Abu-Tair, W. Naeem, 2013. " An automatic COLREGs-compliant obstacle avoidance system for an unmanned surface vehicle,". Journal of Engineering for the Maritime Environment 228(2), pp. 108-121.

[2] Proceedings, J. Larson, M. Bruch, J. Ebken, 2006. "Autonomous navigation and obstacle avoidance for unmanned surface vehicles". Defense and Security Symposium, Orlando, Florida, United States. Unmanned Systems Technology, VIII.

[3] Proceedings, H.K. Heidarsson, G.S. Sukhatme, 2011. "Obstacle detection and avoidance for an autonomous surface vehicle using a profiling sonar". IEEE International Conference on Robotics and Automation, pp. 731–736.

[4] Proceedings, M. Blaich, S. Köhler, J. Reuter, A. Hahn, 2015. "Probabilistic collision avoidance for vessels," IFAC- Workshop 28, pp. 69–74.

[5] J.Y. Zhuang, L. Zhang, S.Q. Zhao, J. Cao, B. Wang, H.B. Sun, 2016. "Radar-based collision avoidance for unmanned surface vehicles". China Ocean Engineering 30, pp. 867–883.

[6] "Multi-Purpose USV, Protector", Rafael Advanced Defense Systems Ltd. See also: http://www.rafael.co.il/Marketing/199-en/Marketing.aspx.

[7] Proceedings, S. Moe, K.Y. Pettersen, 2017. "Set-Based line-of-sight (LOS) path following with collision avoidance for underactuated unmanned surface vessels under the influence of ocean currents". 1st Annual IEEE Conference on Control Technology and Applications, pp. 241–248.

[8] Proceedings, D. Hermann, R. Galeazzi, J.C. Andersen, M. Blanke, 2015. " Smart sensor based obstacle detection for high-speed unmanned surface vehicle" , IFAC-Woekshop 28 , pp. 190–197.

[9] H. Tripp, R. Daltry, 2018. "The Path to Real World Autonomy for Autonomous Surface Vehicles", ASV GLOBAL. See also: https://www.asvglobal.com/wp-content/uploads/2018/01/The-Path-to-Real-World-Autonomy-for-Autonomous-Surface-Vehicles.pdf.

[10] Proceedings, H. Wang, Z. Wei, S. Wang, C.S. Ow, K.T. Ho, B. Feng, 2011. "A vision-based obstacle detection system for Unmanned Surface Vehicle", 2011 IEEE 5th International Conference on Robotics Automation and Mechatronics, pp. 364-369.

[11] T. El-Gaaly, C. Tomaszewski, A. Valada, P. Velagapudi, B. Kannan, P. Scerri, 2013. "Visual Obstacle Avoidance for Autonomous Watercraft using Smartphones" See also: http://repository.cmu.edu/robotics.

[12] Proceedings, H. Allaka, D. Levy, T. Treibitz, M. Groper, 2018. " Vision-aided speed modulation system to enhance seaworthiness of autonomous planing crafts". 15th Workshop on Positioning, Navigation and Communication WPNC 2018, pp. 1–6.

[13] Master Thesis, H. Allaka, 2017. "Motion Assessment of Planing Craft in a Seaway". M.Sc Thesis, University of Haifa, Israel.

[14] In Review, H. Allaka, M. Groper, 2019. "Validation and verification of a planing craft motion prediction model based on experiments conducted on full size crafts operating in real sea". Journal of Marine Science and Technology.

[15] S. Campbell, W. Naeem, G.W. Irwin, 2012. "A review on improving the autonomy of unmanned surface vehicles through intelligent collision avoidance manoeuvres". Annual Review Control. 36 , pp.267–283.

[16] Proceedings, U.D. Nielsen, M. Bjerregard, R. Galeazzi, T.I. Fossen, 2015. " New concepts for shipboard sea state estimation". OCEANS 2015, pp. 1–10.

[17] Proceedings, A. Rosén, E. Begovic, M. Razola, K. Garme, 2017. "High-speed craft dynamics in waves : challenges and opportunities related to the current safety philosophy". 16th International Ship Stability Workshop, Belgrade, Serbia, pp. 1–10.

[18] Proceedings, U.D. Nielsen, R. Galeazzi, A.H. Brodtkorb, 2016. "Evaluation of shipboard wave estimation techniques through model-scale experiments". OCEANS 2016 - Shanghai, pp. 1-8.

[19] Proceedings, Jian Cui, R. Bachmayer, Weimin Huang, B. deYoung, 2017. " Wave height measurement using a short-range FMCW radar for unmanned surface craft," OCEANS 2015 - MTS/IEEE Washington, pp. 1–5.

[20] F. Bergamasco, A. Torsello, M. Sclavo, F. Barbariol, A. Benetazzo, 2017. " WASS: An open-source pipeline for 3D stereo reconstruction of ocean waves". Computers and Geosciences 107, pp. 28–36.

[21] Proceedings, L. Corgnati, L. Mazzei, S. Marini, B. Isoppo, E. Ottaviano, G. Besio, M.G. Magaldi, 2015. " High resolution stereo imaging of sea waves for validation and optimization of wave modelling". OCEANS 2015 MTS/IEEE - Genova: Discovering Sustainable Ocean Energy for a New World.

[22] E.E. Zarnick, 1978. " A non-linear mathemathical model of motions of a planning boat in regular waves". Technial Report DTNSRDC- 78/032, Taylor Naval Ship Reasearch and Development Center, Bethesda, Md, USA.

[23] K. Garme, 2005. "Improved Time Domain Simulation of Planing Hulls in Waves by Correction of the Near-Transom Lift". International Shipbuilding Progress 52, pp. 201–230.

[24] P.R. Payne, 1994. " Recent developments in added-mass planing theory". Ocean Engineering. 21, pp. 257–309.

[25] Proceedings, J. Revaud, P. Weinzaepfel, Z. Harchaoui, C. Schmid, 2015. " EpicFlow : Edge-Preserving Interpolation of Correspondences for Optical Flow", IEEE Conference on Computer Vision and Pattern Recognition, pp. 1164–1172.

[26] Proceedings, S. Farber, H. Allaka, I. Klein, M. Groper, 2018. "Estimating Sea State Using a Low Cost Buoy" IEEE International Conference on the Science of Electrical Engineering in Israel, pp. 1-5.

[27] CAMERI . "Real Time Wave Measurements Haifa". See also: http://www.israports.co.il/SiteAssets/Waves/haifaw-ipa.html.

Proceedings of the 6th International Conference on Ship and Offshore Technology
ICSOT 2019
November 7-8, 2019, IIT Kharagpur, India

ICSOT INDIA 2019-28

EXPERIMENTAL STUDY OF GREEN WATER OCCURRENCE ON KCS

Ravindra Babu Kudupudi
Research Scholar
Kharagpur, West Bengal, 721302
India
ravindra.kudupudi@iitkgp.ac.in

Anirban Bhattacharyya
Assistant Professor
Kharagpur, West Bengal, 721302
India
ab@naval.iitkgp.ac.in

Ranadev Datta
Assistant Professor
Kharagpur, West Bengal, 721302
India
ranadev@naval.iitkgp.ac.in

ABSTRACT

The green water occurrence for KRISO container ship (KCS) model for different speeds and head wave conditions is experimentally investigated in a towing tank. The main focus of the present work is to study the dependency of the green water phenomena on several physical geometrical parameters such as wave steepness, ship forward speed, wave length etc. A series of regular head waves with wavelength/ model length in the range of 0.8 – 1.5 were generated for the model tests. Measurements were recorded for parameters like wave height, motions of the vessel, and impact pressure on the deck and superstructure. The results highlight the influence of parameters like wave length, wave steepness, and ship speed on the green water loading for the vessel.

Keywords:
Green water, impact pressure, wavelength, speed

INTRODUCTION

Occurrence of green water phenomena affects ship safety and may cause damages to the structural part around the bow. Since the modeling of green water occurrence is complicated as it is depends on so many parameters, many researcher conducted several experimental tests to understand the behaviour of the green water occurrence and most importantly predict the impact loads on the different types of the vessels. Buchner (1995) conducted an experiment for impact of green water on a FPSO with respect to design purpose by taking into various parameters like wave length, wave period and current velocity. Later Hamoudi et al., (1998), investigated the impact of green water on deck experimentally for offshore vessels by considering vertical relative motion in regular waves. Berhault (1998) investigated the green water effects on FPSO's for both regular and irregular waves for different bow shapes. . In overall, table 1 gives the view of the experimental investigation has done so far for the past literature.

Though many researchers have concentrated on FPSO's, containership with forward speed will also play crucial for the green water phenomena which it is considered in this paper. The modeling of green water loading is a complex task and depends on many physical and geometrical factors. For example, the green water phenomenon primarily occurs in rough sea condition where waves are highly nonlinear. Also, the severity of the green water loads heavily depends on vessel motions and speed, and the wave characteristics. Hence, to develop a numerical model which considers all these aspects is a non-trivial task.

Therefore, in this paper, KRISO container ship (KCS) hull has been considered for the green water analysis and in fact the information presently offered is very minimalistic for ships with forward speed having heave and pitch motions. The foremost aim was to examine the experimental work part of the influence of different parameters over the shipping of water on the deck and on the superstructure for a containership. The influence of different parameters considered are wavelength, wave steepness and Froude number

Table 1. A brief overview of significant published research on green water loading

Sl No	Author	Contributions
1	Buchner (1995)	Control of influencing parameters under the criterion of wave elevation, time period, wave velocity over the impact of green water on FPSO design.
2	Hamoudi et al. (1998)	Under consideration of the vertical relative motion, the investigation green water occurrence and loading on the deck of offshore vessels.
3	Berhault (1998)	Effect of green water over FPSO for regular and irregular waves considering different wave height, ship motions and bow shapes
4	Greco et al. (2000, 2001)	A 2D study of green water loading by making an allowance for geometric and kinematic parameters
5	Cox and Ortega (2002)	Extreme wave kinematics study based on laboratory annotations of green water

		overtopping on a fixed deck.
6	Tanizawa et al. (2004)	The impact of water shipping over running ship foredeck for regular head seas.
7	Pham et al. (2004)	Application of Whaleback forecastle for high-speed container vessels and reduction of green water loading by utilizing a dam break model applied with the condition of the initial velocity.
8	Fonseca and Guedes Soares (2005)	Exploration of shipping of the green water experimentally over the bow portion of a containership with the application of wave head with the regular and irregular condition.
9	Lee et al. (2012)	Optimization of bow shapes in regular head waves for design aspect by experimental investigation of green water on deck for three different FPSO.
10	Silva et al. (2014)	Exposed condition analysis FPSO towards beam and quarter sea conditions to determine green water loads.
11	Ruggeri et al. (2015)	An experimental approach in fundamental green water study for a captive vessel using simplified FPSO over the condition of head, beam and quartering seas.
12	Lee et al. (2018)	Behavioral study of green water and flow kinematics on deck with various flare angles.
*13	*Present study*	Green water occurrence over KCS hull for different incident wave frequencies, wave steepness and forward speeds.

EXPERIMENTAL INVESTIGATION

Experimental setup

The experiments were conducted in the 150 m long towing tank in the Department of Ocean Engineering and Naval Architecture at IIT Kharagpur, India. The regular head waves were generated by a piston-type wave-maker, and the ship model was towed by a carriage at different forward speeds. A 1:65 scale model was chosen for conducting the experiment. The principal particulars of KCS in both model and full scales are mentioned in Table 2. The connection of model with the carriage is made using a transverse beam mounted at the middle portion of the ship with the support of a wired arrangement (Figure 1). The model was attached to tow rope connected to the carriage to ensure the model velocity remained constant during each run. Pressure transducers are placed at two points (P_1 and P_2) on the deck and one point over the superstructure (P_{R1}) in order to measure the green water pressure.

Table 3. Main particulars for a container vessel in dimension (meters)

	Model Scale	**Full Scale**
The length between perpendiculars (L_{bp})	3.5 m	230 m
Cross section Breadth (B)	0.49 m	32.2 m
Draught of the vessel (T)	0.17 m	10.8 m
Depth	0.357 m	19 m
Displacement	819.5 T	52030 T
Block coefficient (C_B)	0.684	0.684

Figure 1. Crow-foot arrangement of the KCS hull set-up

The measured parameters from the green water experiments are the incident wave heights, ship motions and green water pressures over deck and superstructure. Incident waves are measured using a probe installed over the side portion of the tank. A distance of 20m was kept between the probe and wave inlet to evade the effect of the instabilities triggered due to ship motions. The ship model was free to heave and pitch and the motions were measured using an accelerometer. The arrangement of pressure sensors consists of three sensors as discussed above and the particulars of the probe stations and bow deck sites of pressure sensors are given in Figure 2. The data acquisition system assembles all the signals from various components of observation sensors. The sampling frequency of pressure sensors was 100Hz to capture the pressure fluctuations arising on the deck due to green water phenomena in different conditions.

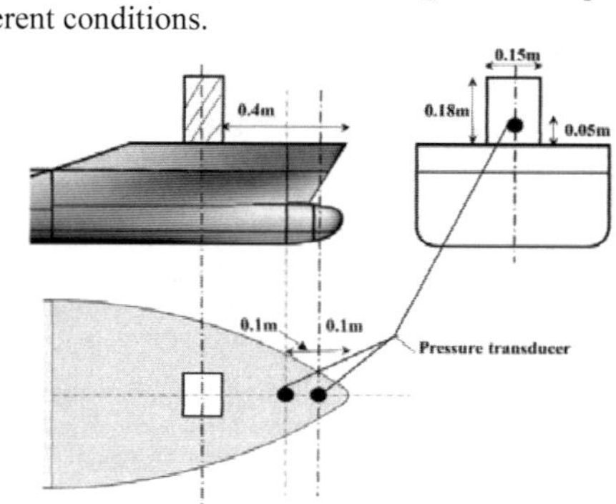

Figure 2: Details of the pressure transducer along with box located on the deck

Test scheme

The experimental test scheme includes measurement of peak pressure due to green water on the deck and bow-mounted superstructure in a series of regular head wave conditions. The choice of wave parameters was based on previous studies on green water occurrence for ships, which indicate the critical wavelength range and steepness resulting in green water incidence on the deck. Four vessel speeds were investigated to understand the dependence of green water phenomena on forward speed, which changes the encounter efficiency in head waves. The entire test matrix consisting of the wave and vessel parameters is presented in Table 3.

Table 3. Complete test matrix for modeling green water experiments

Test No.	Model Speed (m/s)	Froude No.	λ / L	$H(m)$	Comments
1	0.32	0.055	1.0	0.25	Speed variation
2	0.51	0.09			
3	0.95	0.166			
4	0.76	0.13	0.8	0.25	Wavelength variation
5			1.0		
6			1.2		
7			1.5		
8	0.76	0.13	1.0	0.225	Wave height/ steepness variation
9				0.20	
10	0.76	0.13	1.0	0.25	Repeated measurements for Uncertainty Analysis
11					
12					
13					

The measured parameters include the following:

- Pressures at P_1, P_2, and P_{R1}
- Vessel heave
- Vessel pitch, and
- Wave elevation in the towing tank

Results and discussion

Green water evolution

To investigate the accuracy of the generated wave, the time history of the wave profile recorded by a wave probe placed 10m away from the wave maker. The recorded time history of the wave probe is shown in Figure 3. The figure shows a steady sinusoidal wave which confirms the accuracy of the generated wave. Figure 4 presents the time histories of the vessel heave motion at $\lambda / L = 1.0$,- $F_n = 0.13$. A periodic but nonlinear response is observed. Using Fast Fourier Transform (FFT), the amplitude of the first harmonic for different wave frequency is computed and plotted in Figure 5. The amplitude is non-dimensionalised by dividing the response by input wave amplitude.

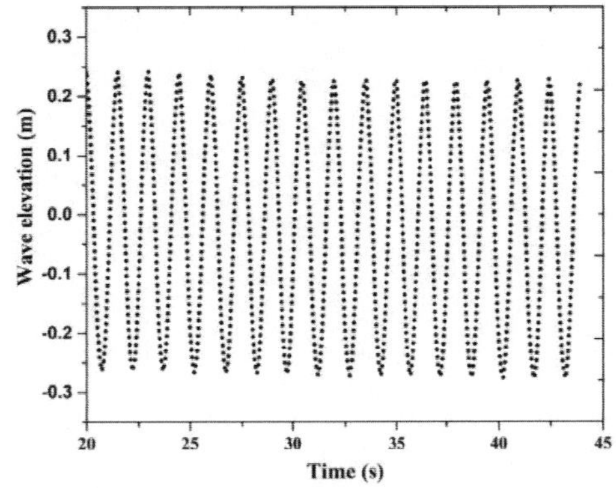

Figure 3. Wave elevation at 10m from the wave maker

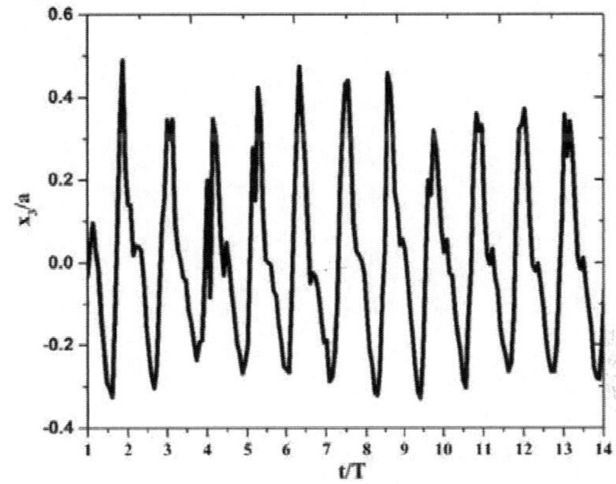

Figure 4. Time histories of heave motion for $\lambda / L = 1.0$

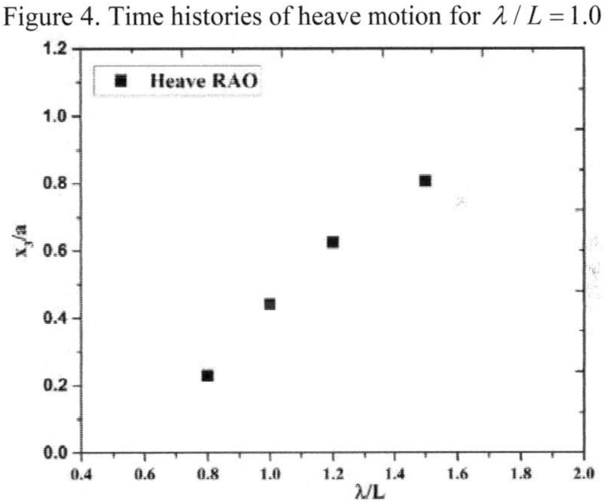

Figure 5. Heave transfer function (1st harmonic), $F_n = 0.13$

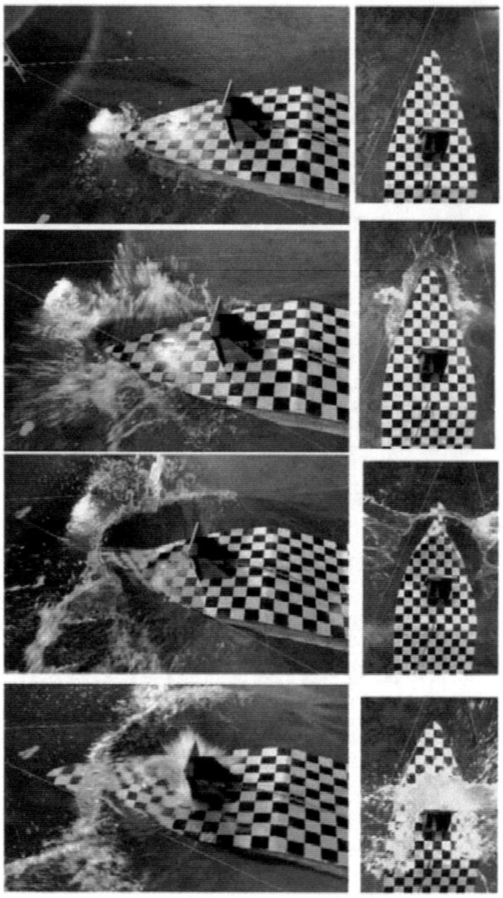

Figure 6. Flow visualization over the deck ($F_n = 0.13$, $\lambda / L = 1.0$)

Figure 6 shows the visualization of the flow evolution over the deck for $F_n = 0.13$ for $\lambda / L = 1.0$. It clearly shows the process of green water flow over the deck and its impact and splashing over the vertically mounted structure. Initially, the green water entered the deck from the bow of the model, and gradually with the sufficient increment of the wave height near the bow region, water started entering the deck from all the directions leading to the 'tongue formation'. As shown in Figure 6, a tongue of water has been formed at the front of the container vessel similar to other experiments (for example Buchner, 2000). Then the flow moved forward like a water jet and climbed up along the vertically mounted square plate.

To understand the uncertainty of the measured pressures, an uncertainty analysis was performed considering the precision errors involved. For this purpose, five repeated measurements were taken for the critical case of $\lambda / L = 1.0$ at $F_n = 0.13$. Thereafter, an estimate of uncertainty was made using Students 't' distribution. The precision limit $P_e^N(CI = C_x)$ for the confidence level C_x is calculated using the following equation:

$$P_e^N(CI = C_x) = \left(\left| \frac{t_{C_x, N-1}^N . S_N}{\sqrt{N}} \right| \times 100 \right) \% \qquad (1)$$

In the above equation, N represents the total number of sampling point, $t_{C_x, N-1}^N$ represents the Student's t-distribution value for the confidence level C_x with $N-1$ degree of freedom, and S_N is denoted as standard deviation. The precision error for 95% confidence interval is 3% for P_1 and 2% for P_2 . It may be noted that, within a 95% confidence interval, the uncertainties are within 3%, showing good consistency of the measured data.

Influence of λ/L ratio

The impact pressure on the deck is investigated for different λ / L ratios for constant model speed to understand the influence of wavelength on the shipping of green water. The model velocity is obtained using Froude scaling of the full-scale ship speed as per standard ITTC practice in ship model testing.

$$F_n = \frac{U}{\sqrt{gL}} \qquad (2)$$

Here, U is the vessel speed, L is the length of the vessel, and acceleration due to gravity is denoted by g . The non-dimensional pressure time history is shown in Figure 7 for the points P_1 , P_2 and P_{R1} at $F_n = 0.13$. It is observed in Figures 7 - 9, that the green water peak pressure is maximum at $\lambda / L = 1.2$ for all the points located on the deck whereas on the superstructure the pressure is maximum at $\lambda / L = 1.0$.

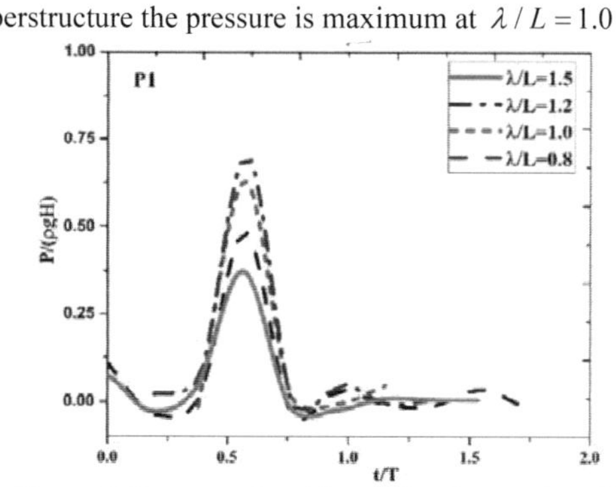

Figure 7. Pressure time history for three points P_1

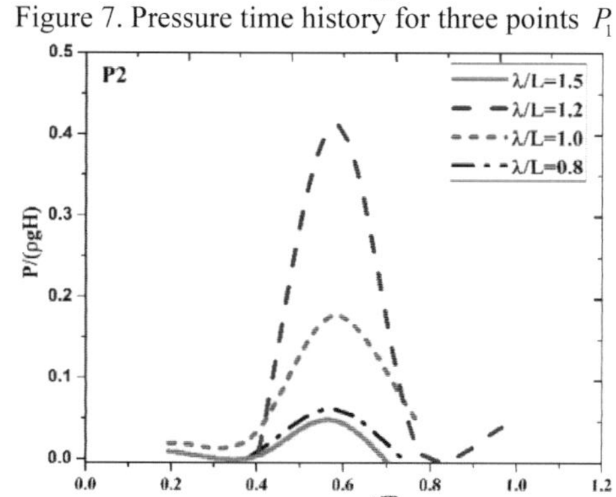

Figure 8. Pressure time history for three points P_2

The variation of the non-dimensional peak pressure with respect to λ / L ratio is plotted in Figure 10. From the figure, it is observed that the pressure decreases at lower and higher λ / L values. This shows the dependency of green water loading on wavelength since the same wave height is used in the experiments. A similar trend was reported by Fonseca & Guedes Soares (2005), where higher green water pressures are observed at $\lambda / L = 1.2$ for the same height of water on deck. The vessel motions are high in this wavelength range. Also, further increasing the wavelength keeping the wave height same implies that the wave steepness is decreased, which might reduce the magnitude of green water loading.

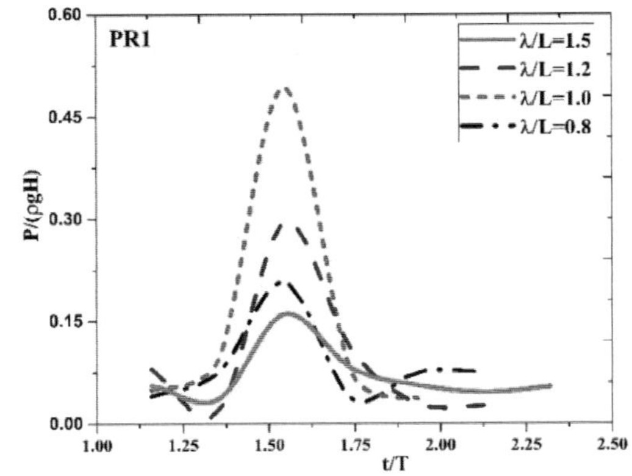

Figure 9. Pressure time history for three points P_{R1}

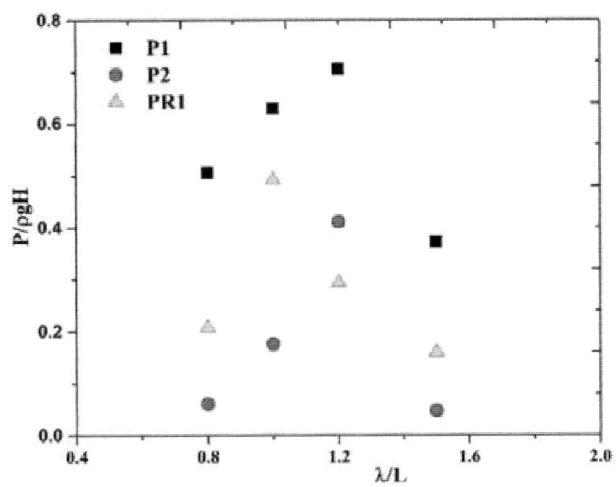

Figure 10. Influence of λ / L on peak pressure

Influence of wave steepness

In order to understand the effect of wave steepness (H / λ), the peak pressure has been measured for different wave steepness ratios. The non-dimensional pressure time history is shown in Figure 11 to 13 for the points P_1, P_2 and P_{R1} at $F_n = 0.13$. From the Figure, it is observed that the pressure increases with the increase in H / λ ratio.

Figure 14 shows the variation of the peak pressure with respect to steepness. The peak pressure is normalized by hydrostatic pressure $\rho g H$. Interestingly, the rate of change in non-dimensional peak pressure for the points P_1 and P_2 is nearly constant. However, for P_{R1}, it is linear. This means that the variation of the dimensional peak pressure at the location of P_{R1} is quadratic. However, the same is linear for the other two points on the deck. This can be explained from the point of view of components of the green water load. According to the Buchner (1995) dam break model, the green water load can be roughly divided into two components, one is hydrostatic and the other is dynamic. In view of that, the present result shows that dynamic part is more dominant at the superstructure as this part is proportional to V^2, i.e. square of the velocity. On the other hand, the static part plays a dominant role at the points P_1 and P_2 on the deck, as this part is proportional to wave steepness.

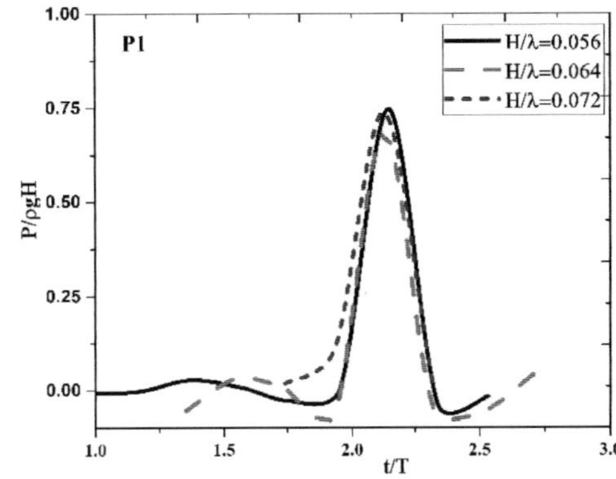

Figure 11. Pressure time history at point P_1

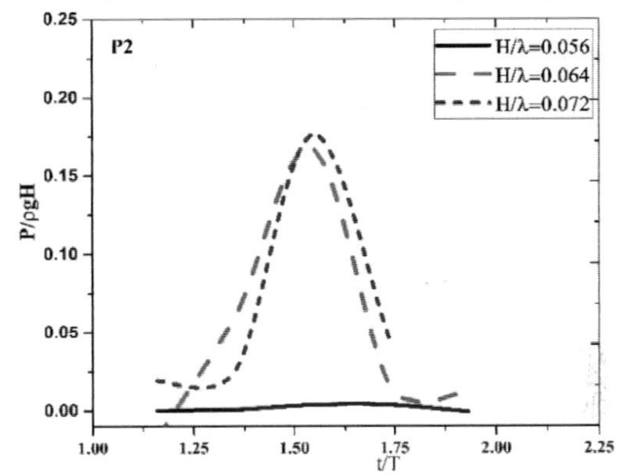

Figure 12. Pressure time history at the point P_2

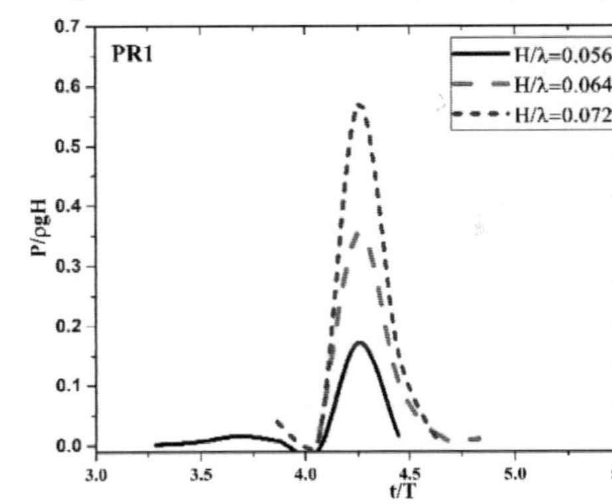

Figure 13. Pressure time history at points P_{R1}

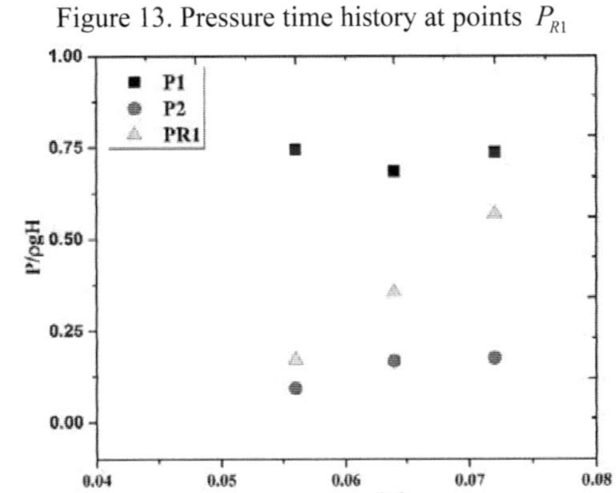

Figure 14. Influence of wave steepness on peak pressure

Influence of ship speed (F_n)

In this section, the influence of Froude number (F_n) on green water loading over the deck for the KCS hull is investigated for $\lambda / L = 1.0$. The different Froude numbers used are F_n = 0.055, 0.09, 0.13 and 0.162 respectively. Figures 15 and 16 represents the time history of the impact pressure at points P_1 and P_{R1} which are located on the deck and superstructure. These results show that the pressure increases with the increasing speed of the vessel for both the locations. The magnitude of impact pressure is higher on the deck compared to the point located on the superstructure.

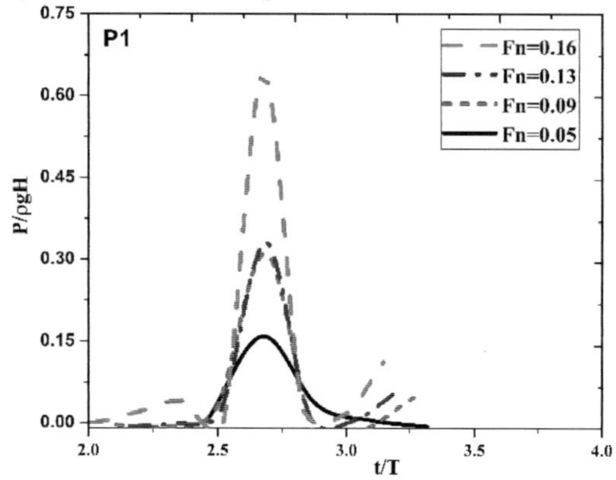

Figure 15. Influence of ship speed at P_1

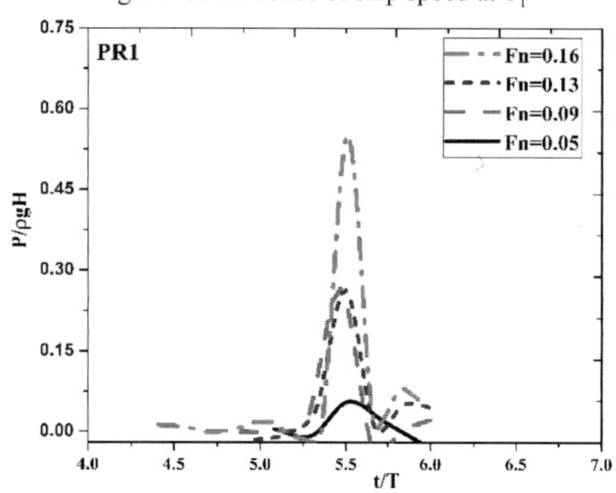

Figure 16. Influence of ship speed at P_{R1}

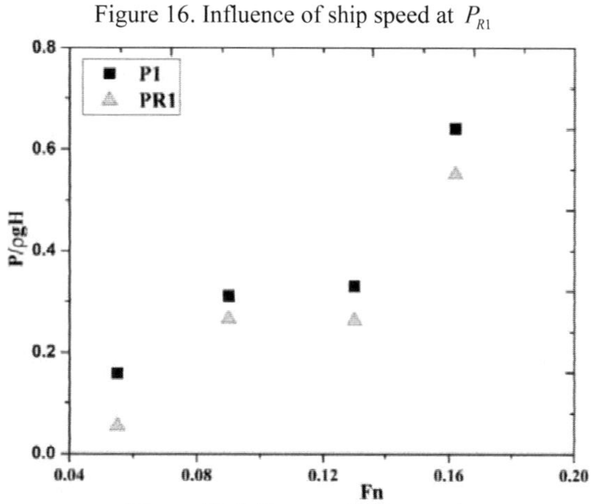

Figure 17. Influence of ship speed

Figure 17 shows the variation of the non-dimensional peak pressure with respect to the Froude number. In Figure 17, it may be noted that there is a sudden rise in the magnitude of the impact peak pressure from $F_n = 0.13$ to $F_n = 0.16$, which is observed from $F_n = 0.05$ to $F_n = 0.09$ also. It means that there is pressure jump when speed varies from low to moderate and moderate to high. At this point, it is difficult to analyze and hence demands more detailed investigations to understand this phenomenon more clearly. However, the obtained green water pressures indicate a strong dependency of the green water loading on the vessel speed.

CONCLUSIONS

In the present paper, green water occurrence is investigated experimentally for the container vessel- KCS with forward speeds. The primary focus of this paper is to understand the effect of λ / L ratio, wave steepness, and F_n on green water loading for a vessel experimentally as well as numerically. The green water impact is computed at three different locations near the bow region- two points on the deck, and one point on the superstructure. The results indicate that green water loading significantly increases with the increase of ship speed, and this is consistent for the selected range of wavelengths and vessel parameters. Also, it is seen that the green water impact largely varies with λ / L ratio and attains its maximum in the range of $\lambda / L = 1 - 1.2$.

Green water pressures for different wave steepness show that for the points on deck the green water load varies linearly, while for the superstructure, the load increases in quadratic order. This shows for higher steepness, damage due to green water on superstructure would be more severe than that of the deck structure. Therefore, one may conclude that green water loading is considerably affected by the vessel speed, wavelength to ship length ratio and wave steepness and one should not ignore this effect while calculating the design load for green water impact.

REFERENCES

[1] Berhault, C., Guérin, P., Martigny, D., & Gueret, R. (1998, January). Experimental and numerical investigations on the green water effects on FPSOs. In The Eighth International Offshore and Polar Engineering Conference. International Society of Offshore and Polar Engineers.

[2] Buchner, B. (1995). The impact of green water on FPSO design. In Offshore technology conference. Offshore Technology Conference.

[3] Buchner, B., and Voogt, A. (2000). The effect of bow flare angle on FPSO green water loading. Proceedings of ETCE/OMAE2000 Joint Conference Energy for the New Millenium, Neworleans, LA.

[4] de Carvalho e Silva, D., and Rossi R. R. (2014). Green water loads determination for FPSO exposed to beam sea conditions. In ASME 2014 33rd International Conference on Ocean, Offshore and Arctic Engineering, San Francisco, United States, pp. V08BT06A012-V08BT06A012.

[5] Faltinsen, O. M., Greco, M., and Landrini, M. (2002). Green water loading on a FPSO. Journal of Offshore Mechanics and Arctic Engineering, 124(2), pp. 97-103.

[6] Fonseca, N., Soares, C. G. (2005). Experimental Investigation of the Shipping of Water on the Bow of a Containership. Journal of Offshore Mechanics and Arctic Engineering, 127(4), pp. 322-330.

[7] Greco, M. (2001). A Two-dimensional Study of Green-Water Loading (Doctoral dissertation, Norwegian University of Science and Technology).

[8] Greco, M., Faltinsen, O. M., & Landrini, M. (2001). Basic studies of water on deck. In Twenty-Third Symposium on Naval Hydrodynamics Office of Naval Research Basin d'Essais des Carenes National Research Council, Val de Reuil, France.

[9] Greco, M., Landrini, M., Faltinsen, O. M. (2004). Impact flows and loads on ship-deck structures. Journal of Fluids and Structures, 19(3), pp. 251-275.

[10] Kleefsman, K. T., Loots, G. E., Veldman, A. E., Buchner, B., Bunnik, T., and Falkenberg, E. (2005). The numerical simulation of green water loading including vessel motions and the incoming wave field. In ASME 2005 24th International Conference on Offshore Mechanics and Arctic Engineering, pp. 981-992.

[11] Kudupudi, R. B., Bhattacharyya, A., & Datta, R. (2019). A parametric study of green water impact on a container ship. Ships and Offshore Structures, pp. 1-7.

[12] Nielsen, K. B., Mayer, S. (2004). Numerical prediction of green water incidents. Ocean Engineering, 31(3), pp. 363-399.

[13] Pham, X.P., Varyani, K.S. (2005). "Evaluation of green water loads on high-speed containership using CFD". Ocean Engineering, 32(5), pp. 571-585.

[14] Tanizawa, K., Sawada, H., Tsujimoto, M. and Koshizuka, S. (2004). Experimental and numerical study of shipping water impact on running ship foredeck in regular head seas. In Hydrodynamics VI: Theory and Applications: Proceedings of the 6th International Conference on Hydrodynamics, Perth, Western Australia, 24-26 November 2004 (p. 125). CRC Press.

[15] Theilen, L., Detlefsen, O., Abdel-Maksoud, M., Bohm, M. (2015). Efficient Simulation Method to Predict Green Water Loads on Superstructures. In ASME 2015 34th International Conference on Ocean, Offshore and Arctic Engineering.

[16] Voogt, A., & Buchner, B. (2004). Prediction of wave impact loads on ship-type offshore structures in steep fronted waves. In The Fourteenth International Offshore and Polar Engineering Conference. International Society of Offshore and Polar Engineers.

[17] Voogt, A.J., Buchner, B. and Garcia, J.L.C. (2004). Wave impact excitation on ship-type offshore structures in steep fronted waves. In Proceedings OMAE Specialty Conference on Integrity of Floating Production, Storage & Offloading (FPSO) Systems (pp. 04-0062).

[18] Zhang, X., Draper, S., Wolgamot, H., Zhao, W., & Cheng, L (2018). Numerical investigation of effects of bow flare angle on green water overtopping a fixed offshore vessel. In ASME 2018 37th International Conference on Ocean, Offshore and Arctic Engineering, Madrid, Spain.

[19] Zhao, X., Ye, Z., Fu, Y. (2014). Green water loading on a floating structure with degree of freedom effects. Journal of Marine Science and Technology, 19(3), 302-313.

[20] Zhu, R., Miao, G., Lin, Z. (2009). Numerical research on FPSOs with green water occurrence. Journal of Ship Research, 53(1), 7-18.

Proceedings of the 6th International Conference on Ship and Offshore Technology
ICSOT 2019
November 7-8, 2019, IIT Kharagpur, India

ICSOT INDIA 2019-29

SEAKEEPING AND SLOSHING ASSESSMENT METHODOLOGY FOR FLOATING LNG TERMINALS

Jai Ram Saripilli
Research and Innovation Centre
Indian Register of Shipping
Mumbai, India
jairam.saripilli@irclass.org

Sharad S Dhavalikar
Research and Innovation Centre
Indian Register of Shipping
Mumbai, India
ss.dhavalikar@irclass.org

Debabrata Sen
Dept. of Ocean Engineering and
Naval Architecture
Indian Institute of Technology
Kharagpur, India
deb@iitkgp.ac.in

ABSTRACT

A methodology for prediction of life-time design slosh-induced loads for floating liquefied natural gas (LNG) terminals is presented. The methodology is in general common to slosh-induced load assessment of all LNG vessels. Additional considerations specific to FLNGs and FSRU terminals are discussed. Vessel motion responses and slosh loads are computed for strength assessment of pump tower and containment system structure. Present paper demonstrates the numerical simulation based methodology by considering two design conditions, a transit LNG carrier mode and a site-specific terminal mode. Slosh coupling effects are accounted in determining the vessel motion responses. Slosh-induced loads are determined using multiphase finite volume method (FVM) based open-source CFD solver OpenFOAM. Dependency of statistical analysis on design basis, assessment methodology and its consequences on derived sloshing loads are discussed. Finally, application of these loads for structural analysis is discussed.

KEYWORDS

Sloshing, slosh-coupled ship motions, design slosh loads

INTRODUCTION

Strength assessment is one of the primary objectives of classification society. Prediction of vessel responses and environmental loads i.e. external wave induced loads and internal sloshing loads is an important prerequisite for structural design assessment of floating LNG terminals. Ship motions which govern the severity of sloshing loads are in turn effected by internal sloshing flow. This strong coupling behaviour of sloshing on ship motions is important to be modelled adequately. Slosh coupling behaviour is even more sensitive for LNG vessels with prismatic containment tanks, operating within and outside barred filled limit.

Sloshing is highly nonlinear and irregular phenomenon. Unlike other wave induced loads (such as bending moment, shear forces etc.), determination of suitable transfer function for slosh-induced loads (typically impact pressure) is not possible. Frequency domain or spectral analysis are hence not applicable for slosh loads evaluation of LNG vessels. And a probabilistic assessment based on the time-domain data is preferred. Probabilistic based assessment framework is proposed in several works (see e.g. [7,9,11,15]), which are typically based on model experiments. Standard guidelines and best practices are also well established by several classification societies [1,2,4,6,8,12,13]. However, it is to be noted that sloshing assessment procedures are still under investigation and there still exists no fully satisfactory direct computation methodology that solves the slosh-induced load assessment problem fully consistently [14].

In this paper, we present a detailed methodology for estimating the life-time extreme motion responses and slosh-induced loads for structural assessment of a Floating LNG terminal. All the major steps involved in the seakeeping and sloshing load assessment, i.e. (a) analysis of wave data, (b) seakeeping analysis with slosh-coupling, (c) sloshing analysis and (d) short-term and long-term statistical analysis are briefly discussed. Methodology for the estimation of slosh design loads relevant to the structural strength assessment of containment system and pump tower is presented. Several aspects of present methodology are described in some of our previous works [18-20, 23]. In this work we lay our emphasis on additional considerations related to environment and subsequent definition of design scenarios that are required for floating liquified natural gas (FLNG) terminals and floating storage and regassification units (FSRU. Further application of these loads for structural analysis is discussed.

A case study is presented, showing the application of the methodology for a floating LNG terminal. Transit and site-specific conditions are considered for the case study. All the design considerations specific to these two conditions are discussed. All the analysis tasks involved in the case study are performed using in-house developed tools, such as SHiPx (an integrated time-domain 3D forward speed seakeeping solver) and its modules such as *shipXslosh*, *sloshpost* (for slosh coupling, spectral and statistical post-processing etc.). The sloshing analysis is performed based on open-source CFD solver OpenFOAM.

Numerical simulations are cost effective and reasonably fast due to the current computational resources. Besides, scaling uncertainties do exist in scaling pressures form model experiments to full scale. In this aspect, present methodology based on full-scale numerical simulations would be advantageous and help designers and classification bodies in strength assessment of LNG vessels.

DESIGN BASIS

The methodology adopted for the estimation of design slosh-induced loads is classified primarily based on design basis. According to which a (i) comparative assessment approach can be employed vessels with minor design change (for which geometrically similar proven reference design exists) and a (ii) detailed assessment approach which is applicable for all designs and typically mandatory for novel designs (new hull lines and/or tank geometry). The second case to be applied where no reference safe design exists. Sloshing is highly stochastic and a small change in probability level will lead to a large difference in slosh pressure. A long-term analysis is recommended considering the stochastic nature of sloshing pressures. However, a quick and simplified comparative assessment approach based on fewer number of design load scenarios can be employed for design checks. Where, we determine the design loads based on short-term statistics of extreme weather conditions. Definition of design load scenario is deferred till subsequent sections. For detailed design assessment, we use long-term approach based on short-term statistics of several or large number of design loads scenarios.

METHODOLOGY

The methodology for sloshing analysis can be categorized into three major steps (i) seakeeping analysis, (ii) sloshing analysis and (iii) statistical analysis. The analysis tasks are performed based on two prerequisite steps, selection of loading conditions and environmental conditions. Besides an intermedia stage for filtering/short-listing load scenarios for detailed sloshing assessment is used to avoid unnecessary large number of simulations. The steps involved in sloshing load assessment methodology are subsequently discussed with the help of a case study.

Case Study

A floating LNG terminal with a capacity of 180000m^3 is considered as shown in fig. 1. The 286.5m long vessel measures 48.0m in breadth with a design draft of 11.5m. The vessel is equipped with four prismatic tanks with MARK III type containment system. The forward most tank (cargo tank no.1) is significantly small in dimensions while the remaining three tanks (cargo tanks no. 2, 3 and 4) are identical. Sloshing is severe in tanks which experience severe motions. It is expected that the forward most tank would experience severe motions. However, considering small dimension of the tank sloshing would be less severe. The aft most tank (cargo tank no. 4) which experiences maximum relative motions among the rest is considered for sloshing load assessment. Subsequently the results relevant to this tank under consideration are discussed.

The approximate values of main dimensions of tank are given in table 1. The density and kinematic viscosity of LNG and its vapor are considered as follows: ρ_{LNG} = 470 kg m^{-3}, ρ_{vapour} = 1.0 kg m^{-3}, υ_{LNG} = 2.5E-7 m^2 s^{-1} and υ_{LNG} = 3.0E-6 m^2 s^{-1}. Generally, a barred fill limit is imposed on LNG carriers. The barred fill limit is defined as 10%H to 70%H (H or H_T represents tank height), i.e. no filling allowed over 10%H and less than 70%H for. Unlike LNG carriers the LNG terminal may be expected to encounter seas with partial filling within this fill range in their design life. To ensure adequate safety, partial filling across the barred fill limits would also be necessary for sloshing assessment of these LNG terminals.

Figure 1. 3D MODEL OF FLNG VESSEL WITH TANKS

Table 1. MAIN DIMENSIONS (APPROX.) OF TANK

Length	L_T	[m]	46.240
Breadth	B_T	[m]	42.160
Height	H_T	[m]	29.245
Lower chamfer height	C_l	[m]	5.910
Upper chamfer height	C_u	[m]	9.650

Design Load Scenarios

We define a load scenario or load case as the combination of relevant irregular short-term sea condition, heading and speed for each loading condition or fill level in tank. And those which are finally used for detailed sloshing tests and contribute to the final slosh-induced load for design assessment are termed as design load scenarios. The choice of these design load scenarios depends on severity of vessel's motion responses subject to severe environment which the vessel is likely to encounter in its life-time for specified relative heading, speed and fill. These are also governed by the operational limits imposed for various types of operations. For instance, loading conditions with fill levels between maximum and minimum barred fill limits is conventionally avoided in transit condition. However, these fill levels become relevant in-case of LNG terminals and FSRUs. Other operational limits governing the definition of load scenarios are the severity of environment for various heading and speed conditions. The procedure adopted in the selection of load scenarios at various stages of analysis is subsequently discussed with the aid of above-mentioned case.

ENVIRONMENT MODELLING

The North Atlantic wave scatter data (IACS Rec. 34 [10]) is generally used to design vessels intended for worldwide operations. Where the individual sea-states can be modelled according to Pierson-Moskowitz (PM) spectrum. For fixed or site-specific operations such as FSRUs or LNG terminals, a site-specific wave scatter data with appropriate spectral definition of individual sea states can be used. A directional wave scatter data is required in addition to establish the probability of relative heading. Similarly, the information for wind and current definition will also be required in case of second order motion responses (accounting mooring or dynamic position system). However, these have negligible influence on slosh induced loads and are not considered here.

A long-term analysis approach estimates slosh-induced loads based on several short-term sea states which included non-extreme but rather more frequently occurring ones as well.

The long-term distribution of loads depends on the probability of occurrence of individual sea conditions based on vessel's operating conditions. The probability of occurrence of sea states used for world-wide operation (transit condition) and site-specific operation considered for present case study is shown in figs. 2 and 3.

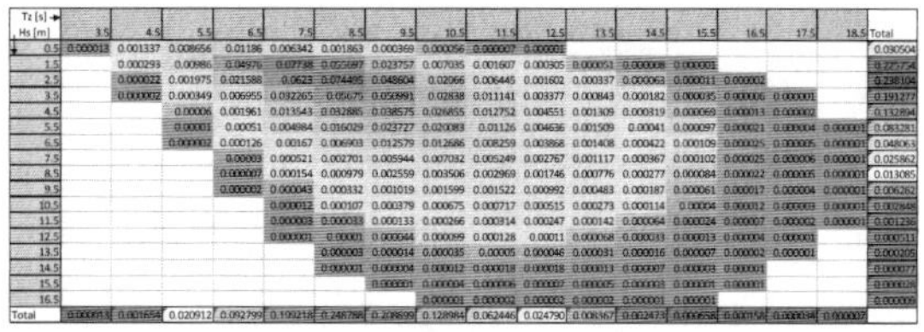

Figure 2. PROBABILITY OF SEA STATES FOR WORLD WIDE OPERATING CONDITION

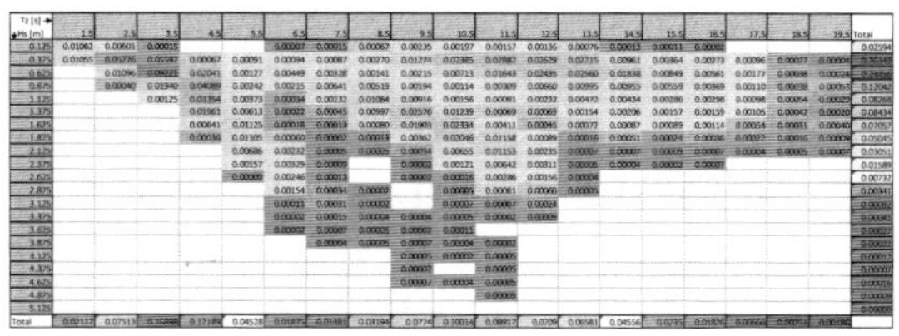

Figure 3. PROBABILITY OF SEA STATES FOR SITE-SPECIFIC OPERATING CONDITION

In the short-term analysis approach (typically used for simplified comparative design assessment approach), slosh-induced loads are computed based on extreme sea states. The extreme seas likely to occur in the design life can be derived from the long-term wave climatology data. Where the significant wave height for desired return period (T_R) at any zero-crossing period (T_z) in the scatter table can be derived as:

$$H_{S_{T_R}} = Q^{-1}\left(\frac{1}{n\,T_R}\right) \quad (1)$$

here, n is number of sea-states recorded per year and $Q(H_s)$ is the exceedance probability function of significant wave height (H_s) at each zero-crossing period T_z. The extreme design sea conditions are derived with a return period equal to design life, i.e. 40-year. Envelope curves for 40-year return period is shown in fig. 4.

Operational Limits

The probability for the vessel to experience 40-year extreme waves is very remote in beam sea condition. Which may happen in the event of failure of control equipment. Considering this imposed operational limit, it is assumed that the vessels encounter extreme waves corresponding to 1-year return period in case of beam sea condition. Thus, the extreme waves for 180° heading are taken at 40-year return level and the those for 90° heading are taken at 1-year return level. The extreme wave envelopes with 40-year and 1-year return period for both world-wide and site-specific operation is shown in fig. 4. The significant wave height for extreme sea conditions for all other heading (χ) is interpolated as:

$$H_{S_\chi} = H_{S_{40-yr}} \times w_{f_{40-yr}} + H_{S_{1-yr}} \times w_{f_{1-yr}} \quad (2)$$

The weight factor (w_f) used for interpolation in this study is given in fig. 5. Linear interpolation can also be used, but here we bias 1-year waves for beam and beam quartering waves. Further operational limits associated to severity of sea states in relation to speed (V) and heading are considered as described in table 2. This is a standard practice and similar assumptions are considered in several sources (e.g. [4,13]).

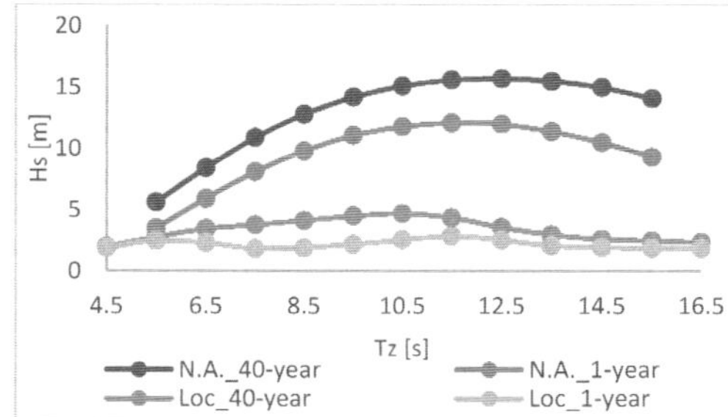

Figure 4. 3D MODEL OF FLNG VESSEL WITH TANKS

In case of terminals, the relative heading is required to be considered to decide the vessels orientation in site. The probability of relative heading can be derived from the site-specific long-term wave climatology data. The marginal probabilities for the site-specific condition is shown in fig. 6. Assuming zero speed for the terminal operation with centreline symmetry, the probabilities of heading conditions for long-term analysis are finally taken as per table 3.

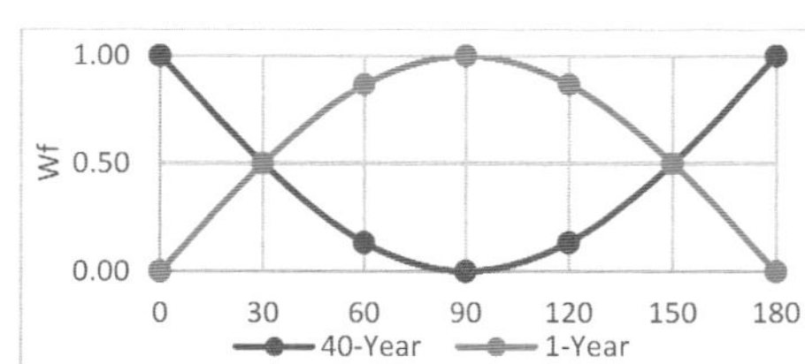

Figure 5. SITE-SPECIFIC PROBABILITY OF RELATIVE WAVE HEADING

Table 2. SEVERITY OF WAVE IN RELATION TO SPEED

Wave Height [m]	Heading [°]				
	0-45	45-60	60-120	120-135	135-180
Hs < 5	V				
5 ≤ Hs < 9	linearly interpolate				
Hs ≥ 9	V/2	5 knots			V/2

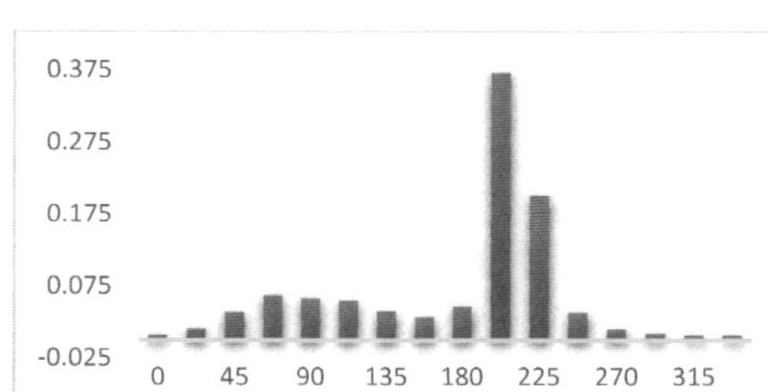

Figure 6. SITE-SPECIFIC PROBABILITY OF RELATIVE WAVE HEADING

Table 3. HEADING PROBABILITY FOR FLNG TERMINAL

0°	0.74%
22.5°	2.20%
45°	4.49%
67.5°	7.07%
90°	7.25%
112.5°	9.25%
135°	24.03%
157.5°	40.38%
180°	4.59%

Additional Considerations : Further, accounting for wind, current and effect of mooring or dynamic positing (DP) system on motion prediction become important in case of LNG terminals. Tank excitations are required to be reproduced based on quadratic transfer functions (QTF) computed accounting these effects. These effects are assumed to have less influence on sloshing and are not accounted in the present work. It is reasonable to say that motions responses obtained without considering these effects will be on a conservative side.

SEAKEEPING ANALYSIS

Tank excitations for every design load scenario are required to be determined to compute slosh loads using CFD simulations or scaled model tests. The ship motion transfer functions are computing using SHiPx, an in-house 3D forward speed boundary element method (BEM) solver based on transient Green function technique (for more details see [19,20]). Weakly-nonlinear approach is adopted in the seakeeping computations. Viscous damping effect is modelled in terms of potential damping.

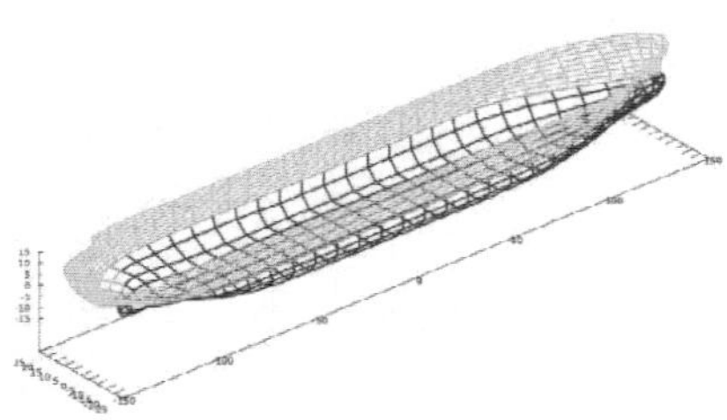

Figure 7. PANEL MESH USED BY SHiPx

Slosh-Coupling

Sloshing is caused or initiated by external ship motions and the slosh forces acting on tank in-turn effect global ship motions (typically sway and roll) [17-22]. From previous studies it is observed that slosh loads derived considering slosh-coupled ship motions is significantly different when compared to those without accounting for slosh coupling effect [17,18]. Where, for the considered cases it was found that not accounting coupling would result in over estimation of slosh loads leading to penalisation of structure. This however may not be the case always. To ensure safe and optimum structural design it is necessary to determine ship motion with slosh coupling.

Slosh coupled ship motion computations are performed using *shipXslosh,* in-house developed module of SHiPx tool. This slosh coupled ship motion solver adopts a hybrid solution algorithm that combines time-domain BEM based ship motion solver with viscous FVM based solver - `interDyMFoam` of OpenFOAM for sloshing (for more details refer [16-20]). The slosh coupled ship motion RAOs are determined over a sufficiently large range of frequencies based on single time-domain simulation run in irregular wave [19].

Spectral Analysis

Ship motion responses to different wave conditions is required to compute the inertial and internal sloshing loads. For sloshing loads the ships motions are required to define the relative tank excitations. Spectral analysis is performed to determine the motion response spectrum for each environmental condition of a load scenario. Spectral analysis is performed for two scatter tables, one for site-specific condition and other for transit world-wide operations. The long-term values of motions and accelerations computed from spectral analysis is shown in table 4. The long-term values are computed for a probability level of 10^{-8}.

Table 4. LONG-TERM MOTIONS

Motions			Accelerations		
Heave [m]	Roll [°]	Pitch [°]	a_x [m s^{-2}]	a_y [m s^{-2}]	a_z [m s^{-2}]
14.88	41.91	9.07	0.059	3.973	4.003

Besides, from the computed response spectra, those load scenarios which result in large responses and whose resonant frequencies are close to slosh natural frequency (for that fill in load scenario) are selected for sloshing assessment. Each such load scenario is now termed as design load scenario. For each of the selected design load scenario, the time history of tank excitations is generated in Monte Carlo approach using precomputed response spectrum using inverse fast Fourier transforms (IFFT) technique.

SLOSHING SIMULATIONS

Slosh loads are computed for each selected design load scenario. Conventionally slosh-induced pressures are evaluated using scaled model experiments. And CFD simulations (based on FVM, FDM, SPH etc.) are used to evaluate slosh-induced loads on pump-tower. Present study, we employ CFD simulations for estimation of loads on both containment system and pump tower. The multiphase interface capturing volume of fluid (VOF) method-based solver (same as slosh-coupled ship motions solver) is used. The in-house developed *shipXuslosh* module of SHiPx (which is based on OpenFOAM) is used to perform sloshing simulations for the selected design load scenarios. The *shipXuslosh* module first generates long duration (3-5 hour) time history of tank excitations (using ship motion response spectrum precomputed based on slosh-coupled ship motion transfer function that was obtained from *shipXslosh*) and runs sloshing simulation using `interDyMFoam` solver of OpenFOAM.

Unlike slosh-coupled ship motion simulations, in case of sloshing simulations much finer grid and time-step size is required. This is very much essential to quantify slosh impact pressures. Dynamic probes are used to monitor the pressures at critical locations and fluid particle velocity and acceleration at pump tower member locations. A schematic of probes where the pressure and fluid particle velocity and accelerations are extracted from CFD is shown in fig. 8. The corner and knuckle locations are considered as hot-spot locations and pressure probes are located at all corners and chamfer locations as shown in fig. 8. The forces acting on pump tower members are modelled according to Froude Krylov assumption. Sufficiently large number of probes are considered on each pump tower member to ensure proper resolution of forces.

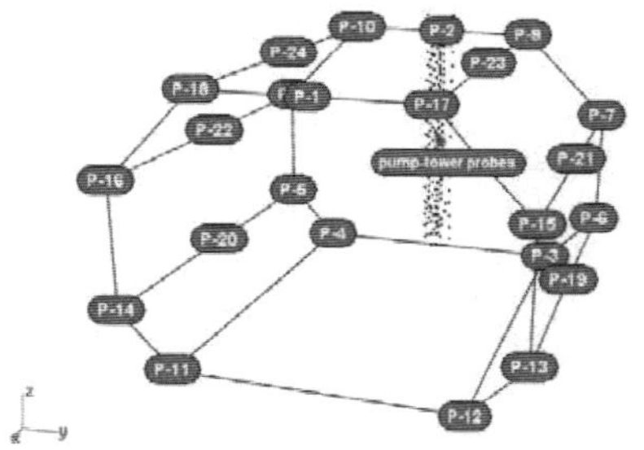

Figure 8. DYNAMIC PROBE LOCATIONS FOR PRESSURES AND PUMP TOWER

Grid and time step sensitivity studies are important to ensure accuracy of the CFD solution. Sensitivity and convergence are not presented here. Based on best practices (e.g. [4,8,13]) and previous experience, sufficient care is taken to create good quality girds with adequate time steps size combined with adjustable time stepping model of OpenFOAM. Long duration (3hours or more) sloshing simulations are performed to ensure statistical convergence within every short-term design load scenario. Fig. 9 shows snapshot of a selected case with 70% filling.

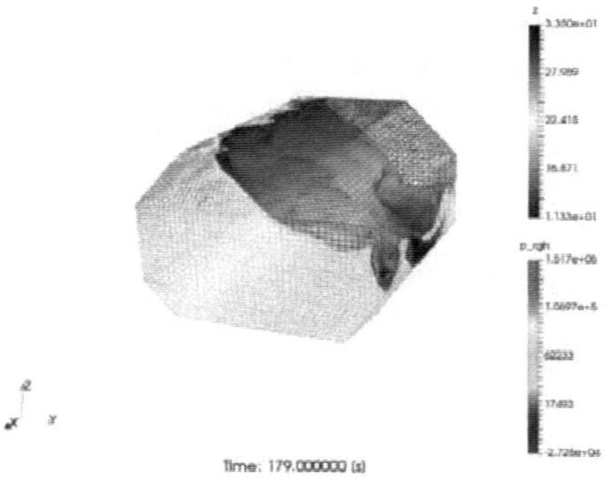

Figure 9. SNAPSHOT OF A SLOSHING SIMULATION

STATISTICAL ANALYSIS

Statistical inferences from each three-hour load scenario is required to mathematically combine the loads of all design load scenarios and then extrapolate to a desired probability level (defined by return period) for the estimation of design loads. A theoretical statistical model is used to describe pressures or pump tower forces for every short-term design load scenario.

Containment System

Processing Pressure Data : Sloshing impact pressure is considered as the governing load for containment system. The pressure data obtained from sloshing tests (CFD simulations for present study) is first processed to identify the pressure peaks corresponding each critical location. A peak over threshold method in combination with moving window technique is adopted to identify the pressure peaks (for more details see e.g. [4,7,8,19,20]). A high pass filter (≈4Hz) is applied to eliminate hydrostatic and low/wave frequency components of pressure. A maximum per event method can be adopted to determine a single description of pressures peaks of all locations. This way a single statistical evolution of representative envelope for all impacts can be drawn for given load scenario. This is done by identifying the maximum pressure peak among all locations within a every time window. Care has to be taken in fixing the time window to account for likely phase lag between impacts at various locations (i.e. impact at lower chamfer would occur prior to that at upper chamfer). Alternatively, location specific design loads are determined and the maximum of all probe locations is finally applied on to the structure. It is important to study the sensitivity of maximum per event approach compared to location specific approach in deriving design loads. This not considered in the present study and would to be studied in future.

A histogram of identified pressure peaks is determined by sorting the magnitude into large number of bins. An exceedance probability function (EPF) is used for visualisation of statistical description, as the main interest is on the extreme values of the probability distribution. The in-house code *sloshPost* (another module of SHiPx) is used to process the dynamic probe data for statistical description of short-term pressures. Time history plots of pressure and their extracted peaks for a sample selected case with 30%H filling is shown in fig 10a (impacts associated to upper chamfer) and 10b (impacts associate to lower chamfer).

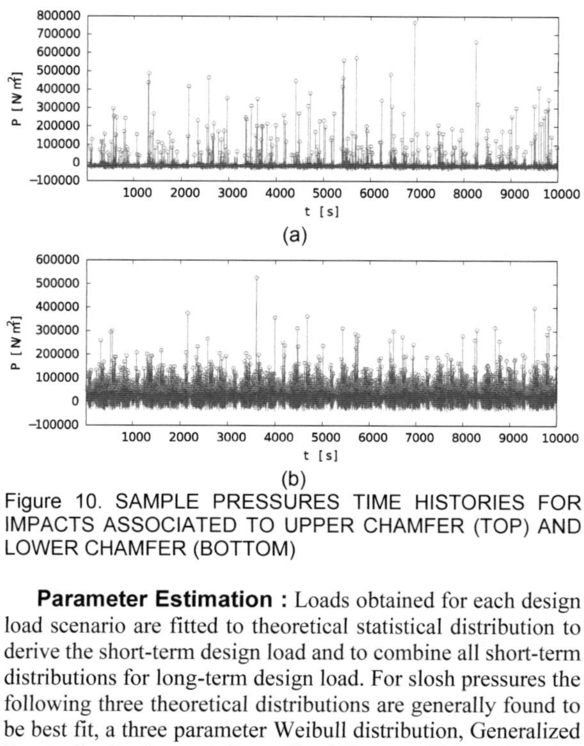

Figure 10. SAMPLE PRESSURES TIME HISTORIES FOR IMPACTS ASSOCIATED TO UPPER CHAMFER (TOP) AND LOWER CHAMFER (BOTTOM)

Parameter Estimation : Loads obtained for each design load scenario are fitted to theoretical statistical distribution to derive the short-term design load and to combine all short-term distributions for long-term design load. For slosh pressures the following three theoretical distributions are generally found to be best fit, a three parameter Weibull distribution, Generalized Pareto distribution and Generalized Extreme Value. Note that there may be no theoretical basis related to best fit distribution, and one may choose any other model as well. The parameter estimation is performed using the method of L-moments and L-moment ratios. Goodness-of-fit studies using Cramer-von Mises (CvM) test is performed to identify the acceptance or rejection of fitted distribution. Note that the statistic of goodness-of-fit test only indicate the acceptance or rejection of theoretical model and does not give any information on accuracy or consequence on predicted value of load. Hence, identification of best fit theoretical model is accompanied by graphical examination. In the present study, the non-rejected theoretical distribution which results in conservative value of load is selected for estimation of design loads.

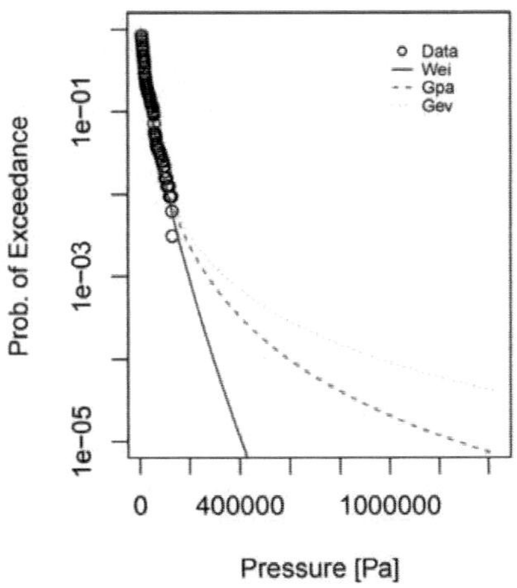

Figure 11. SAMPLE EXCEEDANCE PROBABILITY PLOT OF PRESSURE FOR A SELECTED CASE

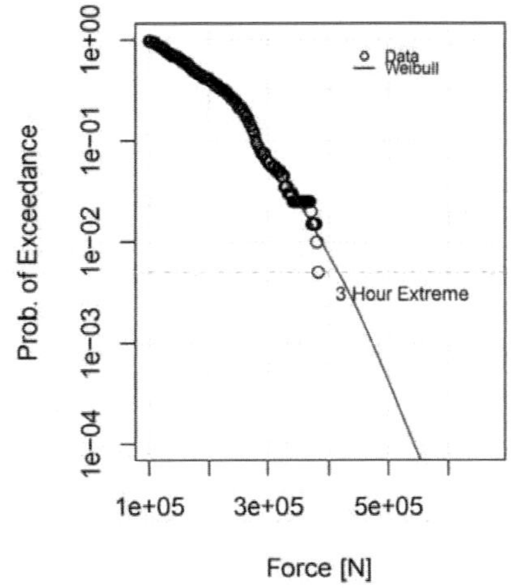

Figure 12. SAMPLE PLOT OF EXCEEDANCE PROBABILITY OF PUMP TOWER FORCE (F_y) FOR A SELECTED CASE

The statistical fit studies are done using the `lmomco` [3] package of `R` programming language. Fig. 11 shows the graphical representation of exceedance probability function of pressure peaks along with fitted theoretical distribution for a sample case. For the case shown in fig. 11, the CvM test statistic indicated that the hypothesis that the distribution follows a Generalized Extreme Value distribution (Gev) is rejected. And the hypothesis that the distribution follows the three parameter Weibull (Wei) and Generalized Pareto (Gpa) distribution is accepted. It can be seen that among the non-rejected Weibull and Generalized Pareto distributions, the Generalized Pareto distribution would result in more conservative value of pressure for higher return periods and hence is finally selected for design load estimation.

Pump Tower

The loads acting on pump tower can be decomposed into components such as hydrodynamic (sloshing), inertial, thermal and other internal loads. The hydrodynamic sloshing and inertial loads are considered for the present study. The inertial loads are computed based on the relative tank/ship accelerations. The hydrodynamic loads on pump tower are characterised using Morrison's equations. For which the fluid velocity and accelerations at pump tower members is computed from CFD simulations. Drag and inertia coefficients are dependent on parameters such as Reynolds number and Keulegan-Carpenter number. These values are also sensitive to local physical behaviour of flow such as wave impact zones and shielding effect. These effects are assumed to be less significant and not considered. Standard definition of these coefficients can be found in several source such as [5,8]. Present study we assume a drag coefficient of 0.7 and inertia coefficient of 2.0.

To avoid discretisation errors, large number of nodes/elements similar to the discretisation used for finite element analysis (used in structural assessment) is used to monitor the velocity and acceleration on pump tower members. Dynamic probes are located at the centreline of each such elements. The volume fraction information is combined to ensure sharp representation of free-surface. The elemental forces are thus computed as:

$$d\mathbf{F} = \left\{\rho_{LNG}\alpha_{LNG}\left(\frac{1}{2}C_D\mathbf{V}_n|\mathbf{V}_n|D + C_M\dot{\mathbf{V}}_nA\right) + \rho_{vapour}\alpha_{vapour}\left(\frac{1}{2}C_D\mathbf{V}_n|\mathbf{V}_n|D + C_M\dot{\mathbf{V}}_nA\right)\right\}dl \quad (3)$$

The time dependent pump tower load is obtained by integrating the load components of all the member elements. Statistical analysis of pump tower loads is performed in a fashion similar to that of pressure on containment system as described in previous section. Unlike pressures, the pump tower forces do not exhibit flat tail nature of the exceedance probability function. And a three parameter Weibull distribution is generally found to be best fit for determining short-term or long-term value of loads. For a sample case, graphical representation of probability of exceedance of lateral component of slosh force (F_y) is shown in fig. 12. It can be observed that the exceedance probability curve for pump tower force (fig. 12) do not exhibit flat tail nature unlike slosh pressure (see fig. 11).

DESIGN LOADS

Short-term

Based on the fitted theoretical distribution for each load scenario, the short-term load for that load scenario can be estimated as the load exceeding short-term probability level. The short-term probability level is defined by the number of events likely to occur in that load scenario. The probability level is defined by the number of events or impacts (in case of slosh pressure). The short-term load (pressure) can be obtained from theoretical fitted exceedance probability function $Q_s(p)$ with N_s number of impacts for the given amount of time spent in that sea state as:

$$Q_s(p_s) = \frac{1}{N_s} \quad (4)$$

Fig. 13 shows a sample short-term exceedance rate of pressure along with fitted distribution, the probability level at which short-term load is evaluated is plotted along for illustration. Note that in the above we show the exceedance probability function in terms of exceedance rate. For this case, the long duration simulations have revealed that event rate is approximately 26. Note that event in this case is the slosh impact on the probes associated to upper chamfer. Assuming the short-term sea with duration of storm as 3 hours, the short-term load for this load scenario is thus estimated as the 3-hour extreme value using the non-rejected conservative distribution as illustrated in fig 13. The maximum pressure for this short-term load scenario is found to be 0.18MPa, while the design load for this load scenario which is derived using Generalized Pareto distribution (Gpa) is found to be 0.20MPa. The short-term design load can finally be estimated as the maximum of all such 3-hour extreme (or based on duration of storm) of all load scenarios.

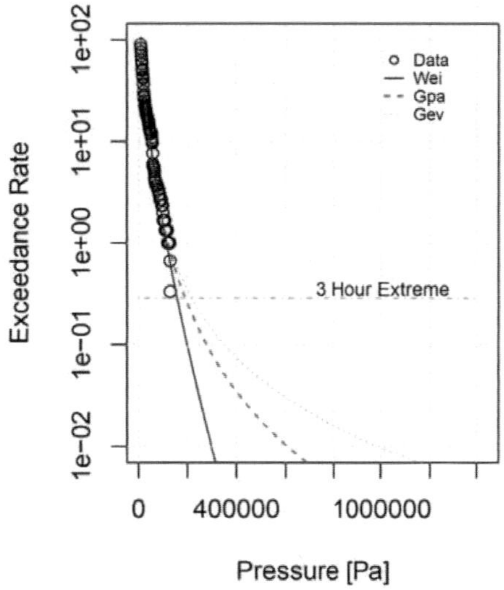

Figure 13. SAMPLE ILLUSTRATION FOR DERIVING SHORT-TERM LOAD OF A SLECTED LOAD SCENARIO

Long-term

The long-term probability of exceedance function of load (consider pressure) can be derived using the fitted short-term theoretical distribution as:

$$Q_L(p) = \sum_{i=1}^{N_{LS}} \alpha_i w_i Q_{s_i}(p) \quad (5)$$

where,

- $Q_{s_i}(p)$ is the short-term exceedance probability function of ith load scenario
- N_{LS} is the total number of load scenarios
- α_i is the probability of ith load scenario (combined probability of wave, heading and filling)
- w_i is the weighting factor defined as $E_i/\bar{E}$ for i^{th} load scenario
- E_i is the event rate of load i^{th} scenario
- $\bar{E}$ is the average event rate over all scenarios

The long-term design load can thus be predicted from the long-term exceedance probability function $Q_L(p)$ as the load (pressure) which exceeds the probability of total number of impacts ($N_{life-time}$) which are expected to occur in the design life, given as:

$$Q_L(p_L) = \frac{1}{N_{life-time}} \quad (6)$$

The short-term approach is generally applicable to extreme events. The extreme events are the extreme sea conditions which the vessel may encounter in its design life. The short-term approach is not ideal and is not recommended for sloshing assessment [9] as this is based on fewer number of sea conditions which do not represent the entire operating scenarios. A long-term approach is recommended as it accounts for all the sea conditions (including more frequent and less sever) or all likely design scenarios representing entire life. We demonstrate this by considering few selected design scenarios (see table 5). Note that that in actual large number of load scenarios are considered, we consider a few among them for illustration purpose.

Table 5. SUMMARY OF SHORT-TERM AND LONG-TERM DESIGN LOAD (PRESSURE) ESTIMATIMATION

Load Scenario	Max. Press. [MPa]	Event rate (E_i)	Fitted dist.	Prob. of L.S. (α_i)	Short-term value [MPa]	Long-term value [MPa]
Case-1	0.017	1.67	Wei	6.5%	0.019	
Case-2	0.052	16.50	Gpa	6%	0.059	
Case-3	0.094	23.33	Gpa	5%	0.106	
Case-4	0.035	4.75	Gpa	12.5%	0.038	
Case-5	0.015	1.78	Wei	20%	0.016	0.32
Case-6	0.056	1.00	Wei	20%	0.058	
Case-7	0.031	5.17	Wei	12.5%	0.033	
Case-8	0.181	26.58	Gpa	5%	0.200	
Case-9	0.044	12.50	Gpa	6%	0.049	
Case-10	0.065	0.60	Wei	6.5%	0.066	
	Max.	Max.	Mean event rate ($\bar{E}$)	Design load:	Short-term	Long-term
	0.181	26.58	**9.388**		**0.20**MPa	**0.32**MPa

For each load scenario, the maximum pressure and event rate observed from the long duration simulation is reported along. The best fit theoretical distribution for each load scenario is also shown. The short-term value of pressure for each scenario is obtained from the fitted distribution. For the considered sample cases, it can be seen that the design short-term pressure is found to be 0.2 MPa. The probability of exceedance for short-term load is determined based on event rate within each load scenario assuming a storm duration of 3 hours. This value approximately ranges in the order of 10^{-2} to 10^{-3}. While the probability of exceedance for long-term value of load assuming the design life of 40-years may be approximately in the order of 10^{-6} to 10^{-8}. For this case considering a mean event rate of 9.388, the probability of exceedance for 40-year life turns out to be approximately 3.7×10^{-7}. The long-term value of pressure from considered cases is thus found to be 0.32MPa. The design loads other than pressures, i.e. pump tower force components and global sloshing loads acting on supporting hull structure can be computed in a similar fashion. Note that in actual we consider large number of load scenarios, in this section we have simplified the cases for illustrative purpose.

APPLICATION OF LOADS ON STRUCTURE

The application of global and inertial loads on structure for finite element analysis (FEA) is trivial and is no different from strength assessment of other general ship types. However, care is required in the application of slosh related loads for strength assessment of containment system and pump tower structure.

Load Application on Pump Tower

The magnitude of the pump tower load components i.e. reaction forces (F_x, F_y, F_z) is quantified based on the long-term analysis as discussed before. Note that long-term statistical analysis of time history data of pump tower is performed on total reaction forces acting on pump tower and the long-term value does not give any idea about the loading profile to be applied. It is thus suggested to consider several loading profiles identified from initial time domain simulations. As a minimum for present study we consider two loading profiles per fill condition for each force component. The two loading profiles correspond to the maximum instances of force components with flow in opposite directions, i.e. in case of motion in transfer mode (say roll) flow from port is considered as one profile and flow from starboard is considered as another profile. Likewise, we select several sets of loading profiles for different situations and scale the loading profile such that the total force is equal to the long-term value. In the FEA, the computed pump-tower sloshing loads are applied to all the columns and braces located below the liquid dome according to the loading profile. In addition, the gravity, inertia and thermal loads are applied to all elements.

Load Application on Containment System

Slosh loads are impulsive in nature and the dynamic response of containment system depends on the time variation of the pressure impulse. The above procedure describes only the statistical evolution of magnitude of pressure however it is important to characterize the dynamic nature of impact. The impact is characterised based on the time taken to reach the maximum magnitude termed as rise time and the time taken for the decay part of the pressure signal termed as decay time. The ratio between the decay time to rise time is termed as skewness of the impact or pressure impulse. From literature, it is found that one of the standard guidelines [13] recommends to choose large decay time longer than rise time as this will increase the structural response. While another standard guideline [8] recommends the rise time as the key parameter for describing dynamic amplification of structure while the decay time is be of secondary importance. On the other hand, few other guidelines (e.g. [4]) do not ascertain precedence over the other and mention it to be discussed and agreed with the society. The influence of dynamic characterisation of impact on structural dynamic response is not studied by the authors at present and a simplified approach is adopted for the present work. In the present study we initially stick to application of several sets of actual impacts of large magnitude observed from the long duration simulations, scaled to match the peak amplitude equal to design long-term value. Further based on scatter of rise time to decay time the most probable impacts are considered. It is observed that the majority of impacts generally exhibit a skewness greater than 1.0 and less than 2.0. For simplification the raise and decay time and the magnitude are used to model the slosh load as triangular impulse and applied to FEA model. Note that the loads obtained from CFD simulations both for containment system and pump tower correspond to full scaled simulations and no scaling is involved.

ADDITIONAL CONSIDERATIONS & WAY FORWARD

Modelling seas based on long-crested assumption may lead to inadequate resolution of critical pressure load in containment system [9]. It is thus important to model short crested seas for each load scenario using adequate spreading function. This would be more critical for site specific floating LNG terminals such as FLNGs and FSRUs which may be required to be designed for fill levels across barred filling range. Further, the assumption of duration of individual storm effects the final design long-term pressure. The standard approach in naval architectural domain assumes the duration of storm to be typically of three hours. But can be extended up to nine or twelve hours. This may be even more significant for floating LNG terminals which are stationary in site.

Though the methodology adopted in this paper is relatively sophisticated, uncertainness's do exist in load estimation when compared to reality. These uncertainties can be regarded to hydroelastic effects, turbulence and compressibility effects etc., among others [12]. Note that as of now, it is impossible to account for all the parameters influencing sloshing and efforts are being laid towards more accurate strength assessment. More accurate reliability based and effective pressure approach (see e.g. [9,10]) based methodologies along with studies on sensitivity of compressibility and local hydroelasticity will help minimising these uncertainties in sloshing assessment.

CONCLUSIONS

In this paper, the methodology for seakeeping and sloshing assessment is presented with the aid of a sample case study. Considerations and assumptions involved in the methodology specific to LNG terminals are pointed out. Later, application of predicted loads for strength assessment is discussed. Slosh coupling effects are incorporated in seakeeping analysis. Extreme event based short-term approach and long-term approach for estimating design slosh-induced loads is presented. A long-term approach accompanied with detailed modelling of sea conditions for load scenarios is recommended for sloshing assessment. For quick design checks the extreme event based short-term load can be used for comparative assessment. The present full numerical simulations-based approach is cost effective and quick. With adequate validation of the CFD models, the full numerical simulation based approach would help designer and classification bodies to study and minimise the uncertainties involved in current assessment standards.

REFERENCES

[1] ABS, 2014. *Sloshing and structural analysis of LNG pump tower*. Guidance Notes, American Bureau of Shipping.

[2] ABS, 2014. *Strength assessment of membrane-type LNG containment system under sloshing loads*. Guidance Notes, American Bureau of Shipping.

[3] Asquith, W., 2018. "L-moments, censored l-moments, trimmed l-moments, l-comoments and many distributions. *R Package*, version 2.3.2.

[4] BV, 2011. *Design sloshing loads for LNG membrane tanks*. Guidance Note. Bureau Veritas NI 554.

[5] Champagnac, M., and Gervaise, E., 2014. "Pump Tower – GTT Methodology for Load Determination of Strength Assessment". In Proceedings International Ocean and Polar Engineering Conference, Vol. 3, pp. 143-149.

[6] ClassNK, 2011. *Guidelines of strength assessment of membrane-type LNG containment systems under sloshing loads*. Guidelines, Nippon Kaiji Kyokai.

[7] Diebold, L., Derbanne, Q., and Gazzola, T., "Statistical behaviour of global and local sloshing key parameters". In Proceedings International Ocean and Polar Engineering Conference, pp. 306-312.

[8] DNV-GL, 2016. *Sloshing analysis of LNG membrane tanks*. Class Guideline, DNV-GL CG-0158.

[9] Gervaise, E., Seze, de P-E., and Maillard, S., 2009. "Reliability-based Methodology for Sloshing Assessment of Membrane LNG Vessels". In Proceedings International Offshore and Polar Engineering Conference, pp. 183-191.
[10] IACS, 2001. *Standard Wave Data, Recommendation No. 34*, International Association of Classification Societies.
[11] Kuo, J. F., Campbell, R. B., Ding, Z., Rinehart, A. J., Sandstrom, R. E., Yung, T. W., Greer, M. N., and Danaczko, M. A., 2009. "LNG Tank Sloshing Assessment Methodology-The New Generation". *International Journal of Offshore and Polar Engineering*, 19(4), pp. 241-253.
[12] LR, 2008. *Procedure for analysis of pump tower and pump tower base*. Procedures, Lloyd's Register.
[13] LR, 2009. *Sloshing assessment guidance document for membrane tank LNG operations*. Additional Design Procedures, Lloyd's Register.
[14] Malenica, S., Diebold, L., Kwon, S. H., and Cho D-S., 2017. "Sloshing assessment of the LNG floating units with membrane type containment system where we are? *Marine Structures*, 56, pp. 99-116.
[15] Paik, J. K., Lee, S. E., Kim, B. J., Seo, J. K., and Ha, Y. C., 2014. "A new method for determining the design sloshing loads for LNG FPSOS". In Proceedings International Conference on Ocean, Offshore and Arctic Engineering, pp. 15.
[16] Saripilli, J. R., and Sen, D., 2017. "Numerical studies of coupling effect of sloshing on 3D ship motions". *International Journal of Offshore and Polar Engineering*, 27(1), pp. 27-35.
[17] Saripilli, J. R., and Sen, D., 2017. "Numerical studies on sloshing loads using sloshing coupled ship motion algorithm". In Proceedings International Ocean and Polar Engineering Conference, pp. 1093-1102.
[18] Saripilli, J. R., and Sen, D., 2018. "Numerical studies on effects of slosh coupling on ship motions and derived slosh loads". *Applied Ocean Research*, 76, pp. 71-81.
[19] Saripilli, J. R., and Sen, D., 2018. "Slosh-coupled ship motion algorithm for estimation of slosh-induced pressures". *Journal of Marine Science and Application*, 17(3), pp. 312-329.
[20] Saripilli, J. R., 2019. "Numerical studies on slosh-coupled ship motion and slosh-induced loads". PhD Thesis, Indian Institute of Technology, Kharagpur.
[21] Sen, D., 2002. "Time domain computation of large amplitude 3D ship motion with forward speed. *Ocean Engineering*, 29, pp. 973-1002.
[22] Sen, D., and Saripilli, J. R., 2017. "Numerical studies on slosh-induced loads using coupled algorithm for sloshing and 3D ship motion". In Proceedings International Conference on Ocean, Offshore and Arctic Engineering Conference, pp. 9.
[23] Saripilli, J. R., and Sen D., 2019. "Estimation of design slosh-induced loads using numerical simulations". In Proceedings International Symposium on Practical Design of Ships and Other Floating Structures, pp. 22.

Proceedings of the 6th International Conference on Ship and Offshore Technology
ICSOT 2019
November 7-8, 2019, IIT Kharagpur, India

ICSOT INDIA 2019-30

EVALUATION OF HULL GIRDER CAPACITY CONSIDERING THE EFFECTS OF LATERAL PRESSURE AND TRANSVERSE STRESSES

Karan Doshi
Indian Register of Shipping
Mumbai, Maharashtra, 400072
India
Karan.doshi@irclass.org

Yogendra Parihar
Indian Register of Shipping
Mumbai, Maharashtra, 400072
India
Yogendra.singh@irclass.org

Saikat Dan
University of Johns Hopkins
Baltimore, United States of
America
sdan3@jhu.edu

ABSTRACT

Hull Girder Capacity evaluation is mandated by Classification Society Rules. Such evaluation is performed using the Smith method by incremental-iterative approach. Load Shortening curves are generated for each plate panel within the hull girder longitudinal section based upon whether the plate is subject to tensile or compressive stresses. The load shortening curves do not account for the effects of lateral pressure and transverse stresses. This could be an important factor for ships with wide beams and large draughts where the lateral pressure and transverse stresses may be significant.

The present paper aims to evaluate the hull girder capacity considering the effect of lateral pressure and the transverse stresses. The load shortening curves consider the biaxial interaction curves of ultimate capacity of the plate using both longitudinal and compressive stresses. The load shortening curves are obtained using a series of finite element analyses considering initial imperfections and lateral pressure effects. The semi-empirical formulae are developed using regression analyses. Proposed formulation are implemented in simplified incremental-iterative approach for ultimate capacity evaluation. The paper concludes that transverse stresses and lateral pressure could lead to further decrease in ultimate capacity of the hull girder.

NOMENCLATURE

a	Frame spacing (mm)
b	Stiffener spacing (mm)
b_f	Width of Stiffener flange (mm)
E	Elastic modulus of steel (N/mm^2)
h_w	Depth of Stiffener web (mm)
N_{stiff}	Number of Stiffeners
p	Lateral pressure (N/mm^2)
Q	Ultimate Strength Ratio (σ_u/σ_{Ye})
r_e	Radius of gyration of section of stiffener with attached plating (mm)
t	Plate thickness (mm)
t_f	Thickness of stiffener flange (mm)
t_w	Thickness of stiffener web (mm)
β	Plate slenderness ratio ($b/t\sqrt{(\sigma_Y/E)}$)
λ	Slenderness ratio of Stiffener with attached effective plating ($a/(\pi r_e)\sqrt{(\sigma_{Ye}/E)}$)
μ	Web Slenderness ratio of Stiffener ($h_w/t_w\sqrt{(\sigma_{Ys}/E)}$)
σ_E	Euler Buckling stress (N/mm^2)
σ_x	Applied longitudinal axial stress (N/mm^2)
σ_{xu}	Ultimate longitudinal axial stress (N/mm^2)
σ_y	Applied transverse axial stress (N/mm^2)
σ_{yu}	Ultimate transverse axial stress (N/mm^2)
σ_{Ye}	Equivalent yield stress of stiffener with attached plating (N/mm^2)
τ	Edge shear stress (N/mm^2)
τ_u	Ultimate edge shear stress (N/mm^2)
FL	Full load condition
BL	Ballast load condition
CSR	IACS Common Structural Rules for Bulk Carriers and Oil Tankers
PIF	Plate Induced Failure
SIF	Stiffener Induced Failure

INTRODUCTION

One of the important aspects of safe design of hull girder is the determination of longitudinal strength and the ultimate strength of the hull girder. Ship structure is composed of stiffened plate panels. The strength of hull girder depends on the stiffening arrangement of panels, scantlings and mechanical properties of the steel. It is also influenced by the imperfections induced at the time of fabrication. Ship hull is typically subjected to loads such as static and dynamic pressures, global bending moments and shear forces and other local loads (if any). It is important to evaluate the ultimate capacity of the hull girder to check its integrity for extreme load conditions. To assess the ultimate capacity of hull girder in bending, various approaches/methodologies were developed. The Simplified Smith's method is popularly used to evaluate the hull girder's ultimate strength.

Smith [1], [2] proposed a simplified method to evaluate the progressive collapse of hull girder under longitudinal bending. To account for the progressive structural strength reduction (with increasing load), the cross-section is discretized into constituent elements, consisting of stiffener with attached plating. Thereafter, buckling or yield failure of each individual member of a section is evaluated by incorporating the average load shortening of plate elements to calculate the ultimate strength of the hull girder, considering the axial rigidity, flexural rigidity of each individual element under uniaxial loads. The accuracy of the calculated results depends on the accuracy of the load shortening curves of individual elements [3]. It may also be noted that the Smith's method is based upon load shortening curves which consider the minimum of the ultimate capacities of individual panels under different failure modes. However, it is not able to account for interaction between the failures modes.

In 1974 and 1984 Ueda and Rashed [4], [5] presented the Idealized Structural Unit Method (ISUM). To simplify and to reduce computational time, a comparatively larger structural element is considered in this approach. This is the stiffened plate flanked by longitudinal girders and transverse frames. The stiffness matrix in incremental form is derived for this unit to consider the effect of yielding or buckling as applicable. The original method of ISUM simulates the buckling and plastic behaviour of element under bi-axial and compression/tension and shear loads. Hence the ISUM method is considered to be more accurate compared to the Smith's simplified method although ISUM method takes higher computational time and resources.

Finite element method (FEM) is also a useful and realistic tool for progressive hull girder collapse analysis. Ultimate capacity of hull girder can be evaluated by incorporating the geometrical imperfections, and material non-linearity. Further, the method can account for a combination of loads such as biaxial compression, shear, torsion etc (and thus interaction of various failure modes). Chen et al [6] evaluated the ultimate hull girder capacity using FEM by considering the elasto-plastic properties of material and geometrically non-linear behavior of elements and their buckling and post buckling behaviour. The FEM based calculations however required longer computation time. They also developed the orthotropic plate elements which simulate the stiffened panel by considering the 'smearing' of the stiffeners over the plate. But in this approach, initial imperfections, local buckling and loss of stiffness were not considered [7]. Later on, Valsgaard and Steen [7] developed a similar FEM based approach by considering the initial imperfections and post buckling behavior to predict the ultimate strength of hull girder.

Various studies were performed using FEM and results were compared with the experimental and other available methods [8]–[10]. FE based analysis provided results with higher accuracy however modelling and analysis were quite complex and time consuming. Such methods are hence considered not very suitable for designers.

In regards to classification rules, IACS published Harmonized Common Structural Rules for Bulk Carriers and Oil Tankers (CSR) [11], in which the simplified Smith's method is adopted to evaluate the ultimate hull girder capacity. For container ships, IACS UR-S11A [12] specifies the same approach. Load shortening curves in various collapse modes are provided e.g. beam-column buckling, torsional buckling, web-buckling, plate buckling and yielding. The effect of initial distortion and residual stress during to fabrication is implicitly incorporated. CSR/ UR S11A based simplified method considers the elasto-plastic behaviour of the steel in tension and compression and influence of shear stresses is neglected [11]. This approach is unable to account for the effect of combined loads like bi-axial compression and lateral pressure. Plating which is subject to lateral pressure loads (wave, cargo or ballast) can fail in Plate Induced Failure mode (PIF) or Stiffener Induced Failure mode (PIF) depending upon the element which is in compression. Various authors in literature [13]–[17] established that lateral pressure combined with transverse/longitudinal compression loads further reduces the capacity of the stiffened panels. Thus, for ships with large beams which could have significant transverse stresses and lateral pressures, the load shortening curves may not be able to account for these. This was demonstrated by the results presented in [18] as compared to FEM.

Considering the above aspects, it is intended to provide an efficient approach which can consider the biaxial compression and lateral pressure for ultimate hull girder capacity determination. It has been shown by Doshi and Vhanmane [19], [20], that uniaxial compressive loads combined with lateral pressure reduce the ultimate capacity of stiffened plate panels. They also proposed a formulation to predict the ultimate capacity of panels by performing a series of FE analyses of stiffened panels of Oil Tankers. The present paper aims to extend this work to consideration of biaxial compression with lateral pressure within the ultimate hull girder capacity evaluation. Formulae for prediction of ultimate capacity under biaxial loads and lateral pressure are developed and used within the load shortening curves for the ultimate hull girder capacity evaluation. Practical initial imperfections (local and global) are considered in form of initial deflections within the stiffened panel. These are aligned with the IACS Rec. 47 [21] which specifies the construction tolerances for shipbuilding.

METHODOLOGY

Simplified method is well recognized to estimate the capacity of hull girder. The proposed method is following simplified method by adopting the incremental-iterative approach as specified in CSR and UR S11A [11], [12]. The progressive collapse of a section between transverse frames is performed by discretising into plate-stiffener elements. Capacity of members under compression is evaluated using the σ-ε curves/ load shortening curves. These σ-ε curves are selected based upon the possible modes of failure for stiffened panels and plates as listed in Table 1. Elements of section are assumed to be independent of each other. The approach presented in this paper replaces the load shortening curves used in CSR and UR S11A [11], [12] with those accounting for biaxial compression and lateral pressure. The method of obtaining the load shortening curves is further introduced in the following sections. Flow chart shown in Figure 1 highlights the overall procedure for ultimate capacity evaluation of hull girder.

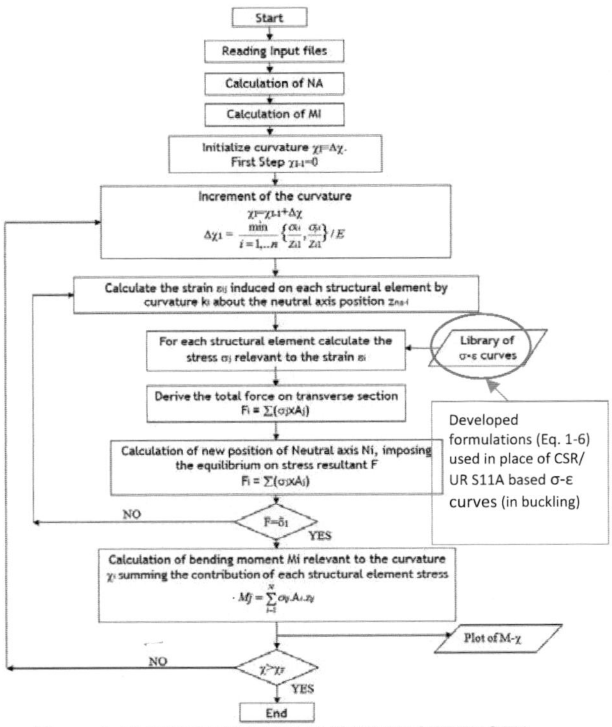

Figure 1: FLOWCHART FOR ULTIMATE STRENGTH EVALUATION

Table 1: Library for σ-ε curves

SN	Element	Mode of failure	σ-ε curves UR S11A	σ-ε curves **(Present)**
1	Lengthened elements or stiffeners	Elasto-plastic collapse	Annex-3: [2.3.2]	UR S11A Annex-3: [2.3.2]
2	Shortened Stiffened elements	Beam-column buckling	Annex-3: [2.3.3]	Refer Eqn. (7)
		Torsional buckling	Annex-3: [2.3.4]	
		Web local buckling of flanged profiles	Annex-3: [2.3.5]	
		Web local buckling of flat bars	Annex-3: [2.3.6]	
3		Plate buckling	Annex-3: [2.3.7]	

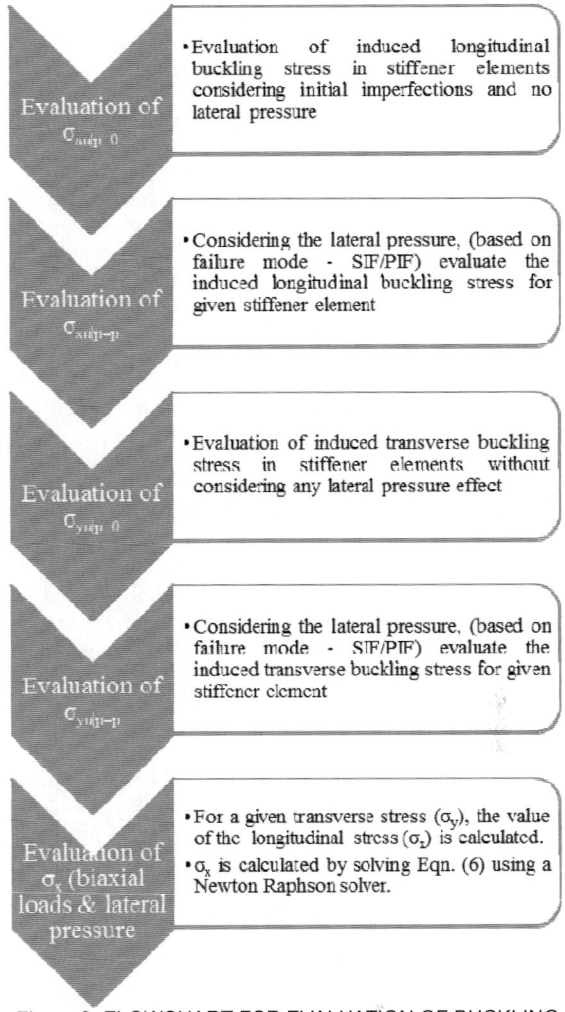

Figure 2: FLOWCHART FOR EVALUATION OF BUCKLING CAPACITY USING DEVELOPED LOAD SHORTENING CURVES

Figure 2 demonstrates the procedure of computing the buckling capacity of a structural member. The first step is to evaluate the longitudinal and transverse buckling capacities of the plating without consideration of lateral pressure. The next step is to calculate the longitudinal and transverse capacities accounting for the presence of lateral pressures. Once, these stresses are obtained, the ultimate longitudinal capacity under last step (see Figure 2) is to be obtained by substituting the actual level of transverse stress in Eqn. (7). (Equation (7) is solved using Newton-Raphson method). Further, the obtained ultimate longitudinal capacity is utilized for evaluation of ultimate capacity of hull girder using incremental-iterative approach (see Figure 1).

BACKGROUND OF DEVELOPED LOAD-END SHORTENING CURVES

Background

To develop the formulae for ultimate capacity prediction, it is important to consider effect of various imperfections, axial loads and lateral pressure. This was accomplished by performing a series of Finite element analyses considering various parameters (as listed in following sections). Practical Ship plating was considered for the analyses. Based upon the results, formulae were derived. The complete work was carried out in an internal project [22]. An advantage of using FEM is that the interaction between the failure modes is also able to be captured in the final ultimate capacity results.

The aspects related to FE modelling, boundary conditions, and imperfections are briefed as follow:

- An extent of ½ + 1 + ½ Frames were modeled in order to correctly assess the stiffness at the transverse frame and stiffener connections, see FIGURE 3. Girders and Transverse floors were generally not modeled as they are assumed to be very stiff when compared to the stiffened panel [17].
- Simply supported boundary conditions were applied on model as shown in Table 2.
- Three levels of initial imperfections were considered in the present study: (i) Local panel imperfection, (ii) Overall stiffened panel imperfection, (iii) Stiffener sideways angular distortion. The initial deflections at different level are describe in FIGURE 4. Panel shape after application of initial deflections is also shown in FIGURE 5. The initial imperfection shape is expressed through eqn. (1):
- Initial imperfections considered were consistent with the shipbuilding tolerances prescribed in IACS Recommendation 47 [21]. For the material non-linearity, a bilinear isotropic hardening model was considered with the tangent modulus taken to be zero.
- Loading is applied at the centroid of the section considering the stiffener and the attached plating.

$$
\begin{aligned}
W_{local-panel} &= W_{opl} \sin\left(\frac{m\pi x}{a}\right) \sin\left(\frac{\pi y}{b}\right) \\
W_{panel} &= W_{oc} \sin\left(\frac{\pi x}{a}\right) \sin\left(\frac{\pi y}{B}\right) \qquad (1) \\
W_{stiff} &= W_{os} \frac{z}{h_w} \sin\left(\frac{\pi x}{a}\right)
\end{aligned}
$$

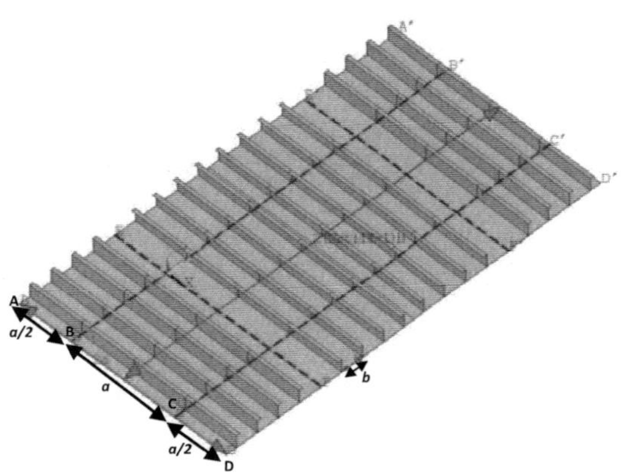

FIGURE 3: SCHEMATICS OF THE STIFFENED PLATE – ½ + 1 + ½ BAY MODEL

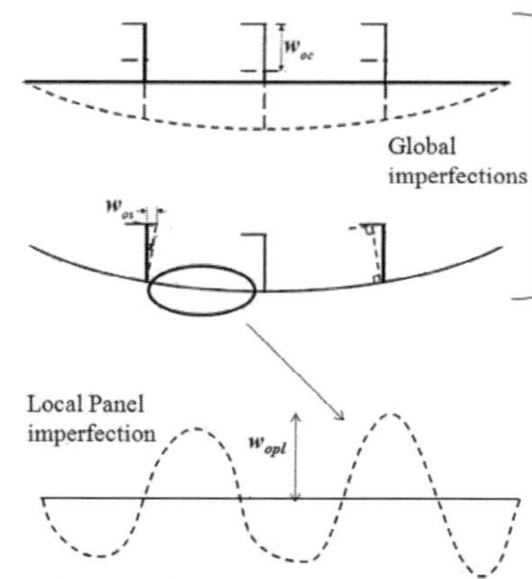

FIGURE 4: SCHEMATICS OF THE INITIAL IMPERFECTIONS

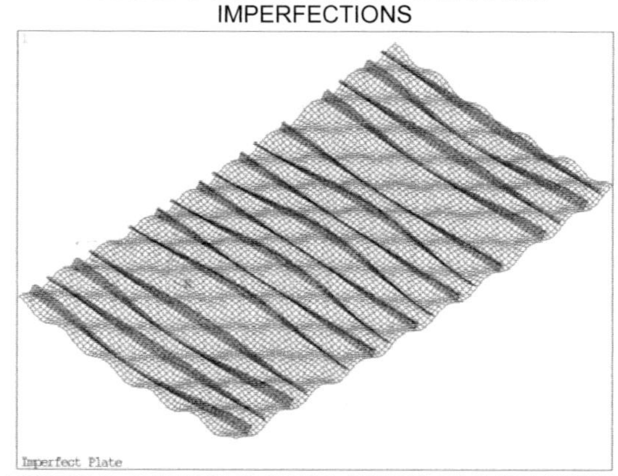

FIGURE 5: STIFFENED PANEL MODELED WITH INITIAL IMPERFECTIONS

Table 2: Boundary conditions

Edge	Constraints
Along Edges ABCD and A'B'C'D': Symmetry conditions:	ROTZ = ROTY=0 and couple UY of all nodes
BB' and CC'; EF and E'F'	U_Z =0
AA' and DD'	Symmetry conditions – ROTZ = ROTX = 0. Couple UX of all nodes

Selection of Analysis Parameters for derivation of the load shortening curves

It may be noted that the selected plate panels are representative of the general range of stiffened plate dimensions recorded within existing ships. Amongst the entire stiffener profiles and the stiffened panels, the plate slenderness ratio β falls in the range of 1.17 – 2.30. The stiffener (with attached plating) slenderness ratio λ lies between 0.14 – 0.62. The actual yield strengths of the plates and stiffeners were considered while performing the analyses

Sets of thirty stiffened plate panels used in practice were considered for the analysis. With 6 lateral pressures and 9 biaxial stress combinations, the total no of cases were become 1620 for each failure mode (SIF and PIF). A range of lateral pressures were considered from 0 – 0.25 for analyses. Biaxial stress combinations (σ_x/σ_y) were varied from 0-1. A series of FE analyses were performed to understand the behavior of stiffened plate panels. The Task was divided into three category:

- Uniaxial longitudinal compressive stress prediction under lateral pressure
- Uniaxial transverse compressive stress prediction under lateral pressure
- Biaxial compressive stress prediction under lateral pressure

FE ANALYSES RESULTS AND CORRELATION ESTABLISHMENT

Formula development

It is envisaged that the ultimate strength degradation has to be benchmarked with the case where lateral pressure is absent. Hence the factor of decrease of ultimate strength can be modelled rather than the actual ultimate strength. This is summarized in Equation 1.

$$\frac{\sigma_{xu}^{p=P}}{\sigma_{xu}^{p=0}} = 1 - f(p, scantling) \qquad (2)$$

With respect to the biaxial interaction, the interaction plots need to be observed so that an equation can be developed. The idea would be to have the equation normalized as shown in Eqn. (2).

Longitudinal Compression

Computation of uniaxial stress in x-direction at pressure p=0 can be predicted using given formulations:

$$\frac{\sigma_{xu}}{\sigma_{Yeq}} = a_1 f_1 f_2 f_3 f_4 f_5 + a_2 \qquad (3)$$

where:

$$\sigma_{Yeq} = \frac{A_{st}\sigma_{Ys} + bt_p\sigma_{Yp}}{A_{st} + bt_p}$$

$$f_1 = 1 - e^{\left[-a_3\left(A_{st}/bt_p\right)^{a_4}\right]}$$

$$f_2 = e^{[-a_5\beta^{a_6}]}$$

$$f_3 = e^{[-a_7\lambda^{a_8}]}$$

$$f_4 = e^{[-a_9\mu^{a_{10}}]}$$

$$f_5 = \left(\frac{\sigma_{Ys}}{\sigma_{Yp}}\right)^{a_{11}}$$

a_1 to a_{11} are the coefficients estimated through regression analysis in internal project [23].

Transverse Compression

The general formula for the ultimate strength of the stiffened panels in transverse compression is shown in Eq. (4):

$$\frac{\sigma_{yu}}{\sigma_{Yeq}} = b_1 f_1 f_2 f_3 + b_2 \qquad (4)$$

where:

$$\sigma_{Yeq} = \frac{A_{st}\sigma_{Ys} + bt_p\sigma_{Yp}}{A_{st} + bt_p}$$

$$f_1 = 1 - b_3\frac{b}{a} + b_4\left(\frac{b}{a}\right)^2$$

$$f_2 = e^{[-b_5\beta^{b_6}]}$$

$$f_3 = 1 + b_7\frac{A_{st}}{bt_p}$$

b_1 to b_7 are the coefficients estimated through regression analysis in internal project [23].

Uniaxial Longitudinal Compressive Stress Prediction under Lateral Pressure

Based on two categories (PIF and SIF) of the induced failure the uniaxial longitudinal compressive stress under lateral pressure can be predicted using following formulations:

$$\frac{\sigma_{xu|p}}{\sigma_{xu|p=0}} = 1 - c_1(\beta)^{c_2}(\lambda)^{c_3}|p|^{\left(1+c_4\frac{h_w^2}{ab}\right)}\left(\frac{h_w^2}{ab}\right) \qquad (5)$$

where:

c_1 to c_4 are the coefficients for the longitudinal ultimate strength of stiffened plates under lateral pressure in PIF and SIF mode, estimated in internal [22].

Uniaxial transverse compressive stress under lateral pressure

Based on two categories (PIF and SIF) of the induced failure the uniaxial transverse compressive stress under lateral pressure can be predicted using following formulations:

$$\frac{\sigma_{yu|p}}{\sigma_{yu|p=0}} = 1 - d_1(\beta)^{d_2}(\lambda)^{d_3}|p|^{\left(1+d_4\frac{h_w^2}{ab}\right)}\left(\frac{h_w^2}{ab}\right) \qquad (6)$$

where, coefficients d_1 to d_4 for the transverse ultimate strength of stiffened plates under lateral pressure were estimated in internal [22].

Biaxial Stress Prediction under Lateral Pressure

In the biaxial case, the failure modes have also been categorized as PIF and SIF. However in this case, a two dimensional region in the σ_x-σ_y space exits. This region is marked with a function of the form:

$$e_1\left(\frac{\sigma_x}{\sigma_{xu|p}}\right)^{e_4} + e_2\left(\frac{\sigma_y}{\sigma_{yu|p}}\right)^{e_5} + e_3\left(\frac{\sigma_x}{\sigma_{xu|p}}\right)\left(\frac{\sigma_y}{\sigma_{yu|p}}\right) = 1 \qquad (7)$$

Where (σ_x, σ_y) is the failure state of stress of a biaxial loaded panel; σ_{xu} and σ_{yu} are the ultimate strengths in the respective uniaxial directions which are computed using Eqn. (5) and (6). Eqn. (7) is generalized, the coefficients e_1-e_5 are expected to be more or less invariant and are obtained using regression analyses in internal project [22].

Due to the considerable scatter of the curves for different stiffened plates with different lateral pressures, it was not possible to fit a single curve for the data sets. However, it is possible to establish the upper and lower bounds of the curves for all data. Using the upper and lower bound curves, it is thus possible to estimate the upper and lower bounds of the ultimate hull girder capacity. The nonlinear curve fitting results are shown in FIGURE 6 to FIGURE 9. Coefficients e1-e5 were derived from curve fitting of three different bounds with respect to type of failure (PIF and SIF).

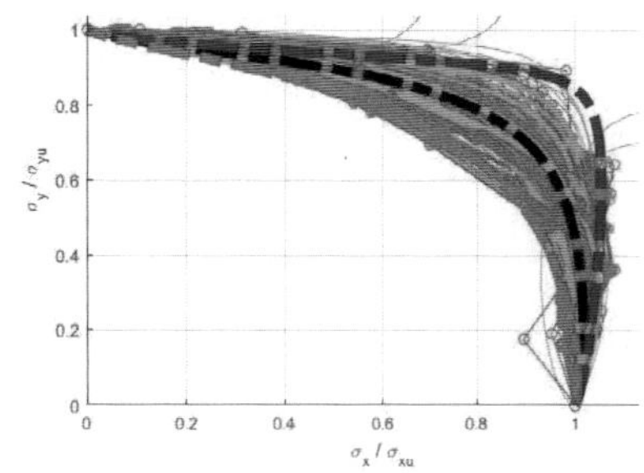

FIGURE 6: CURVE FIT FOR PIF

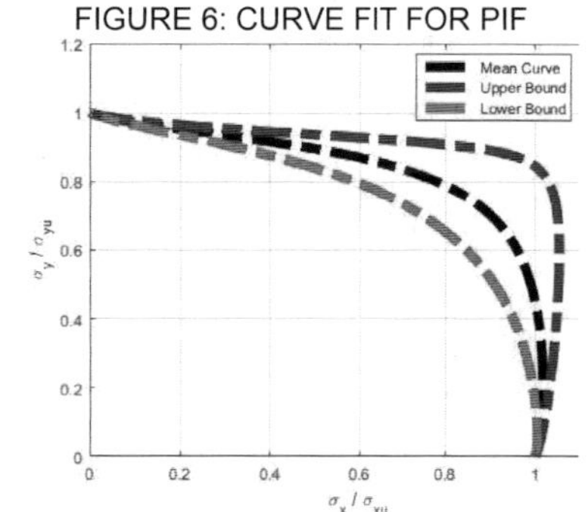

FIGURE 7: UPPER, MEAN AND LOWER BOUNDS FOR PIF

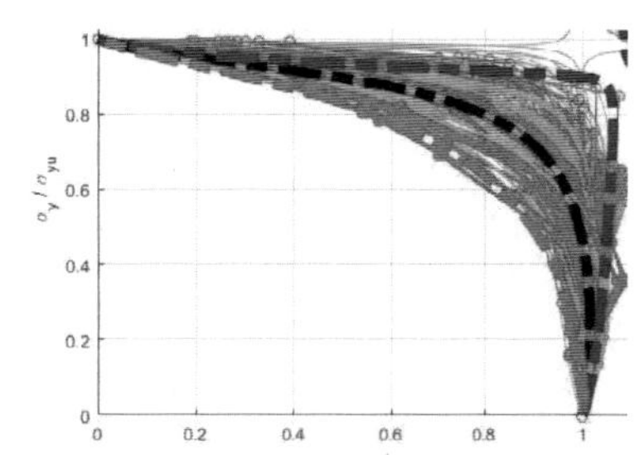

FIGURE 8: CURVE FIT FOR SIF

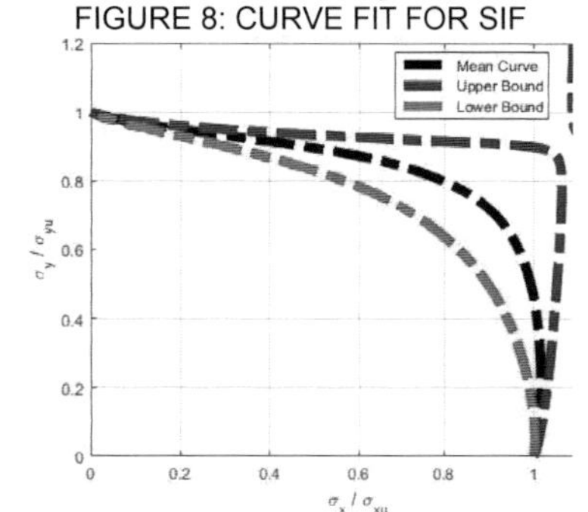

FIGURE 9: UPPER, MEAN AND LOWER BOUNDS FOR SIF

Table 3: Stiffener/plating details used for comparison

Stiffener	t_p (mm)	s (mm)	A (mm)	σ_{Ye} (MPa)
580×12+	22	910	5140	315
200×20	P(FL)	P(BL)	σ_y(hog)	σ_y(sag)
	(MPa)	(MPa)	(MPa)	(MPa)
	0.221	-0.201	100	0

LOAD SHORTENING CURVES

The impact of utilizing the derived load shortening curves is demonstrated in FIGURE 10, which highlights the importance of considering lateral pressure and transverse stresses. The illustration is shown for one sample ship stiffened panel as shown in Table 3. The type failure mode/present formulae considered for comparison purpose are tabulated in Table 4.

Table 4: Details of formulae considered for comparison

SN	Type of failure/mode	Reference
1.	Beam-column buckling (UR S11A)	Annex-3: [2.3.3]
2.	Torsional buckling (UR S11A)	Annex-3: [2.3.2]
3.	Web local buckling of flanged profiles (UR S11A)	Annex-3: [2.3.2]
4.	Buckling (uniaxial)	Eqn. (3), (4)
5.	Buckling (uniaxial with pressure)	Eqn. (5), (6)
6.	Buckling (biaxial with pressure)	Eqn. (7)

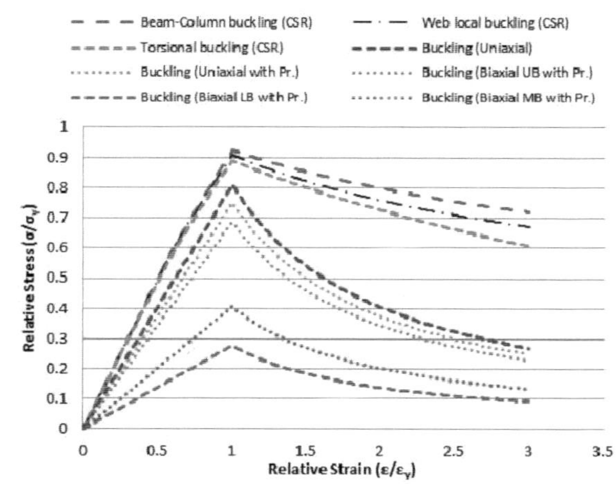

FIGURE 10: COMPARISON OF LOAD SHORTENING CURVES (UR S11A VS DEVELOPED)

ULTIMATE STRENGTH PREDICTION RESULTS USING DEVELOPED FORMULAE

The methodology described in the preceding paragraphs was applied for a post panamax containership. The results were derived for cases using Smith's method, UR S11A, and the present approach. The results are presented in FIGURE 11. A tabular comparison is shown in Table 5. The transverse stresses in each panel were obtained from finite element analyses carried out to check the integrity of the cargo holds as is mandated by the Rules of the Indian Register of Shipping [24].

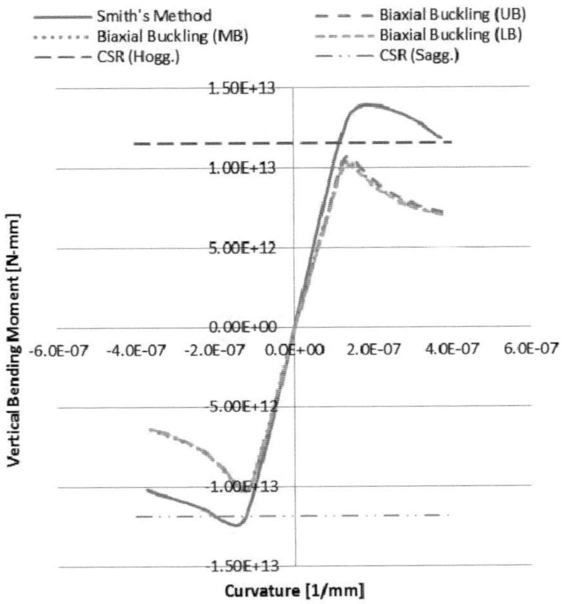

FIGURE 11: PREDICTION OF ULTIMATE STRENGTH OF A CONTAINER SHIP

Table 5: Ultimate capacity of container ship

SN	Applied Method	Hogging (ton-m)	Sagging (ton-m)	% diff w.r.t. UR S11A (Sr.no. 2)
1	Smith's method, M_U	1.391E+07	-1.238E+07	-
2	UR S11A (M_U/(γ_M γ_{DB})) [12]	1.152E+07	-1.179E+07	-
3	Uniaxial formulation (*P*=0)	1.181E+07	-1.072E+07	-2.6(hog), 9.0(sag)
4	Biaxial formulation (upper bound)	1.063E+07	-1.029E+07	7.7(hog), 12.6(sag)
5	Biaxial formulation (mean bound)	1.034E+07	-1.028E+07	10.2(hog), 12.8(sag)
6	Biaxial formulation (lower bound)	1.019E+07	-1.029E+07	11.5(h), 12.7(s)

where:
γ_{DB} = Partial safety factor for the hull girder ultimate bending moment capacity, covering the effect of double bottom bending, to be taken = 1.15 (hogging) and 1.0 (sagging).
γ_M = Partial safety factor for uncertainties in material geometry and strength, to be taken = 1.05

The results of Ultimate Hull Girder capacity are compared with those obtained using Smith Method and UR S11A. It may be noted that UR S11A applies a γ_{DB} and γ_M factor on the moment obtained using Smith's method.

From the comparison, it is observed that the consideration of transverse stresses and the lateral pressure provides an ultimate capacity reduced by 7% – 13% for hogging and sagging conditions. Notably, the difference using the lower bound and upper bound of the formulae is not very significant, this may be attributed to the fact that the transverse stress level may not be very significant (< 0.2 σ_y) in most panels.

CONCLUSION

The following conclusions are drawn from the present work:

- Lateral pressure and biaxial compression have a notable effect on the ultimate hull girder capacity.
- The reduction in hull girder ultimate capacity considering the effects of lateral pressure and transverse compression was 7 % – 13 % when compared to the Ultimate Moment obtained using the Rules.

The present method provides a quick and efficient way of evaluating the hull girder ultimate capacity accounting for presence of lateral pressure and transverse stresses.

ACKNOWLEDGEMENTS

The authors express their earnest thanks and gratitude to the Indian Register of Shipping for supporting the reported work and sponsoring the presentation of this work at ICSOT 2019. The views expressed in this paper are those of the authors and may not necessarily represent those of the Indian Register of Shipping.

REFERENCES

[1] Smith, C., 1977. "Influence of local compression failure on ultimate longitudinal strength of a ship's hull," *Proceeding of International Symposium on Practical Design in Ship Building*, pp. 73–79.

[2] Smith, C., and Dow, R., 1986. "Ultimate Strength of Ship's Hull under Biaxial Bending," in ARE TR86204

[3] ISSC, 2000. "Committee VI.2 - Ultimate hull girder strength," Nagasaki, Japan.

[4] Ueda, Y., and Rashed, S., 1974. "An ultimate transverse strength analysis of ship structure," *J. Soc. Nav. Archit. Japan,* 136, p. 309–324 (in Japanese).

[5] Ueda, Y., and Rashed, S., 1984. "The idealized strucutral unit method and its application to deep girder structure," *Comput. Struct.,* 18 (2), pp. 277–293.

[6] Chen, Y., Kutt, L., Piaszczyk, C., and Bieniek, M., 1983. "Ultimate strength of ship structures," *SNAME Trans*., pp. 49–68,

[7] Valsgaard, S., and Steen, E., 1991 "Ultimate Hull Girder Strength Margins in Present Class Requirements," *Marine Structural Inspection, Maintenance and Monitoring Symposium*, pp. 1–19.

[8] Grodin, G., Elwi, A., and Cheng, J., 1999 "Buckling of Stiffened steel plates - a parametric study," *J. Constr. Steel Res*., 50, pp. 151–175.

[9] Qi, E., and Cui, W., 2006. "Designed-Oriented Methods of Ultimate Hull Girder Strength," *International Conference on Offshore Mechanics and Arctic Engineering*, pp. 1–5.

[10] Mohammed, E., Benson, S., Hirdaris, S., and Dow, R., 2016. "Design safety margin of a 10000 TEU container ship through ultimate hull girder load combination analysis," *Mar. Struct*., 46, pp. 78 101.

[11] IACS, 2018. "Common structural rules for bulk carriers and oil tankers," London, UK.
[12] IACS, 2015. "UR S11A - Longitudinal Strength Standard for Container Ships," London, UK.
[13] Timoshenko, S., and Gere, J., 1963 *Theory of Elastic Stability*, 2nd Edition. McGraw Hill,
[14] Byklum, E., and Amdahl, J., 2002. "A simplified method for elastic large deflection analysis of plates and stiffened panels due to local buckling," *Thin Walled Structures*, 40, pp. 925–953.
[15] Dongqi, Z, 2004. "Stiffened panels subjected to in plane and lateral pressure," Phd. Thesis, National University of Singapore, 2004.
[16] Paik, J., Kim, B., and Seo, J., 2008. "Methods for ultimate limit state assessment for ships and ship-shaped offshore structures: Part I – Unstiffened plates," *Ocean Eng.*, 35, pp. 261–270.
[17] Paik, J., Kim, B., and Seo, J., 2008. "Methods for ultimate limit state assessment for ships and ship-shaped offshore structures: Part II – Stiffened panels," *Ocean Eng.*, 35, pp. 271–280.
[18] Choung, J., Nam, J., and Ha, T., 2012. "Slenderness ratio distribution and load-shortening behaviors of stiffened panels," *Mar. Struct.*, 26, no. 1, pp. 42–57.
[19] Doshi, K., and Vhanmane, S., 2013. "Ultimate strength of unstiffened and stiffened plates subject to combined in-plane compression and lateral pressure," in *International Conference on Ship & Offshore Technology*, 2, pp. 63–71.
[20] Doshi, K., and Vhanmane, S., 2014. "Ultimate Strength of Stiffened Plates Subjected to Longitudinal Compression and Lateral Pressure," in Volume 4B: Structures, Safety and Reliability,.
[21] IACS, "REC. 47 - Shipbuilding and Repair Quality Standard," 2017.
[22] IRS, "RRD-15036-RR - Ultimate strength evaluation of hull girder considering effect of transverse strength and lateral pressure," Mumbai, 2015.
[23] IRS, "RRD-11022-RR - Formula development for ultimate strength assessment of unstiffened and stiffened panels," Mumbai, 2011.
[24] IRS, "Rules and Regulation for the construction and classification of steel ships, Part 3 - General hull requirement," Mumbai, 2015.

Proceedings of the 6th International Conference on Ship and Offshore Technology
ICSOT 2019
November 7-8, 2019, IIT Kharagpur, India

ICSOT INDIA 2019-31

AN EVALUATION OF THE PRESENT SAFETY REGIME OF SHIPPING

Karan Doshi
Indian Register of Shipping
Mumbai, Maharashtra 400080
India
karan.doshi@irclass.org

Apurba Ranjan Kar
Indian Register of Shipping
Mumbai, Maharashtra 400080
India
kar@irclass.org

ABSTRACT

Safety and Environmental Protection of Global shipping are regulated by the International Maritime Organization. IMO provisions in form of Conventions and Codes stipulate the requirements to be adhered to. Such requirements concern the design and construction of the ship and equipment, safe operation of the ship, competency requirements of the crew and maintenance of all of these through the service life of the ship. The ultimate goal is to ensure safety of the life and property as well as protection of the environment. There are safety regimes established to monitor the effective implementation of the IMO Codes and Conventions. These regimes are in form of certification of the ship which include its periodic surveys, inspections by the Flag State and the Port State Control Authorities. The present paper identifies the available data (in form of accidents and casualties, port state control inspections etc.) and presents an analysis to identify the important factors within the present safety regime of Shipping. The present work aims to bring out the salient factors which ensure maritime safety as well as identify potential steps to improve it further.

NOMENCLATURE

DWT	Deadweight
FSI	Flag State Inspection
GT	Gross Tonnage
IMO	International Maritime Organization
ISM	International Safety Management
MoU	Memorandum of Understanding
MLC	Maritime Labour Convention
PSC	Port State Control
RO	Recognized Organization
VDR	Voyage Data Recorder

INTRODUCTION

Shipping is a truly global mode of transportation of goods and people. A Ship typically can be categorized as below to prove this point. Typically, a Ship could be built in Country A, flagged with Country B, owned by a party in Country C, managed by a company in Country D, classified and certified by a body in Country E, manned by crew of Countries F, G, H etc. Thus, multiple nations could be involved in the design, operation and management of a Ship. This requires it to be globally regulated so as to ensure consistent application of standards and codes globally.

Shipping is regulated by the International Maritime Organization (IMO) which develops regulations to be enforced by its member states. These regulations pertain to maritime safety, security and environment. The regulations pertain to various aspects in relation to Ship Design, Construction, Operation, Manning and Management. With the correct application and enforcement of the IMO regulations, it is expected to achieve acceptable standard for seagoing ships in terms of safety, security and impact on the environment. Classification Societies and Recognized Organizations play crucial roles in correct implementation of the IMO regulations in a consistent manner globally.

In the present work, the focus is on the existing mechanisms in place to ensure correct implementation of IMO regulation to achieve maritime safety globally.

CONCEPT OF MARITIME SAFETY

In this paper, Maritime Safety is defined as the absence of dangerous conditions for persons and the Ship in all Maritime Operations. Dangerous conditions are defined as conditions which cause or have potential to cause loss of life, Injuries and Damage to property. It is thus important to track and monitor Maritime Safety, hence, there need to be defined goals/benchmarks against which safety parameters obtained from the service can be compared. Depending upon the objective of the authority monitoring the safety, the goals would also differ.

Ship Level and Ship Fleet Level Safety Goals/ Objectives (Shipowner/Shipmanager authorities)

- Minimize Lost Time due to Injury

- Minimize downtime of the Ship impacting cargo delays, ship repairs, possible detentions etc.
- Prevent loss of reputation of the ship/ship fleet so as can affect its future market value or potential for charters and also the insurance premiums for the ship, crew and cargo. (This can affect the overall profitability. Thus the outlook is mainly commercial)

National Level and International Level Safety Objectives (National Administration/IMO authorities)

- Appropriate implementation of regulations to ensure safety of people & property. Interest in the growth of the national economy may be a lookout, however the main focus remains in the Safety and Protection.

Thus the focus at a ship level or a ship fleet level is influenced by commercial pressures, while at the National and International level, the influence is due to societal pressures. Also, at a ship or ship fleet level, the operational practices can be improved to enhance the safety, at the National/International Level, the focus is on improvement of the regulations being complied by the ship fleet.

The present paper focuses the wider picture – regulation of safety by the National Administrator/IMO or in short, the "Regulator".

A Regulator would require parameters to evaluate the status of prevalent Maritime Safety. Based upon these parameters being measured in the service, the Regulator can adjudge whether the regulations need further tightening to ensure less lapses in safety. Such set of parameters are termed as Performance Indicators of Safety.

SAFETY NETS – REGULATORY EFFECTIVENESS

IMO has adopted various mandatory codes and conventions [1-6] which aim to set an acceptable level of safety. It would be natural to presume that adherence to the codes and conventions is almost akin to an assurance of safety. Any accidents/incidents occurring despite adhering to the Codes and conventions would indicate need for revision to these instruments or creation of a new regulatory instrument.

Thus, monitoring the consistent implementation of codes and conventions is important for maritime safety. The codes and conventions regulate the following marine aspects throughout the life of the ship:

- Ship Design & Construction
- Crew competency
- Safe Operations

To ensure that the Codes and Conventions are being implemented consistently through the lifetime of the ship, IMO has various mechanisms to monitor the same. These are as follows:

- Approval of Ship's plans & drawings by the Flag/Recognized Organizations/Classification Society – To ensure the design is in accordance with the applicable requirements
- Survey of the Ship during construction followed by certificate on completion of Construction of the Ship with its machinery and equipment in accordance with the approved plans & drawings by the flag/RO/Class – To ensure that the actual design has been implemented exactly in reality
- Certificates of competency for the crew – To ensure that the persons onboard are right persons qualified to operate the ship.
- Renewal of the certificates of the ship by periodic inspections – Annual, Intermediate and Renewal Surveys. These are planned and scheduled inspections
- Flag State Inspection – Unscheduled inspection by the Flag to check whether the ship is compliant with the Codes and Conventions
- Port State Inspection – Unscheduled inspection by the Port State to check whether the ship is compliant with the Codes and Conventions.
- Flag State Audits (by IMO) – To ensure that the certificates issuing authority is competent for the job.

The above framework is termed as a 'Safety Net' because its purpose is to effectively detect any anomaly or non-compliance before it can lead to a potential accident(s).

DATA – SAFETY NETS

The evaluation of the regulatory framework and its performance therein is derived from the data relevant to the safety nets. The data may prove the efficacy of safety net in removing deficiencies. The following data may be very relevant in this regard.

Shipboard Incidents

These are typically not of very serious nature (near misses typically) and which may lead to minor injuries or minor damage to the ship of a generally less serious nature. These may perhaps be available with the Ship Operator/Owner but however may not be available in public domain. Some of these accidents could be precursors to potential serious accident(s))

Marine Accidents

These are accidents of a serious nature (fatal injury, death or severe damage to the Ship) typically reported on IMO GISIS MCI module [7] and commercial websites such as IHS database [8], Clarksons, Tradewinds etc. However, the accident investigation and reporting by flag may not be consistent and/or complete. Such data may be typically available in terms of very basic form as below:

- Accident Date
- Ship details (Type, Age, GT, DWT, Principal Particulars, Flag etc.)
- Cargo Status
- Type of Accident/Initial Event
- Consequence of the Accident (whether total loss)
- Location of accident
- Weather conditions
- Brief textual description
- Number of fatalities and missing
- Severity (Very Serious, Serious)
- Other data

Inspections by Authorities

These check the non-Compliances of the Ship with applicable IMO Instruments. This is captured via the inspection by Flag State and the Port State Control. Typically, an inspection may result in detection of deficiencies. Deficiencies are defined as non-compliance with an applicable regulation. If the deficiencies are very serious in nature and pose danger to the ship, life and/or environment, then the ship is detained by the Inspection Authority (Port State or Flag State).

Port State Control Inspections

Port State Control Inspections are conducted by the authority at the Port of Call. These are performed in accordance with established procedures [9].

PSC inspection results are available on the respective MoU websites (and the GISIS PSC module). These typically provide the following data

- Ship details (Type, Age, GT, DWT, Principal Particulars, Flag, and Risk Profile etc.)
- Type of Inspection (initial or follow-up/detailed)
- Results of Inspection (Detained /Identified Deficiencies)
- Number and Types of Deficiencies
- Number and Types of Detainable Deficiencies
- Other data

However, PSC inspection data may have certain bias or limitations due to factors as mentioned below. These factors should hence be considered while drawing a conclusion.

- Type of MoU (MoUs may have different accords or criteria; many ships depending upon size, type etc. may not frequently call at ports of a given MoU (e.g. Capesize bulk carriers and VLCCs may not call very frequently at ports in the Paris MoU, hence may not be sampled adequately)
- Targeting of Ships for Inspections (e.g. based upon their risk factors, high risk ships are targeted more)
- Concentrated Inspection campaigns
- Training and competency of the PSC Officers
- 'Ports of Convenience' (it is alleged in some places that certain ports are lax in detaining the ships, so as to encourage more ships to call)
- Possible Corruption Cases
- Political influence (Ships of one country being detained in the ports of another country with an ulterior motive)
- Misinterpretation (A Ship detention by a Port State may be incorrect due to wrong interpretation of the Regulations, these may be challenged)

Flag State Inspections

This data may be similar to PSC data but not easily available. Ships may not often call at the Ports of the Flag, hence it may not be possible for the Flag to oversee. Also, data is not readily available at a central location/database.

Member State Audits

Member State Audits are performed to verify whether the State's maritime authority is able to implement the IMO Codes and Conventions appropriately/effectively.

This could help identify important areas where member states could improve in implementation (However this may not be direct indicator of safety as compared to the Accident or the PSC data).

Since Audit is done on sampling basis, this may not be very reflective of the actual safety level (rather they just convey the extent and efficacy of the implementation systems of the member state).

DISCUSSION ON THE VARIOUS DATA

The data on Shipboard Incidents and Marine Accidents is typically revealing of the following:

- Cases of the non-compliances (of regulations) which escaped detection by safety net (e.g. approval of incorrect design /defective arrangement/ construction. Incompetent crew were carrying out operations, wrong operations being carried out; even though regulations existed, they were not complied with and the certification and survey regime including the FSI and PSC inspections failed to detect these)

- Cases where there was NO non-compliance with regulations; rather there was no regulation to address the underlying factors or the regulation was not robust enough (e.g. new failure mode which was unknown before, hence there was no provision in the regulations to address this and hence the ship design/operation was inadequate.

The data on Shipboard Incident and Marine Accidents is quite important as it can help identify the lacunae/shortcomings in the Codes and Conventions. Further, it can help reveal failure modes previously unknown.

The data from PSC and FSI typically reveals the cases of Non-compliances (including serious non-compliances which are a threat to the ship, life and environment) with the applicable regulations as adjudged by the PSC Officer.

Statistical Study of the non-compliances from the PSC and Flag State inspection would help identify the regulations which are violated the most and whether further steps need to be taken (either in form of additional guidance or clarification or enhancement of existing guidance) to reduce these. It is of course a good sign that non-compliances are detected before they led to events culminating into a serious accident. This data is hence of importance but considered one level lower compared to the Accident Data, i.e. Secondary Importance.

It is further important to consider the above data for each individual ship types as these may reveal issues uniquely.

CASE STUDY

A case study is performed relevant to the discussion in the preceding sections so as to obtain an insight as to which particular data are more indicative of the safety. The case study and its scope are described as below:

All Bulk Carriers above 500 GT (SOLAS Applicable) are considered which traded between 01-01-2018 to 31-12-2018 (annual data). A total of 11415 ships were identified by the above criteria.

Analysis of the Bulk Carrier Fleet

It would be very helpful to understand the fleet in terms of size and age before proceeding towards detailed analysis. Figure 1 shows the distribution of the fleet size. The fleet has been categorized in terms of the type (handysize, handymax, etc.) for a more lucid description. It may be noted here that there is no distinct definition of these sizes, hence the definition considered by the authors is marked in the plots in terms of the deadweight. It can be observed that the fleet is more or less evenly composed. The capsize category however constitutes more than 25% of the fleet. These are large bulk carriers. Even considering the panama ships, it can be observed that 45% of the fleet is composed of very large ships.

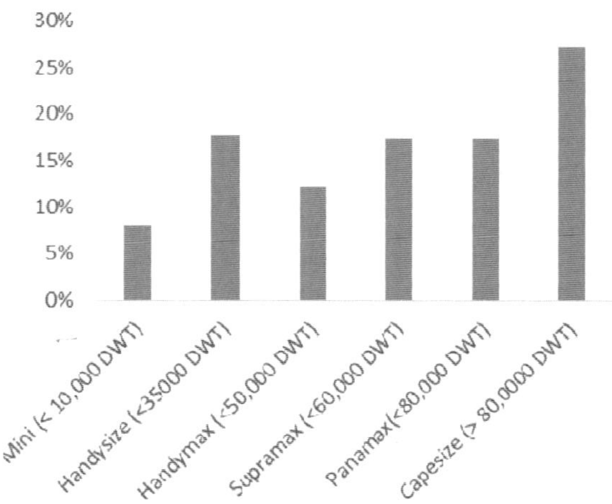

Figure 1: RELATIVE DISTRIBUTION OF THE FLEET SIZE

The fleet age distribution is shown in figure 2. It can be well observed that the fleet is very young. More than 65% of the fleet was composed of vessels with less than 10 years age.

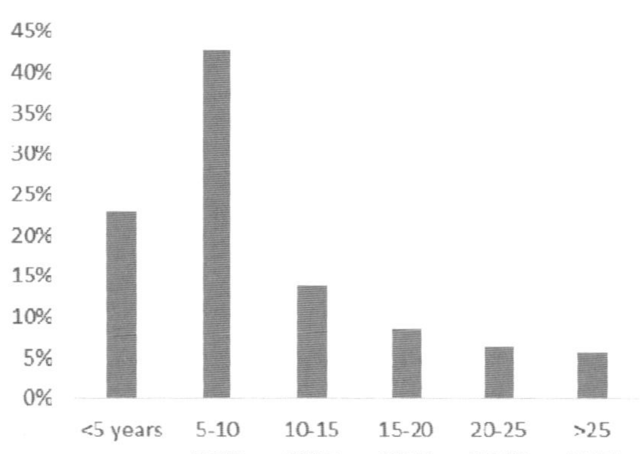

Figure 2: RELATIVE DISTRIBUTION OF THE FLEET AGE

The fleet joint distribution is plotted in Figure 3 for a better projection of the age and the size.

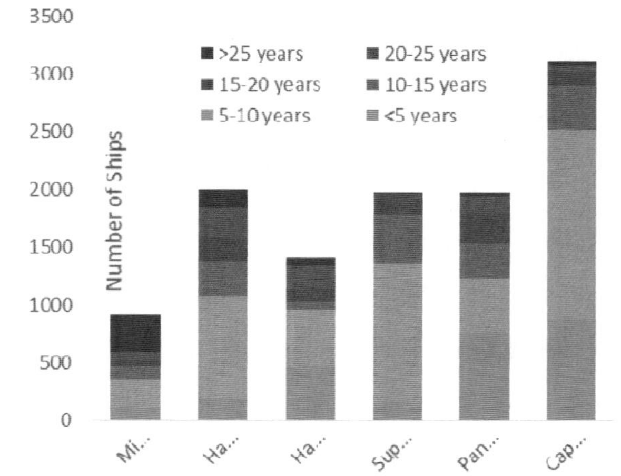

Figure 3: JOINT DISTIBUTION OF THE AGE AND THE SHIP TYPE

It can be seen from the figure 3, that the large size ships (panamax and capesize) form a major portion of the age group less than 10 years.

Analysis of Marine Accident Data

Here, the marine accident data available on IHS Maritime Database is considered. Details of the Shipboard Incidents are not available commonly on commercial databases, hence this data has not been taken for investigation.

The Accident Statistics are depicted in Figures 4 – 5.

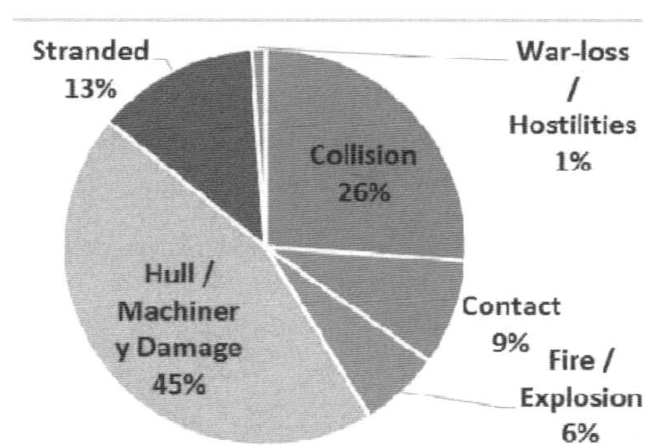

Figure 4: DISTRIBUTION OF SERIOUS ACCIDENT TYPES

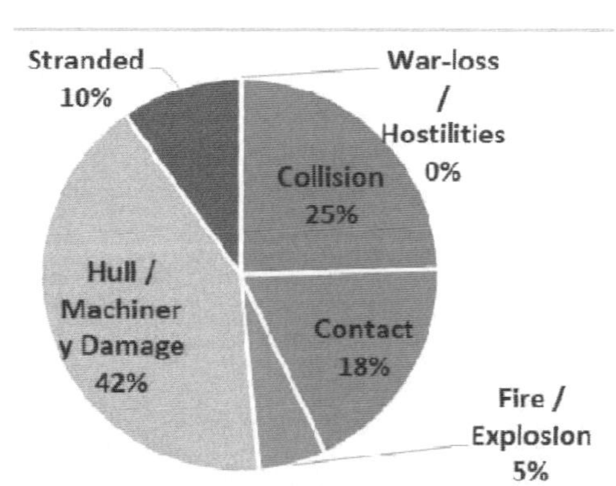

Figure 5: DISTRIBUTION OF LESS SERIOUS ACCIDENT TYPES

It is observed from figures 4 and 5, the pattern of Accidents is more or less unchanged between Serious and Less Serious Accidents. Hull/Machinery Damages account for 45% of the serious accidents. While the Collision, Stranding and Contact accidents contribute more than 50% of all accident causes.

The proportion of serious and less serious accidents is shown in terms of Ship age and Ship size in figures 6-7. It is observed that the accidents (both serious and non-serious) are very less for the large bulk carriers (capsize). This is understandable, since these are expected to be well managed/looked after in terms of maintenance, crew competency and operations. The number of accidents is distinctively more amongst the smaller size bulk carriers.

In terms of age, the bulk carriers above 25 years of age exhibit large number of accidents. The number of accidents in the younger bulk carriers is quite low (<1% of the relative fleet size in that category).

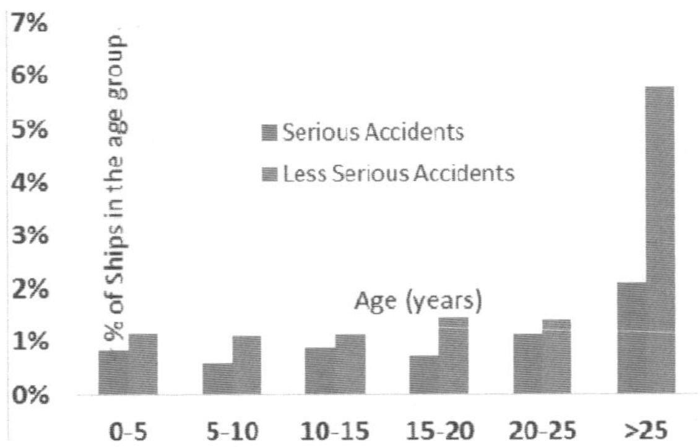

Figure 6: ACCIDENTS IN RELATIVE PROPORTION TO THE SHIP AGES

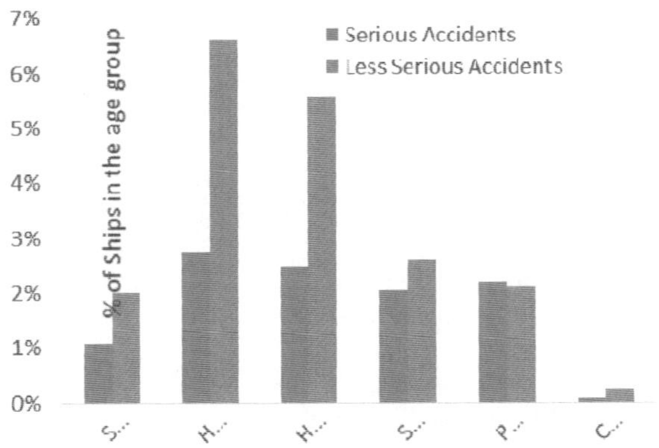

Figure 7: ACCIDENTS IN RELATIVE PROPORTION TO THE SHIP SIZE

Amongst the serious category Hull/Machinery damage, majority (65%) cases occurred with small bulk carriers (Mini and Handysize). Notably, no cases of capsizing accidents were recorded in the year 2018.

Finally, the fatalities in the year 2018 are plotted on the F-N Curve to observe the status/acceptability of risk. The five year moving average is also plotted to gain a better perspective. The results are shown in Figure 8. It can be observed that the trend falls in the ALARP Zone (between the Tolerable and Intolerable risk bounds). The data for the year 2018 had 2 fatalities from 2 accidents. It falls just within the tolerable range.

Analysis of Port State Control Inspections data

Here data from Paris MoU [10] and Tokyo MoU [11] which together account for more than 50% of the PSC inspections world-wide are shown. Not all data available on Paris MoU website are readily available on Tokyo MoU website. Hence the data is compared between the MoUs wherever possible.

It is to be understood that the results of inspections will be dependable based upon the number of inspections. Too less inspections would be questionable as regards confidence in the inference, while too large number of inspections may not be feasible in terms of cost and the trading schedules of the ships.

The sampling (inspection rate) for Paris and Tokyo MoUs is shown in Table 1.

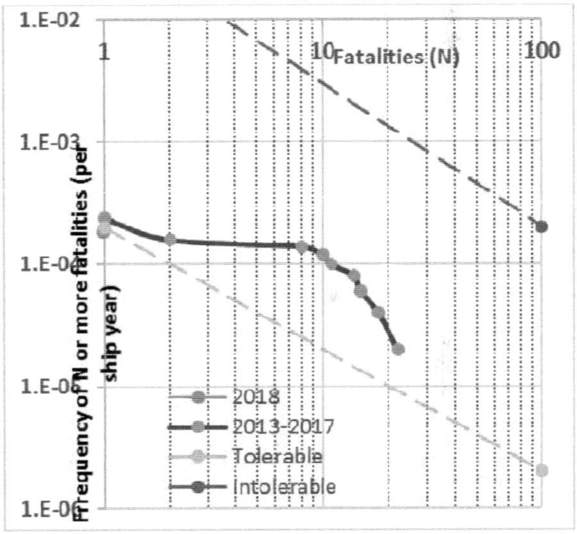

Figure 8: F-N CURVE FOR BULK CARRIERS

Table 1 demonstrates that the sampling rate for inspections is quite substantial i.e. more than 30% of the global fleet is inspected. Here, it must be impressed that the Paris MoU ports may not be frequented by Bulk Carriers as commonly as Tokyo MoU (especially by very large size bulk carriers) due to the bulk cargo being processed mainly outside Europe. These type of data biases must be duly accounted for. On the other hand, Countries in the Tokyo MoU (e.g. China, Australia) are more likely to be frequented by Bulk Carriers due to the trade and the availability/use of pertinent cargoes.

It is remarkable that 48% of the inspections in the Paris MoU resulted in detection of deficiencies (i.e. instances of non-compliance with the regulations). Typically close to 4 deficiencies per inspection and 12 deficiencies per detention were recorded.

The detention rate of approximately 2.6% of the inspections (considering the good sampling rate and considering the same to be representative to the world fleet) would tend to indicate that 2.6% of the global fleet of bulk carriers have serious non-compliances which could lead to endangerment of life and property. However, the credit also lies with the inspections which can detect such anomalies in time, thereby possibly preventing a serious accident.

A comparison of detentions by Ship sizes for the Paris MoU is shown in figure 9. A similar comparison with the Ship's age is shown in figure 10. It is seen that smaller bulk carriers are detained more frequently as well as the younger aged ships. However, the numbers should be taken cautiously as these are judged in relation to total number of detentions and not the total number of ships in the given category (size or age).

Table 1: Paris MoU PSC Inspection Details

Item	Paris MoU	Tokyo MoU[1]
Number of Ships	3350	N.A
Number of Inspections	3711	15241
Number of Detentions	97	313
Number of Deficiencies	7628	N.A
Total Bulk Carriers above 500 GT which operated or operating in global fleet in 2018	11415	11415
Sampling Rate (Inspections per global fleet ships)	32.5%	133%
Detention rate (per inspection)	2.61 %	2.05%
Deficiencies per Inspection	2.05	N.A
Number of Inspections with deficiencies	48.69%	67.33% [2]
Deficiencies per inspection for all inspections with deficiencies	3.82	N.A
Deficiencies per inspection for all inspections leading to detention	11.72	N.A

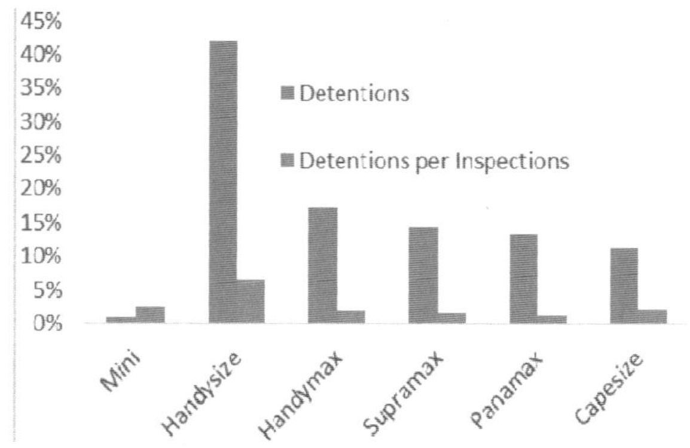

Figure 9: DETENTION RATES BY FLEET SIZE (PARIS MOU)

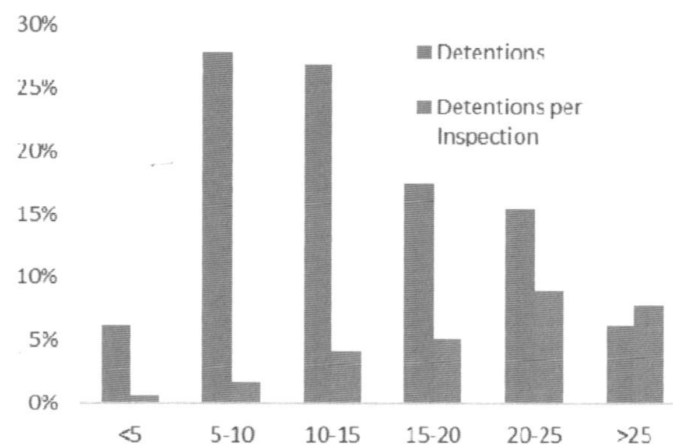

Figure 10: DETENTION RATES BY FLEET AGE (PARIS MOU)

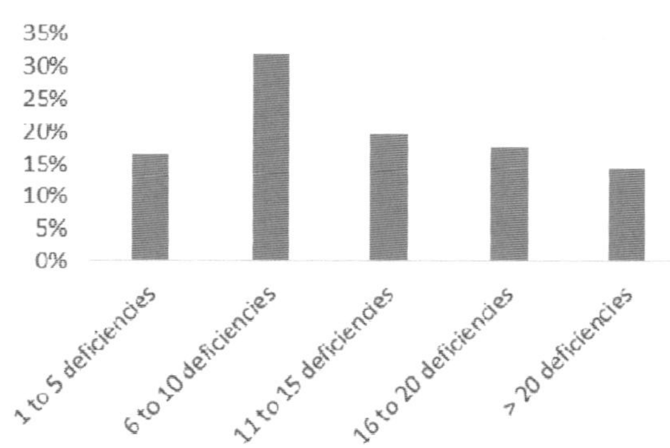

Figure 11: DEFICIENCIES PER DETENTION (PARIS MOU)

A comparison of major deficiency categories (including detainable deficiencies) is shown in Figure 12. It can be seen that Fire Safety, Safety of Navigation, Life Saving Appliances and Emergency Systems form the major share of deficiencies related to technical factors (MLC and Certificates/Documentation are not considered here).

[1] The website does not provide facilities to search by convention size ships (i.e. > 500 GT). The results are therefore for all Bulk Carriers and must be accordingly interpreted.

[2] The website does not permit segregation of deficiencies by type. Hence the number also contains deficiencies pertaining to pollution prevention, Ship security and miscellaneous. The number is thus on a higher side.

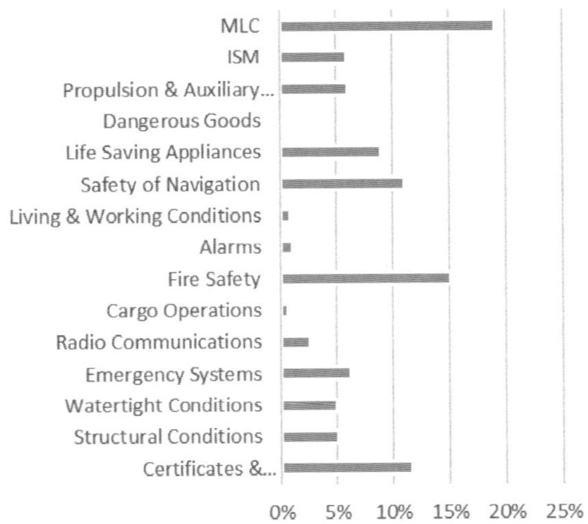

Figure 12: MAJOR DEFICIENCY CATEGORIES

Table 2: Top 20 Detainable Deficiencies

ISM	Seafarers' employment agreement (SEA)
Fire drills	Fire pumps and its pipes
Fire doors/openings in fire-resisting divisions	Abandon ship drills
Fire-dampers	Charts
Covers (hatchway-, portable-, tarpaulins, etc.)	Nautical publications
Lifeboats	Ballast, fuel and other tanks
Voyage data recorder (VDR) / Simplified Voyage data recorder (S-VDR)	Auxiliary engine
Emergency source of power - Emergency generator	Propulsion main engine
Sanitary Facilities	Operation of Life Saving Appliances
Evaluation of crew performance (fire drills)	Ventilators, air pipes, casings

Table 2 provides the top 20 detainable deficiencies commonly detected in the fleet. These 20 account for more than 40% of the all the detainable deficiencies. It can be observed that deficiencies pertaining to Fire Safety are quite dominant in the top 20. The others include items related to navigation (charts and publications), propulsion (auxiliary machinery and main propulstion, emergency source of power etc.). The high number of fire safety related deficiencies being detected may be construed as a possible reason for the relatively less number of fire accidents as seen in the Serious accidents plots above.

Similar exercise with the Tokyo MoU data is undertaken. The results are plotted in Figure 13. The figure shows the detentions in terms of both age and size. (It may be noted that it was not possible to juxtapose Paris and Tokyo MoU results on one plot due to difference of the data formats and sources)

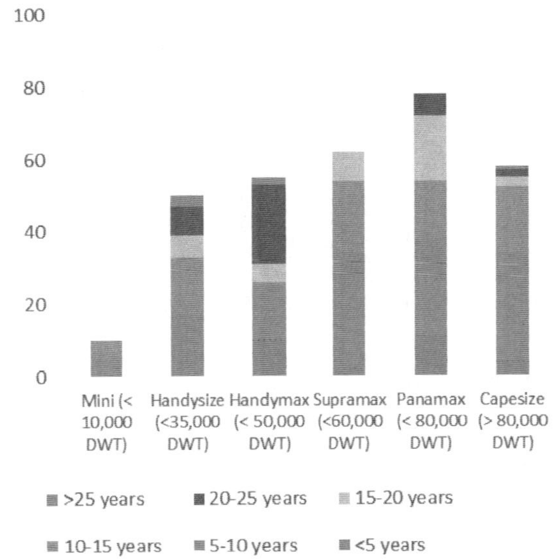

Figure 13: NUMBER OF DETENTIONS OF BULK CARRIERS IN THE TOKYO MOU

The trends in Figure 13 are somewhat different from those observed for the Paris MoU. For instance, the larger ships (Panamax and Capesize) are more frequently detained. Further, the younger aged ships are detained more compared to the older ships. This can be attributed to the fact that Tokyo MoU inspections are four times those of the Paris MoU. Further, the number of large bulk carriers calling at Tokyo MoU ports would be larger as compared to Paris MoU where ports do not receive as much bulk cargo.

The top 20 detainable deficiencies in the Tokyo MoU are shown in Table 3. In the case of Tokyo MoU, the top 20 detainable deficiencies account for more than 65% of all detainable deficiencies. It can be observed that similar to the Paris MoU, the deficiencies related to Fire Safety, Life Saving Appliances, Emergency Systems and ISM form the core of detections in Tokyo MoU (highlighted in greyscale in both the tables).

Table 3: Top 20 detainable deficiencies in Tokyo MoU

Fire Dampers	Ventilators, Air Pipes, Casings
Emergency Source of Power	Charts
ISM (Others)	Unmanned Machinery Space - Ship
Lifeboats	Remote Control (Fire Safety)
ISM (Maintenance of Ship& Equipment)	Launching of rescue boats
Rescueboats	Others
Fire Doors/Openings in fire resisting divisions	Nautical Publications
Fixed Fire Fighting Installation	Fire Detection & Alarms
Emergency Fire Pump	ISM (Resources)
ISM (Shipboard Operations)	Jacketed HP Lines and Alarms

DISCUSSION ON THE CASE STUDY AND PROPOSALS FOR THE FUTURE

The preceding sections demonstrated various ideas which could be used for monitoring of marine safety. At the same time, it should be well noted that such monitoring is heavily dependent on the quantity and quality of data received from Accident Reporting and Investigations as well as the PSC inspections. The preceding analyses reveal the overall trend. Further study would be obviously necessary to uncover more detailed and underlying trends which would be of course depend upon the data as described above.

It may be noted that the above analyses were apolitical, i.e. emphasis was not given to the flag of the ship considering that irrespective of the flag the ship flies, the safety must be uniform. However, it is also acknowledged that this is not the real case. Some flags are quite diligent in their implementation and enforcement of regulations, while some flags are not so. Targeting by flag however should be considered if trends reveal issues with ships flying a particular flag. This could then be considered by the IMO during the audit of that respective flag.

To have an overall view of the safety regime (nets) and its effectiveness, the following data is proposed to be monitored to gain a sense of the pulse of the state of safety.

Primary Indicators (for a ship type)

- Gauging the fleet (in terms of size and age) of a particular ship type under study. The study should essentially consider ship types separately as depending upon the type of cargo carried (dry bulk, liquid bulk, gases in bulk, chemicals in bulk, containers, general cargo). The issues faced as regards safety may be markedly different for each of them.
- Serious Accidents rates (annual with three & five year averages)
- Less serious Accidents rates (annual with three & five year averages)
- Serious Accidents distribution by various categories (e.g. capsizing, Fire, Collision etc.)
- Less Serious Accidents distribution by various categories
- F-N curve (annual, three years and five years)

Secondary Indicators (for a ship type)

- PSC Inspections sampling frequency
- Detention Rates
- Detention rate distribution by age & size.
- Rates of detecting deficiencies per inspections
- Top 20 detainable deficiencies (should at least account for more than 60% of all detainable deficiencies recorded. If not, then more deficiencies should be considered)

Based upon the indicators, the general health of the fleet both in terms of asset and operations can be ascertained. If the safety is being found to lapse, the above indicators may reveal the same and provide idea of the general nature of the problem. It could then be further zoomed upon to understand the nature of the problem in detail.

ACKNOWLEDGEMENT

The authors express their gratitude to Indian Register of Shipping (IRS) for sponsoring the work done within this paper and granting the resources to present the same at ICSOT 2019. The views expressed in this paper are those of the authors and do not necessarily represent those of IRS.

REFERENCES

[1] International Maritime Organization., 1974. "International Convention for Safety of Life at Sea (SOLAS)" (as amended).

[2] International Maritime Organization., 1966. "International Convention on Load Lines" (as amended) (ICLL).

[3] International Maritime Organization., 1972. "Convention on the international regulations for preventing collisions at sea" (as amended) (COLREGS).

[4] International Maritime Organization., 1995. "International Convention on Standards of Training, Certification and Watchkeeping for Sea-farers (STCW) (as amended)".

[5] International Maritime Organization., 1994. "International Code for Safety of High Speed Craft" (as amended).

[6] International Maritime Organization., 2000. "International Code for Fire Safety Systems" (as amended).

[7] International Maritime Organization., 2019. "Global Integrated Shipping Information System (GISIS)". https://gisis.imo.org/Public/Default.aspx

[8] IHS Markit., 2019. "Seaweb: Marine Online Database". https://ihsmarkit.com/products/sea-web-maritime-reference.html

[9] International Maritime Organization., 2017. "Procedures for Port State Control 2017" (as amended).

[10] Paris Memorandum of Understanding (Paris MoU)., 2019. "Inspection Database". https://www.parismou.org/inspection-search/inspection-search

[11] Tokyo Memorandum of Understanding (Tokyo MoU)., 2019. "Inspection Database". http://www.tokyo-mou.org/inspections_detentions/psc_database.php

Proceedings of the 6th International Conference on Ship and Offshore Technology
ICSOT 2019
November 7-8, 2019, IIT Kharagpur, India

ICSOT INDIA 2019-32

FEASIBILITY STUDY OF USING MACHINE LEARNING TOOL FOR EVALUATION OF 3D TIME DOMAIN FREE SURFACE GREEN'S FUNCTION FOR THE SHIP MOTION PROBLEM

K. P. S. Durga Raju
Dept. of Ocean Engineering and
Naval Architecture
IIT Kharagpur, Kharagpur
India
Prudhvikintada@iitkgp.ac.in

Ranadev Datta
Dept. of Ocean Engineering and
Naval Architecture
IIT Kharagpur, Kharagpur
India
ranadev@naval.iitkgp.ac.inl

ABSTRACT
The present paper study the feasibility of using machine learning technique for the evaluation of the 3D time domain free surface Green's function. In order to do that, initially a large dataset for Green's function and its derivatives are generated by using in-house developed deterministic numerical tool for the training. Once the data set is prepared, the same is used to compute the numerical values for the Green's function's and its derivatives. Several critical zones are selected to validate the Random Forest Regressor based machine learning algorithm with numerical data for the feasibility study. from the comparison, it may be concluded that machine learning may be useful tool to compute the Green's function and its derivatives.

INTRODUCTION
The prediction of wave induced loads and motions for sea-going vessels are one of the most fundamental problem in the domain of ocean engineering. Under potential theory based solution, based on the choice of the Green's function, many solution exists. All these methods have several merits and demerits as well, for example frequency domain free surface Green's function based method are very popular in the domain of offshore industries, however, found difficult to apply in forward speed problems. Rankine panel methods are popular in both zero and forward speed problem, however, Rankine panel method is sensitive and often leads to numerical instability due to inaccurate meshing. In this regards, for the forward speed problem, 3D time domain free surface green's function is found to be more robust and easy to implement for these class of problem. Several numerical methods are developed to evaluate the value of 3D time domain green's function and its derivatives. For example, [1] divided the computational domain into several subdomain and apply different series solution for calculating the value of the Green's function and its derivatives. [2] further improved the method by [1] using some more subdomain and by introducing asymptotic expansion. [3] used the series solution, asymptotic solution and Filon quadrature for the evaluation of the green function. [4] introduce one more region where Bessel function expansion is used. Later on [5] showed that the wave part of the time domain Green's function can be obtained by solving a forth order differential equation. [6] proposed a Taylor series expansion method for solving the 4th order differential equation given by [5]. Later on many different solution approaches [7,8] are developed for the solution of 4th order differential equation given by [4]. Bingham [9] pointed out that, for the evaluation of the free surface Green's function, the method propose by [1-4] is better than the method by [6-9]. However, the major problem for adopting the method given by [1-4] is the requirement of huge computational power and memory. Furthermore, due to convolution nature of the integral, computation time during each time step for the evaluation of the green's function is moderately high and therefore performing a long time simulation becomes a tedious job. Recently [10] proposed an efficient algorithm based on residual function for the evaluation of the 3D time domain free surface green's function.

Application of artificial intelligence, machine learning or deep learning in various engineering/medical fields are very common. However, use of machine learning in the field of domain of ocean engineering is limited. Some application is found in prediction of waves [11-12], however not much work is reported in core ocean engineering application.

In the present paper, machine learning tool is used to predict the value of the green's function and its derivatives. Initially, the value of the Green's function is obtained using the method proposed by [1] for the generation of training data. later on, value of the green's function and its derivative is obtained using this training data. random forest based method is used as a machine learning tool. Several critical zones are taken and compare the value of green's function that obtained through the deterministic numerical method and by machine learning based method to check the feasibility of using machine learning for evaluation of green's function.

The Hydrodynamic Problem

Consider a three dimensional arbitrary rigid body floating on a free surface and undergoing arbitrary six degrees of freedom motion in an incompressible, inviscid fluid medium of infinite depth. A linear (sinusoidal) incident wave is approaching making an angle α to the positive x axis. The fluid motion can be defined by the velocity potential:

$$\phi_T(\vec{X};t) = \phi_I(\vec{X};t) + \phi(\vec{X};t) \quad (1)$$

where ϕ_I is the incident wave potential and $\phi = \phi_T - \phi_I$ is the total disturbance potential. Sine ϕ_I is known the problem is therefore to find the solution of the disturbance potential ϕ. This disturbance potential ϕ, for the linearised problem, satisfies the following governing equation, boundary conditions, far field conditions and the initial condition as follows:

$$\nabla^2\phi(\vec{X};t) = 0 \text{ on } \vec{X} = (x,y,z) \in \Omega \quad (2)$$

$$\frac{\partial^2\phi}{\partial t^2} + g\frac{\partial\phi}{\partial z} = 0 \quad \text{on } z = 0 \quad (3)$$

$$\frac{\partial\phi}{\partial n} = V_n - \frac{\partial\phi_I}{\partial n} \quad \text{on } S_0 \quad (4)$$

$$\nabla\phi \to 0 \quad \text{as } R_H \to \infty, \text{ on } z = 0 \quad (5)$$

$$\phi, \partial\phi/\partial t \to 0 \text{ as } t \to 0 \quad (6)$$

In the above, S_0 is the mean wetted body surface. i.e $S_0 = S_B(\vec{X};0)$. $R_H = \sqrt{(x^2+y^2)}$, $\vec{n}$ denotes the unit normal vector which in the present

Expression for 3D Time Domain Green's Function

To solve the above problem, the solution is formulated by introducing a transient Green's function which is a transient free-surface step-function source below the free surface. The Green's function for this problem is given by:

$$G(p,t;q,\tau) = G^0 + G^f \text{ with } t \geq \tau \quad (7)$$

with

$$G^0 = \frac{1}{r} - \frac{1}{r'}$$

$$G^f = 2\int_0^\infty [1 - \cos(\sqrt{g\gamma}(t-\tau)]e^{\gamma(z+\zeta)}J_o(\gamma R)d\gamma$$

$$r = |p-q| = \sqrt{(x-\xi)^2 + (y-\eta)^2 + (z-\zeta)^2}$$

$$r' = |p-q'| = \sqrt{(x-\xi)^2 + (y-\eta)^2 + (z+\zeta)^2}$$

$$R = \sqrt{(x-\xi)^2 + (y-\eta)^2}$$

J_0 = Bessel function of the first kind of order zero. (8)

In the above, G^0 and G^f represent the Rankine part and the free-surface memory part of the Green's function respectively, $p(x,y,z)$ and $q(\xi,\eta,\zeta)$ are the field and the source points, $q' = (\xi,\eta,-\zeta)$ is the image of q above the mean free surface $z = 0$, and γ is the wave number.

Let us assume

$$\beta = (t-\tau)\sqrt{g/r'} \; ; \; \mu = \{z(p(t)) + z(q(\tau))\}/r' \quad (9)$$

Then G^f of equation (8) may be expressed as

$$G(p(t), q(\tau), t-\tau) = \bar{G}^f(\mu,\beta) \quad (10)$$

The computational zone for $\bar{G}^f$ thus ranges in the μ,β plane for $0 \leq \mu \leq 1$, $T_M \leq \beta \leq 0$, where T_M is some large non-dimensional time (theoretically infinity, but practically upto some large value). The largest part of the computational task is the evaluation of $\bar{G}^f(\beta,\mu)$ and its gradient $\nabla\bar{G}^f(\beta,\mu)$. The spatial derivatives of $\bar{G}^f$ can be expressed in terms of $\bar{G}^f_\beta$, $\bar{G}^f_\mu$ using simple chain rule of differentiation. The main computational burden is therefore associated with evaluations of $\bar{G}^f$, $\bar{G}^f_\beta$, $\bar{G}^f_\mu$, since they need to be computed numerous times due to the convolution integral.

Random Forest Regressor

Random Forest (RF) is a flexible, easy to use ensemble machine learning technique that produces considerably effective results time with minimum time spent on hyper-parameter tuning and capable of performing both regression and classification tasks using multiple decision trees and a statistical technique called bagging. Bagging along with boosting are two of the most popular ensemble techniques which aim to tackle high variance and high bias. The detailed discussion on bagging and boosting may be found in many literature (for example 13, 14) and therefore not discussed here. A RF instead of just averaging the prediction of trees it uses two key concepts that give it the name random:

1. Random sampling of training observations when building trees
2. Random subsets of features for splitting nodes

In other words, Random forest builds multiple decision trees and merge their predictions together to get a more accurate and stable prediction rather than relying on individual decision trees.

The fundamental idea behind a random forest is to combine the predictions made by many decision trees into a single model. Individually, predictions made by decision trees may not be accurate but combined together, the predictions will be closer to the true value on average.

Working procedure of Random Forest Regression

The working process of RF may be broadly divided into following steps:

Step 1: Samples are taken repeatedly from the training data so that each data point is having an equal probability of getting selected, and all the samples have the same size as the original training set. Now Bootstrap samples are taken with replacement from the data set.

Step 2: A Random Forest Regressor model is trained at each bootstrap sample drawn, and a prediction is recorded for each sample.

Step 3: Now the ensemble prediction is calculated by averaging the predictions of the above trees producing the final prediction.

Preparation of the Sample Data For Training

Efficient and accurate computation of $\bar{G}^f$ and its derivatives can be performed by subdividing the computational domain $(\mu-\beta)$ into several sub-domains and applying different schemes over the different domains. e.g. asymptotic series, ascending series, economized Chebyshev polynomial approximation etc. (see e.g. [1] - [4]). In the present work, a similar strategy is used, which is arrived at after exploring different schemes available in the literature, and after considerable numerical experimentation. $\bar{G}^f$ and $\nabla\bar{G}^f$ are calculated using a combination of the series expansion and its asymptotic form ([1]), the asymptotic expression given in [4], Filon quadrature and the asymptotic form of Filon quadrature, etc. The resulting algorithm is very robust, and efficient. The different application zones for the various evaluation schemes are determined such that a uniform five digit accuracy is maintained for all the evaluated values of $\bar{G}^f$, $\bar{G}^f_\beta$, $\bar{G}^f_\mu$. Further details on the schemes used for evaluation of these terms are given in [4], [15]. using this numerical model, a set of 1 million data for $\bar{G}^f$, $\bar{G}^f_\beta$ and $\bar{G}^f_\mu$ are prepared for the training purpose. The RF model is trained through this data set.

Results and Discussions

In this section, the computed value of $\bar{G}^f$, $\bar{G}^f_\beta$, $\bar{G}^f_\mu$ using Random Forest Regressior model is compared with that of the deterministic numerical model. It is well known that the Green's function and its derivatives behaves badly near $\mu = 0$, therefore the simulation for $\bar{G}^f$, $\bar{G}^f_\beta$, $\bar{G}^f_\mu$ are carried out for $\mu = 0, 0.05, 0.025, 0.5$. Figure 1(a-d) shows the results of $\bar{G}^f$ at $\mu = 0, 0.05, 0.025$ and 0.5 respectively The numerical results are marked as "orig" (indicating original value) and the results generated by using machine learning tool is marked as "pred" (indicating predicted value).

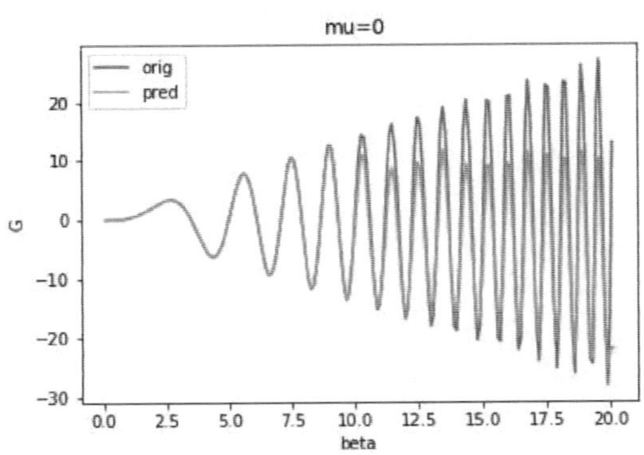

Figure 1a) Comparison of $\bar{G}^f$ at $\mu = 0$

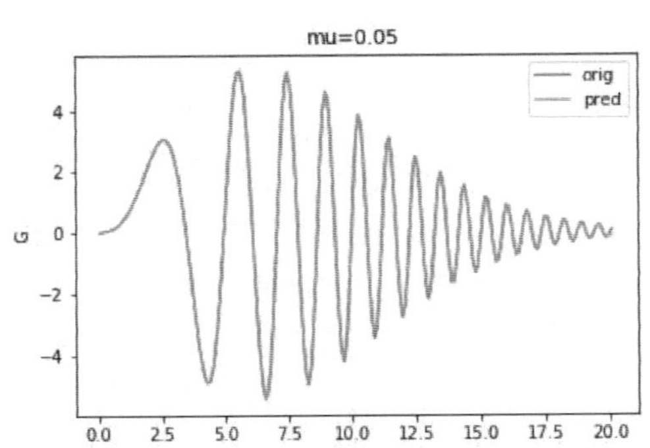

Figure 1b) Comparison of $\bar{G}^f$ at $\mu = 0.05$

From the figures, it may be noted that apart from $\mu = 0$, in all other cases, the predicted results ("pred") shows excellent agreement with "orig" results. Therefore, it may be concluded that, apart from , the RF algorithm works efficiently for evaluating the value of Green's function.

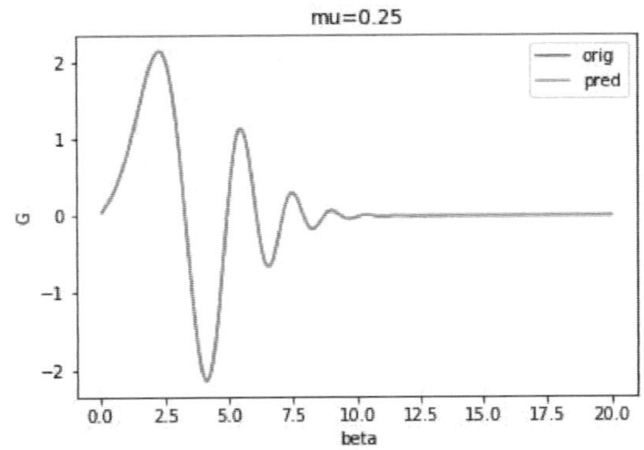

Figure 1c) Comparison of $\bar{G}^f$ at $\mu = 0.25$

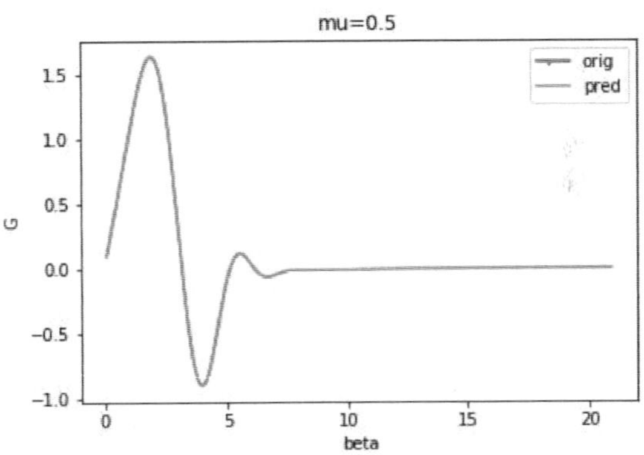

Figure 1d) Comparison of $\bar{G}^f$ at $\mu = 0.5$

The similar observation may be seen in case of $\bar{G}^f_\beta$ and $\bar{G}^f_\mu$ also. Figure 2(a-d) and 3(a-d) show the comparison between "orig" and "pred" at $\mu = 0, 0.05, 0.025, 0.5$ for $\bar{G}^f_\beta$ and $\bar{G}^f_\mu$ respectively.

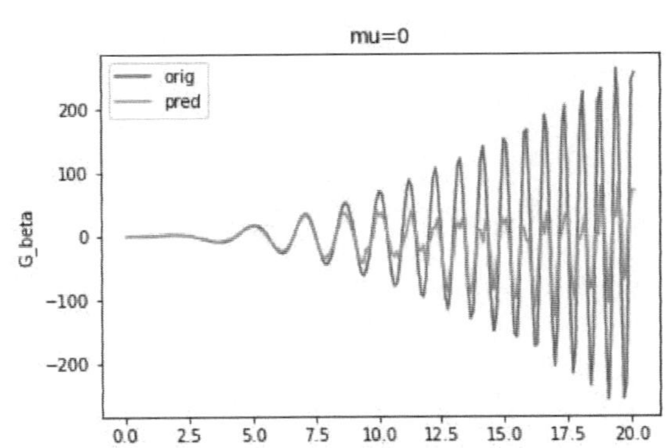

Figure 2a) Comparison of $\bar{G}^f_\beta$ at $\mu = 0$

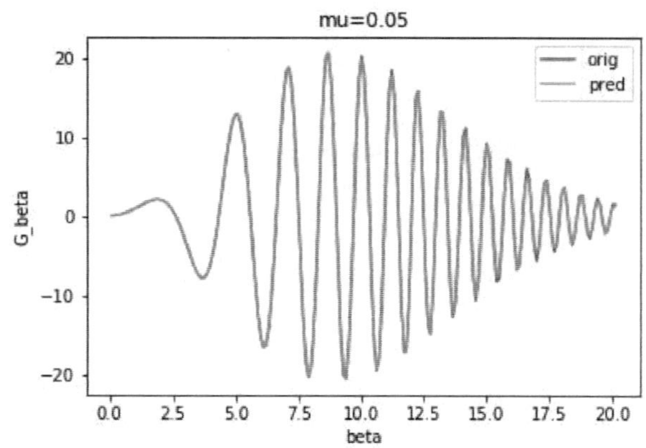

Figure 2b) Comparison of $\bar{G}^f_\beta$ at $\mu = 0.05$

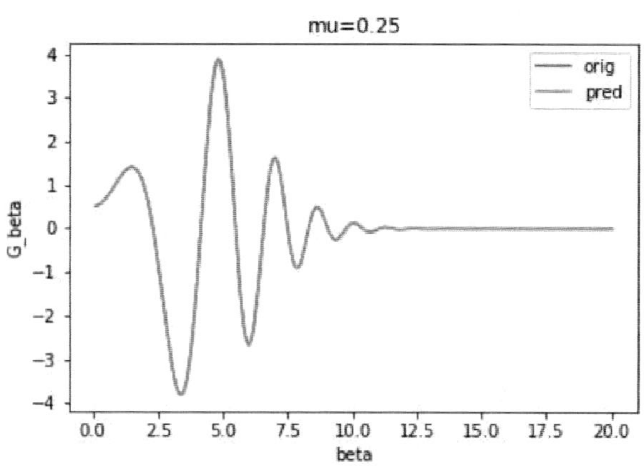

Figure 2c) Comparison of $\bar{G}^f_\beta$ at $\mu = 0.25$

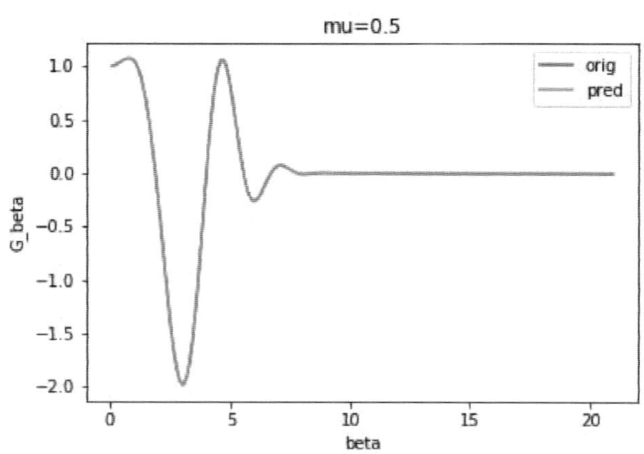

Figure 2d) Comparison of $\bar{G}^f_\beta$ at $\mu = 0.5$

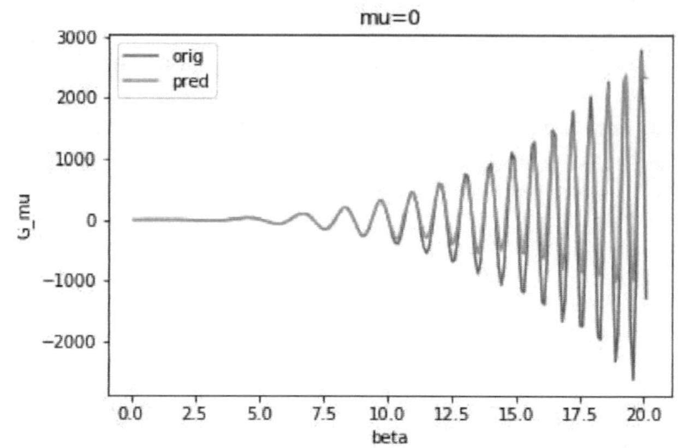

Figure 3a) Comparison of $\bar{G}^f_\mu$ at $\mu = 0$

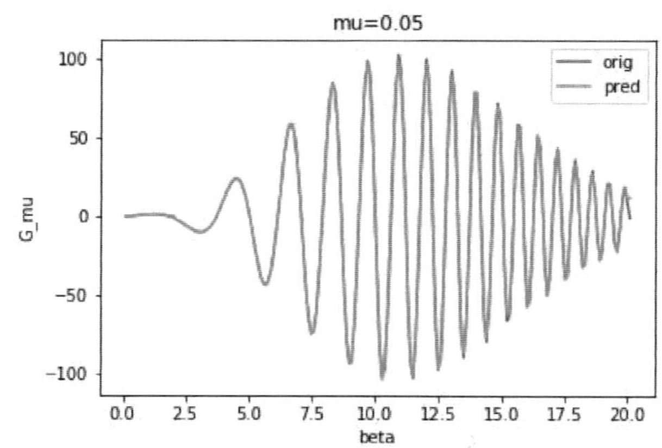

Figure 3b) Comparison of $\bar{G}^f_\mu$ at $\mu = 0.05$

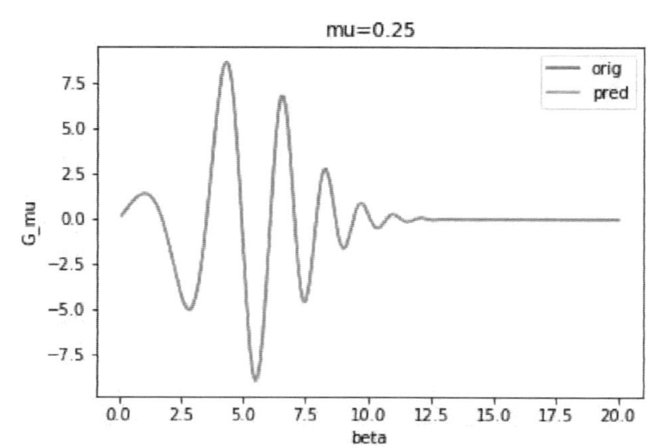

Figure 3c) Comparison of $\bar{G}^f_\mu$ at $\mu = 0.25$

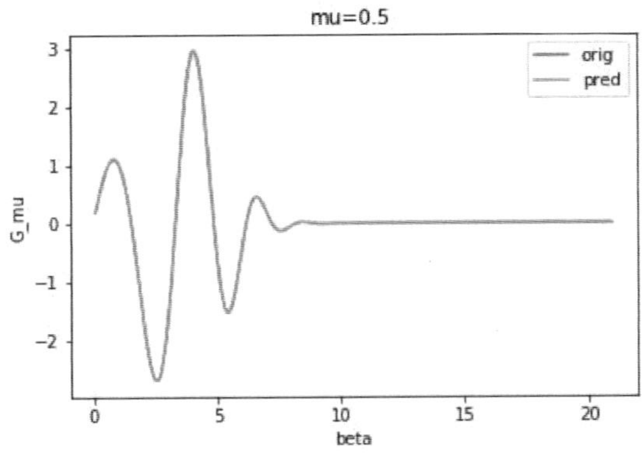

Figure 3d) Comparison of $\overline{G}^{f}_{\mu}$ at $\mu = 0.5$

from the comparison, it may be noted that, apart from $\mu = 0$, in all other case, the RF algorithm evaluate the Green's function and its derivative with high accuracy. Even if the value of μ marginally away from zero , i.e. at $\mu = 0.05$, the RF compares excellently with deterministic numerical results. It is well known fact that, the behaviour of Green's function and its derivative near $\mu = 0$ is highly oscillatory and diverging which leads to numerical instability in many cases. Datta et al [16] studied this problem and observed that, by making artificial vertical panel near free surface, one can avoid this numerical instability, which means, keeping the value of μ away from zero. In such cases, the present machine learning tool also may evaluate the green's function with required accuracy, as it is seen from the above plots that, RF compares very well with the "orig" for $\mu = 0.05$, which is very close to $\mu = 0$.

Conclusion

In the present study, a machine learning tool based of Random Forest Regressor is developed to compute the value of 3D time domain free surface Green's function and its derivatives. The computed results are then compared with in-house developed deterministic numerical technique. The results are obtained for various critical zones. It is found that, apart from $\mu = 0$, the results obtained from machine learning tool compare excellently with that of the numerical model. The value of becomes zero when the ship is having a flare or a transom stern. This indicate that, if a ship does not have flare or transom stern, the present methodology may be useful. Even for the ship with flare or transom, use of artificial panel near free surface ay solve the instability issue. Therefore, it may be concluded that use of machine learning is feasible for evaluating the Free Surface Green's function and its derivative. The present study is focuses only to compare the value of Green's function and its derivative with numerical methods at different zone, however, comparison of memory and time requirement is required to analyze rigorously before implementing the present machine learning based into ship motion codes.

REFERENCES

[1] Newman, J. N., 1985, Algorithms for the free-surface Green function. *Journal of Engineering Mathematics,* vol. 19, pp. 57-67.

[2] Lin, W. M. and Yue D, 1990, Numerical solution for large amplitude ship motions in the time domain. *Proceedings of 18th ONR Symposium on Naval Hydrodynamics*, National Academy Press, Washington D.C, pp. 41-66

[3] Beck, R.F., Liapis, S.J., 1987 Transient motion of floating bodies at zero forward speed. *Journal of Ship Research* 31, 164–176.

[4] King, B., 1987, Time domain analysis of wave exciting forces on the ships and bodies *Ph.D. Thesis,* Dept. of Naval Arch. and Marine Eng, The University of Michigan, USA

[5] Clement, A.H., 1998 An ordinary differential equation for green function of time-domain free-surface hydrodynamics, *Journal of Engineering Mathematics,* 33, 201–217.

[6] Chuang J.M., Qiu, W., Peng H., 2007, On the evaluation of the Green's function., *Ocean Engineering*, 34, p.p. 962-969

[7] Li, Z.F., Ren, H.L., Tong, X.W., Li, H., 2015. A precise computation method of transient free surface Green function, *Ocean Engineering*, 105, 318–326

[8] Zhu, R.C., Zhu, H.R., etc, Shen L., 2007, Numerical treatments and applications of the 3D transient green function. China *Ocean Engineering*. 21 (4), 637-646

[9] Bingham, H.B., 2016. A note on the relative efficiency of methods for computing the transient free-surface Green function. *Ocean Engineering*. 120, 15-20

[10] Shan P., Wang, Y., Wang F, Wu J., Zhu R, 2019, An efficient algorithm with new residual functions for the transient free surface green function in infinite depth. *Ocean Engineering*. 178, 435-441.

[11] Gopinath D. I., Dwarakishb G. S., 2015 "Wave prediction using neural networks at New Mangalore Port along west coast of India, Aquatic Procedia, 4, 143-150.

[12] Tom T, , Ikemoto A, Mase H, 2019, Wave prediction in the sea of Japan by deep learning using meteorological data, journal of Japan. doi : https://doi.org/10.2208/kaigan.75.I_145

[13] https://pubs.acs.org/doi/abs/10.1021/ci0256438

[14] http://www.jocpr.com/articles/a-study-of- english-reading-ability-based-on-multiple-linear-regression-analysis.pdf

[15] Sen, D., 2002, Time domain computation of large amplitude 3D ship motion with forward speed, *Ocean Engineering,* vol. 29, pp. 973-1002.

[16] Datta R., Rodrigues J. M., Guedes Soares C., 2011, Study of motions of fishing vessels by a time domain panel method. *Ocean Engineering,*38, p.p.782-792

The Royal Institution of Naval Architects
8-9 Northumberland Street
London, WC2N 5DA
United Kingdom

ISBN 978-1-7138-0867-1

CURRAN ASSOCIATES INC.
proceedings.com